mathematics for Information Technology

Alfred Basta

Stephan DeLong

Nadine Basta

DELMAR
CENGAGE Learning

Australia • Brazil • Japan • Korea • Mexico • Singapore • Spain • United Kingdom • United States

DELMAR
CENGAGE Learning

Mathematics for Information Technology
Alfred Basta, Stephan DeLong, and Nadine Basta

Vice President, Careers & Computing: Dave Garza

Director of Learning Solutions: Sandy Clark

Associate Acquisitions Editor: Kathryn Hall

Director, Development-Career and Computing: Marah Bellegarde

Managing Editor: Larry Main

Senior Product Manager: Mary Clyne

Editorial Assistant: Kaitlin Murphy

Brand Manager: Gordon Lee

Market Development Manager: Erin Brennan

Senior Production Director: Wendy Troeger

Production Manager: Mark Bernard

Content Project Manager: Christopher Chien

Senior Art Director: David Arsenault

Media Editor: Deborah Bordeaux

Cover Image: © kentoh/Shutterstock

Library of Congress Control Number: 2012931207

ISBN-13: 978-1-111-12783-1

ISBN-10: 1-111-12783-2

Delmar
5 Maxwell Drive
Clifton Park, NY 12065-2919
USA

Cengage Learning is a leading provider of customized learning solutions with office locations around the globe, including Singapore, the United Kingdom, Australia, Mexico, Brazil, and Japan. Locate your local office at:
international.cengage.com/region

Cengage Learning products are represented in Canada by Nelson Education, Ltd.

To learn more about Delmar, visit **www.cengage.com/delmar**

Purchase any of our products at your local college store or at our preferred online store **www.cengagebrain.com**

Printed in the United States of America
1 2 3 4 5 6 7 16 15 14 13 12

Contents

Chapter 10

Chapter 11

Chapter 12
Probability 401

Chapter 13
Statistics 440

Preface

Mathematics for Information Technology is written to help students develop the specific math skills and understanding they need to succeed in electronics, computer programming, and information technology (IT) programs. With topical coverage tailored to important IT applications, this text delivers easy-to-understand and balanced mathematical instruction for students in 9- to 12-week college courses. A wealth of illustrations, examples, applications, and exercises will guide students toward an understanding of the content from a number of different angles.

The authors' combined experience teaching this material in live classrooms and in online/distance learning formats has uniquely qualified them to develop this text for a variety of learning environments. Whether students are learning in a classroom or online, in a 9-week course or a 12-week semester, or in an electronics, computer programming, or IT department, they will find *Mathematics for Information Technology* an invaluable resource throughout their studies.

This Book's Approach

Beginning with basic concepts of sets and logic, the authors have taken care to build students' knowledge and skills step-by-step throughout each chapter. The authors begin each chapter with an application, an outline of topics, and a list of learning objectives. They present the material with clear, student-friendly explanations and illustrative examples drawn from IT applications as well as from everyday life. Each chapter closes with a summary of relevant concepts, a glossary of key terms, and a variety of practice exercises to build problem-solving skill and confidence.

Organization

Mathematics for Information Technology begins with a fundamental discussion of sets, logic, and number systems. Having built a solid foundation in these essential concepts, students will progress through topics in algebra and trigonometry to finish with probability, statistics, and graph theory.

Features of the Text

The authors use a clear, conversational writing style to present mathematical concepts and apply them to situations that students will understand from everyday living as well as to problems they will encounter in their careers. The following suite of text features will help guide students toward mastery of each chapter's knowledge and skill set:

Chapter openers introduce each chapter with interesting and motivational applications, illustrating the real-world nature of the chapter topics.

Chapter objectives outline the knowledge and skills students will master in each chapter.

Key terms are highlighted and carefully defined to help students increase technological literacy.

Dozens of detailed examples in every chapter help students master a step-by-step approach to problem solving. Geared toward IT majors, examples provide both technical and everyday scenarios to help students see how math concepts are applied in the real world.

More than 1,700 section review problems provide frequent problem-solving practice throughout each chapter.

Chapter summaries and glossaries assist students as they review the material and prepare for tests.

Exercise sets at the end of each chapter provide hundreds of problems for skill-building practice and test prep.

Help for Teaching and Learning

A robust supplements package accompanies this text to help students and teachers maximize their learning advantages.

Instructor Resources

The Instructor Companion Web site provides the following support for teachers:

- Solutions to *all* text problems
- Computerized test banks in ExamView® software
- PowerPoint® presentations
- An Image Gallery including all text figures

Applied Math CourseMate

Mathematics for Information Technology includes Applied Math CourseMate®, Cengage Learning's online solution for building strong math skills. Students and instructors alike will benefit from the following CourseMate Resources:

- An interactive eBook, with highlighting, note-taking, and search capabilities
- Interactive learning tools including the following:
  ✓ Quizzes
  ✓ Flash cards
  ✓ PowerPoint slides
  ✓ Skill-building games
- An appendix of answers to odd-numbered chapter exercises.
- And more!

Instructors will be able to use Applied Math CourseMate to access the Instructor Resources and other classroom management tools. Go to login.cengagebrain.com to access these resources, and look for this icon CENGAGE brain.com to find resources related to your text in Applied Math CourseMate.

A Message to Students

Mathematics textbooks for students of information technology curricula are often steeped in complicated terminology and lacking in explanations. We hope you will find this textbook a pleasant change from the others available in the breadth of the examples, the caliber of the mathematical instruction, and the relevance of the content. We've restricted the content to the material you'll find most useful in your future studies and presented the material in such a way that (even if you are at a distance from your instructor) the information is accessible and understandable and will lay the foundations for your future academic success.

Whether you are studying in a traditional "brick and mortar" environment or in a virtual "online" setting, we urge you to choose the most effective supporting materials possible. For instance, if you do choose to use Web resources to serve as study aids, select them carefully and ensure that they are reputable and reliable. Your instructor will undoubtedly offer guidance in the choice of resources, but a general rule of thumb is to eschew sites that are open for modification by contributors who do not submit their work through a process of peer review. Generally it is safe to rely upon sites with the suffix .org or .edu, but there are other reliable resources to which your instructor can guide you.

You will need a calculator to solve many of the problems presented in this book. The calculator you select need not be overly expensive or complicated to handle the mathematics in this text. Many students find graphing calculators useful but not essential in this course. We have found that even a scientific calculator is sufficient for this material.

Acknowledgments

A text of this complexity could not happen without assistance from knowledgeable experts.

Content Review

The publisher wishes to acknowledge the following individuals who carefully reviewed the content at manuscript stage:

Robert Gallante
Shawsheen Valley Technical High School, Billerica, Massachusetts

James McCallum
YTI Career Institute, York, Pennsylvania

Mark Schwind
YTI Career Institute, York, Pennsylvania

Francisco Soto
Career Center for Texas

Charulata Trivedi
Quinsigamond Community College, Worcester, Massachusetts

Technical Review

The publisher acknowledges our series adviser, John C. Peterson, for his review of technical accuracy during the manuscript stage and his tireless dedication to generating the solutions manual. In addition, we thank Linda Willey for her steady and detailed review of every step of the solutions.

Authors' Acknowledgments

In addition, the authors wish to acknowledge the assistance and support of a dedicated community of family, friends, and colleagues:

To my wife, Nadine: It is the continuing symphony of your loving thoughts, caring actions, and continuous support that stands out as the song of my life.

To my mother: You are a never-ending melody of goodness and kindness. You are without equal in this world.

And to the memory of my father: If one is weighed by the gifts one gives, your values given are beyond estimation.

—*Alfred Basta*

First, I would like to thank God for giving me the chance to complete this work. Every day I thank Him for my three precious gifts: Alfred, Becca, and Stavros.

To my beloved husband, Alfred: Thank you for your continuous love and support throughout our wonderful 17 years together.

To our children, Rebecca and Stavros: You are the true joy of our lives and our greatest blessing. We pray for you every day to live a life that honors and glorifies God. Fix your hearts upon Him, and love Him with all your strength.

—Nadine Basta

To my dear wife, Debbie, our cats Alex, Daphne, and Freddy, and my parents Bonnie and Dave, all of whom supported me through the process of preparing this book. Also, to the many kind and supportive colleagues at Cengage Learning, particularly Mary Clyne, whose kindness and understanding to a novice author places me forever in her debt.

—Stephan DeLong

About the Authors

Alfred Basta is a professor of mathematics, cryptography, and information security. He is a professional speaker on topics in Internet security, networking, and cryptography. Alfred is a member of many associations, including the Mathematical Association of America.

Stephan DeLong is a professor of mathematics with over 20 years of teaching experience on campus and more than six years online. He received a master of science degree in mathematics from Lehigh University and is a member of many mathematical associations. In addition to teaching and speaking about mathematics, Stephan is an avid cyclist and writer of fiction.

Nadine Basta teaches information systems security, risk management, and cryptography. She holds a bachelor of science degree in special mathematics and a master of science degree in computer science. Nadine is a member of the Mathematical Association of America.

Chapter 1 Sets

This chapter introduces some preliminary concepts you may have studied before: the concepts and notations of sets and subsets, representation of sets by the visual tool called a Venn diagram, and the idea of infinite sets. These notions will prove useful in the chapters that follow, and therefore we will be well served by establishing a solid foundation in these topics. If the idea of a set is new to you, don't be concerned because we will be developing the concept from early principles, and you will be up to speed in no time.

Initially, we'll use illustrative examples taken from a general context, with the thought that familiarity with the concepts can be gained best through the use of nontechnical examples. Consequently, you'll find that (at the outset) we'll anticipate no knowledge of computer programming or preexisting information technology–specific knowledge. This material will be presented in a general context, using concepts accessible to all.

Objectives

By the time you have successfully completed the materials of this chapter, you will be able to:

- Recognize the fundamental notions and notations of sets.
- Understand the subset relation between sets.
- Construct and interpret Venn diagrams.
- Understand applications that employ sets.
- Recognize and distinguish between infinite and finite sets.

1.1 Set Concepts

Terminology of Sets

A **set** is a well-defined, unordered collection of objects having no duplicate members. When we use the term **well-defined set**, we mean that membership in the collection is unambiguous, not open to interpretation, and strictly capable of determination through an investigation of the facts. This concept, when viewed with the greatest possible generality, is a truly fundamental notion for us, since all of mathematics is constructed on it.

As an illustration, "the set of all legal residents of your town" is a well-defined set because legal stipulations dictate who is and who is not a legal resident of your town. Confronted by any particular individual you meet, direct examination of the facts will enable you to determine whether that person is a legal resident of your town and hence considered to be an object within the set. Alternatively, "the set of all smart people in your town" is not well defined because the perception of what it means to be a smart person is open to debate, and may not be interpreted the same way by different people. Consequently, whether John Smith is a legal resident of your town can be definitively established, while whether John Smith is a smart person in your town cannot.

Before advancing further, let's get our feet wet by confirming our understanding of the notion of a well-defined set.

EXAMPLE 1.1

Are the following sets well defined?

1. The set of tall people in your class NO
2. The set of letters in the English alphabet YES
3. The set of warm days in the past year NO
4. The set of libraries in New Jersey having more than 1000 library card holders. YES

SOLUTION

1. For the first set, we have a concept that is not well defined. The reason for this is that the notion of what constitutes "tall" is ambiguous. Is a six-foot-tall man "tall"? Or do we reserve that designation for someone who is seven feet tall and plays in the National Basketball Association?
2. The set of letters in the English alphabet, on the other hand, *is* well defined. There are precisely 26 letters in the English alphabet, and thus it is clear what is in the set of letters in the English alphabet and what is not.
3. The third set is not well defined for the same reason as the first set was not: ambiguity of the membership criteria. What is considered "warm" may differ from person to person, and therefore the criterion for membership in the set is not definitively established.
4. The last set is a well-defined set. If we consider any particular library in the state of New Jersey, it is a matter of investigation to determine whether that library has more than 1000 library card holders. Consequently, we can definitively establish membership within the set, and so the set is well defined.

Note

A set is considered well defined if, given an arbitrary object, you can tell definitively if the object is in the set or is not!

Let a be an object in set A. The notation $a \in A$ is used to indicate that the object a is a **member** of set A or that a is an **element** of A. The notation $a \notin A$ is used to say that the object a is not a member of set A. It is conceivable, of course, that a set could have no members, and the symbol $\varnothing$ is used to denote such a set, which we refer to as the **empty set** or the **null set**.

Membership in a set can be specified in at least three ways. The first way is through a **verbal description**, wherein the members of the set are described in an unambiguous manner. An illustration would be "M is the set of all individuals currently on the roster of the New York Mets baseball team." That the set is well defined is apparent. If I ask, for instance, whether Reggie Jackson is a member of set M, the answer can be definitively determined through investigation of the facts.

The second way of expressing the membership of a set is called **roster notation**, wherein the members of the set are listed for us. Roster form is particularly useful if the membership of a set is fairly limited in size or if the members follow an easily recognizable pattern. An example of roster notation would be $A = \{1, 2, 3, 4\}$. Should the list of members be rather substantial, we can use an **ellipsis** to shorten the roster, provided that we can present the roster in such a way that membership in the set is clear. An ellipsis is a succession of three "dots" that indicate that the demonstrated pattern of numbers continues, either forever or until a number following the ellipsis is reached. For instance, we could say $B = \{2, 4, 6, ..., 100\}$ if we wanted to indicate the even natural numbers less than or equal to 100, but it would be improper to attempt to give the same set as $B = \{2, ..., 100\}$, since we are not provided with enough information to be certain of the set's membership criteria.

The third way of describing set membership is called **set-builder notation**, which looks like this: $P = \{x \mid x \text{ is an even number}\}$. The information within the braces tells us that "generic" members of the set P will be referred to as x and that the form of the elements x is described by the rule following the vertical line. We refer to that as a **characteristic property** or **characteristic trait** of the members of the set. In this case, the members of set P are those numbers that happen to be multiples of 2.

The set-builder notation form is read in a particular way, which is completely nonobvious on inspection. For example, we would read the preceding illustration, $P = \{x \mid x \text{ is an even number}\}$, as "$P$ is equal to the set of all elements x such that x is an even number." This takes some getting used to, and we encourage you to think carefully about how set-builder notation would be read whenever you encounter it. That practice will provide you with the familiarity you need to make this potentially strange and new notation less intimidating.

Consider the following examples of sets given in set-builder notation:

$A = \{x \in N \mid 5 < x < 6\}$;
$B = \{x \mid x \text{ is any of points common to any two distinct parallel lines in a plane}\}$.

In keeping with our suggestion, let's indicate how the notations would be read:

In the first case, we have "the set A is equal to all elements x within the natural numbers, such that 5 is less than x and x is less than six." In the second case, "the set B is equal to the set of all elements x, such that x is one of the set of points common to any two distinct parallel lines in a plane." The language is cumbersome, perhaps, but we do need to gain familiarity with it, and so we definitely want to practice using it.

It may not be immediately apparent, but the two sets we have just presented are empty. That they *are* empty becomes evident on examination of the criteria for membership. For the set A, it is impossible for a natural number to occur between 5 and 6, and thus there are no elements capable of satisfying the definition for membership in A. That B is empty is a consequence of the definition of parallel lines in Euclidean geometry, where a postulate dictates that distinct parallel lines can share no points. Consequently, we could say $A = B = \varnothing$. Note that the two sets are not obviously equal when first presented but are found to be so only on further examination.

Note

Be sure you know how to interpret each of the different forms in which sets can be presented—they will be used interchangeably throughout this textbook!

Within mathematics, we should become familiar with particular sets of numbers, as they form a rich source of examples. Each of these sets is represented by a particular boldfaced capital letter, and each of those notations should be reserved for its particular set. Here are the sets, in "increasing" order (what we mean by increasing will be seen as we set forth the definitions; each successive set is made up of its predecessor combined with additional members):

Some of the Sets of Numbers

N = the set of natural numbers = $\{1, 2, 3, \ldots\}$

W = the set of whole numbers = $\{0, 1, 2, 3, \ldots\}$

Z = the set of integers = $\{\ldots, -3, -2, -1, 0, 1, 2, 3, \ldots\}$

There are additional sets of numbers: the rational numbers (symbolized by **Q**), the irrational numbers (for which there is no universally accepted set name), the real numbers (symbolized by **R**), and the complex numbers (symbolized by **C**). We will discuss these further as we move through the text and need to understand them.

None of the three forms of specification for membership—verbal description, roster notation, or set-builder notation—is generally superior to the others, but in some cases one is more appropriate than the others. For instance, if your set were finite and relatively small or if the numbers followed an easily detected pattern of progression, roster notation would likely be the most useful form for you to use. If your set was easily described in words, then the verbal description method might be best. If all the members of your set followed a particular pattern, then set-builder notation might be preferable. Often, which form you choose is a matter of expedience or personal preference. We will use the methods interchangeably throughout this text.

EXAMPLE 1.2

For the set $G = \{x \in W \mid x \leq 7\}$, describe the members of the set using roster form and also a verbal description.

SOLUTION

Recalling the definition of the set W, our familiar set of whole numbers, and with an understanding of the $\leq$ symbol in hand, we can list the members of the set G as the roster $G = \{0, 1, 2, 3, 4, 5, 6, 7\}$. The ordering of the numbers could be juggled as long as all those numbers listed are included. The verbal description allows a bit more creativity, since a group of objects can be described in various ways. Often, the best approach is to simply use the membership criterion from set-builder notation, and so we can describe the members of G as "the set of whole numbers less than or equal to 7." We could also describe the members as "the set of whole numbers less than 8," since this description also captures the nature of the members of set G.

When we consider the elements of a set A, several things should be apparent by inspection. First, the description of the requirement for membership in A should be so unambiguous that it is clear whether a particular object is an element of A. This is a necessary quality of a set being well defined: that the criteria for membership should be evident. Second, the members of A should be distinguishable in the sense that it is clear whether two allegedly distinct objects are, in fact, one and the same. Remember, no duplication of members should occur within a set.

That last remark may seem ambiguous. How could it be the case that two elements of a set would be indistinguishable? Consider the following question: are q and Q the same thing? That might depend on something we refer to as the **universal set**.

In the study of sets, an early step is to establish a context within which the set exists. It is customary to define this context by means of an all-encompassing set called the universal set, which we shall symbolize by U. In a given situation, all sets considered must take their members only from that universal set. An example of a universal set could be a particular set of numbers or a particular group of individuals. If no universal set is specified, it may be that its composition is irrelevant for the purposes of the problem, so we can take it to be some larger set within which the members of our set reside.

Note that for a particular set, the universal set can vary from example to example. Suppose that our set was the set of all persons currently on the playing roster of the Philadelphia Eagles. The universal set could (unless specifically stated) be the set of all players in the National Football Conference, the set of all players in the National Football League, or the people in the United States of America. When the universal set is not made explicit by context and when the identity of the universal set is of importance, we will state it specifically.

Once the universal set has been specified, we will consider only members from that set for the duration of the problem under consideration. If, for instance, we define $U = \{1, 2, 3, 4\}$, then for the remainder of the problem, no other objects exist. Thus, using that universal set, if we then wanted to create a set consisting of all even numbers from the universal set and call that set E, we would say $E = \{2, 4\}$. Although there are other even numbers from the broader context of all possible numbers, from the perspective of our universal set, these are the only even numbers in existence.

Returning to our example of the set elements q and Q, if the universal set were the 26 letters of the English alphabet, then we could rightly claim that both symbols represent the same thing and thus are the same. If, on the other hand, the universal set is the 52-member collection of capital and lowercase letters of the English alphabet, then the elements q and Q are distinct and should not be treated as being one and the same.

The moral to that last example is that we must be very careful when we consider the members of sets because it is possible that the specification of the universal set could affect our observations and analysis.

Cardinality of a Set

When examining sets, we will want to consider various features, including the **cardinality** of the set. Cardinality refers to the quantity of members belonging to the set, and it is represented by a symbol that is similar to the familiar notation of absolute value: $|S|$. An alternative notation used in some texts is the symbol $n(S)$, which also represents the cardinality of set S.

Cardinality may be **finite** (that is, membership in the set is limited to a particular, possibly large quantity of members) or **infinite** (there are an unlimited number of members in the set). We will consider infinite sets more closely in Section 1.5 but may refer to them in some examples in the meantime. The cardinality of the empty set is zero, since that set has been defined to have no members.

EXAMPLE 1.3

Determine if the following sets are finite or infinite:

1. The set of players currently on the roster of the New York Mets baseball team
2. The set of grains of sand existing on all the beaches of the planet Earth
3. The set of blood cells in your body at any given time
4. The set of numbers between 0 and 2

SOLUTION

1. Rosters of baseball teams (and of the New York Mets in particular) are limited by league rules and are thus finite.
2. The number of grains of sand on the beaches of the planet Earth is quite large, but it is, nonetheless, finite. There is, ultimately, only so much sand on the planet.
3. At any given time, your body contains only a finite number of cells, although that number may be large.
4. The set of numbers between 0 and 2 is infinite. This may be surprising to you, but keep in mind that there exist all manner of fractions lying between 0 and 2, and their number is unlimited.

Equality and Equivalence of Sets

When we consider two separate sets, we can establish the relationships of equality and equivalence between them. Two sets A and B are said to be **equal**, symbolized $A = B$, if their membership is entirely identical (keeping in mind that the order of the elements in the sets need not be the same; the criteria for equality is that the members of the two sets be identical, but they may be listed in any order we wish!). Two sets A and B are said to be **equivalent**, symbolized $A \sim B$, if they share the same cardinality. Note that if two sets are equal, they are also equivalent, but the converse is not true. There exist equivalent sets that are not equal!

For instance, $A = \{1, 2, 3, 4\}$ and $B = \{a, b, c, d\}$ are equivalent sets, since both have cardinality 4, but they are not equal, since they do not have the same members.

Note

Cardinality of a set (the quantity of members possessed by the set) can be either finite or infinite. If the set is named S, then the notation for the cardinality of a set is either $n(S)$ or $|S|$.

EXAMPLE 1.4

Are the following sets equal, equivalent, both, or neither?

1. The set of all letters in the English alphabet and the set of whole numbers between 0 and 25, inclusive EQUIVALENT
2. The sets $A = \{1, 2, 3, 4\}$ and $B = \{3, 4, 2, 1\}$ BOTH
3. The sets $C = \{3, 1, 4, 5\}$ and $D = \{a, b, c, d, e\}$ NEITHER

SOLUTION

1. In the first case, the sets are clearly not equal because one has members that are letters, while the other has members that are numbers. They are, however, equivalent sets because the cardinality of each set is 26. For the set of numbers, that may not be obvious at first glance, but think it over carefully!
2. For the second example, the sets are equal and hence also equivalent because they share exactly the same members (albeit in a different ordering, which is, as mentioned earlier, irrelevant).
3. For the third example, the sets are neither equal nor equivalent. It is obvious that their members are not identical from set to set, and set C has cardinality 4 while set D has cardinality 5.

The Complement of a Set

Once the universal set has been established, possibly by inference or implication, and a set A described, the universal set is subdivided naturally into two groups: those members of the universal set lying within A and those not lying within A. The members of the universal set that do not fall within the set A are said to form the **complement** of A, denoted A^c or A'.

Let's suppose $U = \{1, 2, 3, \ldots, 10\}$ and $A = \{1, 2, 3, 4\}$. Then $A' = \{5, 6, 7, 8, 9, 10\}$. Note that the set U is completely subdivided by a set and its complement in the sense that all members of U must lie in one of those two sets.

EXAMPLE 1.5

Let $U =$ the set of all whole numbers between 10 and 20 and $A = \{11, 13, 14, 17\}$. What is A'?

$A' = \{12, 15, 16, 18, 19\}$

SOLUTION

The set of elements of U not in the set A would be $\{12, 15, 16, 18, 19\}$. Notice that the explicit language of the description of the set dictates that 10 and 20 are not within the universal set! Had we intended them to be included, we would have used the phrase "between 10 and 20, inclusive."

EXAMPLE 1.6

Let $U =$ the set of all letters in the English alphabet, and $A = \{a, e, i, o, u\}$. What is A'?

SOLUTION

The set A consists of the vowels in the English alphabet, and therefore the complement would be the set of the consonants in the English alphabet. We can, if we wish, produce the membership list in roster form, but this would be unwieldy, and thus we can simply use the descriptive form of membership identification and state that $A' =$ the set of consonants in the English language. Observe that it is perfectly acceptable for the set A to be given in one form, while the complement is given in another.

Exercises

In the following exercises, answer with complete sentences and correct grammar and spelling.

1. What is a set?
2. What does it mean to say two sets are equal?
3. What does it mean to say two sets are equivalent?
4. What is meant by the cardinality of a set?
5. What is the empty set, and how is it symbolized?
6. What does it mean to say a set is finite?
7. What is an ellipsis, and what does it represent?

In the following problems, determine if the sets are well defined. If they are not well defined, state why.

8. The set of paid employees of the U.S. government
9. The set of most efficient computer brands
10. The set of odd integers less than 100

11. The set of well-spoken professors at Harvard University
12. The set of astronauts who have piloted the space shuttle *Atlantis*
13. The set of even integers between 6 and 7

In the following problems, determine if the sets are finite or infinite. If they are finite, state their cardinality.

14. The set of even integers between 10 and 30, inclusive
15. The set of states in the United States of America at the present time
16. The set of digits in the number "1 trillion"
17. The set of digits in the full decimal expansion of pi
18. The set of numbers between 4 and 10

Express the following sets in roster form and state the cardinality of the set.

19. The set of letters in the word "Mississippi"
20. The set of all natural numbers less than 50
21. The set of all states in the United States whose names begin with the letter N
22. $A = \{x \mid 2 - x = 7\}$
23. The set of all cities in Oregon having a population of more than 3 million people
24. The set of all states in the United States that share a border with a foreign country
25. The set of all living persons in the United States who hold or have held the office of president of the United States

Express the following sets in set-builder notation.

26. $A = \{1, 2, 3\}$
27. $B = \{0, 2, 4, ...\}$
28. $C = \{2, 3, 5, 7, 11, 13\}$
29. D is the set of all months in the year having exactly 20 days.
30. E is the set of all odd natural numbers less than 1000.

Give a verbal description of the members of the following sets.

31. $\{3, 6, 9, 12, 15\}$
32. {Fred, Barney, Betty, Wilma}
33. $\{x \in W \mid 2 < x \leq 6\}$
34. {Hawaii, Alaska}
35. $\{1, 3, 5, 7, ..., 19\}$

For the following problems, use the sets $A = \{2, 4, 6, 8, 10\}$, $B = \{3, 4, 5, 6\}$, and $C = \{a, b, c, d\}$.

36. Determine $|A|$
37. Determine $|B|$
38. Determine $|C|$

For the following problems, determine if the given sets are equal, equivalent, both, or neither.

39. $B = \{3, 4, 5, 6\}$, $C = \{a, b, c, d\}$
40. $S = \{1, 2, 3, 4\}$, $T = \{1, 3, 2, 4\}$
41. B = set of letters in the word "pool," C = set of letters in the word "lop"
42. C = the set of U.S. state capital cities, S = the set of U.S. states
43. $A = \{x \in \mathbf{N} \mid x > 2\}$, B = the set of numbers greater than 2

1.2 SUBSETS

In many situations, a particular individual may be a member of several different sets. For instance, if we consider a particular person, whom we can call John Smith, it is certainly true that John is a member of his own family, but his family is, in turn, a member of a larger group, such as a neighborhood community. That community is part of a town, which is part of a county, which is part of a state, and so on. This illustrates that some sets happen to be contained within other sets.

Given two sets, A and B, suppose that every element of A is also an element of B. This condition gives a relationship between the two sets, called the **subset** relationship. This is denoted by the symbol $\subseteq$, which is read as "is a subset of." In symbols, $A \subseteq B$ if $a \in A$ implies that $a \in B$.

Every nonempty set must have at least, trivially, two subsets. The set itself satisfies the definition of a subset, so for all sets A, we must have $A \subseteq A$. Additionally, the empty set is a subset of any set. That this is true is a consequence of the definition of the subset relationship. There are no members of the empty set that are not members of A, so $\varnothing$ is a subset of A.

Proper Subsets

If A is a subset of B, then B is called a **superset** of A, and if the set B contains additional elements that are not elements of A, then we say that A is a **proper subset** of B and use the specific notation $A \subset B$. When it is unclear whether A is a proper subset of B or if we prefer to maintain generality, the regular subset symbol, $\subseteq$, may be used, but if it is known that the subset relationship is proper, we should use the proper subset symbol for purposes of clarity.

Referring back to our known sets of numbers from mathematics, we can establish a chain of subsets as an illustration: $N \subset W \subset Z \subset Q \subset R$. Note that each of the subset relations in the example is a proper subset relation, since each superset contains additional members beyond those of the subset.

EXAMPLE 1.7

For each of the following pairs of sets A and B, determine whether set A is a subset of set B.

1. A = {John, Mary, Bob}, B = {John, Mary, Bob, Scott}
2. $A = \{2, 3, 5, 7, 11, 13, ...\}$, $B = N$
3. A = set of states in the United States, B = {New York, New Jersey}

SOLUTION

1. Yes, A is a subset of B. Note that every element of set A is also an element of set B, and thus A satisfies the definition of what it means to be a subset of B.
2. Yes, A is a subset of B. A consists of all the prime numbers, and each of those numbers also happens to be a natural number.
3. No, A is not a subset of B. In fact, the reverse is true! All the elements of B are members of set A, and therefore B is a subset of A, but A is not a subset of B.

Set notation can be a bit tricky in some cases, and care has to be taken to not confuse the symbols. Let's suppose that $A = \{a\}$ and $B = \{a, b, c\}$. Is it true that $A \in B$? This is a bit unusual because we do see the element "a" within set B, but we should be cautious. It is true that $a \in B$, but it is not true that $\{a\} \in B$ because B does not contain the set whose only member is "a." What *is* true is that $A \subset B$, since A is a proper subset of B.

The Power Set of a Set

Let's consider a particular example of a relatively small set, $A = \{1, 2, 3\}$, and attempt to list all its subsets. To begin, we have already stated that, by assumption, $\varnothing$ is a subset of A and that A is a subset of itself (this notion is so uncomfortable for some people that the set A is said to be an "improper" subset of itself, and this motivates the use of the terminology "proper subset"). Now let's turn to the other subsets. First, consider the one-element subsets. A set with one element is called a **singleton** set. There are three such subsets of A: $\{1\}$, $\{2\}$, and $\{3\}$. Now consider the two-element subsets: $\{1, 2\}$, $\{1, 3\}$, and $\{2, 3\}$. We have completed the list of all possible subsets of A and have found that there are a total of eight: A, $\varnothing$, $\{1\}$, $\{2\}$, $\{3\}$, $\{1, 2\}$, $\{1, 3\}$, and $\{2, 3\}$.

The set we can make whose members are all the subsets of a set A is called the **power set** of A, denoted 2^A or sometimes $P(A)$. Thus, another way of asking you to list all the subsets of a set A is to call for the power set of A.

> **Note**
>
> The power set of a particular set has as its members all possible subsets of that particular set. The cardinality of the power set is always $2^{n(S)}$.

It would be useful for us to know how many subsets a particular set would have so that we would know for sure we had listed them all. It turns out that for finite sets of cardinality n, there are always 2^n subsets (and this, in fact, is the motivation for the notation of the power set!). Note that a set with just one element satisfies this rule, since the set itself and the empty set are its only subsets and that the empty set (which has only itself as a subset) also satisfies the rule, since $2^0 = 1$.

For even relatively small sets, the number of subsets can be quite alarming. Consider the set of letters in the English alphabet, which has 26 members. That set has more than 67 million subsets! If you could write one subset per second, working continuously and not pausing for the inconvenience of sleep, it would take you *more than two years* to produce the complete listing.

Just as we had a symbol to represent nonmembership in a set, we have a symbol to indicate the nonexistence of the subset relationship. The symbol is the same as the subset symbol but with a slash through it: $\not\subset$. This symbol is read as "is not a subset of."

Let's do a few examples to illustrate the concept of the subset relationship.

EXAMPLE 1.8 List the subsets of the set L, whose members are the distinct letters in the word "loop."

SOLUTION

Note that the set is $L = \{l, o, p\}$. Keep in mind that we do not list the repetition here, since the description of the set specifies that the members are the distinct letters involved. Since there are three elements to the set L, our rule for subset cardinality indicates that there are eight subsets we need to find. The best place to start is to list $\varnothing$ and L first, since those are often overlooked when listing the set of all subsets of a given set. We can proceed in any manner we choose, but it may be best to follow the systematic method of listing the singletons first, then the two-element subsets. Since we've already listed the only three-element subset, L itself, we'll then be done. The singletons are {l}, {o}, and {p}, and the two-element subsets are {l, o}, {l, p}, and {o, p}. We now have eight subsets in our hands, and therefore our list is complete.

EXAMPLE 1.9

List all the subsets of the set $E = \{\varnothing\}$.

SOLUTION

Here, we must be very careful. The set E consists of the symbol for the empty set and therefore has one member. A set with one member has two subsets, itself and the empty set, so the subsets of E are $\varnothing$ and $\{\varnothing\}$.

EXAMPLE 1.10

List all subsets of the set C = {cat, dog, fox, bat}.

SOLUTION

There are four elements to set C, and therefore there are $2^4 = 16$ subsets. Again, the first two we can list are $\varnothing$ and C itself. Now we systematically list all other subsets:

{cat}, {dog}, {fox}, {bat}

{cat, dog}, {cat, fox}, {cat, bat}, {dog, fox}, {dog, bat}, {fox, bat}

{cat, dog, fox}, {cat, dog, bat}, {cat, fox, bat}, {dog, fox, bat}

A count establishes that we have shown 16 subsets, and therefore the list is complete.

Applications of Subsets

There are many interesting applications of subsets, some of which are presented in the exercises. One of those, which we will discuss as a means of closing this section, is the voting coalition problem. The premise is that a particular committee or electorate is charged with a task, and the action of the committee is determined by the outcome of a vote in which each member will either have a vote equal in strength to all other voting members or in which the votes are "weighted" in such a manner that some voters have more voting strength than others. Some predetermined quantity of votes, perhaps a simple majority, is required for a decision to be made, and the question might be how many "winning" coalitions exist within the voting structure. Such a question leads one naturally to the subject of probability, which we will investigate later in this text.

For now, let's suppose that four students form the Student Government Association and that one, the president, has two votes, while the other members have one vote each. If a proposal comes before the association, it is put to a vote by the four students, and a total of three votes (simple majority) is needed for approval of the proposal. How many coalitions of votes will produce approval?

Let's identify the four students as P (the president), A, B, and C, and construct the set of all voting outcomes where a particular individual has voted to approve the proposal. This set would consist of elements such as PAC, which is intended to indicate that the president and students A and C voted in favor. All possible outcomes are as follows:

P, A, B, C, PA, PB, PC, AB, AC, BC, PAB, PAC, PBC, ABC, PABC

Now, which of those outcomes would generate approval of the proposal? A brute-force consideration of each such vote, remembering that the president's voting strength is double that of his colleagues, reveals the winning coalitions to be {PA, PB, PC, PAB, PAC, PBC, ABC, PBC}. This subset of the set of all possible voting outcomes reveals that there are eight winning coalitions of votes that would allow the proposal to be passed by the Student Government Association.

Exercises

In the following exercises, answer with complete sentences and correct grammar and spelling.

1. What do we mean by the term "subset"?
2. What do we mean by the term "proper subset"?
3. What do we mean by the term "superset"?
4. What is the power set of a set?
5. Is it possible to have a set that has no subsets at all? Explain.

For the following sets, determine if $A \subset B$, $B \subset A$, $A \subseteq B$, $B \subseteq A$, $A = B$, or if none of these relationships exist. It is possible that more than one relationship could hold, and if that is the case, list all applicable relationships.

6. $A = \{a, e, i, o, u\}$, B = set of letters in the English alphabet
7. A = set of positive even integers less than 20, $B = \{0, 2, 4, ..., 20\}$
8. $A = \{x \mid x \in N, 10 < x < 20\}$, $B = \{x \mid x \in N, 11 \leq x \leq 19\}$
9. $A = \{x \mid x$ is a retired professional basketball player$\}$, B = {Michael Jordan, Larry Bird, Julius Erving}
10. A = set of players in the starting lineup for the New York Yankees in the first game of the 2009 World Series, B = set of players in the starting lineup for the Philadelphia Phillies in the first game of the 2009 World Series
11. $A = \varnothing$, $B = \{x \mid x \in N, x < 5\}$
12. A = set of letters in the word "Mississippi," B = set of letters in the word "sip"

For the following problems, identify the cardinality of the power set of A.

13. $A = \{a, e, i, o, u\}$
14. $A = \varnothing$
15. A = set of days in the work week
16. $A = \{x \mid x \in N, 3 \leq x \leq 5\}$
17. $A = \{1, 2, 3, 4, 5, 6, 7, 8, 9, 10\}$

For the following problems, determine if the statements are true or false. Explain your answer or give an example to illustrate.

18. There exists a set A such that $A \subset A$.
19. There exists a set A having exactly 20 subsets.
20. If a set A has 40 members and you can list one subset per second and work continuously for seven days, it is possible to list all the subsets of A in that time.
21. If A and B have the same cardinality, then so do 2^A and 2^B.
22. $\varnothing \subset \varnothing$
23. For sets A, B, and C, if $A \subset B$ and $B \subset C$, then it must be the case that $A \subset C$.

The following problems are applications using subsets and may require some "out-of-the-box" thinking.

24. If a sandwich shop allows you to pick any combination of ingredients from its menu to build a Panini sandwich and the menu shows 18 ingredients, how many different sandwiches are possible?
25. Four voters are going to vote yes or no (Y or N) on an upcoming town council issue. The measure will pass if three or more of the voters vote yes. How many different voting combinations exist, and how many of these will be results that allow the measure to pass?
26. A tyrannical historical commission has the power of decision over permission for home owners in a neighborhood to repaint their homes a proposed color. The commission has four members, each having a particular number of votes: president (four votes), vice president (three votes), architectural committee leader (two votes), and fence height overseer (two votes). If a simple majority of votes is needed to approve a resident's application, how many winning "coalitions" of votes exist?

1.3 Venn Diagrams

Venn Diagrams

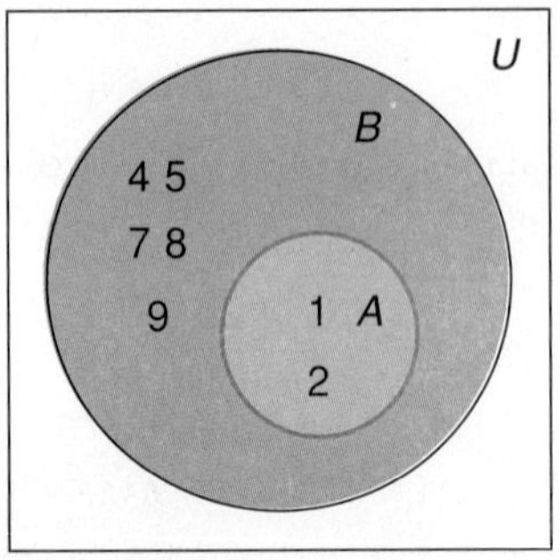

Figure 1.1 Venn diagram depicting set A as a proper subset of set B

There are times when it proves useful or desirable for us to represent sets and the relationships among them in a visual manner. This can be beneficial for a variety of reasons, among which is the possibility that a pictorial representation may reveal relationships that were unclear through other descriptive styles.

The method we are going to use is due to John Venn and was named in his honor. The process is called a **Venn diagram** construction. A Venn diagram uses a series of closed curves, usually circles or ellipses, to depict sets. The circles or ellipses are placed within a rectangular box that is intended to depict the universal set, and the relationships (overlapping, nonoverlapping, or inclusion) between the circles/ellipses will indicate corresponding relationships among the sets.

Figure 1.1 shows a typical Venn diagram representing two sets, A and B, within an arbitrary universal set. In this situation, the set $A = \{1, 2\}$, and the set

$B = \{1, 2, 4, 5, 7, 8, 9\}$. There are two significant features of this particular Venn diagram that we should observe and remark on. The first is that the elements 1 and 2 are possibly not obvious as members of set *B*. This is somewhat disguised by the fact that they lie within the circle depicting set *A*, but since that circle is contained wholly within set *B*, the elements are also members of *B*. The second observation is that there are no indicated elements outside of set *B*, and that would suggest that *B* contains all elements of the universal set.

The relationship between the two circles exhibits the same status shared between the sets *A* and *B*. *A* is a proper subset of *B*, since *B* contains all the elements of *A* as well as other elements not in *A*. Note that the circle depicting *A* is contained entirely within the circle depicting *B* but that there are other elements within the *B* circle and outside the *A* circle.

The Intersection of Two Sets

The circles representing two sets within a Venn diagram can appear in a variety of configurations. The circles may be positioned in such a manner that one is entirely contained within the other, as we saw in the first example of this section. The two circles may have a partial overlap, or they may not overlap at all.

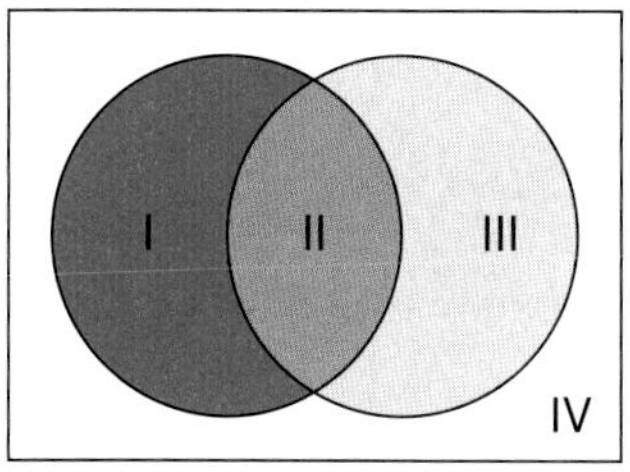

Figure 1.2 Sample Venn diagram showing areas of universal set

In the case where the circles had a partial overlapping, the situation would look something like Figure 1.2. Note that we have not identified the sets by name, but this is only a pictorial convenience to avoid overly cluttering the diagram with additional letters. We may assume that each of the circles represents a distinct subset of the universal set. We have also labeled various parts of the diagram using Roman numerals for the specific reason that we wish to discuss the various regions of the diagram produced by the occurrence of this partial overlap.

Let's begin our assessment of the situation by directing our attention to the center part of the diagram, where the two circles overlap one another. That part of the diagram, labeled with Roman numeral II, represents the members that are common to both sets, and this is an informal definition of **intersection**. Let's formalize this notion, along with the symbol for the intersection relation:

DEFINITION: $A \cap B$, the intersection of sets *A* and *B*, is the set of all elements that are simultaneously elements of set *A* and set *B*.

In what situation might sets "overlap" in this manner? Think of any two organizations to which you belong; examples might be your math class and your family. The set *A* might be the set of all students enrolled in your math class and the set *B* the set of all persons in your family. The sets *A* and *B*, by construction, have at least one common member: *you*. Thus, there is an overlap, or intersection, of the two sets that we would describe as the set of all people who are both in your math class and in your family. The word *and* is a word typically used to signify intersection.

We should mention, parenthetically, that the expression "by construction" is a rather common mathematical phrase. It means "by design," "on purpose," or "intentionally." The two sets *A* and *B* described above were designed in such a manner that we knew there was going to be an intersection because we started from the notion that you were in two particular, distinct groups and then established those groups as the sets.

EXAMPLE 1.11 Let A = the set of all whole numbers less than or equal to 10 and B = the set of even integers. Find the intersection of sets A and B.

SOLUTION

Since set A is fairly small, we can list its membership through the use of a roster, which would be $A = \{0, 1, 2, 3, 4, 5, 6, 7, 8, 9, 10\}$. The intersection of set A with set B would be those elements common to the two sets. Consideration of the elements of A suggests the intersection to be $A \cap B = \{0, 2, 4, 6, 8, 10\}$.

EXAMPLE 1.12 By using a Venn diagram, produce the prime factorizations of the numbers 42 and 30, and from that Venn diagram determine the greatest common factor of those numbers.

SOLUTION

This illustration will allow us not only to explore the notion of the intersection of two sets but also to remind ourselves of the process of prime factorization. The prime factorization of 42 is (2) (3) (7), while that of 30 is (2) (3) (5). We can take the universal set to consist of all natural numbers (since no universal set is specified, we can take it to be any convenient superset of the numbers involved in our problem) and the sets A and B to be the prime factors of 42 and 30, respectively. Therefore, the situation, visually, would resemble Figure 1.3. Note that the factors of 2 and 3 occur in both sets and hence form the intersection of sets A and B. The product of the prime factors within the intersection is the greatest common factor of 42 and 30. In this case, it is fairly trivial to compute that the greatest common factor of the numbers is 6.

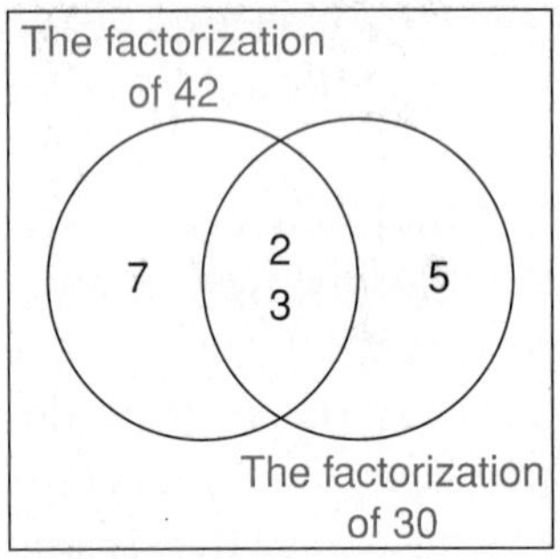

Figure 1.3 Venn diagram showing prime factorizations of 42 and 30

The Union of Two Sets

Returning to the matter of Venn diagrams, we can also view the diagram as a way to depict the result of combining the two sets together to form one larger set. When the two sets are combined, we say that we are creating the **union** of the two sets; we obtain the combination of the regions marked I, II, and III on the diagram.

DEFINITION: $A \cup B$, the union of the sets A and B, is the set of all elements that are elements of set A, set B, or both. Building on our previous example of your class and your family, the union of the two sets would be the set of all the members of your math class, together with the members of your family.

Just as the word "and" signaled intersection, the word "or" references union. We would say, in the previous example, that the union is the set of all persons who are in your math class or who are in your family. There might be some debate about this phrasing because some might wonder if that statement does not intend to include those individuals who happen to be in both sets. However, the understanding in mathematics and in logic is that the word "or" is meant to represent the **inclusive or**, which means "either or both." So, when we say that

x is an element of set A or set B, then this accepts the possibility that x might be an element of both of the sets as well.

EXAMPLE 1.13

By using a Venn diagram, produce the prime factorizations of the numbers 42 and 30 and use them to determine the least common multiple of 42 and 30.

SOLUTION

We have already produced the prime factorizations of these numbers as well as the Venn diagram depicting the situation (see Figure 1.4). The least common multiple of the two numbers is produced by forming the product of all prime factors of the two numbers, where the common prime factors are used only once in the formation of the product. In this case, the least common multiple of the numbers 42 and 30 would be (2) (3) (5) (7) = 210.

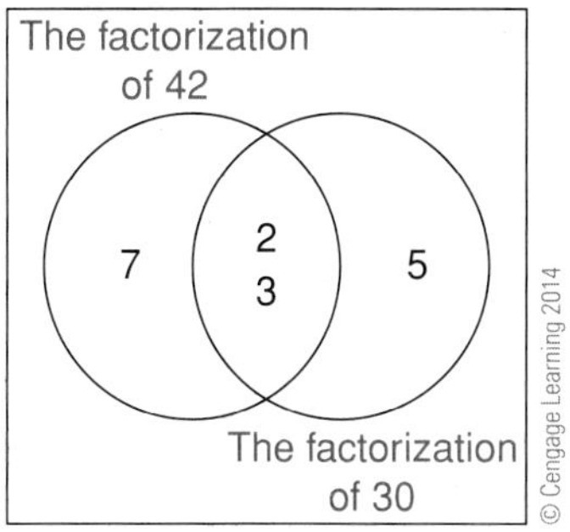

Figure 1.4 Venn diagram showing prime factorizations of 42 and 30

Disjoint Sets

Another possible orientation of two sets in a Venn diagram would be the case where the two sets were situated in such a way that there was no overlap of their representative circles. In such a case, we say the intersection of the two sets is empty or that the sets share no common members. The two sets are said to be **disjoint**, and the circles depicting them in the Venn diagram would have no overlap, as seen in Figure 1.5.

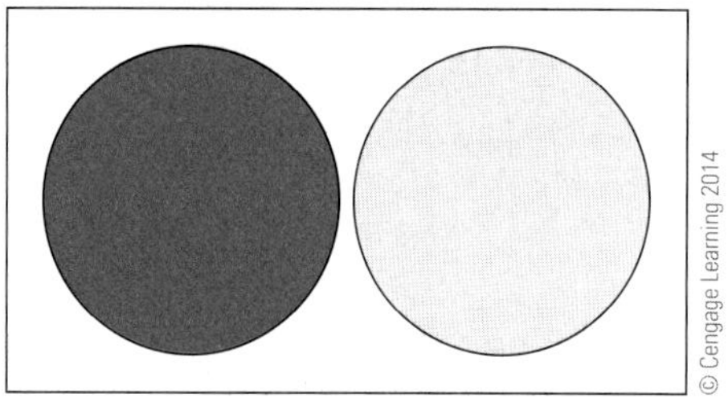

Figure 1.5 Venn diagram showing disjoint sets

An example of disjoint sets might be the set A = all even natural numbers and the set B = all odd natural numbers. Since nothing is both an even natural number and an odd natural number, the two sets are disjoint.

Before moving ahead, let's consider several more examples to illustrate the concepts introduced to this point.

Suppose we have two sets of objects, relatively well understood, so that the example will be fairly self-evident as an illustration. For the universal set, we will choose the student body at a public high school. The two individual sets involved will be A = {football players on the varsity team} and B = {baseball players on the varsity team}.

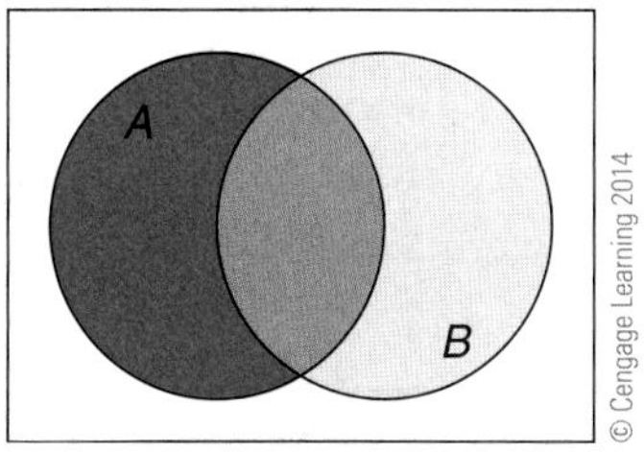

Figure 1.6 Venn diagram relating varsity football players and baseball players

What would the Venn diagram of the situation look like? In our experience, it is common for varsity football teams and varsity baseball teams have at least some athletes in common, since there are what we refer to as "dual sport athletes." It is unlikely that, in most schools, all of the athletes participating in one of the sports would be participating in the other, so we can assume that the proper subset relationship does not hold in this situation. Therefore, we can expect the sets to not be disjoint. The Venn diagram would probably look like Figure 1.6. How would we describe the regions in the diagram in words? The region we had labeled in the general diagram as region I is the part of set A that does not overlap in any way with set B. This would be the set of students in the student body who are on the varsity football team but not on the varsity baseball team. The region we had labeled as II is where the two sets intersect. This would be the set of students in the student body who are on both the varsity football team and the varsity baseball team. The region labeled as III would be those students in the student body who are on the varsity baseball team but not on the varsity football team.

If we consider the combination (union) of the two sets, $A \cup B$, that would be the set of all students in the student body who are on the varsity football team or the varsity baseball team (keep in mind that we are using the inclusive *or* here, so that includes the students who are on both teams).

The region outside the two sets, the complement of $A \cup B$, is the set of students in the student body who are not on the varsity football team or the varsity baseball team.

Can we imagine a nonnumerical situation that would lead to a Venn diagram incorporating disjoint sets? An example would suit if we could find two sets that were necessarily having no overlap. Let's suppose the universal set is the set of all household pets in the city of Plains, Georgia. We can let A = pet cats in the city of Plains and B = pet dogs in the city of Plains, and then the Venn diagram would look like Figure 1.7. It is clear that there are no pets that are both a cat and a dog at the same time, and therefore the sets A and B have no overlap.

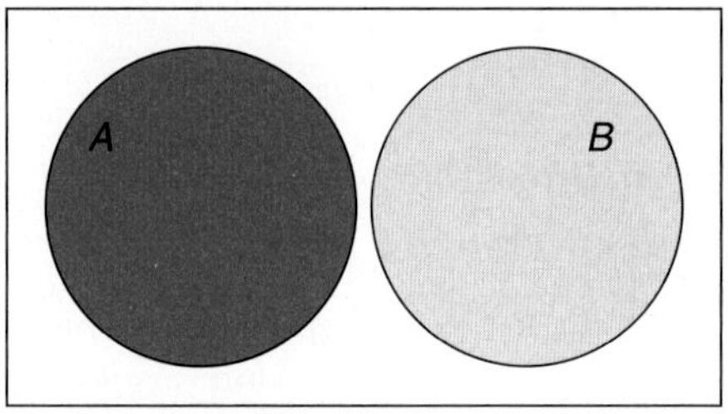

Figure 1.7 Venn diagram showing disjoint nature of sets of pet cats and pet dogs in Plains, Georgia

Set-Theoretic Expressions and DeMorgan's Laws

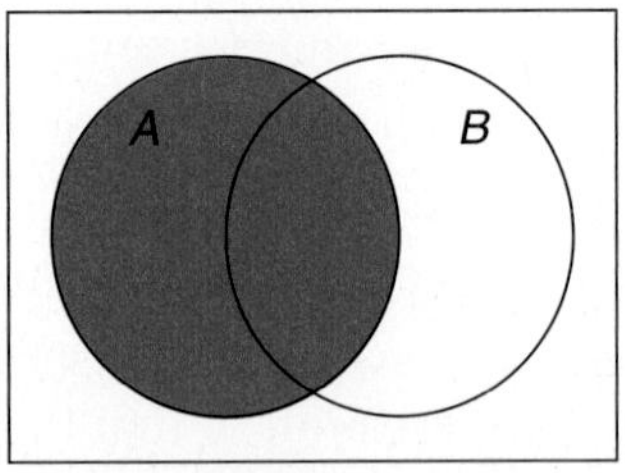

Figure 1.8 Venn diagram showing set A as a shaded region

Venn diagrams have another useful function, the confirmation of equality for set-theoretic expressions. When we use the term "set-theoretic expression," we refer to expressions where sets are associated using the operations of intersection, union, or complementation. An elementary expression with which we are already familiar would be $A \cap B$.

If you were given two set theoretic expressions, such as $A \cap B^c$ and $A \cup (B \cap A)^c$, you might wonder whether they are they equal to one another. There are various methods we could employ to assess the situation, but one useful and timely method employs Venn diagrams. The statements are equal if their Venn diagram representations are identical.

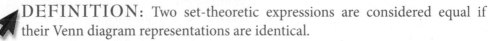

DEFINITION: Two set-theoretic expressions are considered equal if their Venn diagram representations are identical.

Let's consider the first statement, $A \cap B^c$. We want to take the intersection of A with the complement of B, so we consider the two sets independently and then construct their intersection (see Figures 1.8 and 1.9). The overlap of the

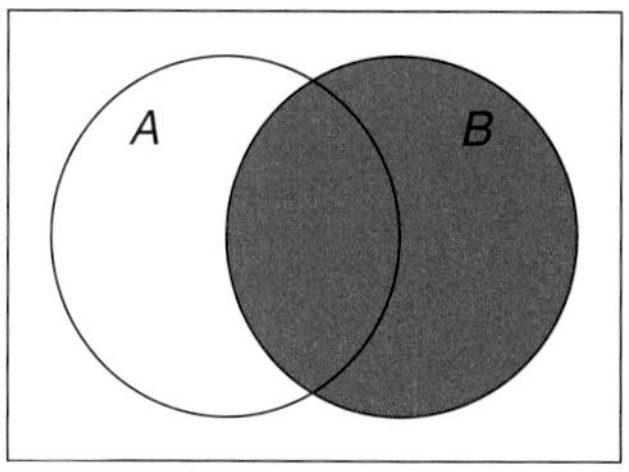

Figure 1.9 Venn diagram showing set B^c as a shaded region

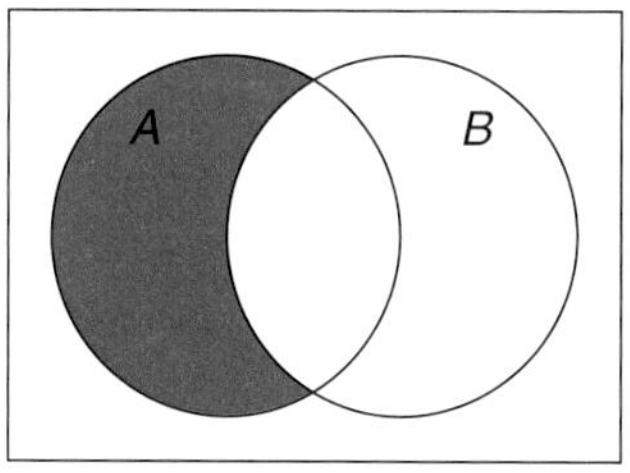

Figure 1.10 Venn diagrams showing set A, and set B^c as shaded regions

two regions is the part of the Venn diagram shaded simultaneously in both graphs (see Figure 1.10).

Now consider the other expression, $A \cup (B \cap A)^c$ and repeat the process. Recalling that $(B \cap A)$ is the "football-shaped area" where the circles overlap, we see that its complement is everything outside that part of the diagram. When combined with set A, the entire Venn diagram is filled (see Figure 1.11). The two expressions do not yield the same Venn diagram, and hence the expressions are not equal to one another.

There are two famous set-theoretic results we shall ask you to confirm in the exercises named in honor of the logician who first defined them. **DeMorgan's laws**, first formalized by Augustus DeMorgan in the nineteenth century, state that the operations of union and intersection interchange under the operation of complementation. Symbolically,

$$(A \cup B)^c = A^c \cap B^c, \text{ and } (A \cap B)^c = A^c \cup B^c$$

The result was actually known in antiquity (similar results were employed, for example, by Aristotle), but through what is known as the development of algebraic logic, performed by the logician George Boole in the nineteenth century, it became and is still credited to DeMorgan.

We will establish the validity of the first of DeMorgan's laws here and leave the other to the exercises for you to confirm.

EXAMPLE 1.14

Verify the set theoretic equality $(A \cup B)^c = A^c \cap B^c$

SOLUTION

We will produce the Venn diagram representing the expression on each side of the equation and demonstrate through comparison of the results that the Venn diagrams are identical. This will establish the first of DeMorgan's laws.

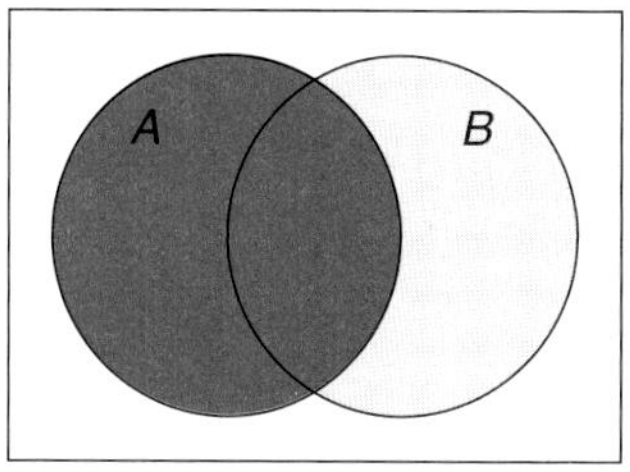

Figure 1.11 Venn diagram showing $A \cap B^C$

First we note that $A \cup B$ consists of the combined interior regions of sets A and B and that the complement of that union would be the region of the universal set lying outside both of the circles (see Figure 1.12).

When we consider A^c, this is the part of the universal set lying outside set A, and similarly B^c corresponds to the part of the universal set lying outside set B (see Figure 1.13). If the intersection of these two regions is taken, we obtain the result shown in Figure 1.14 and see that our result is the same diagram we obtained earlier. Hence, the two expressions are equal.

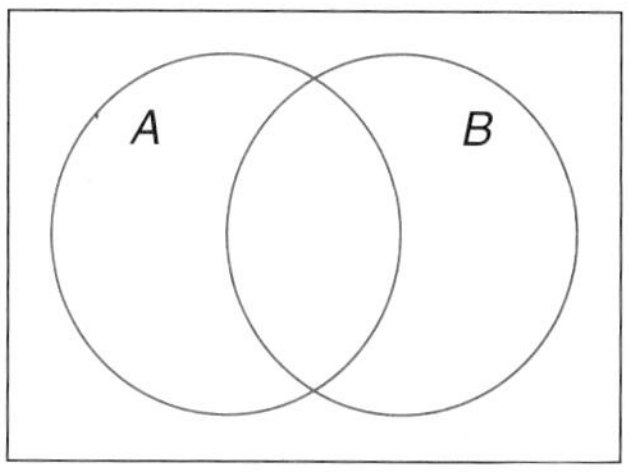

Figure 1.12 Venn diagram showing $(A \cup B)^c$

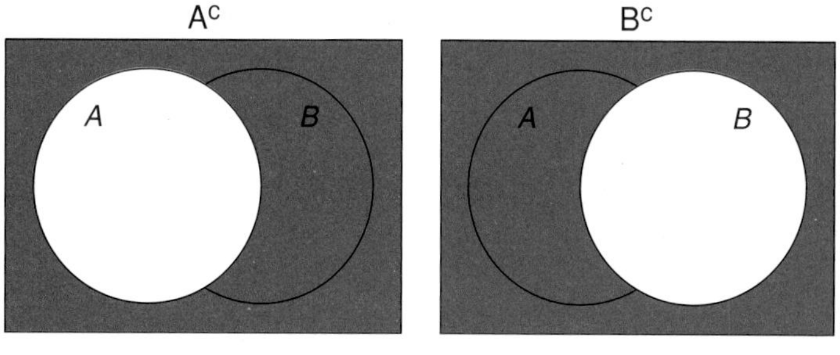

Figure 1.13 Venn diagrams showing A^c and B^c

In the examples we have seen so far, we have limited our Venn diagrams to have two sets depicted within the universal set. This is not necessary, however,

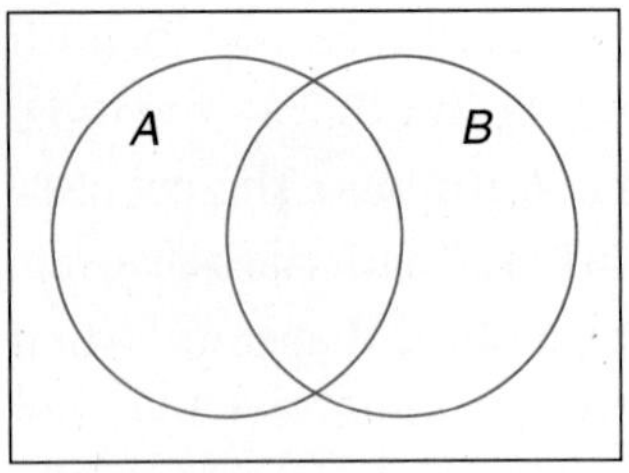

Figure 1.14 Venn diagram showing $A^c \cap B^c$

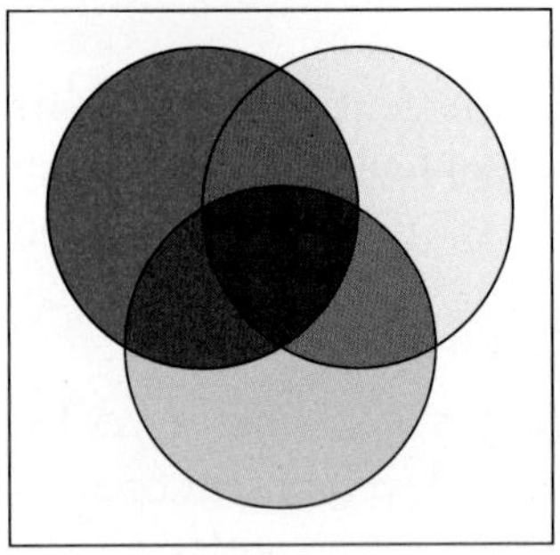

Figure 1.15 Venn diagram showing three sets

and there could be as many sets as desired. For practical reasons, we will limit our examples to those having three sets within them. Such a Venn diagram would look something like Figure 1.15.

The situation becomes slightly more complicated with three sets in the Venn diagram, but there are no real additional concepts involved. Let's see if we can, in words, describe what the regions might mean, just using the ideas we have developed to this point. We'll number the pieces of the Venn diagram (see Figure 1.16) and then proceed.

The sets will be *A* in the upper left, *B* in the upper right, and *C* in the lower middle (see Figure 1.17). Let's think of what the individual regions would mean and describe them.

The region marked 5, the central region, is the area common to all three sets. This, as has been described before, is the intersection of the sets. We would designate this region of the Venn diagram as $A \cap B \cap C$.

Consider the football-shaped region that consists of the areas marked 4 and 5. This is the overlap of sets *A* and *C*, which we would describe as $A \cap C$. Do you see how the regions are described using the same concepts as was done when only two sets were involved? Each region of the three-set Venn diagram can be interpreted in a similar fashion.

EXAMPLE 1.15

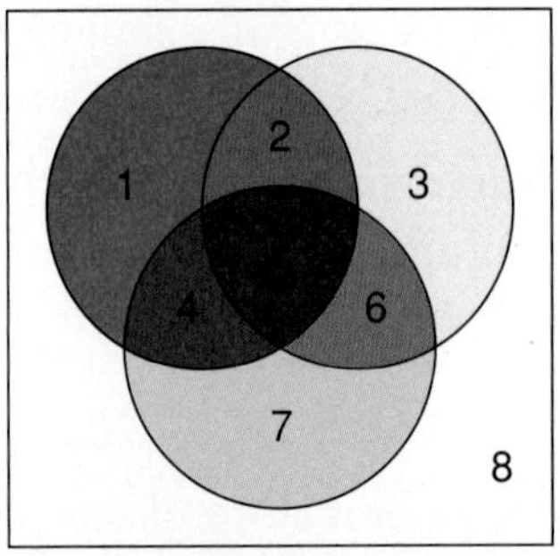

Figure 1.16 Venn diagram showing three sets, with distinct regions of the diagram indicated by number

Let the universal set be the set of students in a local college. Set *A* will represent the students enrolled in an astronomy class, set *B* will represent the students enrolled in an biology class, and set *C* will represent the students enrolled in a chemistry class. Recalling the labels we had placed on the Venn diagram, interpret the regions labeled 1, 2, and 6 of the diagram.

SOLUTION

How would we interpret region 1? These students are enrolled in an astronomy class, since they are contained in set *A*, but they are not part of either of the other two sets and therefore are not enrolled in either a biology class or a chemistry class.

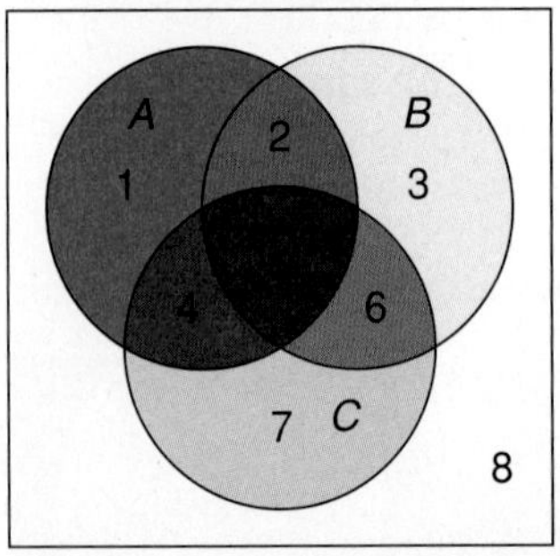

Figure 1.17 Venn diagram showing three sets, with distinct regions of the diagram indicated by number and letter

What about region 2? The students in this region would be in both the astronomy class set and the biology class set and therefore are taking both of those types of classes. They are not, however, within the chemistry class set, and therefore we would describe them as being students who are enrolled in an astronomy class and a biology class but not in a chemistry class.

Region 6 is the set of students that are within the biology set and the chemistry set but not the astronomy set. These students would be taking astronomy and biology but not chemistry.

Notice that there is a logical interpretation we can make about each region and a particular, specific, unambiguous way that we can describe members of the universal set that fall within each region. This will play a significant role when we look at applications of Venn diagrams in the next section of the text.

Intersections, Unions, and Cardinality

Whenever we contemplate sets or parts of sets, it's natural to wonder about their cardinality. What about the cardinality of one of the combinations of sets, such as the intersection of two sets, or the union of two sets?

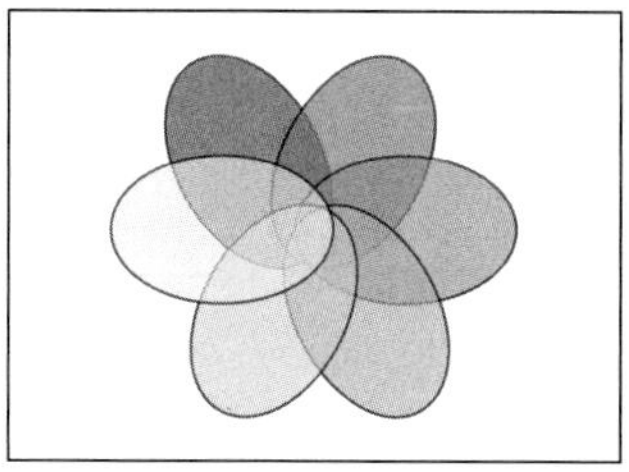

Figure 1.18 Venn diagram showing six sets within the universal set

In order for us to understand this, let's consider cases. If the two sets are disjoint, then clearly the cardinality of their union must be the sum of the individual cardinalities as a consequence to the definition of disjoint sets. If they are not disjoint, then we can reason as follows. Suppose the cardinality of set A was added to the cardinality of set B. We would obtain a total that was in excess of the cardinality of the union because those elements that were in both sets would have been counted twice. In order to compensate for this, we can subtract the cardinality of the intersection, which will remove the double counting. That is,

$$n(A \cup B) = n(A) + n(B) - n(A \cap B)$$

We could reason in a similar manner to obtain a formula for the cardinality of the intersection, but there is no need. The formula given above, being algebraic and involving the cardinality of the intersection as it does, can be solved for that cardinality, and our second formula will be obtained. That is, the formula for the cardinality of the intersection of sets A and B is given by

$$n(A \cap B) = n(A) + n(B) - n(A \cup B)$$

EXAMPLE 1.16

Suppose there we have the following set of numbers: $A = \{1, 3, 5, 7, 9\}$, $B = \{2, 3, 4, 5\}$. Verify the formulas given for the cardinality of the intersection and of the union by producing the relevant sets and determining their cardinality directly.

SOLUTION

We can see the intersection of the two sets is $\{3, 5\}$, and hence its cardinality is 2. The union of the two sets is $\{1, 2, 3, 4, 5, 7, 9\}$, and it has cardinality 7. By observation, $n(A) = 5$ and $n(B) = 4$, and the derived formula tells us that the cardinality of the intersection should be $5 + 4 - 7 = 2$, which is precisely what we found.

On the other hand, the cardinality of the union was 7, and the formula predicted that the cardinality of that set would be $5 + 4 - 2 = 7$, so the example does illustrate the validity of the formulas.

Earlier, it was mentioned that we would only consider three-set Venn diagrams in this development. Before moving on, we should at least mention what happens when more than three sets occur because the outcome is visually fascinating.

We could use any shapes, in principle, to depict the sets, but the common practice is to use circles or ellipses. When there are more than three sets, the circular approach is generally abandoned, and we appeal to ellipses, which allow us to fit more sets into our diagram in an aesthetically pleasing manner. Suppose there were six such sets. The Venn diagram would look something like Figure 1.18.

The images generated by Venn diagrams with more than three sets display a feature that is, perhaps, not so evident with the diagrams having only two or three sets. By constructing the diagram using ovals of similar size and rotation determined by dividing 360° by the number of sets, we generate beautiful images possessing rotational symmetry. A figure possessing such symmetry is unchanged by rotation through a fixed number of degrees (in this case, 60°). Although they are striking images, they can be difficult to work with in practice.

Exercises

In each of the following, construct a two-set Venn diagram that depicts the described region.

1. $A \cap B^c$
2. $A^c \cup B$
3. $(A \cap B)^c$
4. $(A \cap B)^c \cup B$
5. $(A \cup B^c)^c$
6. $A^c \cap A$
7. $(A^c \cap A)^c$

Use Venn diagrams to determine if the following equalities are true or false.

8. $(A \cup B)^c = A^c \cap B^c$
9. $(A \cup B)^c = A^c \cup B^c$
10. $A^c \cap B = B$
11. $(A^c \cup U)^c = A$
12. $(A \cap B) \cup (A \cup B^c)^c = B$
13. $(A \cap B) \cup (A \cap B^c) = A$
14. $(A \cup B) \cap A = A$
15. $(A \cap B)^c = A^c \cup B^c$

For the following sets of Venn diagrams, use set notation to describe the shaded area, using intersection, union, or complementation if needed. The answers are not necessarily unique.

16. In the readings of this section, we argued for the formulas for the cardinality of the intersection and for the union of two sets. By similar process of reasoning, generate the formula for the cardinality of the union of three sets.
17. Following up on Problem 16, generate the formula for the cardinality of the intersection of three sets.

1.4 Applications of Sets

Now that we have developed some of the terminology and concepts of sets, we will turn our attention to an examination of some of the applications in which sets might prove useful. There are many such applications, and we will touch on a few of them here.

The Survey Problem

In a survey problem, we have information about the cardinality of various sets and their intersections or unions with other sets. The problem is, typically, to determine some fact about the situation, such as the total number of individuals involved or the number of individuals possessing some particular trait. Such problems are often encountered in the leisure press in the form of "logic puzzles."

Consider the following situation. A group of 500 students are asked if they are taking a math course, a philosophy course, both, or neither. The poll finds that 178 students are taking both types of courses, 88 students are taking neither type of course, and 308 are taking a philosophy course. With that information in hand, we are now asked the following questions: How many students are taking *only* a philosophy course but not a math course? How many students are taking a math course? How many students are taking *only* a math course but not a philosophy course?

You have probably seen such problems in the past and recognize that it boils down to assembling the data in an organized fashion and using the known information to piece together the facts you require in order to answer the questions.

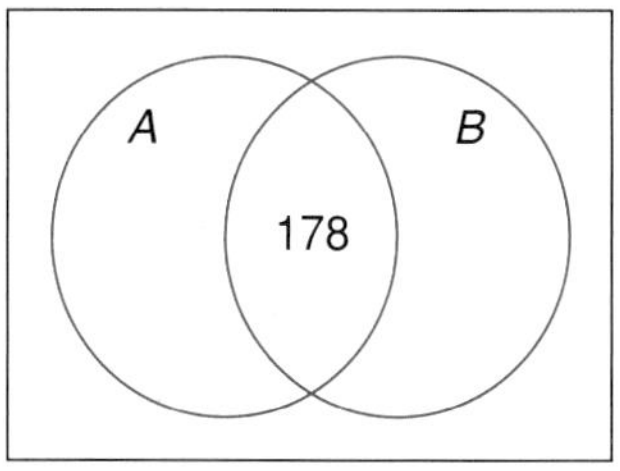

Figure 1.19 Venn diagram showing initial completion of cardinality of region of intersection

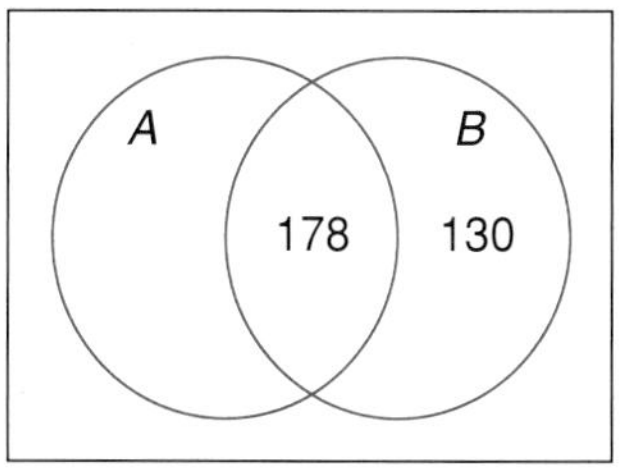

Figure 1.20 Venn diagram showing second region's cardinality

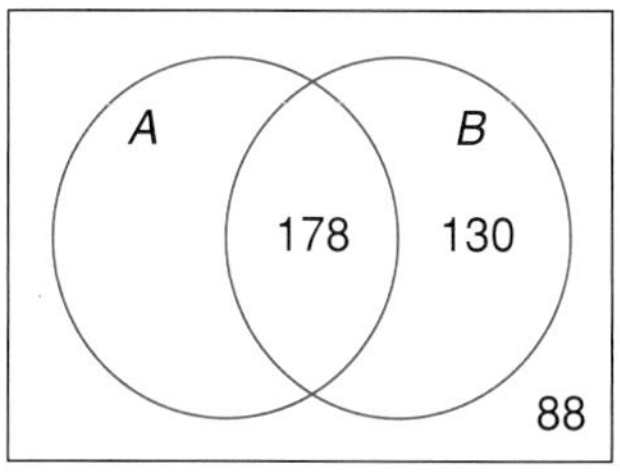

Figure 1.21 Venn diagram showing cardinality of that portion of the universal set outside the two sets

The process can be systematized by construction of a Venn diagram, as we will see, and this is particularly useful when the problems become more complex, as will be the case in the next example.

There are two sets involved here: A = {students taking a math class}, and B = {students taking a philosophy class}. The universal set is U = {500 students being surveyed}. We will construct a two-set Venn diagram and use the given information to fill in the various components of the diagram with the cardinality of the set depicted by each region. It is typically best, if possible, to start in the middle of the diagram and work your way outward, but it is not necessary, and we may not have enough information to do so.

We'll start with the general diagram (see Figure 1.19) and observe that we are told that 178 students are taking both course types. The diagram will be gradually "filled in" with the information as we consider it.

There is no specific series of steps we must follow in order to reach our conclusion, since every problem will provide us with different combinations of information. For this reason, filling in the rest of the diagram must be performed in an ad hoc manner where we attempt to deduce the cardinality of each region of our graph.

We note that the problem told us 308 students were taking a philosophy course. Since 178 students are already accounted for in the overlap of the math and philosophy sets, this implies that 130 students must be within the philosophy class circle yet outside the overlap with the math class circle. Thus, we can fill in a second part of the diagram, as shown in Figure 1.20.

Finally, we are told 88 students took neither course and thus lie within the universal set but outside of both of the individual sets, as seen in Figure 1.21.

There is just one region of the Venn diagram whose cardinality is yet to be determined. This is the group of students who are taking a math course yet not taking a philosophy course. The problem gives no direct data about that collection of students, but we can make an inference based on all available data. The universal set was known to have 500 members, and 396 are now accounted for in the already-known areas. This means 104 students are unaccounted for and therefore must lie in the remaining region (see Figure 1.22).

The situation is slightly more complicated when three sets are involved but not oppressively so. We will consider an example of such a problem before looking at a different sort of application.

EXAMPLE 1.17

Suppose 200 households are surveyed to determine how many cats, dogs, and ferrets are kept as pets. It is found that three households have all three types of pet, five have a cat and a ferret, 11 have a dog and a ferret, 20 have a cat and a dog, 66 have only a cat, 46 have only a dog, and 46 have no pets at all. Construct a Venn diagram to describe this situation.

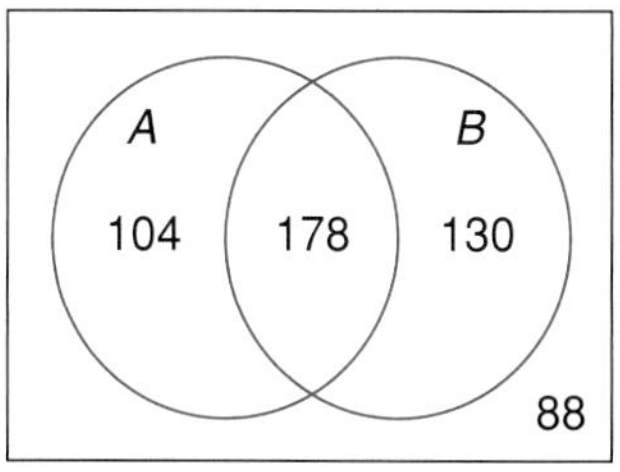

Figure 1.22 Completed Venn diagram

SOLUTION

As before, we begin with a blank Venn diagram and attempt to fill in all regions with cardinalities. The sets will be A = {households having a cat}, B = {households having a dog}, and C = {households having a ferret}. We will suppress some of the work, suggesting where the values come from, and see if the logic is apparent.

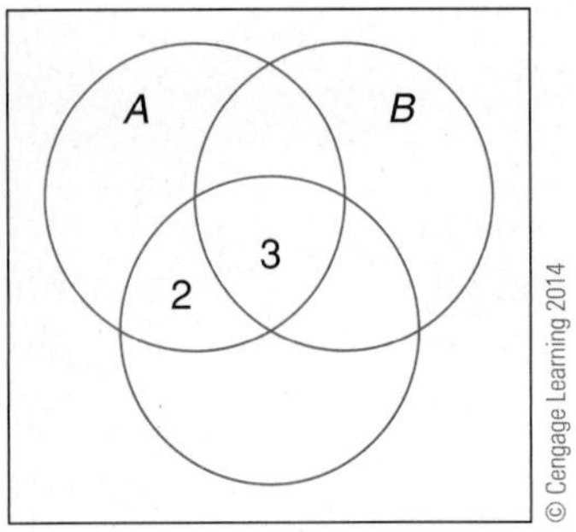

Figure 1.23 Venn diagram showing cardinalities of regions in intersection of set of families with cats with set of families with ferrets

First note that three households have all three types of pets and that five have a cat and a ferret. This tells us that the overlap of all three sets has three households in it and that of the five households having a cat and a ferret, only two are such that the household has no dog. Thus, we can begin our diagram as shown in Figure 1.23.

We are also told that 11 have a dog and a ferret and that 20 have a cat and a dog. Two more regions can be filled in for the table (see Figure 1.24).

Can you see from where the values of the numbers are being obtained? It may take some consideration for you to be able to follow the reasoning, but be sure that you do before moving on to the rest of the explanation!

We know that 66 families had only a cat, 46 had only a dog, and 46 had no pets at all. This allows us to complete nearly all of the remaining fields in Figure 1.25.

All that remains is the families who owned only a ferret, and since there were 200 families altogether and 188 are now accounted for, the remaining region has 12 families, and the problem is finished.

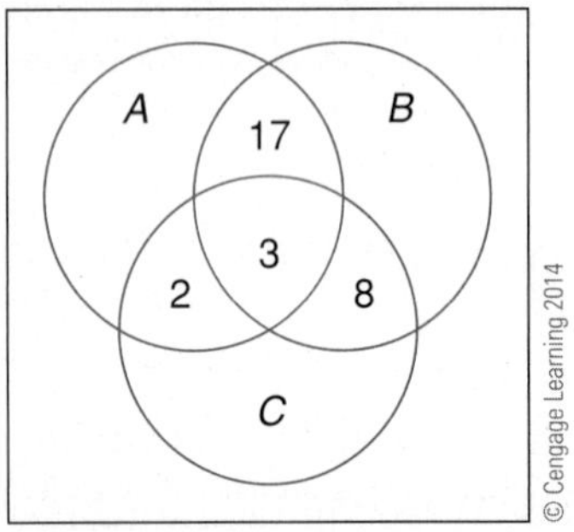

Figure 1.24 Venn diagram with remaining intersection cardinalities included

Establishment of Correct Reasoning

Venn diagrams can also be used to assess reasoning. The term "reasoning" here refers to a determination if a series of evidential facts necessarily leads to a particular conclusion. When used in this manner, the Venn diagram is referred to as an **Euler diagram**, named after the famous Swiss mathematician Leonhard Euler (pronounced "oiler").

As an illustration, consider the following series of statements:

- All bachelors are reclusive.
- Some reclusive people are strange.
- Therefore, some bachelors are strange.

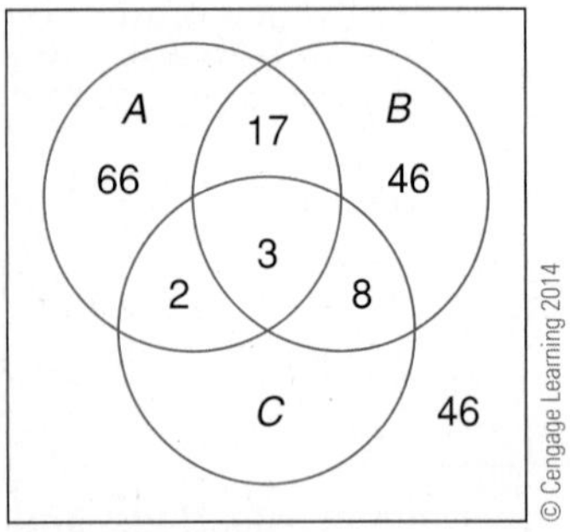

Figure 1.25 Venn diagram showing all explicitly provided information

Does this sound convincing to you? The statements are rather bold ones, making what we call "universal" and "existential" claims about persons and drawing a conclusion based on them. The term "universal" refers to the statement employing the word "all" and indicates that the assertion holds for every single bachelor. The term "existential" refers to the statements having the word "some" and purports that there are at least a few people who are strange and a few bachelors who are as well.

The reasoning is what we may describe as "sound" (or "valid" if we can establish that); assuming that we accept the first two statements (which we will come to know as premises) as being true, then the third statement, following the word "therefore," must be true as well. That final statement we will refer to as the conclusion of the argument. These terms will be explored more fully in Chapter 2, but with Venn diagrams we can investigate this problem with no further knowledge in our hands.

What we will do here is to draw an Euler diagram that captures the precise meaning of the premises and then assess whether the resulting diagram supports the conclusion. Recall that the premises were as follows:

- All bachelors are reclusive.
- Some reclusive people are strange.

We have three types of people being described: bachelors, reclusive people, and strange people. We will name those sets as *B*, *R*, and *S* so that the names of the sets reflect their membership. This is a common practice, we should mention, to choose set names in such a way that (when possible) the name of the set relates directly to the membership. It is not necessary, however, and you could always use the names *A*, *B*, and *C* for your sets if you wish. The universal set is not specified, but we can assume it to be any convenient superset of all three sets described in the problem. We will take the superset to be the set of all male humans and designate that as *U*.

Consider the first statement. "All bachelors are reclusive" tells us that the set of bachelors is entirely contained within the set of reclusive people or that (in our specific language) the set of all bachelors is a subset of the set of all reclusive people. Our experience tells us that some recluses are married, and therefore the set of bachelors must, in fact, be a proper subset of the set of reclusive people.

The second premise, "Some reclusive people are strange," indicates that there exists an overlap between the set of reclusive people and the set of strange people but that *not all* reclusive people are within the set of strange people. Remember, the premise merely states that *some* reclusive people are strange.

The logical argument is valid if there is no possible way in which we could produce an Euler diagram in contradiction with the content of the conclusion, "Some bachelors are strange." The conclusion indicates that there is definitely an overlap between the set of bachelors with the set of strange people. Must this be the case?

We can envision an Euler diagram construction where the premises would be satisfied, but the conclusion is not required to be true. Consider, for instance, the diagram in Figure 1.26. Here, the set of bachelors is clearly a subset of the set of reclusive people, and there are some reclusive people who are within the set of strange people. Notice, though, that there is, in this orientation, no overlap between the set of bachelors and the set of strange people.

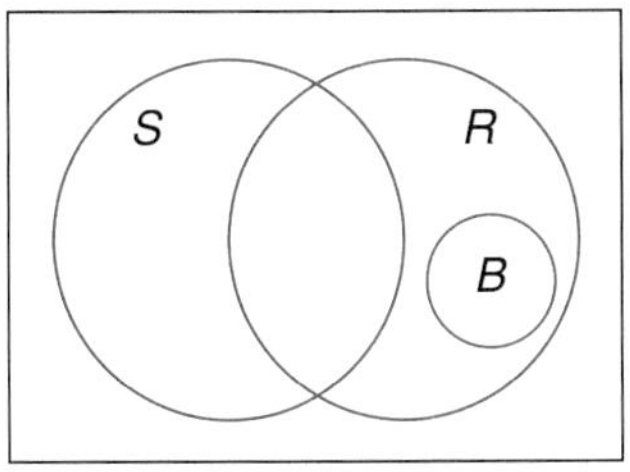

Figure 1.26 Euler diagram showing possible relationships between sets *B*, *R*, and *S*

You may want to suggest that the orientation we have depicted is not the only possibility and that it is conceivable that the circle depicting set *S* *could* overlap with set *B*. This is precisely the point, however! The point is that it is not *necessary* that an overlap between sets *B* and *S* exists, and therefore we say that the argument lacks validity and is disproven using an Euler diagram analysis.

Let's take a look at another example. "All children are graceful. Sarah is not graceful. Therefore, Sarah is not a child." Keep in mind that what we are not doing is assessing the truth or falsity of the individual statements. It is not relevant, in this situation, whether the statements are factually true. Our question is this: if we *assume* that the premises are true, *must* the conclusion be true as well, or is it *conceivable* that under the given assumptions the conclusion could be false?

We will assign names to the sets involved: *C* will be the set of children, *G* the set of graceful people, and *S* the singleton set consisting of Sarah. The universal set *U* shall be the set of all people.

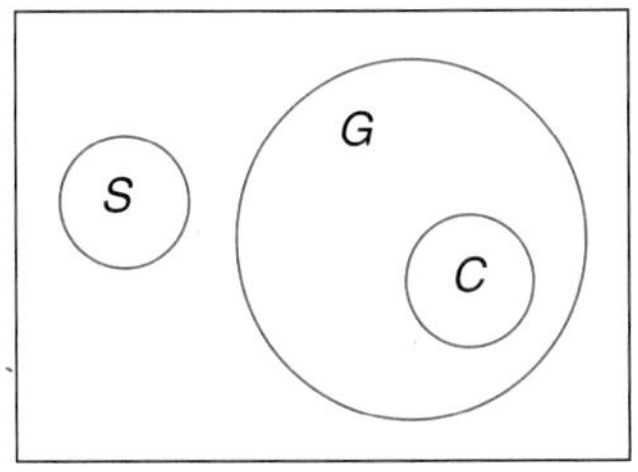

Figure 1.27 Euler diagram showing relationships between sets *S*, *G*, and *C*

The first premise, "All children are graceful," indicates that *C* is a proper subset of set *G*. The second premise, "Sarah is not graceful," tells us that the singleton set *S* is not contained in the set *G* and so is in G'. The Euler diagram for this situation strictly limits our positioning of the set *S* in such a way that it must lie outside the set *G* and therefore at a distance from the set *C* (see Figure 1.27). Consequently, it is impossible that Sarah could be a child, and the argument is valid.

Arguments of the form demonstrated in the previous two examples are referred to as **syllogisms**, which are sets of premises followed by a conclusion. They are special forms of arguments that contain **quantifying** words, such as *some*, *all*, and *none*.

EXAMPLE 1.18

Determine whether the following syllogism is valid: All ID cards are made of plastic. My credit card is made of plastic. Therefore, my credit card is an ID card.

SOLUTION

Let's call the sets involved I = {set of all ID cards}, P = {set of all things made of plastic}, and C = {set consisting of my credit card}. The universal set will be an arbitrary set containing all these sets as subsets, possibly the set of all manufactured things.

The first premise tells us that the set I is a proper subset of the set P. The second premise tells us set C is a proper subset of set P as well. The question now becomes, Must it be the case that set C lies within set I, or can we envision a scenario where set C could be outside of set I?

Consider the Euler diagram in Figure 1.28. Since there is a potential orientation of the sets that would allow set C to exist outside of set I, we conclude that the reasoning is invalid.

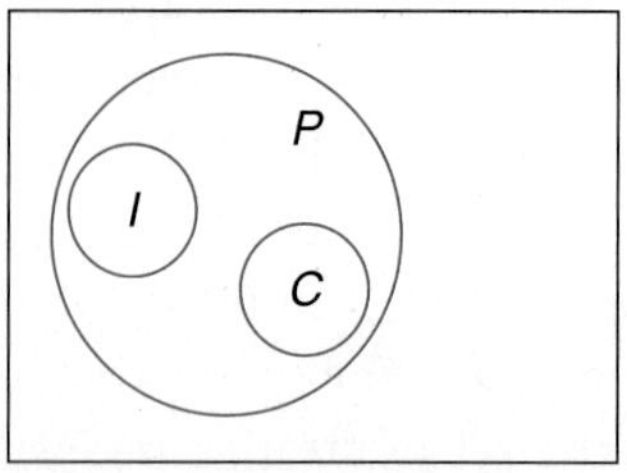

Figure 1.28 Euler diagram showing possible orientation of sets I, P, and C

Exercises

1. Suppose that your college conducts a survey to determine where students are residing. Of 175 students surveyed, it was found that 79 lived on campus, 93 lived in an apartment, and 44 lived in apartments on campus. Of those interviewed, how many lived in apartments off campus? How many lived on campus but *not* in an apartment? How many students lived neither on campus nor in an apartment?
2. In a small midwestern town, a survey was conducted of the 125 residents of a particular neighborhood. It was found that 88 of the residents owned a car, 59 owned a truck, and 21 owned no vehicle at all. Of those surveyed, how many owned both a car and a truck? How many owned only a truck? How many owned only a car?
3. At a particular cat rescue, a set of 50 cats was examined. It was found that 12 had only black fur, 11 had only orange fur, 9 had only gray fur, 24 had some black fur, 6 had some black and some orange fur but no gray fur, 5 had some black fur and some gray fur but no orange fur, and no cats had both gray and orange fur but no black fur. Find the number of cats who had at least some one of the three colors in their fur, the number that had all three colors in their fur, the number that had none of the colors in their fur, and the number that had exactly two of the colors in their fur.
4. A police officer studied 63 cars passing the local high school. He observed that 19 cars were driven by what appeared to be teenagers, 37 were driven by females, 13 were driven by what appeared to be teenage boys, 6 were driven by what appeared to be teenage girls, and 31 were driven by

women who were past teenage years. How many drivers appeared to be men who were past teenage years?

5. A State Patrol officer sampled cars crossing the border from Colorado into Wyoming. In his report he indicated that of 95 cars sampled, 45 cars were driven by men, 63 cars were driven by Colorado residents, 53 cars had three or more passengers, 37 cars were driven by men who were Colorado residents, 35 cars were driven by men and had three or more passengers, 30 cars were driven by Colorado residents and had three or more passengers, and 25 cars were driven by men who were Colorado residents and had three or more passengers. His supervisor read the report and determined that the report was in error. Explain how the supervisor knew this.
6. A survey was taken of students at State College. If 75 students were surveyed and it was found that 25 were majoring in engineering, 26 were majoring in biology, 30 were majoring in chemistry, 8 were double majoring in engineering and chemistry, 11 were double majoring in engineering and biology, 7 were double majoring in biology and chemistry, and no one was triple majoring. How many students were majoring in engineering only, biology only, and chemistry only?
7. A survey of 427 farmers showed that 135 grew only beets, 120 grew only radishes, 100 grew only turnips, 210 grew beets, 50 grew beets and radishes, 45 grew beets and turnips, and 37 grew radishes and turnips. Find the number of farmers who grew at least one of the three, grew all three, did not grow any of the three, or grew exactly two of the three.
8. Finally, 150 children were surveyed in a particular town, and it was found that a total of 35 played hockey, 71 played baseball, 30 played soccer, 10 played all three, 3 played only soccer, 17 played only hockey, 6 played only soccer and hockey, 48 played only baseball, and 53 played none of the three sports. Find how many played soccer and baseball only, and how many played hockey and baseball only?

1.5 Infinite Sets

Infinite Sets and the One-to-One Correspondence

In this section, we will attempt to develop an understanding of a rather elusive topic: infinity. Among the first to extensively study infinite sets was the German mathematician and logician Georg Cantor, who created a furor in the late nineteenth century with his exploration of the infinite. Cantor, considered by some to be the father of what we now call set theory, believed that his notions of infinity were communicated to him directly by God. His results were considered shocking and divisive by his peers and were ridiculed by such giants as the legendary philosopher Ludwig Wittgenstein (who dismissed some of Cantor's ideas as laughable and absurd) but praised by equally giant figures, such as the analyst David Hilbert (who described the landscape explored by Cantor as a "paradise").

Before proceeding, we should establish the terminology lying at the root of the intellectual conflict that raged during Cantor's time. First, we must understand what is meant by infinity. As has already been stated, sets can be categorized broadly into two classes: the finite and the infinite. Finite sets have a limited membership, although that membership might well be immense. Infinite sets have unlimited membership.

For instance, the set of all stars in the Milky Way galaxy at any particular moment is finite. Although there are an incredible number of stars in the galaxy, that number (at any given time) is limited. The set of grains of sand on the planet Earth is finite. The number of students in your class is also finite and a much smaller number, undoubtedly, than the number of stars in the Milky Way galaxy.

The set of all integers, on the other hand, is infinite. There is no limit to the number of integers because whatever integer you name, we can always add one to any positive integer (or subtract one from any negative integer) to produce yet another integer. The set of rational numbers is infinite by similar reasoning. Were you to name a rational number, we could always, for instance, halve that number to find yet another rational number.

As we begin our consideration of the infinite, you should realize that intuition has the tendency to break down when attempting to cope with notions dealing with the concept. As an elementary example, consider the following question: when you add together infinitely many positive numbers, is the sum necessarily infinite? The natural thought is that this assertion is true, but it is, in fact, not necessarily true. Suppose that the numbers being combined by addition follow the progression 1, ½, ¼, ⅛, and so on. It can be shown that the sum is limited by the finite value 2.

Returning to our illustrations of infinite sets given previously, we uncover the root of the trouble Cantor discovered when exploring infinite sets: clearly, the integers are contained within the rationals, and therefore the set of rationals would appear to be larger than the set of integers. But how could this be the case, since both of them are infinitely large sets, unless there were different "types" of infinity?

We will return to this question shortly but first will introduce Cantor's concept for defining infinity. The definition he provided for us relies on a preliminary notion we must introduce: **one-to-one correspondence.**

Two sets are said to be in one-to-one correspondence if we can pair the elements of the two sets in such a manner that each element of one set is associated with one and only one member of the other set and vice versa. There are precise mathematical definitions for one-to-one correspondence, but they would be out of place here, so we will suppress them. A truly informal description of a one-to-one correspondence would be that two sets are in one-to-one correspondence if they have the same cardinality.

For instance, set $A = \{1, 2, 3, 4\}$ is in one-to-one correspondence with set $B = \{a, b, c, d\}$, since we can form (among others) the pairings (1, a), (2, b), (3, c), and (4, d). We have established a pairing wherein each member of set A is associated with one and only one element of set B. Of course, sets A and B are obviously finite, so now we will use the concept of one-to-one correspondence to define the notion of infinite sets:

A set is infinite if it can be placed in a one-to-one correspondence with a proper subset of itself.

Cantor used the representative letter c to denote the cardinality of a set whose cardinal number exceeds f_0. The letter c is chosen to represent the word "continuum."

Cantor's Exploration of Infinity

Let us close the discussion of the infinite sets and their cardinality by returning to Georg Cantor and his controversial exploration of the infinite. Cantor proposed that there was an infinite number of infinite cardinal numbers, first of which we called $\aleph_0$. He suggested that the power set of the natural numbers had cardinality $\aleph_1$, which is obviously greater than $\aleph_0$, and speculated that this number, $\aleph_1$, was one and the same with the value c but was unable to prove the result. The result became known as the *continuum hypothesis* and remained open until the early 1960s, when a mathematician at Princeton University, Paul Cohen, established that it is impossible to either prove or disprove the conjecture. This spawned the divergence of set theory into two directions: "Cantorian set theory," wherein the continuum hypothesis is accepted, and "non-Cantorian set theory," which rejects the hypothesis.

Both approaches have led to consistent but significantly different areas of mathematics.

Exercises

Answer the following, using complete sentences and correct spelling.

1. What is meant by an infinite set?
2. What is meant by the term "one-to-one correspondence"?
3. What is meant by "countable"?

Show that the following sets are infinite by constructing a one-to-one correspondence between the set and a proper subset of itself.

4. $\{1, -1, 2, -2, 3, -3, \ldots\}$
5. $\{100, 200, 300, 400, \ldots\}$
6. $\{2, 4, 6, 8, \ldots\}$
7. $\left\{1, \frac{1}{2}, \frac{1}{3}, \frac{1}{4}, \frac{1}{5}, \ldots\right\}$

Show that there exists a one-to-one correspondence between the given set and N.

8. $\{5, 10, 15, 20, \ldots\}$
9. $\{-1, -2, -3, \ldots\}$
10. $\{10, 14, 18, 22, \ldots\}$
11. $\{3, 6, 9, 12, \ldots\}$

Are the following sets infinite or finite? Explain.

12. The set of seconds that have elapsed since the year 0 AD.
13. The set of digits in the full decimal representation of e, the base of the natural logarithm.
14. The set of points in the Cartesian plane.
15. The set of seconds in one day.
16. The set of prime numbers divisible by 2.

Summary

In this chapter, you learned about:

- The terminology and notation of sets.
- Cardinality, equality, equivalence, and complement.
- Proper and improper subsets and the power set.
- Venn diagrams: intersection and union of sets and disjoint sets.
- Some applications of sets using Venn and Euler diagrams.
- Infinite and finite sets: countability and one-to-one correspondence.

Glossary

cardinality (of a set): The quantity of members within a set; represented as $n(S)$.

characteristic property (characteristic trait): A quality or feature of an object that makes it both distinctive and identifiable.

complement (of a set S**):** Those elements of the universal set that are not elements of S.

countable set (countability): A set that is finite or that can be placed in one-to-one correspondence with the set of natural numbers.

DeMorgan's laws: $(A \cup B)^c = A^c \cap B^c$ and $(A \cap B)^c = A^c \cup B^c$.

disjoint sets: Sets whose intersection is empty.

element: An object that is a member of a set.

ellipsis: A series of three dots, indicating a progression of numbers continues according to a demonstrated pattern.

empty set: A set having no members.

equal sets: Two sets having precisely the same set of members.

equivalent sets: Two sets having precisely the same number of members.

Euler diagram: A Venn diagram used to determine validity of reasoning.

finite set: A set having a limited number of members.

inclusive or: That usage of the word "or" taken to mean "either or both."

infinite set: A set having an unlimited number of members.

intersection (of sets): The elements common to two given sets.

member: An object belonging to a set.

null set: A set having no members.

one-to-one correspondence: A relationship between two sets that associates to each member in one set a unique element in the other set and vice versa.

proper subset: A subset containing some but not all of the members of a particular set.

power set (of a set): The set of all subsets of a given set.

quantifying: A word (such as "all," "none," or "some") that expresses the portion of a set possessing a particular property.

roster notation: A listing of all the members of a set.

set: A well-defined, unordered collection of objects having no duplicate members.

set-builder notation: Representation of set membership, including an arbitrary variable used as a generic set member, together with a characteristic property/trait describing the set members.

singleton: A set whose cardinality is 1.

subset: A collection of objects taken from a particular, specified set.

superset: A larger set containing a given set as a subset.

syllogism: A set of premises followed by a conclusion.

union (of sets): A combination of two sets.

universal set: An overall set specified in a particular problem within which other sets are specified.

Venn diagram: A visual tool used to demonstrate the relationships between sets within a universal set.

verbal description (of a set): A nonambiguous written description of the members of a set.

well-defined set: A collection of objects membership within which is unambiguous.

End-of-Chapter Problems

1. Is the set of congresspeople of the United States well defined? If it is not well defined, state why.
2. Is the set of cats in this room well defined? If it is not well defined, state why.
3. Is the set of all cats in the world well defined? If it is not well defined, state why.

In the following problems, determine if the sets are finite or infinite. If they are finite, state their cardinality.

4. The set of odd integers between 20 and 40, inclusive.
5. The set of digits in the full decimal expansion of all real numbers.

Express the following sets in roster form and state the cardinality of the set.

6. The set of letters in the word "Oklahoma."
7. The set of all whole numbers less than 30.

Express the following sets in set-builder notation and determine their cardinality.

8. $A = \{2, 3, 4, 5\}$
9. B is the set of all rational numbers less than 100.

Give a verbal description of the members of the following sets.

10. $\{4, 8, 12, 16, 20\}$

For the following problems, determine if the given sets are equal, equivalent, both, or neither.

11. $B = \{10, 11, 12, 13, 14, 15\}$, $C = \{a, b, c, d, e, f\}$
12. $S = \{5, 4, 3, 2\}$, $T = \{2, 4, 3, 5\}$

For the following sets, determine if $A \subset B$, $B \subset A$, $A \subseteq B$, $B \subseteq A$, $A = B$, or if none of these relationships exist. It is possible that more than one relationship could hold, and if that is the case, list all applicable relationships.

13. A = {Massachusetts, Connecticut, Rhode Island, Vermont, New Hampshire, Maine}, B = the set of all states in the United States
14. $A = \{x \mid x$ is or was a president of the United States$\}$, B = {Bill Clinton, Jimmy Carter, Ronald Reagan}
15. A = set of positive odd integers less than 21, $B = \{1, 3, 5, ..., 21\}$

For the following problems, identify the cardinality of the power set of A.

16. $A =$ {Massachusetts, Rhode Island, Connecticut, Vermont, New Hampshire, Maine}
17. $A = \{x \mid x \in W, 2 < x < 3\}$

For the following problems, determine if the statements are true or false. Explain your answer or give an example to illustrate.

18. There exists a set A such that $A \supset A$.
19. There exists a set A having exactly 16 subsets.

The following problems are applications using subsets and may require some out-of-the-box thinking.

20. Suppose that a committee is going to vote on an item. The item will pass if a simple majority of the votes cast are in favor of it. The committee has a chairman, who gets four votes, and three other members, each of whom gets one vote. How many winning "coalitions" of votes exist?

Determine if the following equalities are true or false.

21. $(A \cup B^c)^c = A^c \cap B$
22. $A^c \cup B^c = (A \cap B)^c$
23. $A^c \cap U = A \cap U$
24. $A^c \cup U = B^c \cup U$
25. $A \cup U^c = A$
26. $(A^c \cap B) \cup (A \cap B) = B$
27. $(A \cap B) \cup (A \cup B) = A \cap B^c$

For the following sets of Venn diagrams, use set notation to describe the shaded area, using intersection, union, or complementation if needed. The answers are not necessarily unique.

28.
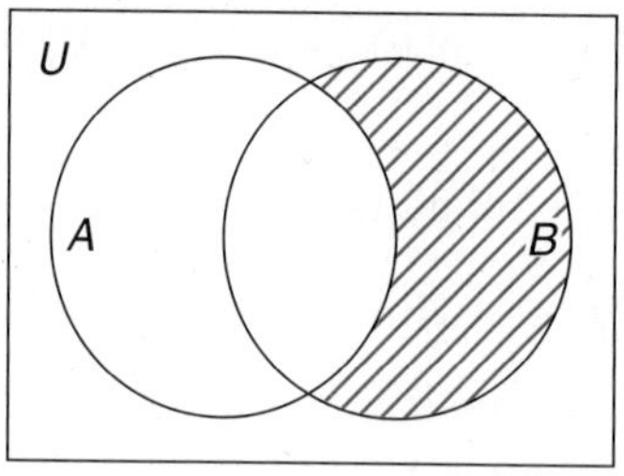

29.
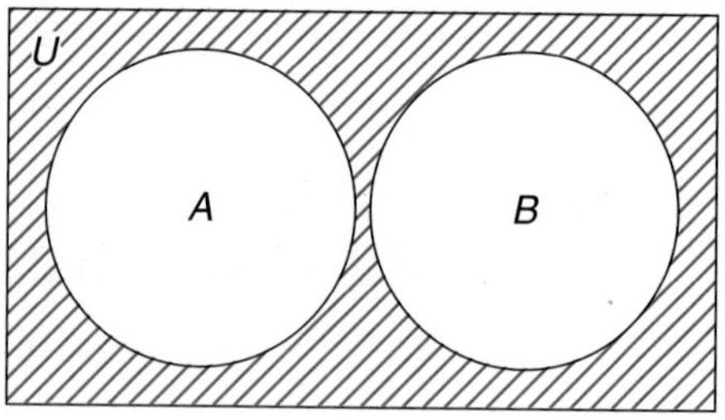

30.
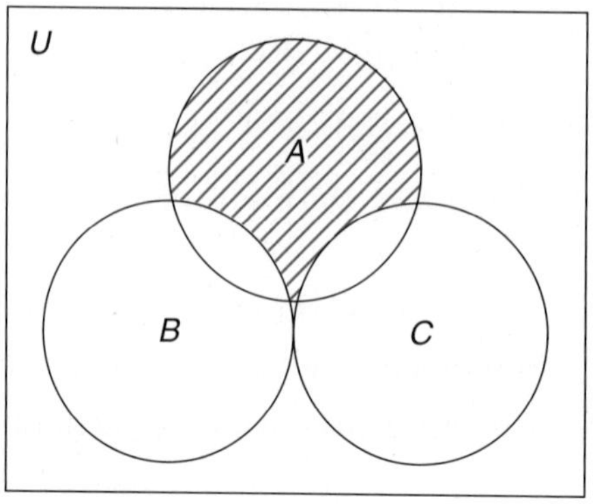

31. The U.S. Census Bureau did a study of 283 married couples with children living in poverty. It was found that 127 of them lived with one of the couple's parents, that 227 of them lived in apartments, and that 95 lived in an apartment with one of the couple's parents. How many of them lived in an apartment without parents, how many of them lived with parents but not in an apartment, and how many of them lived neither with parents nor in an apartment?
32. In a suburban community, a survey was conducted of the 732 residents of a particular neighborhood. It was found that 645 owned computers, that 534 had swimming pools, and that only 37 had neither a computer nor a swimming pool. Of those surveyed, how many had both a computer and a swimming pool? How many had only a computer? How many had only a swimming pool?
33. At a Portuguese Water Dog National Specialty dog show, it was observed that there were 342 dogs entered. Of these dogs, 158 were wavy coated, while the rest were curly coated; 37 were brown, while the rest were black; and 162 of them were female. There were 8 brown wavy females, 78 wavy males, and 160 black females. How many curly females were there? Is there enough information to determine how many brown wavy males there were? If so, how many?
34. A library has 15,832 books and 169 movies. At a particular point in time, 453 of the items were new, and 1832 of them were checked out. There were 32 new movies checked out, and 302 of the books were new. None of the old movies were checked out. There were 1700 old items checked out. How many new books were checked out? How many old books were checked out?
35. A lecture class has 512 students, 263 of them being women; 412 of them have declared a major, and 368 are freshmen; 130 of the freshmen women have declared a major. All the nonfreshmen have declared a major. There are 182 freshmen men in the class. How many freshmen women are there? How many of the freshmen men have declared a major?

Are the following syllogisms valid?

36. All professors are human.

    All humans get sad sometimes.

    Therefore, all professors get sad sometimes.
37. All dogs bark sometimes.

    Some dogs eat too much.

    Therefore, all dogs that bark eat too much.
38. All cats are independent.

    Some independent things like people.

    Therefore, some cats like people.

39. All books are printed.
    Some printed things are wrong.
    Therefore, some books are wrong.
40. All lamps give off light.
    Some lamps are bright.
    All bright things are hard to look at.
    Some bright things are bronze.
    Therefore, some bronze things are lamps.

Show that the following sets are infinite by constructing a one-to-one correspondence between the set and a proper subset of itself.

41. $\{1, 3, 5, 7, \ldots\}$
42. $\{1, 2, 4, 8, 16, 32, \ldots\}$
43. $\left\{1, \frac{1}{3}, \frac{1}{5}, \frac{1}{7}, \frac{1}{9}, \ldots\right\}$

Show that there exists a one-to-one correspondence between the given set and N.

44. $\{3, 8, 13, 18, \ldots\}$
45. $\{7, 8, 9, 10, 11, \ldots\}$
46. $\{9, 12, 15, 18, \ldots\}$
47. $\left\{1, \frac{1}{4}, \frac{1}{9}, \frac{1}{16}, \frac{1}{25}, \ldots\right\}$

Are the following sets infinite or finite? Explain.

48. The set of all meals eaten by all human beings since there were human beings.
49. The set of digits in the full decimal representation of $\sqrt{2}$
50. The set of prime numbers divisible by 15.

Chapter 2 Logic

Logic is an instrument for analysis of reasoning, whose object is to distinguish good, or valid, reasoning from bad, or invalid, reasoning. It is not a mechanism for determining if particular statements are true or false; that is a matter for experimentation, consideration, or research. With logic, we consider a collection of supportive statements and ponder the following question: if it is assumed that each of the collection of statements is true, then does a purported consequent statement *have* to be true?

You might be wondering what the study of logic would have to do with information technology, but be assured the answer is that the connection between the two is both intimate and essential. Early in the twentieth century, the work of Alan Turing and Alonzo Church spurred the development of a new field of inquiry referred to as "computability." Turing's inspiration was the desire to solve the **decision problem** posed by Hilbert, asking whether a standard procedure could be developed to determine whether a particular statement could be proven. This is, as you can appreciate, one and the same as the challenge of whether a particular computer algorithm can be devised to perform a particular task.

Turing's study led to his development of theoretical computation devices now known collectively as **Turing machines**, which have evolved into what is referred to as a compiler, an essential component for the processing of any programming language. The behavior of a Turing machine is, by construction, based entirely on the application of logic, and as such devices are the progenitors of the modern computer, it goes without saying that a solid foundation in logic will prove invaluable in our studies.

Repeating the conceptual framework of Chapter 1, we'll continue (for the time being) to present the concepts from a general standpoint and not presuppose knowledge of information technology applications. The time will come—and very soon—that our attention will become more focused on application-specific concepts, but that moment is not yet here.

Objectives

By the time you have successfully completed the materials of this chapter, you will be able to:

- Recognize simple and compound statements.
- Understand truth-values.
- Recognize and work with logical connectives.
- Construct and interpret a truth table.
- Recognize logically equivalent statements.
- Construct symbolic representations of arguments.
- Recognize and use Euler diagrams.

2.1 Statements and Logical Connectives

Statements

In the study of logic, we begin with the introduction of **statements**. A statement, from the logician's point of view, is a declarative sentence that is either true or false. A word that is often used synonymously with "statements" is "propositions." In this work, we will choose to use the former over the latter, but be aware that others may choose to use "propositions" out of personal preference. Examples of statements would be "a triangle is a four-sided polygon" or "George Washington was the first president of the United States." Each of these has the property that they are declarative (that is, they make a presentation about a condition) and also that they are (by virtue of empirical evidence) either true or false.

The two examples given above express a single idea, but this is not necessary. A statement could convey a more complicated combination of ideas, such as "Mark Twain was an author and Cleopatra was an actress." When a statement expresses a single idea, we call it a **simple statement**; otherwise, we call it a **compound statement**.

Before proceeding, we should point out that the quality of being true or of being false is what we refer to as a statement's **truth-value**. Shortly, we will adopt the practice of designating true statements with the abbreviation T and false statements with the abbreviation F. This will come into play in the next section of the book when we introduce truth tables, but it is expedient to introduce the notions at this point.

A statement cannot present a position that is ambiguous or open to interpretation; this is part and parcel of the definition of what it means to *be* a statement. For instance, "Raquel Welch is beautiful" is not a statement in the sense already described because the notion of what is beautiful is open for debate, and thus the remark cannot be judged on its face to be true or false. Nonsensical remarks, such as "Groknell is a nutritious slumping," are not statements because they can, by construction, be neither true nor false.

The typical practice in the examination of an argument is to associate statements with capital letters, such as A and B, which need not have any relationship to the content of the statement. For instance, I can specify

A: A triangle is a four-sided polygon.

while the letter A has no relationship to any part of the statement. Note that, recalling our convention of identifying T and F with the truth-values of "true" and "false," respectively, it would be a tragic choice of notation to choose identifying names of T or F to represent statements. This should be avoided at all costs!

We have already given one sentence that would not qualify as a statement because of its ambiguity. There are other sentences that would not be considered statements, and it would be beneficial to see examples of those.

- What time is it? (this is not a statement because it is a question and cannot be true or false)
- I am lying. (this is what we consider a paradox because it can be neither true nor false)

There are other examples we could create, but these will do for the moment.

The second example given above is an interesting sentence to consider and is famously known as the **liar's paradox**. Consider if the statement were true: the truth of the assertion would negate itself. The same interpretation can be made if the statement were considered to be false.

EXAMPLE 2.1

Are the following phrases "statements," or are they not?

1. The Empire State Building is in New York City.
2. The Soviet Union was an evil empire.
3. There are 10,560 feet in one mile.
4. Is it time for dinner?

SOLUTION

1. This is a statement. Whether or not the Empire State Building is in New York City is a matter of definitive fact, which can be determined without ambiguity.
2. This is not a statement but rather an expression of opinion. There is no method through which we could establish the truth or falsity of the phrase, and hence it does not satisfy the specific definition of what it means to be a statement.
3. This is a statement. Do not be misled by the fact that it is, in fact, false. That is not the point of the question. A statement is a declarative sentence that can be either true or false, and which is the case is a matter left for exploration. There either are or are not 10,560 feet in a mile, but that is not relevant here. The phrase is a declarative sentence whose truth-value can be established.
4. This is not a statement. Recall that questions can not be statements, as they are not declarative sentences.

The Logical Connectives

Logical connectives are those words that join one simple statement to another while not serving any other grammatical purpose. Common examples of logical connectives would be words such as "not," which represents negation; "and," which represents conjunction; "or," which represents disjunction; "if-then," which in combination represent implication; and "if and only if," which represents what we call a biconditional relationship.

Naturally, when first presented, technical terms, such as conjunction and disjunction, are likely to have little meaning for you. Each of these will be clarified in their meaning, and the associated symbols used to abbreviate the logical connectives will be spelled out. You will want to memorize such symbols quickly, possibly making a study chart to keep them straight in your mind, since the symbolic representation of statements and the logical connectives will play a major role in what is to follow.

Let's first consider the logical connective of "not," or "negation," which we shall abbreviate with the symbol ~, which is called a "tilde," pronounced "til-da." You should always be clear, by the way, about the context in which a symbol appears. In logic, the tilde represents negation, while in other fields the symbol can have different meanings. For instance, in geometry, the tilde is often used to symbolize similarity of objects.

Maintaining our first statement representation, A: A triangle is a four-sided polygon, we can create the negation, ~A, read as "not A," and interpret that to mean "It is not the case that a triangle is a four-sided polygon." Common practice is to use the phrase "It is not the case that" to commence the negation of a statement.

Note the following interesting observation: the truth of a statement and its negation *must be* reversals of one another. That is, if A is true then ~A must be false and vice versa. It is impossible that a statement and its negation should have the same truth-value.

As a second example, let's revisit our second statement from earlier, and identify it as

B: George Washington was the first president of the United States.

Observe that ~B can be interpreted as "It is not the case that George Washington was the first President of the United States." One could, equivalently, phrase this as "George Washington was not the first president of the United States," but this tends to muddy the waters a bit as to what statement B actually declared. For this reason, we will, generally, maintain our use of "It is not the case that" to precede a statement we are negating.

Question: If our statement is "It is raining outside," is the related statement "It is sunny outside" its negation?

When you first read that, you may think that this is true. However, note that if it is not raining, that does not mean that it is sunny outside. It could be very cloudy yet not raining. The negation is simply the statement "It is not the case that it is raining outside" or possibly "It is not raining outside." The moral here is to be very cautious when you construct negations of statements because your supposed negation could be far too broad in its scope.

The second logical connective we mentioned was "and." This is symbolized as $\wedge$, and it allows us to combine two different statements together. The symbolic representation would be as follows: $A \wedge B$. We can create many possible examples, and here is one:

A: The New York Yankees are a baseball team.
B: The Philadelphia Flyers are a hockey team.

If we use these as our statements, then the conjunction of the statements would be "The New York Yankees are a baseball team, and the Philadelphia Flyers are a hockey team." The presence of the connective "and" makes this a compound statement.

It is worth mentioning at this point that some logical connective concepts can be captured by different wordings. For instance, a conjunction need not be created solely through the use of the word "and" but might also be signified by another equivalent word or phrase. Examples of such words or phrases would include "however," "but," and "moreover." For instance, if I say "It is raining and the sun is shining," this is equivalent to the statement "It is raining, but the sun is shining." Both phrases capture the idea that two distinct conditions exist.

Whenever we introduce a logical connective, it makes sense to discuss the truth-value associated with that connective. A conjunctive statement is true precisely when each of the individual statements are true. So, if I say, "It is raining and the sun is shining," that compound statement is true if *both* individual statements are true, and it is false if one or both is not true.

The third logical connective is "or," which is the disjunction. The symbol for "or" is an upside-down version of the "and" symbol, which would be $\vee$. If we use the statements

A: A triangle is a four-sided polygon.
B: George Washington was the first president of the United States.

then the disjunction would be $A \vee B$, which we would interpret as "A triangle is a four-sided polygon, or George Washington was the first president of the United States."

The truth-value of a disjunction can be a bit more slippery to understand. There are two possible interpretations of "or," which are typically referred to as the "inclusive or" and the "exclusive or." We encountered this distinction in Chapter 1, and there is a similar concern we must address in the study of logic. The "inclusive or" is considered to be true if one or both of the individual statements are true. The "exclusive or" is considered to be true if one, but not both, of the individual statements are true.

Thus, if our statements were

A: The New York Yankees are a baseball team.
B: The Philadelphia Flyers are a hockey team.

then, for the "inclusive or," $A \vee B$ would be true, since both of the statements are true. For the "exclusive or," the disjunction $A \vee B$ would be false, since in the exclusive case we take the disjunction to be true if one but not both of the individual statements are true.

To avoid ambiguity and continue the practice adopted in Chapter 1, in this work we will refer only to the "inclusive or" and will view the disjunction as being true in the event that one or both of the individual statements are true. It is, by the way, the Latin word "vel," which captures the meaning of the inclusive form of "or," and it is this that motivates the notation chosen for disjunction. In the rare case where the exclusive case is required, we will include the phrase "but not both" in our wording.

The fourth connective was implication, often referred to as the "if-then" connective. While there are several symbols in common use for implication, the two most common are a right-pointing arrow, $\rightarrow$, and the symbol $\supset$. The use of the symbols would be $A \rightarrow B$ in the first case and $A \supset B$ in the second. There is no particular reason to prefer one to the other, and so we arbitrarily will select the arrow. The notation $A \rightarrow B$ is read as "If A, then B" or sometimes "A implies B." The statement chosen as A is referred to as the "antecedent," while that chosen as B is referred to as the "consequent."

In terms of truth-value, an implication is true in all cases except one: $A \rightarrow B$ shall be considered false only in the case where the antecedent is true while the consequent is false. This presents the rather unusual condition that an implication whose antecedent is false is, from a logical standpoint, true.

We'll introduce just one more logical connective at this point, the biconditional, or "if and only if" connective. This is symbolized with a double-headed arrow between the related statements, $A \leftrightarrow B$, which is read as "A if and only if B." Such a compound statement is considered true only in the case where both A and B are true statements. As an example, we might assign

A: The elephant is a mammal.
B: The snake is a lizard.

Then the compound statement $A \leftrightarrow B$ would be read as "The elephant is a mammal if and only if the snake is a lizard."

It's important to realize that more than one logical connective can be used at a time in order to make progressively more complicated sentences. For instance, we might identify the following statements:

A: It is raining outside.
B: I forgot to bring my umbrella.
C: My shirt will get wet.

We could create the compound statement $(A \wedge B) \to C$, which would be equivalent to "If it is raining and I forgot my umbrella, then my shirt will get wet." You can imagine that the combinations of statements could become quite complicated, and consequently we will have to be very careful when we symbolize and analyze expressions.

So far, we've started with the identification of statements, followed by their reduction into symbolic form. Suppose we were given the sentence and had to assign the statement names ourselves and then produce the symbolic abbreviation? Let's try an example to see if we can manage it.

> If the Yankees win their division and the Red Sox win the wild card, then the Blue Jays will not appear in the American League playoffs.

There are several ways we could do this, but what we might like to look for first are the logical connectives. Note the presence of the words "if" and "then," which indicate that there exist an antecedent and a consequent we must specify. Also, in the antecedent, we detect the word "and," indicative of a conjunction, and, in the consequent, the presence of the word "not" signifies a negation.

So, we might choose our statements as follows:

A: The Yankees win their division.
B: The Red Sox win the wild card.
C: The Blue Jays will appear in the American League playoffs.

Under these identifications, the symbolization of the compound statement would be as follows: $(A \wedge B) \to \sim C$.

Statements should, in general, be affirmative in nature. It is conceivable that we could create negative statements, such as

A: It is not raining outside.

but this is, in general, bad practice. The reason has to do with the confusion that arises naturally from such assignments, and we will avoid defining our statements in this manner.

Exercises

In the following exercises, identify which sentences are logical statements.

1. The planet Venus is the closest planet to Earth's sun.
2. Let's play racquetball.
3. Do you like French food?
4. Average gas mileage for automobiles in the United States has increased over the past 30 years.

5. You will win the lottery next week.
6. To thine own self be true.
7. Sylvester Stallone is an actor.

The following exercises contain statements. Identify whether they are simple statements or compound statements.

8. My favorite baseball team is the New York Yankees, and my favorite baseball player is Alex Rodriguez.
9. It will snow tomorrow.
10. If I can't have you, then I don't want anybody.
11. Most students never read their math book.
12. Dirty Harry was a fictional police officer.
13. You win some, you lose some.
14. Terry is egotistical and vain.

For each of the following compound statements, assign variables to the individual statements and then symbolize the compound statements using the logical symbols developed in this section.

15. Debbie is a pretty girl or I am a monkey's uncle.
16. If math is not fun, then I don't know what fun is.
17. I will either buy my Christmas gifts with cash, or I will buy them with a credit card.
18. The president will win reelection, or he will be viewed as a failure.
19. Mark is not charismatic, and will not win the debate.

For the following symbolizations, create statements corresponding to each letter, and then write an English sentence that is equivalent to the symbolization.

20. $p \rightarrow (q \wedge r)$
21. $p \vee (q \rightarrow r)$
22. $(p \wedge q) \rightarrow r$
23. $p \rightarrow (q \vee r)$
24. $p \wedge (q \vee r)$

For the following statements, give the negation.

25. I am six feet tall.
26. Scott is intelligent.
27. Snakes are slimy.
28. Red is my favorite color.
29. The Lakers is a soccer team.
30. Apple Corporation manufactures computers.
31. The number of computer users in the United States is doubling every year.
32. China is the largest economy in the world.
33. The Chinese have developed the fastest supercomputer in the world.
34. Australia is the world's largest continent.

2.2 Truth Tables for Negation, Conjunction, and Disjunction

A **truth table** is a device we can use to investigate a statement containing a finite number of simple statements in combination. The table contains all possible truth-values of the statements involved and allows us to examine the consequences of the combinations of the various alternatives. That sounds terribly abstract, we know, but you will find that truth tables are actually nice constructions with which to work, and an example or two will clarify the situation greatly. Be assured, by the way, that the notion of truth tables will recur later in this text when we attempt to optimize circuit pathways.

First, let's start with a peculiar fact: the number of rows of a truth table will always be a power of two. That is, the number of rows a truth table must have is 2, 4, 8, 16, 32, and so on. Why would this be the case? Every statement has two possible truth-values of its own, T or F. Whenever one simple statement is combined with another through the use of one connective, there are four possible combinations to consider: listing the first statement's truth-value first, the combinations are TT, TF, FT, and FF. Every time another statement is appended to the compound statement construction, the number of possible truth-value combinations doubles.

In the construction of a truth table, we will list the statements as the headings of the columns of the table, and the truth-values for the particular statements will occur in the rows of the table. For instance, consider the truth table for $A \wedge B$, where A and B are arbitrary statements. We have already declared that a disjunction of this nature is true only in the event both simple statements are true.

The truth table would look like Table 2.1. Note that the four rows accommodate all possible combinations for truth-values of statements A and B and that the final column reveals the truth-value of the conjunction for those individual combinations.

TABLE 2.1 Example of a Truth Table

A	B	$A \wedge B$
T	T	T
T	F	F
F	T	F
F	F	F

Every connective we have defined has its own truth table. The truth table for negation of a statement is fairly elementary, and we produce that table now. Keep in mind our remark made earlier in this chapter: the truth-value of a statement and the truth-value of its negation are in exact opposition to one another. Therefore, the truth table for the negation of statement A would look like Table 2.2.

TABLE 2.2 Truth Table for ~A

A	~A
T	F
F	T

We have already specified the criteria for truth of the disjunctive compound statement, and as exhibition of that criteria, we present the truth table for disjunction in Table 2.3.

TABLE 2.3 Truth Table for A ∨ B

A	B	A ∨ B
T	T	T
T	F	T
F	T	T
F	F	F

Truth tables for the connectives become progressively more elaborate as the number of statements involves increases. Let us consider a more complicated example where we look at the following compound statement:

John is intelligent and handsome, or I am a monkey's uncle.

This statement needs to be expanded somewhat to reveal all of the involved simple statements. The first fragment, "John is intelligent and handsome," is actually a combination of two simple statements: "John is intelligent" and "John is handsome." Therefore, the compound statement can be stated more explicitly as:

John is intelligent and John is handsome, or I am a monkey's uncle.

We can now associate the simple statements with identifying letters:

A: John is intelligent.
B: John is handsome.
C: I am a monkey's uncle.

Symbolically, the compound statement is now $(A \wedge B) \vee C$. Note that, much like we find in algebraic statements, associated statements are clustered by symbols of grouping. We now will build up a truth table for the compound statement and then interpret our result. With practice, we can construct the entire table in one step, but we should take care not to jump too far too fast.

Recall the truth table for conjunction (Table 2.4). This will play a role in the construction of the new table. Since three simple statements are involved, the table will contain $2^3 = 8$ rows, accounting for all possible truth-value combinations of the simple statements A, B, and C (Table 2.5).

TABLE 2.4 Truth Table for A ∧ B

A	B	A ∧ B
T	T	T
T	F	F
F	T	F
F	F	F

TABLE 2.5

A	B	C	A ∧ B	(A ∧ B) ∨ C
T	T	T		
T	T	F		
T	F	T		
T	F	F		
F	T	T		
F	T	F		
F	F	T		
F	F	F		

Observe the first three columns create all the possible combinations for the truth-values of A, B, and C and are arranged in such a manner that the construction of the table is highly systematic. The first column is halved with T and F values for statement A, with the first four rows assigning truth-value T to statement A. The second column shows the truth-value assignments for statement B, but the string of T and F entries are halved from the organization for the first column. This process continues in the third column and would (in principle) continue for as many simple statements as our problem contained.

Next, we fill in the column for the conjunction (Table 2.6). The truth-values of A and B, individually, can now be ignored, for all that remains is for us to consider the disjunction of statement A ∧ B with statement C (Table 2.7).

TABLE 2.6

A	B	C	A ∧ B	(A ∧ B) ∨ C
T	T	T	T	
T	T	F	T	
T	F	T	F	
T	F	F	F	
F	T	T	F	
F	T	F	F	
F	F	T	F	
F	F	F	F	

TABLE 2.7

A	B	C	A ∧ B	(A ∧ B) ∨ C
T	T	T	T	**T**
T	T	F	T	**T**
T	F	T	F	**T**
T	F	F	F	**F**
F	T	T	F	**T**
F	T	F	F	**F**
F	F	T	F	**T**
F	F	F	F	**F**

We can now assess the truth-value for the overall compound statement given the truth-values of the individual statement. As an illustration, consider the first row of the table. In the case that it is true that John is intelligent and that John is handsome and that I am a monkey's uncle, then the overall compound statement is true. In short, once empirical evidence reveals to us the truth-values of the individual simple statements, the truth-value of the compound statement can be determined.

If a drawback exists to the presentation of these facts in a truth table, it is in the sheer size of the instrument. For comparison purposes, which we shall employ shortly, it is inconvenient for us to have to consider such large constructions. There are, however, instruments we can employ that will consolidate the table into a more compact form that can (at times) prove useful. More will be said about this at the appropriate time.

It is possible that the final column of a truth table would consist entirely of true results. This indicates the compound statement is, under all conditions, exclusively true. Such a statement is referred to as a **tautology**. This is one extreme for the result of a truth table investigation, and it stands in opposition to the possible result where all entries in the final column would be false results. Such a statement is referred to as a **contradiction**.

As we prepare to close this section, it is natural to ponder how truth tables could possibly find application in the "real world" apart from the field of semantics. Each of the logical connectives has an analog in various applications, but for the moment we'll focus our attention on two of the connectives. An application of conjunctions and disjunctions is in the circuitry of computers and the transmission of current.

We know that current will pass through a switch if the switch is closed and that it will not pass through a switch if the switch is open. When we have a series circuit, we have one switch followed immediately by another; this corresponds to a logical conjunction. That is, if we use the representations

A: Switch A is closed.
B: Switch B is closed.

then current will flow through a series circuit precisely in the case where the disjunction $A \wedge B$ is true.

When we have a parallel circuit, we have one switch alongside another; this corresponds to a logical disjunction. For a series circuit, current passes through only if both switches are closed, while for a parallel circuit, current will pass through if at least one of the switches is closed. Consequently, the condition for current to flow matches precisely the situation where $A \vee B$ is true.

Exercises

Construct a truth table for each statement.

1. $p \wedge \sim q$
2. $p \vee p$
3. $p \vee \sim p$
4. $p \wedge \sim p$
5. $\sim p \vee \sim q$
6. $\sim(\sim p) \wedge \sim q$

7. $\sim(\sim(\sim(\sim p)))$
8. $(\sim p \wedge \sim q) \vee (\sim p \vee \sim q)$
9. $(\sim p \vee q) \wedge r$
10. $\sim(p \wedge \sim q) \vee r$

Do the following statement pairs have identical truth tables, or do they not?

11. $\sim(p \vee \sim q)$ and $(\sim p) \vee q$
12. $\sim(\sim p \vee \sim q)$ and $p \vee q$
13. $p \vee (\sim q \wedge r)$ and $p \wedge q \wedge \sim r$
14. $p \wedge (\sim q \vee \sim r)$ and $(p \wedge (\sim q)) \wedge (p \wedge \sim r)$
15. $p \wedge \sim(q \wedge r)$ and $p \wedge (\sim q \vee \sim r)$
16. $\sim(p \vee q)$ and $\sim p \vee \sim q$

In the following exercises, circuits are shown. Symbolize them logically and then (if possible) try to devise an equivalent circuit that uses fewer switches than the one depicted in the problem.

17.
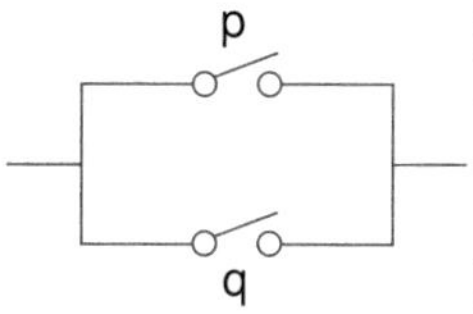

18.

19.
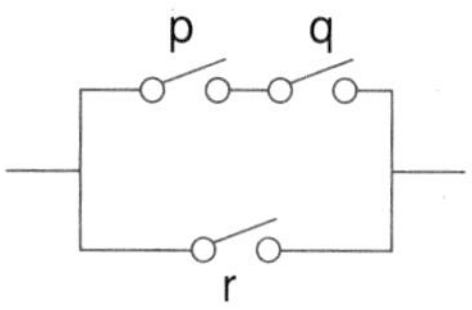

20.
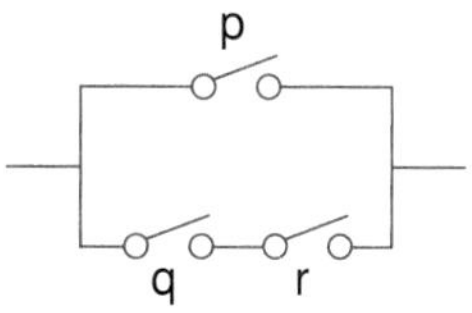

21.
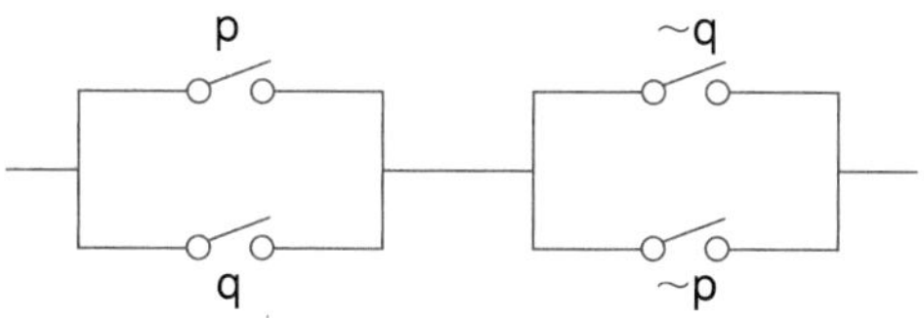

22.
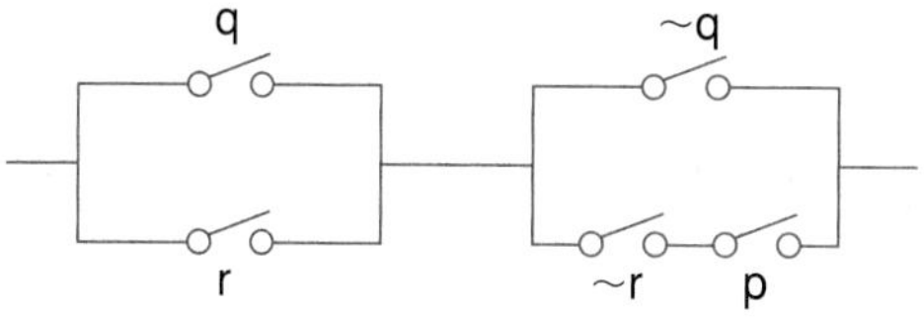

23.
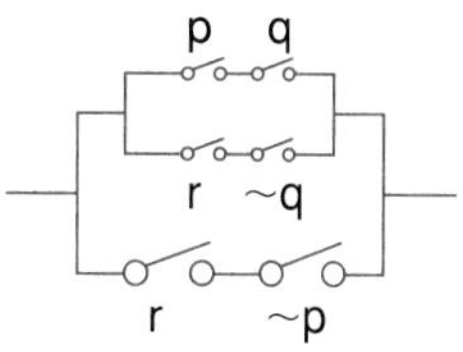

24.
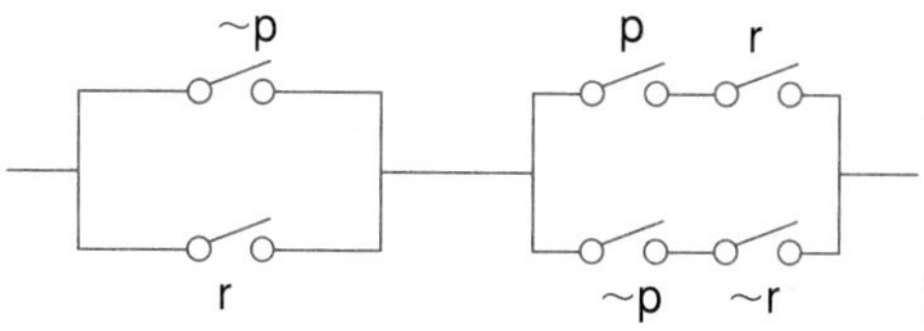

25.
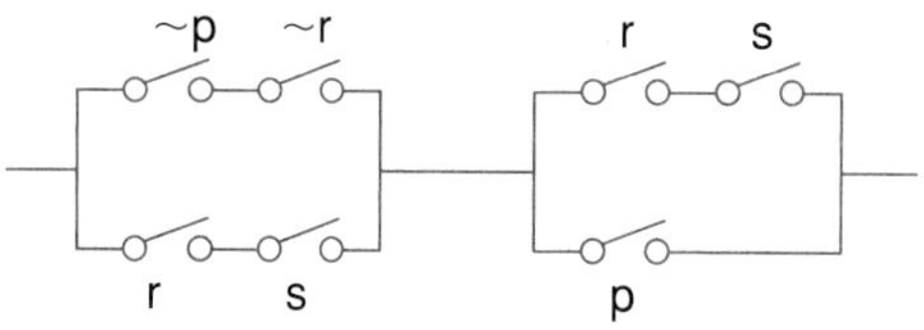

26.
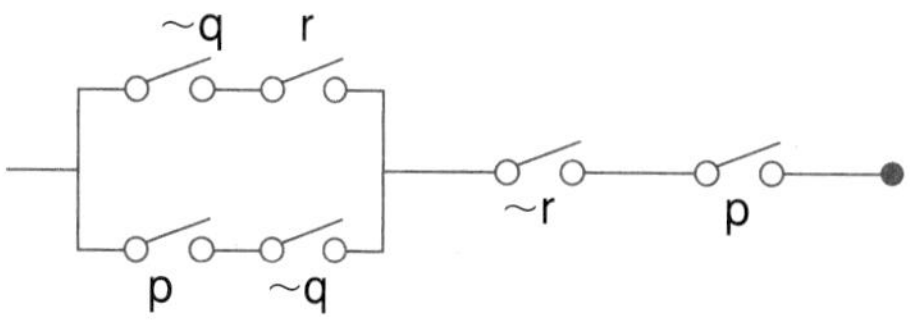

In Problems 27 to 34, construct switching circuits that represent the symbolic statements.

27. $p \wedge \sim q$
28. $p \vee \sim q$
29. $p \vee (q \wedge r)$
30. $p \vee (q \wedge \sim r)$
31. $\sim p \vee (q \wedge \sim r)$
32. $\sim p \wedge (\sim q \vee \sim r)$
33. $(p \vee q) \wedge \sim r$
34. $(\sim p \vee q) \wedge (r \wedge s)$

2.3 Truth Tables for the Conditional and the Biconditional

The compound statement "If I study, then I will be prepared for the exam" is what we have designated a **conditional statement**, signaled to us by the presence of the word combination "if-then." Such statements are also referred to as implications, hypothetical, or implicatives, but we will maintain the use of the term "conditional."

Conditional statements contain two components: an antecedent (occurring between the words "if" and "then") and a consequent (following the word "then"). Conditional statements tell us that if the antecedent is true, then the consequent must be true. That is, the antecedent *implies* the consequent. In order for us to determine the truth-value of an implication, we must consider when such a statement would be false. The only conditions that would create a false outcome would be if the essence of implication were not honored; that is, if the antecedent was true but the consequent was false. That is, *the only instance where A → B would be false would be if the conjunction A ∧ ~B were true.*

Therefore, A → B must have the same truth table as the compound statement ~(A ∧ ~B). Let us construct the truth table for that compound statement, and we will have the truth table for the conditional as shown in Table 2.8. Thus, A → B is false only in the particular case where A is true while B is false.

TABLE 2.8

A	B	~B	A ∧ ~B	~(A ∧ ~B)	A → B
T	T	F	F	T	T
T	F	T	T	F	F
F	T	F	F	T	T
F	F	T	F	T	T

We should note that conditional statements are often presented using the if-then wording format, but that this is not always the case. As is the case with virtually all phrases, alternate wordings exist that capture the same meaning. For the conditional, there are many such variations, such as a simple reversal of the wording. If someone is heard to say "If I do a good job, then I will be paid my salary," it could be stated in an equivalent manner with the phrase "I will be paid my salary if I do a good job." That is, an alternative to the phrase "If A, then B" would be "B if A."

Another possible variation is the phrase "A is sufficient for B." This may not be the natural way to phrase an implication, but it is certainly another way to capture the idea. When we say "A is sufficient for B," we are saying that if A happens, then this is enough to cause B to happen. Consider the mathematical phrase "Having exactly three sides is sufficient for a polygon to be a triangle." This is another way of stating "If a polygon has three sides, then it is a triangle."

Two additional variants on the "if-then" wording are "A only if B" and "B is necessary for A." Admittedly, these are less common than is the "if-then" phrasing, but we should be aware of them in case they should arise.

The **biconditional statement** is also known as the "if and only if" statement. Symbolized by a double-headed arrow, this relationship is also known as the condition of material equivalence and has the property that it is true if each of the individual statements have the same truth-value. That is, the statement "A if and only if B" is true precisely when both statements are true or when both statements are false.

An example of such a compound statement would be "I am alive if and only if I am breathing." Another way to view the biconditional would be to visualize the relationship as the conjunction of distinct conditional statements: $(A \rightarrow B) \wedge (B \rightarrow A)$. From that perspective, the previously given example would be "If I am alive, then I am breathing, and if I am breathing, then I am alive."

Biconditional statements have the following truth table (Table 2.9). Recall we have said that the biconditional can also be viewed as the conjunction of two conditional statements. We establish this by producing the truth table of that expression (Table 2.10) and showing that its truth table is identical to that presented above. Note that the final column of each of the tables is identical to the other, as was indicated.

TABLE 2.9

A	B	$A \leftrightarrow B$
T	T	T
T	F	F
F	T	F
F	F	T

TABLE 2.10

A	B	$A \rightarrow B$	$B \rightarrow A$	$(A \rightarrow B) \wedge (B \rightarrow A)$
T	T	T	T	**T**
T	F	F	T	**F**
F	T	T	F	**F**
F	F	T	T	**T**

Exercises

Answer the following in complete sentences, using correct spelling and grammar.

1. What is meant by a "conditional" statement?
2. What is meant by a "biconditional" statement?
3. What is meant by the term "antecedent?"
4. Under what conditions will the truth table result of a conditional statement be false?

For the following statements, construct a truth table.

5. $A \rightarrow \sim B$
6. $\sim A \rightarrow B$
7. $\sim(\sim A) \leftrightarrow B$
8. $\sim(A \rightarrow B)$

9. $(A \vee B) \leftrightarrow A$
10. $\sim(A \wedge \sim B) \rightarrow (A \vee B)$
11. $\sim(A \vee B) \rightarrow \sim(A \wedge C)$
12. $(A \vee B) \leftrightarrow \sim(A \vee C)$
13. $(A \rightarrow \sim B) \leftrightarrow (B \rightarrow \sim A)$

For every conditional statement A → B, we can form a "reversed" statement called its "converse." Write the converse of the following conditional statements.

14. If I don't pay my taxes, I will go to jail.
15. If you like the car I sell you, you will buy your next car from me.
16. If a polygon is a triangle, it has three sides.
17. If health care is reformed, the public will be happy.

For every conditional statement A → B, we can form a related condition consisting of the negations of the individual statements, ~A → ~B, called the "inverse" of the original conditional statement. Write the inverse of the following conditional statements.

18. If marijuana is legalized, then heroin will ultimately be legalized.
19. If we declare war now, we will ensure peace in the future.
20. If you return the product in its original packaging, you will be given a full refund.
21. If you get a C in this course, then you will get your diploma.
22. If you accumulate 20,000 points on your credit card, you will receive a free flight to Hawaii.

Rewrite the conditional statements using the alternate form "A is sufficient for B."

23. If it rains today, then my shirt will get wet.
24. If I do not read my math book, then I will be in trouble on the test day.
25. If I am convicted of a felony, then I will not be able to join the army.
26. If my bid is accepted, I will buy that house.

Rewrite the following statements using the word combination "if-then."

27. I'll celebrate if I win the election.
28. To lease this car, it's sufficient for your credit to be approved.
29. You will graduate in May only if you file an application by January 30.
30. In order to marry Susan, it is necessary that you divorce Dakota.

2.4 Equivalent Statements

Two statements are said to be **logically equivalent** if they have the same truth functional value. In this sense, the last two tables in Section 2.3 indicate the equivalence of the statements $A \leftrightarrow B$ and $(A \rightarrow B) \wedge (B \rightarrow A)$. We will now explore a variety of statements and attempt, through the use of our truth table construction, to establish logical equivalence.

One might wonder, at first glance, why we would bother? If two statements are equivalent, why would it be preferable to choose one form over the other? The reason is that we would prefer to have one form if it were simpler, or more elegant, in appearance than the other. A complicated expression might mask the content of our statement, which could be apparent if a simpler form were provided.

Let's look at an example: "If I pass this math class, I will be happy." This is obviously a conditional statement, with A: I pass this math class, and B: I will be happy. Is this statement equivalent to "If I won't be happy, then I will have not passed this math class"?

The two compound statements are clearly related, but it is unclear at first glance if they state the same content. For us to determine their logical equivalence, we must determine if they generate the same truth tables. We already know the truth table for the conditional statement (Table 2.11).

TABLE 2.11 Truth Table for the Conditional Statement

A	B	A → B
T	T	T
T	F	F
F	T	T
F	F	T

Let's now symbolize the supposedly equivalent form and construct its truth table. Using the same notions for statements A and B, we see that the symbolization of "If I won't be happy, then I will have not passed this math class" would be ~B → ~A. We set up the table as shown in Table 2.12.

TABLE 2.12

A	B	~B	~A	~B → ~A
T	T	F	F	**T**
T	F	T	F	**F**
F	T	F	T	**T**
F	F	T	T	**T**

Notice the truth table results are identical to one another in each table; this establishes the logical equivalence of the two statements and indicates that they do, in fact, express precisely the same concept. That second conditional, by the way, is a common logically equivalent form to the conditional statement A → B, which we call the **contrapositive** of the conditional statement.

You might remark, and rightly so, that the equivalent statements given above are equally complicated, and therefore it seems to make no difference which one we use. As a consequence, you might suspect that the importance of logical equivalence is marginal at best.

Consider the following statement:

$$(A \wedge B) \vee [(B \vee C) \wedge (B \wedge C)]$$

We can imagine this as being representative of any compound statement involving three simple statements A, B, and C, so possibly A: I take a math class, B: I take an English class, and C: I take a history class. The statement would indicate

I will take a math class and an English class, or I will take an English class or a history class and I will take an English class and a history class.

The mere reading of the statement exhibits that confusion could abound because it is unclear what the speaker is trying to tell us, and the presentation is quite disorganized.

Let's consider the truth table for the compound statement. We will begin with an eight-row table, since there are three statements, and a compound statement involving three variables requires 2^3 rows (Table 2.13). As we can see, the process is quite complicated. Eight columns of truth table values had to be determined, with a series of conjunctions and disjunctions. However, consider the following statement by way of comparison: $(A \vee C) \wedge B$. The truth table for this statement would look like Table 2.14.

TABLE 2.13

A	B	C	$A \wedge B$	$B \vee C$	$B \wedge C$	$[(B \vee C) \wedge (B \wedge C)]$	$(A \wedge B) \vee [(B \vee C) \wedge (B \wedge C)]$
T	T	T	**T**	T	T	**T**	**T**
T	T	F	**T**	T	F	**F**	**T**
T	F	T	**F**	T	F	**F**	**F**
T	F	F	**F**	F	F	**F**	**F**
F	T	T	**F**	T	T	**T**	**T**
F	T	F	**F**	T	F	**F**	**F**
F	F	T	**F**	T	F	**F**	**F**
F	F	F	**F**	F	F	**F**	**F**

TABLE 2.14

A	B	C	$A \vee C$	$(A \vee C) \wedge B$
T	T	T	T	**T**
T	T	F	T	**T**
T	F	T	T	**F**
T	F	F	T	**F**
F	T	T	T	**T**
F	T	F	F	**F**
F	F	T	T	**F**
F	F	F	F	**F**

Observe that this much simpler statement has precisely the same truth table, and thus the statement is logically equivalent to the previous statement! In words, this would be "I will take a math class or a history class, and I will take an English class." Even with our introductory exposure to logic, we can see that the latter statement is much less confusing and certainly more elegant.

A natural question one can raise at this point would be, Where did the alternative, simpler form of the statement come from? That is a justifiable question, and we will address it shortly. For now, we'll content ourselves with addressing the following question: given two statements, determine if they represent logically equivalent propositions.

We'll begin with presentation of known, fundamental logical equivalencies. Our first will be that of **double negation**. It has been presented that a statement and its negation have opposite truth-values. That is, when A is true, ~A is false and vice versa. If the negation of A is, itself, negated, we obtain a statement that must have the same truth-value as statement A and hence be logically equivalent to A.

As an explicit exhibition of this, let's show a truth table that displays ~(~A) (Table 2.15). We see that the columns for A and for ~(~A) are identical, and therefore the statements are logically equivalent.

TABLE 2.15 Truth Table for Double Negation

A	~A	~(~A)
T	F	**T**
F	T	**F**

As a second illustration of logical equivalence, compare the following two statements:

$A \rightarrow B$ versus $\sim A \vee B$

We propose that the two statements are logically equivalent. The truth table for the former is already established, so we will focus our attention on the latter (Table 2.16). Note that, as suggested, the truth tables are identical, and thus the statements are logically equivalent.

TABLE 2.16 Truth Table Demonstrating Logical Equivalence

A	B	~A	~A ∨ B	A → B (known)
T	T	F	T	**T**
T	F	F	F	**F**
F	T	T	T	**T**
F	F	T	T	**T**

In the exercises, you will be asked to confirm or disprove the logical equivalence of statements. The problem amounts to construction of a truth table for each statement and comparison of the results. If the truth tables are identical, the statements are logically equivalent; otherwise, they are not.

EXAMPLE 2.2

Here is a bit of semantics that may create confusion:

1: It is not the case that: the bank is open on Saturday and the mailing center is open on Saturday.
2: The bank is not open on Saturday, or the mailing center is not open on Saturday.

Are the two remarks the same?

SOLUTION

They are clearly similar, but on first glance, it is unclear if they state the same information. In order to determine whether they are equivalent, we create symbolization of the statements and construct the truth tables. Symbolically, let us say A: the bank is open on Saturday, and B: the mailing center is open on Saturday. The statement labeled as 1 above is then symbolized as $\sim(A \wedge B)$, while the second statement is symbolized as $\sim A \vee \sim B$. We now construct the truth tables and make our comparison (Tables 2.17 and 2.18). Noting that the final columns match, we conclude that the statements are logically equivalent and thus express the same meaning.

TABLE 2.17

A	B	A ∧ B	~(A ∧ B)
T	T	T	**F**
T	F	F	**T**
F	T	F	**T**
F	F	F	**T**

TABLE 2.18

A	B	~A	~B	~A ∨ ~B
T	T	F	F	**F**
T	F	F	T	**T**
F	T	T	F	**T**
F	F	T	T	**T**

That last example is one of the logical forms of DeMorgan's laws, and proof of the other form of DeMorgan's laws will be presented to you in the exercises.

We will adopt the symbol ≡ to represent logical equivalence, and that symbol will be read as "is logically equivalent to" or, more briefly, as "is equivalent to." Much like the equals sign in computational mathematics, this symbol is not a logical operator but rather an indicator of a condition of equivalence between two statements.

Exercises

Answer each of the following using complete sentences, proper grammar, and correct spelling.

1. What is meant by "logical equivalence"?
2. How do you establish logical equivalence using truth tables?

Demonstrate that the following statements are or are not logically equivalent to one another.

3. $\sim(A \wedge \sim B)$ and $(\sim A) \vee (\sim B)$
4. $\sim(A \wedge B)$ and $\sim A \wedge \sim B$
5. $\sim(A \vee \sim B) \wedge \sim(A \vee B)$ and $A \vee (A \wedge B)$
6. $\sim(A \vee \sim B) \vee (\sim A \wedge \sim B)$ and $\sim A$

7. $((A \wedge B) \wedge C) \wedge D$ and $((D \wedge C) \wedge B) \wedge A$
8. $A \vee (A \wedge B)$ and A
9. $A \vee B$ and $\sim(\sim A \wedge B)$
10. $(A \to B) \wedge (B \to A)$ and $(A \leftrightarrow B)$
11. $(A \vee B) \vee C$ and $A \vee (B \vee C)$
12. $(A \to B) \wedge (A \to C)$ and $A \to (B \wedge C)$
13. $(A \wedge (\sim C \vee B))$ and $(A \wedge C) \vee (A \wedge B)$

We established in an example of this section the equivalence of two statements that represented one form of DeMorgan's laws, $\sim(A \wedge B) \equiv \sim A \vee \sim B$. The following problem presents the other form of DeMorgan's laws, which you should verify.

14. Show $\sim(A \vee B) = \sim A \wedge \sim B$

Use DeMorgan's laws to write the negation of the following statements.

15. I will be rich and successful.
16. It will rain today or it will be sunny.
17. You will participate or I will have you removed.
18. The Democrat won the election and the Republican is unhappy.
19. The number N is not positive and it is not negative.
20. I will continue to teach mathematics or I will go to law school.
21. I will either purchase a Macbook or I will purchase a Dell laptop.
22. Scott is an actuary, and Donna is a computer programmer.
23. China is a major manufacturer of silicon chips and India is a major producer of software.
24. Pi is an irrational number and *e* is a rational number.

2.5 Symbolic Arguments

In this section, we consider a collection of statements alleged to give support to a logical conclusion. Such a series of statements is what we refer to, in the context of logic, as an **argument**. Note that this is in contrast to the angry outburst you may associate with the concept of an argument, so you will want to maintain awareness of this distinction. In logic, arguments are not combative exchanges but rather a series of statements that allege to imply another statement.

Consider the following, relatively simple, illustration:

If Kyle did well on the GRE examination, then he will be admitted to graduate school.
Kyle did well on the GRE examination.
Therefore, Kyle will be admitted to graduate school.

The first two statements listed here are referred to as the **premises**, and the final statement, prefaced by the word "Therefore," is referred to as the **conclusion**.

This terminology appeared in Chapter 1 when we explored the application of Venn diagrams, but it serves us well to remind ourselves here.

Our concern is whether the argument is **valid**. That is, if the two premises are true, must the conclusion also be true? In other words, does the conclusion necessarily follow from the premises?

We can use Venn diagrams, as was shown in Chapter 1, to assess the validity of arguments, but we can also use truth tables. To do so, we will symbolize the individual premises and then take their conjunction. Following that process, we will form a conditional statement whose antecedent is the conjunction and whose consequent is the conclusion. If that conditional is a tautology, a statement all of whose truth-values are true, then the argument is considered valid; otherwise, it is considered **invalid**.

We realize that the preceding paragraph may be a bit confusing because of its extensive use of terminology. Let's return to the example and illustrate the intent of those instructions.

> If Kyle did well on the GRE examination, then he will be admitted to graduate school.
> Kyle did well on the GRE examination.
> Therefore, Kyle will be admitted to graduate school.

We will symbolize the argument using the following abbreviations:

> A: Kyle did well on the GRE examination.
> B: Kyle will be admitted to graduate school.

With these in mind, our argument takes the form

$A \rightarrow B$

A

$\therefore B$

Note the introduction of a triangular array of dots in the final line of the symbolic representation of the argument. That symbol is the representation of the word "therefore."

The "conjunction of the premises" is the compound statement $(A \rightarrow B) \wedge A$. This is the combination of the two premises by the logical connective "and." We next construct a "conditional" statement having that conjunction as the antecedent and the argument's conclusion as the consequent. By this we mean $((A \rightarrow B) \wedge A) \rightarrow B$.

The problem now boils down to the determination as to whether that last statement represents a tautology, and to this our attention now turns (Table 2.19).

TABLE 2.19

A	B	$A \rightarrow B$	$(A \rightarrow B) \wedge A$
T	T	T	**T**
T	F	F	**F**
F	T	T	**F**
F	F	T	**F**

We interrupt the construction of the truth table at this point to be sure that you follow the development. Carefully consider the table and be sure that each column's content is completely clear to you before moving on to the final table, in which we will produce the conditional statement described in the process earlier. The final column of the truth table is uniformly true, and hence the outcome is a tautology (Table 2.20). Therefore, the argument has been established as valid.

TABLE 2.20

A	B	$A \rightarrow B$	$(A \rightarrow B) \wedge A$	B	$((A \rightarrow B) \wedge A) \rightarrow B$
T	T	T	**T**	**T**	**T**
T	F	F	**F**	**F**	**T**
F	T	T	**F**	**T**	**T**
F	F	T	**F**	**F**	**T**

This example, having the form $[((A \rightarrow B) \wedge A) \rightarrow B]$, is a particular illustration of a known form of argument, referred to as the **law of detachment.** Any argument that can be symbolized into this structure is now known to be valid by that law.

There are other forms of known valid arguments, referred to as the laws of contraposition, syllogism, and disjunctive syllogism. They are shown below:

Law of contraposition:

$A \rightarrow B$

$\underline{\sim B}$

$\therefore \sim A$

Law of syllogism:

$A \rightarrow B$

$B \rightarrow C$

$\therefore A \rightarrow C$

Law of disjunctive syllogism:

$A \vee B$

$\sim A$

$\therefore B$

The proofs that all these are forms of valid arguments will be left to the exercises and are similar to the establishment of the law of detachment. Any time you recognize a particular argument as having one of the four known forms, you may declare them valid by referencing the known form by name.

EXAMPLE 2.3

Show that the following argument is valid:

If John goes to school, then John will do well on the test.
If John does well on the test, then he will pass the class.
Therefore, if John goes to school, then he will pass the class.

SOLUTION

This argument can be symbolized using the following abbreviations:

J: John goes to school
W: John does well on the test
P: John passes the class

$J \rightarrow W$
$W \rightarrow P$
$\therefore J \rightarrow P$

The argument is valid because it takes the form of the law of syllogism, which is known to be valid.

An argument form that is known to be invalid is referred to as a **fallacy**. There are two forms of fallacies we will introduce here: the **fallacy of the converse**, and the **fallacy of the inverse**. Both of these fallacies involve conditionals and are closely related in form to the known valid arguments. Take care that you do not mistake the fallacies from valid argument forms, as this can be disastrous.

Fallacy of the Converse:

$A \rightarrow B$

B

$\therefore A$

Remember, in the original conditional statement of the argument, we are told "If A, then B." Nothing is said in that statement about the power of implication wielded by B. Therefore, the occurrence of B does not imply the occurrence of A. Here is an illustration of the fallacy:

If Mark gets married, then he will be happy.
Mark will be happy.
Therefore, Mark got married.

If you think a bit deeply about this argument, you can realize the nature of the fallacy quite clearly: there are many occurrences that could prompt Mark to be happy, and being married is just one of them. We cannot say that, since he is happy, the one thing that caused that state of happiness was his marriage.

By way of further illustration, let's attempt to prove the validity of the argument (Table 2.21). The final column of the table does not consist of entirely true results, and therefore the conditional statement made up of the conjunction of the premises together with the conclusion is not a tautology. We conclude the argument to be invalid.

TABLE 2.21

A	B	$A \rightarrow B$	$(A \rightarrow B) \wedge B$	$((A \rightarrow B) \wedge B) \rightarrow A$
T	T	T	T	**T**
T	F	F	F	**T**
F	T	T	T	**F**
F	F	T	F	**T**

The other known form of fallacy takes the following form:

Fallacy of the Inverse:

$A \rightarrow B$

$\sim A$

$\therefore \sim B$

Consider the following as an example: "If I do all my chores, I will receive my allowance. I did not do all my chores. Therefore, I will not receive my allowance." It could very well be the case that the individual involved would have parents who were not entirely strict, and therefore the act of not doing chores might not necessarily indicate that an allowance would not be received.

Any time you recognize an argument as having one of the forms of the two known fallacies, you may state that the argument is invalid, provided that you refer to the particular fallacy demonstrated by the argument, using the proper name of the fallacy.

When more than two statements are involved in an argument, the valid argument forms mentioned earlier can save us from spending an inordinate amount of time constructing a lengthy truth table. Consider the following argument:

It is sunny outside.
If it is sunny outside, then I will wear my favorite ball cap.
If it is sunny outside and I wear my favorite ball cap, then the home team will win the game.
If I do not wear my favorite shoes, then the home team will not win the game.
Therefore, I will wear my favorite shoes.

Note that the validity of the argument has nothing to do with whether the remarks made within the argument are factually correct or even believable. The only thing that matters is the form of the argument; the matters of factual evidence are not under consideration in this sort of problem.

Using the following abbreviations,

S: It is sunny outside.
B: I will wear my favorite ball cap.
W: The home team will win the game.
F: I will wear my favorite shoes.

we obtain the following symbolization:

S

$S \rightarrow B$

$S \wedge B \rightarrow W$

$\sim F \rightarrow \sim W$

$\therefore F$

From the first two lines, using the law of detachment, we can determine that B is true. Since S and B are individually true, we conclude that $S \wedge B$ is true. Since $S \wedge B$ is true, the law of detachment applied to the third line of the argument establishes the truth of W.

The fourth line is known to be logically equivalent to its contrapositive, and thus the fourth line can be written as W → F, and hence, by the law of detachment applied to that equivalent form, we conclude that F is true, and therefore the argument is valid. It is worth mentioning that this could also be established by use of truth tables, and it would be an interesting exercise to do so. Because the problem involves four statements, we recall our rule for truth table size determination and recognize that such a table would have 2^4, or 16, rows.

Both approaches (truth tables and known argument forms) are acceptable ways to establish the validity of arguments. The truth table approach is what one might refer to as a "brute-force" style where we plow through the problem by methodically constructing a truth table and filling in column after column until the final result is determined. There is nothing inherently wrong in doing so, but it could be cumbersome and time consuming. Use of known argument forms is a more refined approach, demonstrating a great deal more elegance, but it requires the memorization of known argument forms.

EXAMPLE 2.4

Use known argument forms to assess the validity of the following argument:

If Ashley majors in mathematics, then she will open doors of opportunity.
Ashley did not major in mathematics.
Therefore, Ashley will not open doors of opportunity.

SOLUTION

To remove distraction, we symbolize the statements involved as

A: Ashley majors in mathematics.
D: Ashley opens doors of opportunity.

and represent the argument as

$A \to D$
$\sim A$
$\therefore \sim D$

This argument has the form of the fallacy of the inverse and is known to be invalid. Simply because Ashley does not major in mathematics, we cannot say that she will not open doors of opportunity! That is too broad a conclusion to draw based on the premises.

EXAMPLE 2.5

Use known argument forms to assess the validity of the following argument:

If you place your math book under your pillow at night, you will sleep soundly.
If you sleep soundly, you will do well on your mathematics final examination.
Therefore, if you place your math book under your pillow at night, you will do well on your mathematics final examination.

SOLUTION

Again, resorting to symbolization,

P: You place your math book under your pillow at night.
S: You will sleep soundly.
W: You will do well on your mathematics final examination.

We obtain the following representation of the argument:

$P \rightarrow S$

$S \rightarrow W$

$\therefore\ P \rightarrow W$

This is a syllogism and is therefore valid despite the silliness of the sequence of statements. Remember, we are not interested in whether the information has a basis in fact but merely if the form of argument is valid!

EXAMPLE 2.6

Use known argument forms to assess the validity of the following argument:

If you buy your car from an established dealership, your car will not have mechanical problems.
If you do not buy your car from an established dealership, then you are asking for trouble.
You are not asking for trouble.
Therefore, your car will not have mechanical problems.

SOLUTION

As before, we'll symbolize the individual statements,

B: You buy your car from an established dealership.
P: You will have mechanical problems.
A: You are asking for trouble.

Under these symbolizations, we can express our argument as

$B \rightarrow \sim P$

$\sim B \rightarrow A$

$\sim A$

$\therefore\ \sim P$

The second and third premises, under the law of contraposition, tell us $\sim(\sim B)$, or B. This, together with the first premise, by the law of detachment, implies $\sim P$, which is the conclusion. Therefore, the argument is valid using the laws of contraposition and detachment.

Exercises

Using complete sentences and correct spelling and grammar, answer the following questions.

1. What is an argument, from the point of view of logic?
2. Within an argument, what are the premises?
3. Within an argument, what is the conclusion?
4. What is meant by "a valid argument"?
5. What are the two different methods described in this section of assessing the validity of an argument?

Using truth tables, verify the validity of the following argument forms.

6. The law of contraposition
7. The law of syllogism
8. The disjunctive syllogism

Using truth tables, verify that the following argument forms are invalid.

9. The fallacy of the converse
10. The fallacy of the inverse

Using the known argument forms determine, whether the following arguments are valid.

11. If the price of oil falls, then people will drive more.
    The price of oil is falling.
    Therefore, people will drive more.
12. If a movie features a major star, it will gross a lot of money.
    This movie grossed a lot of money.
    Therefore, it featured a major star.
13. If you do not pay your tuition on time, you will be dropped from your classes.
    You paid your tuition on time.
    Therefore, you will not be dropped from your classes.
14. If you buy antivirus software, your computer will not be infected with a virus.
    You did not buy antivirus software.
    Therefore, your computer will be infected with a virus.
15. If you order the high-performance engine, you will order the sports suspension.
    You ordered the sport suspension.
    Therefore, you ordered the high-performance engine.
16. If you love me, you will believe every word I say.
    You do not believe every word I say.
    Therefore, you do not love me.
17. You will either get a job in town or you will move to New York.
    You did not get a job in town.
    Therefore, you will move to New York.

For the following symbolically represented arguments, determine if they are valid.

18. $\sim A$
    $A \rightarrow B$
    $\therefore \sim B \wedge A$

19. A
    $\sim B \rightarrow \sim A$
    $(A \wedge B) \rightarrow C$
    $\therefore B \rightarrow C$

20. B
    $B \rightarrow \sim A$
    $B \rightarrow (C \wedge A)$
    $\therefore C$

21. A
    $\sim A \rightarrow \sim B$
    $(A \wedge B) \rightarrow C$
    $\therefore \sim B \rightarrow C$

Symbolize the following arguments and then use the method of your choice to determine if the argument is valid.

22. Doris will go to Hawaii.
    If Doris does not go to Hawaii, she will not go on vacation.
    If Doris goes to Hawaii on vacation, then her friends will miss her.
    Therefore, if Doris goes on vacation, her friends will miss her.
23. Alex speaks Russian.
    If Alex speaks Russian, then he will get a job with the CIA.
    He will not live in New York or will not work for the CIA.
    Therefore, Alex will live in New York.
24. If Sue breaks up with Jon, she will not come to the dance.
    If Sue does not break up with Jon or does not come to the dance, then she will go shopping.
    Sue will not go shopping.
    Therefore, Sue will not break up with Jon.
25. Tom is tall and Scott is short.
    If Scott is short, then he can get away with wearing striped pants.
    If Wayne is tall, then he can not get away with wearing striped pants.
    Therefore, Wayne is not tall.

2.6 Euler Diagrams and Syllogistic Arguments

Recall from Chapter 1 that a syllogism is a particular type of argument, consisting of at least two premises and a conclusion. The premises and conclusion may contain what we referred to as "quantifiers," words such as "all," "some," and "none." Moreover, an Euler diagram is a particular instance of a Venn diagram, applied to the analysis of arguments.

Keep in mind that, when attempting this method for verifying the validity of arguments, your examination is directed toward an attempt to produce a possible diagram that would show the syllogism to be invalid. If no such diagram can be produced, you can conclude that the argument is valid. This practice is sometimes referred to as the "counterexample principle."

Here's an illustration:

Some children love cookies.
Scott does not love cookies.
Therefore, Scott is not a child.

Is this argument valid, or is it invalid? Let's construct an Euler diagram to assist in our investigation.

The first premise talks about children and their love of cookies and tells us that some children love cookies. This suggests that there are two sets of individuals present: the children and the individuals who happen to love cookies. That some children love cookies would suggest that the set of children

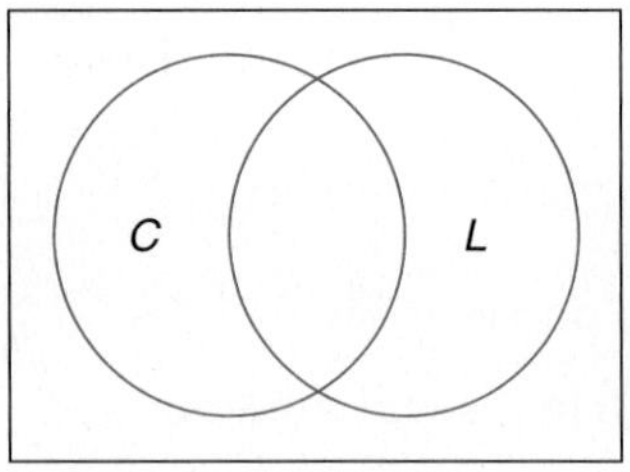

Figure 2.1 Venn diagram comparing set of all children (*C*) with those who love cookies (*L*)

intersects the set of individuals who love cookies but that the set of children is not wholly contained in the set of those who love cookies. Hence, a visualization of this condition might look like Figure 2.1, taking *C* to be the set of children and *L* to be the set of individuals who love cookies.

The second premise tells us that Scott does not love cookies, and therefore Scott is one of those individuals who is located beyond the limits of set *L*. Remember, this merely tells us that Scott is *somewhere* in the universal set, outside of set *L*. His location is ambiguous other than for that fact.

Our question now is this: the conclusion tells us Scott is not a child. Is it *necessary* that Scott be outside of the set of children based on the stated assumptions?

Knowing only that Scott lies outside of set *L*, we have no assurance that Scott is beyond the limits of set *C*, and therefore the syllogism is invalid. Scott could, very well, be within that part of set *C* not intersecting with set *L*.

EXAMPLE 2.7

Use an Euler diagram to determine the validity of the following syllogism:

Some mathematicians love tennis.
Everyone who loves tennis is genuine.
Sebastian is not genuine.
Therefore, Sebastian is not a mathematician.

In this case, we have four concepts: mathematicians, individuals who love tennis, individuals who are genuine, and Sebastian. Let's define the sets and then consider the possible or necessary locations of our friend Sebastian.

M: the set of mathematicians
T: the set of individuals who love tennis
G: the set of individuals who are genuine

Note that set *M* intersects with set *T*, but there is not a subset relationship present, since we are merely told that *some* mathematicians love tennis, not all. On the other hand, set *T* must be a proper subset of set *G*, since all those who love tennis are, evidently, genuine.

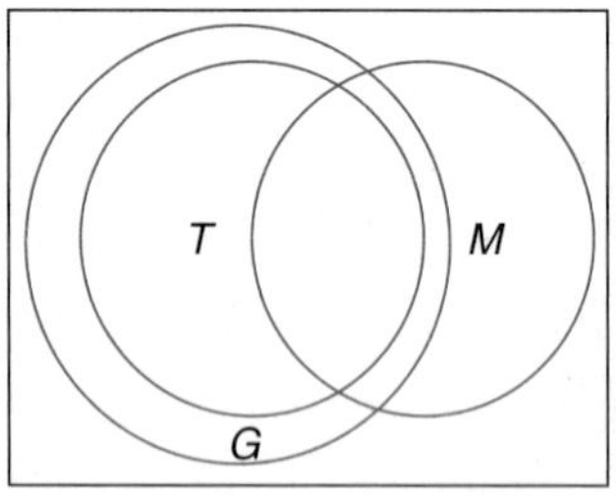

Figure 2.2 Venn diagram for Example 2.7

The Euler diagram for the situation might look like Figure 2.2. Now, we confront the crucial question: the conclusion suggests that it is impossible that Sebastian could be found within set *M*, knowing from the third premise that he does not belong to set *G*. Is this mandatory?

By inspection, we can see that the requirement that Sebastian's location be outside set *G* does not preclude his presence in set *M*, since there is a segment of set *M* that is beyond the limits of set *G*. Consequently, it is not necessary that Sebastian not be a member of set *M*, and hence we conclude the argument to be invalid.

Our examples, to this point, have contained the quantifier "some" but have avoided the use of the concept of "none." Let us close this section with an example capturing that notion, and then our exposition on logic will be complete.

EXAMPLE 2.8

No Europeans are honest.
Scott is not honest.
Therefore, Scott is European.

Is this argument valid? We will construct an Euler diagram to assess the situation, identifying the involved sets as *E*: the set of Europeans, and *H*: the set of honest people.

We are told in the first premise that no Europeans are honest. This indicates the set *E* has no intersection with the set *H*. The second premise tells us that Scott lies somewhere beyond the limit of set *H*. Consider a diagram showing disjoint sets *E* and *H*, as in Figure 2.3. The conclusion tells us that Scott is definitely within set *E*. Must this be the case? The second premise tells us that Scott is not an honest person, but this only requires that he be located outside of set *H*. There is ample room in the universe for Scott's location to be outside of set *H* and still outside set *E*, and therefore we have no assurance that Scott, dishonest though he may be, is European. The argument is invalid.

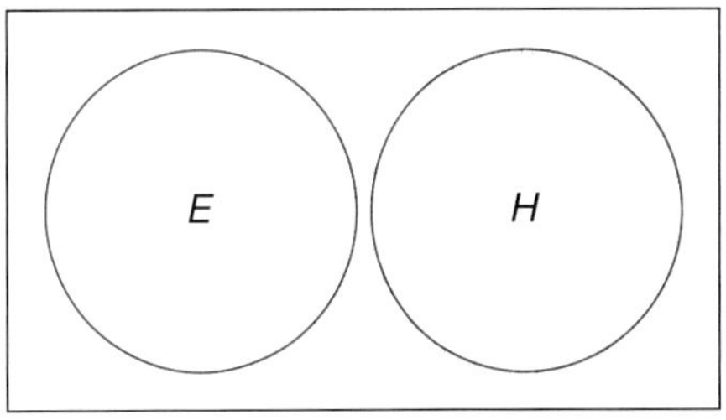

Figure 2.3 Venn diagram for Example 2.8

Exercises

Using complete sentences and proper spelling and grammar, answer the following questions.

1. What is an Euler diagram?
2. What is meant by the term "quantifier"?
3. What does it mean to say a syllogism is valid?
4. What does it mean to say a syllogism is invalid?

Determine whether the following syllogisms are valid or invalid.

5. All college students drink alcohol.
   Scott drinks alcohol.
   Therefore, Scott is a college student.
6. All mathematicians are stodgy.
   Ramanujan was a mathematician.
   Therefore, Ramanujan was stodgy.
7. Some vitamins keep you healthy.
   Milk keeps you healthy.
   Therefore, milk is a vitamin.
8. Some diplomats are corrupt.
   Chamberlain was a diplomat.
   Therefore, Chamberlain was corrupt.

9. All salespeople are annoying people.
   Kelly is a salesperson.
   Therefore, Kelly is an annoying person.
10. Some mathematicians believe in the Axiom of Choice.
    All who believe in the Axiom of Choice are silly.
    Alfred is not silly.
    Therefore, Alfred is not a mathematician.
11. No cars manufactured before 1960 were equipped with seat belts.
    Some MGs were manufactured before 1960.
    Therefore, some MGs were not equipped with seat belts.
12. All volleyball players are tall.
    Some women play volleyball.
    Therefore, some women are tall.
13. All vegetarians eat tofu.
    Kristine eats tofu.
    Therefore, Kristine is vegetarian.

The following syllogisms lack conclusions. Assuming that the syllogisms are to be valid, provide a possible conclusion that follows from the premises.

14. Some Pontiacs are fast cars.
    All fast cars are dangerous.
    Therefore, …
15. All math teachers are saintly.
    Biff is not saintly.
    Therefore, …
16. Some singers have deep voices.
    All people with deep voices are confident.
    Therefore, …
17. No Heisman Trophy winner has led the NFL in interceptions.
    Some linebackers have led the NFL in interceptions.
    Therefore, …

Summary

In this chapter, you learned about:

- Simple and compound statements.
- Logical connectives: "not," "and," "or," "if and only if."
- Truth tables for negation, conjunction, and disjunction; conditional and biconditional statements; equivalent statements; and symbolic arguments.
- Using Euler diagrams to determine the validity of syllogisms.

Glossary

argument: A collection of statements alleged to give support to a logical conclusion.

biconditional statement: A compound statement involving the "if and only if" logical connective form.

compound statement: A statement conveying two or more ideas.

conclusion: A final statement in a logical argument alleged to follow from given premises.

conditional statement: A compound statement involving the "if-then" logical connective form.

contradiction: A compound statement that, under all conditions, is exclusively false.

contrapositive: The logical equivalent form $\sim B \rightarrow \sim A$ to $A \rightarrow B$.

decision problem: A question posed by Hilbert that asked if a standard procedure could be developed to determine if a particular statement could be proven.

double negation: A succession of two "not" symbols in a logical expression.

fallacy: An argument form known to be invalid.

fallacy of the converse: $A \rightarrow B$

B

$\therefore A$

fallacy of the inverse: $A \rightarrow B$

$\sim A$

$\therefore \sim B$

invalid argument: A logical argument that is not valid.

law of detachment: $[((A \rightarrow B) \wedge A) \rightarrow B]$.

law of contraposition: $A \rightarrow B$

$\underline{\sim B}$

$\therefore \sim A$

law of syllogism: $A \rightarrow B$

$B \rightarrow C$

$\therefore A \rightarrow C$

law of disjunctive syllogism:

$A \vee B$

$\sim A$

$\therefore B$

liar's paradox: A classic statement ("I am lying") that cannot be either true or false.

logical connectives: Words that connect one simple statement to another.

logically equivalent statements: Two statements having the same truth-functional value.

premises: Statements within a logical argument that allege to imply a particular conclusion statement.

simple statement: A statement conveying a single idea.

statement: A declarative sentence that is either true or false.

tautology: A compound statement that, under all conditions, is exclusively true.

truth table: A device we can use to investigate a statement containing a finite number of simple statements in combination.

truth-value: The truth or falsity of a statement.

turing machines: A theoretical device that would manipulate symbols on a spool of tape in accordance with a set of predesignated rules.

valid argument: An argument such that, if the given premises are true, the conclusion must be true.

End-of-Chapter Problems

In the following exercises, identify which sentences are logical statements.

1. The sky is blue.
2. The Boston Bruins are a baseball team.
3. Have you eaten dinner?

The following exercises contain statements. Identify whether they are simple statements or compound statements.

4. The Celtics are a basketball team and math is fun.
5. Harry Potter attended Hogwarts.
6. If Hermione Granger attended Hogwarts, then so did Ronald Weasley.

For each of the following compound statements, assign variables to the individual statements and then symbolize the compound statements using the logical symbols developed in this chapter.

7. George owns an iPod, or George owns a Zune.
8. Dinner is good and it is not the case that lunch was good.
9. This text is good if and only if the questions in this text are good.

For the following symbolizations, create statements corresponding to each letter, and then write an English sentence that is equivalent to the symbolization.

10. $P \leftrightarrow (Q \wedge R)$
11. $P \vee (Q \wedge R)$
12. $(P \vee Q) \wedge (P \vee R)$

For the following statements, give the negation.

13. I am a vegetarian.
14. Australia is the only country that is also a continent.

Construct a truth table for each statement.

15. $P \vee (Q \vee R)$
16. $P \leftrightarrow {\sim}Q$
17. $P \rightarrow (P \wedge Q)$

Do the following statement pairs have identical truth tables, or do they not?

18. $P \leftrightarrow Q$ and $(P \wedge Q) \vee ({\sim}P \wedge {\sim}Q)$
19. $P \rightarrow {\sim}Q$ and $Q \rightarrow {\sim}P$
20. $P \vee (Q \vee R)$ and $(P \vee Q) \vee R$

In the following exercises, circuits are shown. Symbolize them logically and then (if possible) try to devise an equivalent circuit that uses fewer switches than the one depicted in the problem.

21.

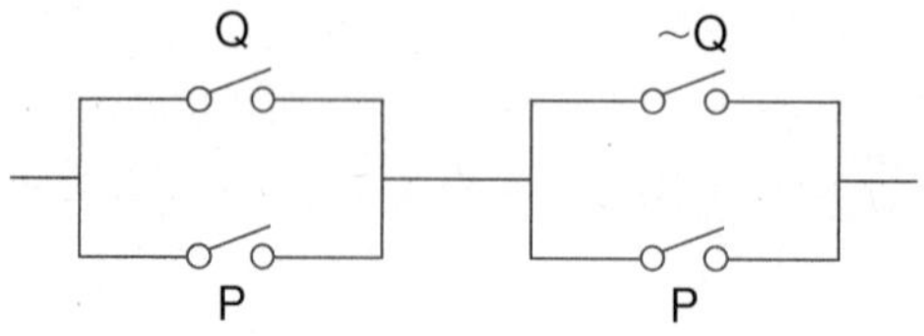

22.

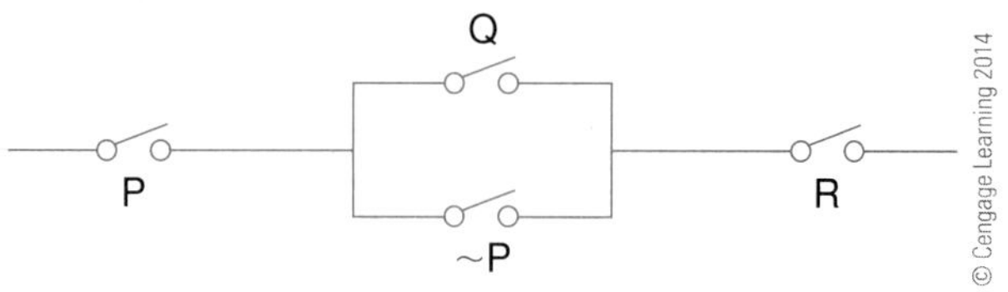

23.

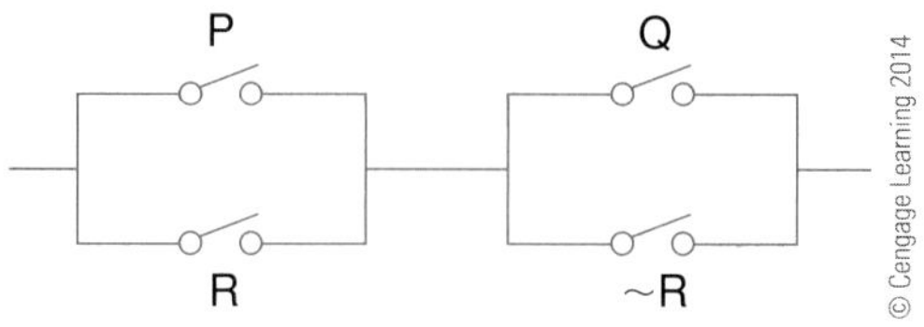

In the following exercises, construct switching circuits that represent the symbolic statements.

24. $(P \vee {\sim}Q) \wedge ({\sim}P \vee R)$

25. $(P \wedge Q) \vee {\sim}R$

For the following statements, construct a truth table.

26. $A \leftrightarrow {\sim}B$

27. $(A \wedge {\sim}B) \leftrightarrow ({\sim}A \wedge C)$

Write the converse of the following conditional statements.

28. If the economy crashes, people will go hungry.
29. If I am hungry, I will cook dinner.
30. If black is white, pigs will fly.

Write the inverse of the following conditional statements.

31. If a wizard needs a wand, spells are easy.
32. If George is tall, it is not the case that George is short.

Rewrite the conditional statements using the alternate form "A is sufficient for B."

33. If there is cake for dessert, then I will be happy.
34. If night falls, the moon comes out.

Rewrite the following statements using the word combination "if-then."

35. I will be happy if class goes well.
36. To see your professor, it suffices to go to her office hours.

Demonstrate whether the following statements are or are not logically equivalent to one another.

37. ${\sim}(A \vee {\sim}B)$ and ${\sim}A \vee B$

38. $(A \rightarrow B) \wedge (B \rightarrow C)$ and $A \rightarrow C$

Use DeMorgan's laws to write the negations of the following statements.

39. The summertime is coming and the leaves are sweetly blooming.
40. George has some money and George will not eat at home tonight.

Using the known argument forms, determine whether the following arguments are valid.

41. If George plays the upright bass, then he has a hatchback or a station wagon.
George plays the upright bass.
Therefore, George has a hatchback or a station wagon.

42. If I eat too much cake, I will have a bellyache.

    I have a bellyache.

    Therefore, I ate too much cake.

For the following symbolically represented arguments, determine if they are valid.

43. $A \rightarrow B$

    $B \rightarrow C$

    $\sim C$

    $\therefore \sim A$

44. $A \rightarrow \sim B$

    $B \rightarrow C$

    $\therefore \sim C$

Symbolize the following arguments and then use the method of your choice to determine if the argument is valid.

45. If Fred speaks Apache, then he will get a job with the Bureau of Indian Affairs.

    If Fred does not live in Washington, then he will not get a job with the Bureau of Indian Affairs.

    Therefore, if Fred speaks Apache, then he lives in Washington.

46. If I hate doing something and I do it anyway, then I am a hero.

    I am not a hero.

    I hate doing something.

    Therefore, I do not do it.

Determine whether the following syllogisms are valid or invalid.

47. All desserts are bad for you.

    All cookies are desserts.

    Therefore, all cookies are bad for you.

48. All mammals suckle their young.

    An ostrich is not a mammal.

    Therefore, an ostrich does not suckle its young.

The following syllogisms lack a conclusion. Assuming that the syllogisms are to be valid, provide a possible conclusion that follows from the premises.

49. All teachers dislike grading.

    George is a teacher.

    Therefore, …

50. Some dogs are Portuguese Water Dogs.

    All Portuguese Water Dogs like the water.

    Therefore, …

Chapter 3 Binary and Other Number Systems

Probably the most common system for representing numbers currently in use is the decimal, or base-10, number system. This system identifies quantities using numerals from the set {0, 1, 2, 3, 4, 5, 6, 7, 8, 9} and forms the foundation for the Hindu-Arabic number system with which we are all likely to be familiar.

A lesser-known fact is that the base-10 number system is not the only numeration system we can use to express quantities, nor is it always the most efficient method. In fact, the underlying numeration system for storing information electronically uses only the numerals {0, 1} and is referred to as the binary, or base-2, system.

Objectives

By the time you have successfully completed the materials of this chapter, you will be able to understand the relationships between the following number systems:

- Binary and decimal
- Binary and hexadecimal
- Decimal and hexadecimal
- Binary and octal
- Binary and 8421 BCD numbers

3.1 The Decimal Number System

The most common "representation" form currently in use for numbers is the **decimal** system. This is the system of numeration we use in our everyday lives and is probably the one with which you are most familiar. Computer programmers use other number systems: **binary**, **hexadecimal**, and (to a lesser extent) **octal**. Why this would possibly be the case may be unclear, but shortly the reasons for the use of alternative systems will be made apparent. Of course, once a new number system is introduced, the question of arithmetic within that system naturally arises, and we will explore that question as well.

As a precursor to those issues, let's provide some motivation for exploration of the first of those systems, the binary numbers, and then devote some time to formalizing the decimal number system and its terminology. As we'll see, the process of computation in both systems uses the same underlying theory, but the limitation of using fewer numerals requires us to take care in the implementation of our familiar arithmetic rules.

The binary system as applied to information assigns the numbers 0 and 1 to represent alternative "states," such as off-on, no-yes, false-true, and zero-nonzero. Alternative states are a matter of interest in any situation in which only one of two mutually exclusive conditions can exist. Through clever combinations of 0s and 1s, any such data can be successfully represented numerically. Computers and other electronic devices use binary numbers, in which integers are represented through the use of the digits 0 and 1 exclusively. Everything in your computer, for instance, is represented within the computer's memory only in the form of 0s and 1s. Digital information stored on DVDs or CDs is also recorded in binary form, and all instructions between a computer's hardware and software are ultimately reduced to a simple "yes or no" determination.

There is some warranted concern that computation by hand with binary and other number systems can be challenging, but computers are capable of working with such numbers with ease. After we have developed an understanding of the procedure, we will be able to assign such tasks to mechanical devices, comfortable with our ability to perform the computations by hand, if need be. At the moment, it is unclear precisely how written information can be expressed numerically using 0s and 1s, but we'll discuss that once the binary system and its operations have been developed.

With those thoughts in mind, we turn to our familiar decimal number system, within the framework of which we will familiarize ourselves with the terminology we will employ in the other number systems. You will find that, as we do so, terms that are familiar in the decimal system will have necessary modifications to analogous concepts within those other systems.

Our development begins with the introduction of the term **base**. An alternative term used in place of the word "base" is **radix**, but we will use the word "base" in this text. The base of any number system is the number of distinct characters employed for numeration when using that system. For instance, in our common decimal system, there are 10 symbols we use for numeration: 0, 1, 2, 3, 4, 5, 6, 7, 8, and 9. All numbers within the decimal system are constructed using those characters. The number of available characters is given as the "base of the system," and hence the decimal system is also referred to as the **base-10 number system**.

Parenthetically, note that the other number systems use a different set of characters that may be a smaller or larger set than that used in the decimal system. The binary system, as mentioned previously, employs only the digits 0 and 1, and therefore the base of that system is 2. It is natural to be puzzled regarding how other numbers could be depicted using only those two digits, but please be patient! That bridge will be crossed in the fullness of time, and how we could represent a decimal number such as 7, for instance, using only 0s and 1s, will be made clear.

When a base-10 digit is used in combination or series with other such digits, producing multidigit numbers such as 318, we recognize that each digit occupies a particular place within the number and that it has a unique **place value**, .capturing its own meaning. The place value is determined by the location of the digit relative to the "decimal point." Here we must be a bit careful, for in different countries different symbols may be used to mark the separation between integral place values of a number and fractional values. In the United States, for instance, a dot is used to represent the decimal point, while in Great Britain a comma is employed. It might be more accurate to refer to the decimal point as a **decimal separator**, to avoid confusion, and we will generally adopt that procedure here. That being said, the term "decimal point" is widely in use, and hence that term will be taken to be synonymous with "decimal separator."

The decimal separator indicates the line of demarcation between the integral component of the decimal number and the fractional component. To the left of the decimal separator we find the integral component of a number, while to the right we find the fractional component. You are likely to be familiar with the place names for the decimal number system, but we will provide an example as a reminder.

Consider the decimal number 174.3609. The 1 at the start of the number is said to be in the "hundreds" place, the 7 following it to be in the "tens" place, and the 4 following that to be in the "ones" place. Digits to the left of the decimal separator have their place names ending in "s" for all locations. To the right of the decimal separator, the 3 is said to be in the "tenths" place, the 6 following it to be in the "hundredths" place, the 0 following that to be in the "thousandths" place, and the 9 that terminates the number to be in the "ten-thousandths" place. Note that digits to the right of the decimal separator have their place names ending in "ths."

From the place name of the digit, we obtain what is referred to as its place value. For instance, in the previously mentioned example, 174.3609, the 1, which occupies the hundreds place, possesses place value 100. The 7 has place value 70 and so on. The digits to the right of the decimal separator have place values such as (for the 6, as an illustration) six hundredths.

Using this notion of place values, we can generate the **expanded form** of a decimal number. This **exploded form** decomposes the original number into a sum of terms consisting of the individual digits within the number multiplied against an appropriate power of 10. In base-10, you are likely to find the method to be nearly self-evident, but the process may be slightly less natural when we turn our attention to the other bases we will use, and so we'll risk a bit of tedium at this point in order to lay the groundwork for the analogous structure in less familiar bases.

Prior to demonstrating the process, consider an illustration of addition of decimal numbers:

$$300 + 80 + 2 + 0.1 + 0.09 = 382.19$$

The addition is fairly straightforward, but we will now view it in a rather unconventional manner. The **symmetric property of equality** tells us that the equation can be reversed, yielding

$$382.19 = 300 + 80 + 2 + 0.1 + 0.09$$

The expression now to the right of the equals sign forms the basis for the expanded form we are intending. Note that the term "expanded" merely refers, in this case, to a horizontal enlargement of the expression.

Each of the terms on the right side of the equation can be expressed as a single digit multiplied by a power of 10, in a manner evocative of scientific notation. That is,

$$\begin{aligned} 382.19 &= 300 + 80 + 2 + 0.1 + 0.09 \\ &= 3 \times 10^2 + 8 \times 10^1 + 2 \times 10^0 + 1 \times 10^{-1} + 9 \times 10^{-2} \end{aligned}$$

This final form is what we refer to as the expanded form. The entire number is written as a sum of terms, each of which uses a digit from the numeration system times a power of the base (which is 10 for the decimal number system).

Returning to 174.3609, we can decompose the number into the following sum:

$$100 + 70 + 4 + 0.3 + 0.06 + 0.000 + 0.0009$$

Then, proceeding as was done before, we can express the individual terms as a digit multiplied by a power of 10, obtaining

$$1 \times 10^2 + 7 \times 10^1 + 4 \times 10^0 + 3 \times 10^{-1} + 6 \times 10^{-2} + 0 \times 10^{-3} + 9 \times 10^{-4}$$

You may be wondering if it is necessary to show the term where the digit is zero. In the fully expanded form of the decimal number, all the digits from the original number should be shown for the sake of completeness and systematization. Thus, it would technically be a mistake to suppress that term from the expansion.

EXAMPLE 3.1

Identify the place values for the digits and produce the expanded form of the decimal number 5,386.724.

SOLUTION

For the digit 5, the place value is 5,000, since the 5 occurs in the thousands place. For the remaining digits, we can list them systematically:

> The place value of 3 is 300, the place value of 8 is 80, the place value of 6 is 6, the place value of 7 is 0.7, the place value of 2 is 0.02, and the place value of 4 is 0.004.

The expanded form of the decimal number, using the digits multiplied by powers of 10, would be

$$5 \times 10^3 + 3 \times 10^2 + 8 \times 10^1 + 6 \times 10^0 + 7 \times 10^{-1} + 2 \times 10^{-2} + 4 \times 10^{-3}$$

We close out this section with a final bit of terminology: in a number represented in the base-10 system, the digit farthest to the left is referred to as the **most significant digit (MSD)**, and the digit farthest to the right is referred to as the **least significant digit (LSD)**. In the preceding illustration, the MSD is 5, while the LSD is 4.

Exercises

Answer each of the following, using complete sentences and proper grammar and spelling.

1. What is meant by the base of a number system, and what alternative terminology can be used for the word "base"?
2. How do we determine the place value of a digit within a number?
3. What is meant by the expanded form of a decimal number?
4. What, if anything, is the difference between the word *base* and the word *radix*?

In each of the following, identify the place name and place value of every digit.

5. 1,387
6. 281.793
7. 0.01379
8. 10,002.00208

In each of the following, give the expanded form of the decimal number.

9. 874.983
10. 16,039.177
11. 201.9938
12. 123,654.92846

For the following, give the decimal number whose expanded form is shown.

13. $3 \times 10^5 + 0 \times 10^4 + 4 \times 10^3 + 3 \times 10^2 + 2 \times 10^1 + 9 \times 10^0$
14. $0 \times 10^0 + 1 \times 10^{-1} + 3 \times 10^{-2}$
15. $3 \times 10^0 + 1 \times 10^{-1} + 4 \times 10^{-2} + 1 \times 10^{-3} + 5 \times 10^{-4} + 9 \times 10^{-5}$
16. $2 \times 10^4 + 9 \times 10^3 + 1 \times 10^2 + 9 \times 10^1 + 5 \times 10^0 + 3 \times 10^{-1} + 7 \times 10^{-2}$

Referencing the problems indicated, for each of the following, identify the LSD and MSD.

17. Problem 5
18. Problem 6
19. Problem 7
20. Problem 8
21. Problem 9
22. Problem 10
23. Problem 11
24. Problem 12

Give examples of decimal numbers satisfying the following conditions (answers will vary).

25. A whole number having MSD is 5 and LSD is 8.
26. A number having 0s in the tens, hundredths, and thousands places.

3.2 The Binary Number System

The decimal system is likely to be familiar to you, but the binary system may be somewhat less so. As was remarked in the initial comments of the chapter, the binary (or base-2) system employs only two digits in all its numerical representations, 0 and 1. Thus, a number expressed in the binary system would look something like this: 100101101.11.

There is legitimate concern that this number would be mistaken for a decimal number, and therefore we must adopt a convention that prevents this confusion. When numbers are given in a base system other than the decimal system, we shall display the base as a subscript following the number, as in 100101101.11_2. This will exhibit clearly that the number shown is a representation in base-2, the binary number system. We could extend this concept to cover the decimal system as well, following a number with a subscript of 10, but we shall agree that if no subscript follows the number, it will be understood that our representation is in base-10. Hence, 3 457.58 will be considered equivalent to $3\,457.58_{10}$. This practice is much like the agreement we make that rational numbers having denominator 1 are typically not expressed in fraction form, and as long as we all agree on the stipulation, no confusion will arise.

The two digits in use in the binary system are sometimes referred to as **bits**. A group of eight bits is called a **byte**, and half a byte (four bits) is called a **nibble**. Different types of computers are capable of processing different lengths of data segments at one time, some able to handle 8 bits, some 16 bits, others 32 bits, and still others 64-bit segments. Our objective now is to develop familiarity with representations of numbers in the base-2 system and to then construct the arithmetic of numbers represented in this way.

We encounter similar terminology within the binary system as that developed for the decimal system. For one thing, numbers represented in the binary system consist of the digits 0 and 1, each having a place value within the representation. Unlike the places in the decimal system, however, the place values of the binary system are powers of 2 rather than powers of 10.

Although they may be familiar to you, we should review some of the smaller powers of 2 for the sake of clarity (see Table 3.1).

When we present a number in the binary system, the symbol that resembles what we commonly refer to as the decimal point and that we identified

TABLE 3.1 Some of the Powers of 2

Power of 2	Decimal Value
2^{-5}	$\frac{1}{32}$
2^{-4}	$\frac{1}{16}$
2^{-3}	$\frac{1}{8}$
2^{-2}	$\frac{1}{4}$
2^{-1}	$\frac{1}{2}$
2^{0}	1
2^{1}	2
2^{2}	4
2^{3}	8
2^{4}	16
2^{5}	32

as the decimal separator in Section 3.1 will now be referred to as the **binary point**, or the **binary separator**. The digit occurring immediately to the left of the binary point has place value 2^0, the digit to its left has place value 2^1, and so on. The digits to the right of the binary point have place values $2^{-1}, 2^{-2}, 2^{-3}$, and so on.

Consider the following illustration of a binary number and its interpretation in our more familiar decimal form: 101101.011_2. Using the stated place values for the binary number and noting that the first digit is in the 2^5 place (we can tell this by counting positions to the left of the binary point, keeping in mind that the initial digit to the left of that point is the 2^0 position), we obtain

$$1 \times 2^5 + 0 \times 2^4 + 1 \times 2^3 + 1 \times 2^2 + 0 \times 2^1 + 1 \times 2^0 + 0 \times 2^{-1} + 1 \times 2^{-2} + 1 \times 2^{-3}$$

Converting to decimal form, this would be equivalent to

$$\begin{aligned} &32 + 0 + 8 + 4 + 0 + 1 + 0 + \frac{1}{4} + \frac{1}{8} \\ &= 32 + 8 + 4 + 1 + .25 + .125 \\ &= 45.375 \end{aligned}$$

Shortly, we will formalize the method for conversion between binary and decimal numbers, but before doing so, we present a bit more terminology. The digit occurring in the highest place position within the number is referred to as the **most significant binary digit**, or **most significant bit (MSB)**. The digit with the lowest place position, on the other hand, is said to be the **least significant binary digit**, or **least significant bit (LSB)**. In the example, both the MSB and the LSB were 1, which is almost always the case in a binary number. The only exception to this situation would be where the MSB, LSB, or possibly both would be zero.

Because it is common to subdivide binary numbers into nibbles, as an aid to readability, the preceding example would more likely be encountered as $10\ 1101.011_2$.

The illustration just considered provides an example of how to convert from binary numbers to equivalent decimal numbers, but we will now formalize that procedure. To convert from the binary to the decimal system, perform the following steps:

1. Express the binary number in expanded binary form, showing each digit of the binary number multiplied by an appropriate power of 2.
2. Convert the fractions obtained from the binary places to the right of the binary point into decimal equivalents (this is not, strictly speaking, necessary, but it is the convention we will adopt here).
3. Add the resulting decimal numbers together.

EXAMPLE 3.2

Convert $11\ 0101\ 0110_2$ to decimal form.

SOLUTION

Since we do not see the binary point present, we assume (as is always the case with decimal numbers) that the binary point occurs at the end of the number and that it has been suppressed.

Noting the MSB occurs in the 2^9 place (we can determine this by counting places from right to left, aware that the first digit represents the 2^0 place, or observing that the power of 2 applicable to the MSB is always one less than the number of digits to the left of the binary point), we proceed as follows:

$$\begin{aligned} &1 \times 2^9 + 1 \times 2^8 + 0 \times 2^7 + 1 \times 2^6 + 0 \times 2^5 + 1 \times 2^4 + 0 \times 2^3 \\ &\quad + 1 \times 2^2 + 1 \times 2^1 + 0 \times 2^0 \\ &= 512 + 256 + 0 + 64 + 0 + 16 + 0 + 4 + 2 + 0 \\ &= 854 \end{aligned}$$

EXAMPLE 3.3

Convert $1.\ 0111\ 01_2$ to decimal form.

SOLUTION

We note that the MSB is in the 2^0 place and recall the binary places to the right of the binary point begin with 2^{-1} and have the powers decrease by 1 with each advancing position to the right. Therefore, our representation expands to

$$\begin{aligned} &1 \times 2^0 + 0 \times 2^{-1} + 1 \times 2^{-2} + 1 \times 2^{-3} + 1 \times 2^{-4} + 0 \times 2^{-5} + 1 \times 2^{-6} \\ &= 1 + 0 + \frac{1}{4} + \frac{1}{8} + \frac{1}{16} + 0 + \frac{1}{64} \\ &= 1 + .25 + .125 + .0625 + .015625 \\ &= 1.453125 \end{aligned}$$

Provided that we count our place locations carefully and compute our powers of 2 correctly, the conversion from binary to decimal form is fairly straightforward.

3. Repeatedly multiply the remaining decimal fraction by 2 at each stage of the process. Continue until the fractional part becomes 0.
4. The integral parts of the products, when read in order and placed after the binary separator, yield the equivalent binary number.

You might perceive that the process would break down if the criteria in the third step were never satisfied. In such an event, one stops the process of repeated multiplication at the point where the number of binary digits produced is three times the number of digits in the decimal form of the number.

The choice of "three times the number of digits in the decimal form of the number" might sound arbitrary, but there is sound support for that decision. Three decimal digits are roughly equal in precision to 10 binary digits. For example, compare the following numbers:

$$.001 = 1/1000 \text{ versus } .000\,000\,000\,1_2 = 1/1024 \approx .0098$$

These values differ by approximately two ten-thousandths and are (for all intents and purposes) equivalent. In approximation, we can use the rule of thumb that we will triple the number of digits in the binary representation from that of the decimal representation and maintain the same level of accuracy.

Let's consider several examples.

EXAMPLE 3.5

Convert the decimal fraction .625 to binary form.

SOLUTION

To keep things organized, we will again employ a tabular method (Table 3.4). Strictly speaking, this is not necessary, but it is convenient and helps us keep track of the little details. The binary form is obtained from the "Integral Part" column, read downward (rather than upward, as was done for the earlier examples), to yield $.101_2$ as the binary representation.

TABLE 3.4 Converting a decimal fraction to binary

Product	Integral Part	Decimal Fraction	
$.625 \times 2 = 1.25$	1	.25	MSB
$.25 \times 2 = .5$	0	.5	
$.5 \times 2$	1	0	LSB

EXAMPLE 3.6

Convert the decimal fraction .4018 to binary form.

SOLUTION

Although the decimal fraction has not become 0, we stop here because the number of bits now obtained is 12, which is three times more than the number of decimal digits in the original number. Consequently, the binary equivalent of .4018 would be $.0110\,0110\,1101_2$.

It is surprising to find that even relatively simple decimal fractions might have unending binary equivalent representations. We consider one such example, the fairly tame decimal number 0.1, as an illustration.

TABLE 3.5 Converting a decimal fraction to binary

Product	Integral Part	Decimal Fraction	
$.4018 \times 2$	0	.8036	MSB
$.8036 \times 2$	1	.6072	
$.6072 \times 2$	1	.2144	
$.2144 \times 2$	0	.4288	
$.4288 \times 2$	0	.8576	
$.8576 \times 2$	1	.7152	
$.7152 \times 2$	1	.4304	
$.4304 \times 2$	0	.8608	
$.8608 \times 2$	1	.7216	
$.7216 \times 2$	1	.4432	
$.4432 \times 2$	0	.8864	
$.8864 \times 2$	1	.7728	LSB

EXAMPLE 3.7

Convert the decimal fraction 0.1 to binary form.

SOLUTION

We have intentionally gone beyond the tripling of the original number of decimal digits to illustrate the point mentioned prior to the example. Observe that the binary digits generated become cyclic, and therefore the procedure will never terminate. In a case such as this, we could adopt the convention for nonterminating, repeating decimal numbers and place a bar over the sequence of repeating digits: $.0\overline{0011}_2$

Should a decimal number contain both integral and fractional parts, those parts should be converted to binary form separately and then combined across the binary point to obtain the complete binary number.

TABLE 3.6 Converting a decimal fraction to binary

Product	Integral Part	Decimal Fraction	
$.1 \times 2$	0	.2	MSB
$.2 \times 2 = .4$	0	.4	
$.4 \times 2$	0	.8	
$.8 \times 2$	1	.6	
$.6 \times 2$	1	.2	
$.2 \times 2$	0	.4	
$.4 \times 2$	0	.8	
$.8 \times 2$	1	.6	
$.6 \times 2$	1	.2	
$.2 \times 2$	0	.4	
$.4 \times 2$	0	.8	

EXAMPLE 3.8

Convert the decimal number 167.625 to binary form.

SOLUTION

The integral and decimal fraction parts of this number were converted to binary form in earlier examples, and thus we recall $167 = 1010\ 0111_2$ and $.625 = .101_2$. Consequently, the complete binary representation would be $1010\ 0111.101_2$.

The Arithmetic of Binary Numbers

With an understanding established for the conversion between the familiar decimal representation of numbers and the binary representation, we now turn our attention to our well-known operations of addition and subtraction. The operations can be performed in the binary system, using rules analogous to those employed in the decimal system.

To add binary numbers, we use the following rules:

1. In binary addition, 0 added to 0 yields 0. That is, $0_2 + 0_2 = 0_2$.
2. In binary addition, 0 added to 1 yields 1. That is, $0_2 + 1_2 = 1_2$.
3. In binary addition, 1 added to 1 yields 0, with a carry of 1 to the place on the left. That is, $1_2 + 1_2 = 0_2$ with a carry of 1_2.
4. In binary addition, adding the binary number 1_2 to itself with a carry of 1_2 from the place on its right yields 1_2 with a carry of 1 to the place on its left.

EXAMPLE 3.9

Add $11\ 0110_2$ to $10\ 1011_2$.

SOLUTION

We will use a table to organize the addition in order to make clear the process of carrying and addition. The numbers shown in the table are understood to be binary, and therefore we will omit the subscript of 2 for this illustration.

Step One: **We add the binary digits in the right-most column, yielding a result of 1, which generates no carried value (Table 3.7).**

TABLE 3.7 Binary Addition process

Carry						0	
		1	1	0	1	1	0
		1	0	1	0	1	1
Sum							1

Step Two: **Continuing the process in the next column to the left, we add the carried digit to the other binary digits. Following rule 3 above, the sum of that column's entries is 0, with a carry of 1 to the next column to the left (Table 3.8).**

TABLE 3.8 Binary Addition process

Carry					1	0	
		1	1	0	1	1	0
		1	0	1	0	1	1
Sum						0	1

Step Three: **The process continues, following the stated rules of binary addition (Table 3.9).**

TABLE 3.9 Binary Addition process

Carry				1	1	0	
		1	1	0	1	1	0
		1	0	1	0	1	1
Sum					0	0	1

Step Four: **(Table 3.10).**

TABLE 3.10 Binary Addition process

Carry			1	1	1	0	
		1	1	0	1	1	0
		1	0	1	0	1	1
Sum				0	0	0	1

Step Five: **(Table 3.11).**

TABLE 3.11 Binary Addition process

Carry		1	1	1	1	0	
		1	1	0	1	1	0
		1	0	1	0	1	1
Sum			0	0	0	0	1

Step Six: **(Table 3.12).**

TABLE 3.12 Binary Addition process

Carry	1	1	1	1	1	0	
		1	1	0	1	1	0
		1	0	1	0	1	1
Sum		1	0	0	0	0	1

Step Seven: **(Table 3.13).**

TABLE 3.13 Binary Addition process

Carry	1	1	1	1	1	0	
		1	1	0	1	1	0
		1	0	1	0	1	1
Sum	1	1	0	0	0	0	1

The sum of the binary numbers is $110\ 0001_2$.

Subtraction of binary numbers is similar to addition, and there is an analogous set of rules we must follow:

1. In binary subtraction, 0 subtracted from 0 yields 0. That is, $0_2 - 0_2 = 0_2$.
2. In binary subtraction, 0 subtracted from 1 yields 1. That is, $1_2 - 0_2 = 1_2$.
3. In binary subtraction, 1 subtracted from 1 yields 0. That is, $1_2 - 1_2 = 0_2$.
4. We cannot subtract the binary number 1 from the binary number 0. We can "borrow" from the digit to the left to perform the subtraction, with the result that $0_2 - 1_2 = 1_2$, with the digit to the left of the 0 reduced by 1. In appearance, this would be equivalent to $10_2 - 1_2 = 1_2$.

The first few rules are probably fairly obvious, but the process of borrowing needs illustration. We'll give a few examples to demonstrate the procedure, including the decimal equivalents as a verification of the process.

EXAMPLE 3.10

Subtract $101\ 1101_2$ from $111\ 1011_2$.

SOLUTION

It isn't obvious what the decimal equivalents for these binary numbers would be, and it would be good practice to convert each number from binary, so we'll begin with that and then turn to the subtraction following the binary rules.

The subtrahend, $101\ 1101_2 = 64 + 16 + 8 + 4 + 1 = 93$, while the minuend $111\ 1011_2 = 64 + 32 + 16 + 8 + 2 + 1 = 123$, and therefore we are (in decimal form) intending to subtract $123 - 93$, and hence the result should be 30, which has binary equivalent $1\ 1110_2$. It is tremendously inconvenient to translate back and forth between binary and decimal form in order to perform what should be a relatively simple computation, so one would hope that the binary subtraction algorithm would not be discouragingly complicated! We'll attempt that now, organizing our work in Table 3.14 for clarity, as we did with addition. Note the first two steps in subtraction follow the rules listed quite readily, since no borrowing is involved. The third step will require borrowing, since we cannot perform the $0_2 - 1_2$ process directly. Keep in mind that the result will be a 1 in the difference, with the number to the left of the 0 in the minuend reduced by 1 (Table 3.15). Note that now, as often is the case in base-10 subtractions, we must borrow again (Table 3.16) and still again (Table 3.17)! The remaining steps require no borrowing, and produce the result shown in Table 3.18.

TABLE 3.14 Binary Subtraction process

Borrow							
	1	1	1	1	0	1	1
	1	0	1	1	1	0	1
Difference						1	0

TABLE 3.15 Binary Subtraction process

Borrow					1		
	1	1	1	~~1~~0	0	1	1
	1	0	1	1	1	0	1
Difference					1	1	0

TABLE 3.16 Binary Subtraction process

Borrow				1			
	1	1	**~~1~~0**	**~~1~~0**	0	1	1
	1	0	1	1	1	0	1
Difference				1	1	1	0

TABLE 3.17 Binary Subtraction process

Borrow			1				
	1	**~~1~~0**	**~~1~~0**	**~~1~~0**	0	1	1
	1	0	1	1	1	0	1
Difference			1	1	1	1	0

TABLE 3.18 Binary Subtraction process

Borrow							
	1	**~~1~~0**	**~~1~~0**	**~~1~~0**	0	1	1
	1	0	1	1	1	0	1
Difference	0	0	1	1	1	1	0

Our difference is the binary number $1\ 1110_2$, as our work in the equivalent decimal representation predicted.

EXAMPLE 3.11

Subtract 1100.11_2 from 1111.01_2.

SOLUTION

Step One:

Borrow							
	1	1	1	1	.	0	1
	1	1	0	0	.	1	1
Difference					.		0

Step Two:

Borrow						1	
	1	1	1	~~1~~0	.	0	1
	1	1	0	0	.	1	1
Difference					.	1	0

Step Three:

Borrow						1	
	1	1	1	~~1~~0	.	0	1
	1	1	0	0	.	1	1
Difference				0	.	1	0

Step Four:

Borrow						1	
	1	1	1	~~1~~0	.	0	1
	1	1	0	0	.	1	1
Difference			1	0	.	1	0

Step Five:

Borrow						1	
	1	1	1	~~1~~0	.	0	1
	1	1	0	0	.	1	1
Difference		0	1	0	.	1	0

Step Six:

Borrow						1	
	1	1	1	~~1~~0	.	0	1
	1	1	0	0	.	1	1
Difference	0	0	1	0	.	1	0

Therefore, $1111.01_2 - 1100.11_2 = 10.10_2$.

You may have observed that in the computations shown to this point, we have scrupulously avoided the appearance of negative numbers. There is a reason for this, which we should discuss, at least to some degree.

It is common knowledge that negative numbers in base-10 are indicated by the presence of a "minus sign" preceding the number, and while we could attempt to introduce similar notation for base-2, this becomes problematic for our work. The purpose for using binary notation is to ultimately represent alternate states of equipment using 0s and 1s, and therefore the use of a third symbol, such as the minus sign, is an extravagance we cannot afford.

One solution, and one that we shall not explore at this time, is to adopt the practice that a "sign digit" of 0 or 1, to indicate positive and negative values, respectively, shall precede every binary number. That is, we could choose to represent the number "-5" using the binary representation of 5, namely, 101_2, preceded by an extra digit of 1 to indicate that the number is negative.

In our work, this possible resolution to the problem muddies the waters to an unacceptable degree, and therefore we shall not consider negative binary numbers at this time. You should simply be aware, for the time being, that there exist creative solutions to the representation of signed numbers in alternative bases that do not require the incorporation of the minus sign symbol.

Up to now, we have presented addition and subtraction of binary numbers, but it is also possible to multiply and divide binary numbers. As we will see, multiplication proceeds in a manner similar to the method of base-10 multiplications, but that it is actually easier because of the limited possible outcomes for multiplications of binary numbers.

There are four rules to remember when multiplying binary numbers (recognizing that these numbers are intended to be base-2 digits):

$$0 \times 0 = 0$$
$$0 \times 1 = 0$$
$$1 \times 0 = 0$$
$$1 \times 1 = 1$$

Consider, using those rules, the multiplication of binary numbers 101_2 and 11_2, following the rules of multiplying and carrying with which we are familiar from base-10 multiplication. Observe that we are actually multiplying the decimal numbers 5 and 3, so the outcome must be the binary form of 15, which is 1111_2, and that we are constructing the example within a reference grid (see Tables 3.19 through 3.22), so it is easier to recognize the placement of the numbers.

TABLE 3.19 Binary Multiplication process

	1	0	1
	×	1	1

Much as is done with decimal number multiplication, we begin with the rightmost digit of the lower number and multiply that quantity times the entire upper number (Table 3.20).

TABLE 3.20 Binary Multiplication process

	1	0	1
	×	1	1
	1	**0**	**1**

To serve as a placeholder, we put a 0 in the rightmost position of the next product line, and then multiply again (Table 3.21).

TABLE 3.21 Binary Multiplication process

	1	0	1
	×	1	1
	1	0	1
1	**0**	**1**	**0**

Now we will add the numbers in the product lines together, obtaining our final result. Keep in mind the rules for addition of binary numbers and carrying (Table 3.22).

TABLE 3.22 Binary Multiplication process

	1	0	1
	×	1	1
	1	0	1
1	0	1	0
1	**1**	**1**	**1**

In a similar way, we can introduce binary division, which again mimics the process in base-10 but is again easier because of the simplified rules of binary addition and subtraction. Recall the base-10 procedure directs us to divide the number constructed from the first sequence of digits in the dividend whose value is at least as great as the divisor, placing the whole number result in the quotient. That whole number is multiplied by the divisor, the product placed beneath the dividend, and we subtract.

Following this, the next digit in the dividend is brought down, and we iterate the process until the outcome of the subtraction becomes less than the divisor and no further digits can be brought down from the dividend, at which time the division process terminates. The final difference is referred to as the remainder of the division.

Consider the following example, which demonstrates that the process in the binary number system is analogous.

EXAMPLE 3.12 Divide 1011_2 by 11_2.

SOLUTION

$11\overline{)1011}$ As with base-10 division, note that it is not possible to multiply 11_2 by an integer to obtain 1_2, or 10_2, so our focus is on what the divisor can be multiplied times to be less than or equal to the binary number 101_2. Keeping in mind that our only true choices are to multiply by 0_2 or 1_2, it is apparent that the only suitable choice would be to multiply by 1_2. We put a 1 in the quotient, multiply by 1_2, and place the product beneath the dividend, and subtract:

$$\begin{array}{r@{}l}
 & 1 \\
11 & \overline{)1011} \\
 & \underline{11} \\
 & 101
\end{array}$$

Note that we have "brought down" the next digit in line.

Repeating the division, we obtain

$$\begin{array}{r@{}l}
 & 11 \\
11 & \overline{)1011} \\
 & \underline{11} \\
 & 101 \\
 & \underline{11} \\
 & 10
\end{array}$$

The "10" in the final line, being less than the divisor, must be our remainder, and hence our result is 11_2, with remainder 10_2.

In decimal form, the problem was to divide 11 by 3, and thus our binary result should be equivalent to a quotient of 3, with remainder 2. A quick conversion of the binary result shows that this is precisely what we obtained.

An Application of Binary Numbers: ASCII

When we began our discussion of the number systems, we indicated that all data stored on your computer and other electronic devices were recorded in binary form, including nonnumerical data. It seems appropriate to now investigate, in an introductory manner, how this is done.

You may be familiar with **ASCII (American Standard Code for Information Interchange)**. This was developed from a foundation of telegraph codes and is a means through which we can encode the English alphabet using binary representation. A total of 95 printable ASCII characters exist and are associated with decimal numbers from 32 through 126, inclusive (the reason for this span of enumeration is not significant to us here, but we should mention for the sake of completeness that there are other "control" characters, now mainly obsolete, that were represented by the preceding whole number values).

How this relates to your computer may not be immediately clear, but consider your computer's keyboard and look at all those keys that generate a "printable" character. There are 47 of them, and with the "Shift" key providing a secondary character assignment to each key, we double the total number of ASCII characters to 94, with the 95th being the space bar (an "invisible" character).

When the printable characters are numbered in this manner and their representation is converted to binary form, we gain the capacity to represent written material using 0s and 1s exclusively. For instance, consider the binary string 0010 0000. This is the binary equivalent of the decimal number 32 and has been assigned as the representation for the invisible character, or space. You may wonder why the number begins with what appears to be a pair of superfluous 0s, but this is done merely so that every character is represented using one byte.

Here is a sample of binary code: 01001101 01111001 00100000 01100011 01100001 01110100 00100000 01101001 01110011 00100000 01101111 01101110 00100000 01101101 01111001 00100000 01101100 01100001 01110000 00101110. While it is totally nonobvious what this means, we can tell you that it represents the phrase "My cat is on my lap." Note that there are precisely as many bytes as characters in the phrase, if one includes spaces and the period.

The key to deciphering this message—or any other phrase represented in binary form—is to have in hand the assignment of letters, numbers, and symbols to their identifying binary numbers, and then the translation is straightforward. For instance, the character "M" has been assigned the binary representation 0100 1101, and that is the first byte of information given in the encryption of the phrase relating my cat's location. Table 3.23 shows the overall representation of the 95 characters in binary form.

EXAMPLE 3.13 Express the statement "It was a dark and stormy night." in binary form.

SOLUTION

We begin by considering the table relating the ASCII symbols and their binary equivalents. We only need to locate the bytes representative of the individual characters, maintaining an awareness that "I" is not to be confused with "i" for the purpose of encryption. As we can see from the chart, the representation of

TABLE 3.23 Binary Codes

Char	Bin	Char	Bin	Char	Bin
(sp)	010 0000	@	100 0000	`	110 0000
!	010 0001	A	100 0001	a	110 0001
“	010 0010	B	100 0010	b	110 0010
#	010 0011	C	100 0011	c	110 0011
$	010 0100	D	100 0100	d	110 0100
%	010 0101	E	100 0101	e	110 0101
&	010 0110	F	100 0110	f	110 0110
‘	010 0111	G	100 0111	g	110 0111
(	010 1000	H	100 1000	h	110 1000
)	010 1001	I	100 1001	i	110 1001
*	010 1010	J	100 1010	j	110 1010
+	010 1011	K	100 1011	k	110 1011
,	010 1100	L	100 1100	l	110 1100
–	010 1101	M	100 1101	m	110 1101
.	010 1110	N	100 1110	n	110 1110
/	010 1111	O	100 1111	o	110 1111
0	011 0000	P	101 0000	p	111 0000
1	011 0001	Q	101 0001	q	111 0001
2	011 0010	R	101 0010	r	111 0010
3	011 0011	S	101 0011	s	111 0011
4	011 0100	T	101 0100	t	111 0100
5	011 0101	U	101 0101	u	111 0101
6	011 0110	V	101 0110	v	111 0110
7	011 0111	W	101 0111	w	111 0111
8	011 1000	X	101 1000	x	111 1000
9	011 1001	Y	101 1001	y	111 1001
:	011 1010	Z	101 1010	z	111 1010
;	011 1011	[	101 1011	{	111 1011
<	011 1100	\	101 1100	|	111 1100
=	011 1101	]	101 1101	}	111 1101
>	011 1110	^	101 1110	~	111 1110
?	011 1111	—	101 1111	(del)	111 1111

"I" is 0100 1001, and from there we can proceed to obtain all the binary representations, with the following result:

01001001 01110100 00100000 01110111 01100001 01110011 00100000
01100001 00100000 01100100 01100001 01110010 01101011 00100000
01100001 01101110 01100100 00100000 01110011 01110100 01101111
01110010 01101101 01111001 00100000 01101110 01101001 01100111
01101000 01110100 00101110

Naturally, producing the representation by hand is highly tedious, and we won't make a practice of doing so. This example is intended only to demonstrate that, in principle, we could express all written documents in binary form.

Exercises

For each of the following problems, answer in complete sentences, using proper spelling and grammar.

1. What is meant by the mathematical term "bit"?
2. What is meant by the mathematical term "byte"?
3. What is the "most significant bit" in a binary number, and what is its abbreviation?
4. What is the "least significant bit" in a binary number, and what is its abbreviation?

For each of the following decimal numbers, convert to binary form.

5. 1,865
6. 62,093
7. 128
8. 4733
9. 193,207
10. 1,038
11. 18.125
12. 213.8
13. 0.15

For each of the following binary numbers, convert to decimal form.

14. $110\ 1011_2$
15. 101_2
16. $1110\ 1001_2$
17. $1\ 1111\ 1101_2$
18. 110.011_2
19. 0.1101_2
20. $1010\ 1101.111_2$
21. $0.0000\ 001_2$

Perform the following binary additions. Check your work by converting the addends and sum to decimal form and adding these to show that the same result is obtained.

22. $1101_2 + 1011_2$
23. $10\ 1101_2 + 1\ 0110_2$
24. $1.0011_2 + 0.111_2$
25. $100.001_2 + 1101.011_2$

Perform the following binary subtractions. Check your work by converting minuend, subtrahend, and difference to decimal form and subtracting these to show that the same result is obtained.

26. $1101_2 - 1011_2$
27. $110_2 - 11_2$
28. $1001\ 1101_2 - 101\ 1100_2$
29. $11.0111_2 - 1.1_2$

Perform the following binary multiplications. Check your work by converting the factors and the product to decimal form and multiplying these to show that the same result is obtained.

30. $(1101_2)(101_2)$
31. $(10\ 1101_2)(110_2)$
32. $(111_2)(111_2)$
33. $(1001_2)(1010_2)$

Perform the following binary divisions. Check your work by converting the dividend, divisor, and quotient to decimal form and dividing these to show that the same result is obtained.

34. $101101_2 \div 1001_2$
35. $1110\ 0111_2 \div 111_2$
36. $1000\ 0100\ 1101_2 \div 1\ 0001_2$
37. $1000\ 0010\ 0000_2 \div 10\ 0000_2$

Convert the following phrases to ASCII representation.

38. Now is the time for all good men to come to the aid of the party.
39. It is fun to write in binary.

Convert the following binary representations into English, using the ASCII conversion chart.

40. 01001000011000010111011001100101001000000110000100100000011 01101011000010111010001101000011001010110110101100001011001 1101101001011000110110000010110110000010000000110010001100000 10 1111001
41. 01001101011000010111010001101000001000000110100101110011001 00000011001100111010101101110

3.3 The Hexadecimal Number System

As we've already discussed, all the information on your computer's hard drive is stored in binary form. We have also mentioned how, using the ASCII system, we can represent English statements using only 0s and 1s. The motivation for the choice of base-2 stems from the fact that it is the easiest way for current digital structures to store, retrieve, and act on information.

The alternative states of off-on, no-yes, false-true, negative-positive, and so forth lend themselves ideally to being associated with either a 0 or a 1. This is significant because digital circuitry, for instance, employs transistors that are electronically switched either on (1) or off (2) to represent the value of a binary variable. Magnetic storage devices, like computer hard drives, represent data with sectors that are magnetically polarized to be in a state that is either positive or negative. Optical media storage, like CDs or DVDs, are etched thermally by a laser so that particular portions of the disk are either reflective or nonreflective. Because of the omnipresence of binary numbers in modern technology, it is essential that electrical engineers, computer scientists, and information technology professionals develop fluency with binary numbers.

However, while computers operate quite efficiently with the binary number system, we humans sometimes encounter difficulty working with representations of numbers in that form because of the number of digits involved in the representation. As an illustration, the decimal number 489 has binary representation $1\ 1110\ 1001_2$, requiring nine digits. For this reason, it is sometimes desirable to use yet another number system, the *hexadecimal* system, sometimes referred to simply as "hex" for the sake of brevity, which humans can read with much greater facility. The name of the system is derived from a combination of the Greek word "hexa," meaning six, and the Latin word "decima," meaning ten.

Hex is a system that uses sixteen characters for number representation, and thus we would refer to it as the **base-16 number system**. Since 16 is the fourth power of 2, we can imagine that some relationship might exist between base-2 and base-16 that was not present for base-10, and therefore the conversion between the two forms might be performed more readily. We will explore that matter shortly, but first we need to learn more about hex.

Because hex has radix 16, we require 16 symbols to depict numbers in the system. The characters employed are 0, 1, 2, 3, 4, 5, 6, 7, 8, 9, A, B, C, D, E, and F. The symbol A corresponds to the decimal number 10, B to the decimal number 11, C to the decimal number 12, D to the decimal number 13, E to the decimal number 14, and F to the decimal number 15. This may seem bewildering, but it merely requires practice to attain familiarity and fluency with the new representation symbols, and the reduction achieved in the length of the representation of numbers will be well worth the effort.

In general, a number written in hex is approximately one-fourth as long as a number expressed in binary form. We have mentioned the binary equivalent to the decimal number 489 as $1\ 1110\ 1001_2$ and now (without proof for the moment) reveal that the hex form of the number is $1E9_{16}$. Of course, you might rightly argue that the equivalence of this number to 489 is no more obvious than was the binary representation, but keep in mind that the underlying purpose for the use of hex is the combination of the shorter representation of the number with the ease of conversion between binary and hex, as opposed to the conversion between binary and decimal.

Just as the binary and decimal systems were based on powers of 2 and 10, respectively, hex is constructed using the powers of 16. It would be surprising (and, to be frank, somewhat disturbing) if you possessed ready knowledge of the powers of 16, and thus we will list some of the powers here in Table 3.24 for reference. The digits of a number written in hex are weighted by position, just as in the decimal and binary systems. The place values express powers of 16, and the **hexadecimal point**, or **hexadecimal separator**, delineates the integral and hexadecimal fraction components of the number. For brevity, we can refer to this as the "hex point," or "hex separator."

TABLE 3.24 Some of the Powers of 16

16^{-3}	.000244140625
16^{-2}	.00390625
16^{-1}	.0625
16^{0}	1
16^{1}	16
16^{2}	256
16^{3}	4096

The Relationship between Binary and Hex

We have hinted that a strong relationship exists between binary numbers and hex numbers, and since that is the premise for the expenditure of labor to learn hex, we will begin by exploring that connection. Let's start with a list of the first 16 whole numbers, their binary expression, and their hex representation (Table 3.25). You can see that we have opted to show each of the binary digits as nibbles in order to create a four-digit representation of each binary number in the table. Observe that each of the hexadecimal digits represents exactly one nibble of binary information. Manual conversion between binary and hexadecimal is very easy: you need only do substitute one hex digit for every nibble of binary information or vice versa.

TABLE 3.25 Decimal Numbers Expressed in Binary and Hex Forms

Decimal Number	Binary Form	Hex Form
0	0000_2	0_{16}
1	0001_2	1_{16}
2	0010_2	2_{16}
3	0011_2	3_{16}
4	0100_2	4_{16}
5	0101_2	5_{16}
6	0110_2	6_{16}
7	0111_2	7_{16}
8	1000_2	8_{16}
9	1001_2	9_{16}
10	1010_2	A_{16}
11	1011_2	B_{16}
12	1100_2	C_{16}
13	1101_2	D_{16}
14	1110_2	E_{16}
15	1111_2	F_{16}

EXAMPLE 3.14

Convert the binary digit 11 0110 1111 0110_2 to hex.

SOLUTION

Using the methods for conversion to decimal numbers, we can show that this number is 14070 in base-10, but this is not relevant for our exercise here. If we expand the first nibble to include two 0s preceding the 1s, we obtain the equivalent form:

$$0011\ 0110\ 1111\ 0110_2$$

Now we'll replace each nibble by the hex number determined from the table above. This yields

$$36F6_{16}$$

This is truly astonishing, assuming that it is true. We will now confirm its accuracy.

As was done with binary numbers, we can use powers of the base to convert hex numbers to our familiar decimal form. That is, if we have a number such as $36F6_{16}$, we can use the powers of 16 to convert the number to base-10. This illustration will serve to verify the result obtained in the previous example:

$$\begin{aligned} &3 \times 16^3 + 6 \times 16^2 + 15 \times 16^1 + 6 \times 16^0 \\ &= 12288 + 1536 + 240 + 6 \\ &= 14070 \end{aligned}$$

This is precisely what we had indicated was the decimal equivalent to the binary number 11 0110 1111 0110_2, and this indicates that our method for conversion is accurate.

Conversion from hex to binary is the exact inversion of the procedure, with the hex digits replaced, in order, by the related nibbles of binary information.

EXAMPLE 3.15

Convert the hex number $BA085_{16}$ to binary form.

SOLUTION

Referencing our table of binary-hex equivalents, we observe the equivalent nibbles are B ~ 1011, A ~ 1010, 0 ~ 0000, 8 ~ 1000, and 5 ~ 0101. Therefore, our equivalent binary number is 1011 1010 0000 1000 0101_2.

We will confirm the equivalence by producing the decimal form of each of the two numbers and showing that they agree:

$$\begin{aligned} BA085_{16} &= 11 \times 16^4 + 10 \times 16^3 + 0 \times 16^2 + 8 \times 16^1 + 5 \times 16^0 \\ &= 720896 + 40960 + 0 + 128 + 5 \\ &= 761989 \end{aligned}$$

$$\begin{aligned} 1011\ 1010\ 0000\ 1000\ 0101_2 &= 2^{19} + 2^{17} + 2^{16} + 2^{15} + 2^{13} + 2^7 + 2^2 + 1 \\ &= 524288 + 131072 + 65536 + 32768 + 8192 \\ &\quad + 128 + 4 + 1 \\ &= 761989 \end{aligned}$$

As in the explanation above, we construct the expanded form of the hex number, utilize the powers of 16, and simplify to obtain our result:

$$\begin{aligned} BA085_{16} &= 11 \times 16^4 + 10 \times 16^3 + 0 \times 16^2 + 8 \times 16^1 + 5 \times 16^0 \\ &= 720896 + 40960 + 0 + 128 + 5 \\ &= 761989 \end{aligned}$$

The Relationship between Decimal and Hex

We essentially glossed over the method for transitioning from hex to decimal form. The procedure simply involved multiplication of the hex digits by the appropriate power of 16 and simplifying the result.

Conversion from decimal to hex, however, is slightly more complicated. Naturally, one possible approach is to convert decimal numbers to binary, from which the translation to hex is a triviality. Naturally, this is unsatisfactory, since it would involve a superfluous intermediate step that we would much rather avoid.

The procedure is much as we saw in the case of the binary number system, and the method for converting is analogous. As we did in the binary case, we will address the integral and decimal fraction cases separately and then combine the results to handle mixed numbers.

For integral valued decimal numbers, we divide repeatedly by 16, maintaining a record of our quotients and remainders, proceeding until a quotient of 0 is attained. The hex number is then constructed using the remainders of the successive divisions, read in reverse order.

EXAMPLE 3.16

Convert the decimal number 936 to hex form.

SOLUTION

The hex equivalent is constructed using the remainders, read upward through the column: $3A8_{16}$ (Table 3.26).

TABLE 3.26 Converting a decimal number to hex

Division	Remainder as Decimal Number	Remainder as Hex	Quotient
$936 \div 16$	8	8	58
$58 \div 16$	10	A	3
$3 \div 16$	3	3	0

EXAMPLE 3.17

Convert the decimal number 1876923 to hex.

SOLUTION

Having reached a quotient of 0, we obtain our number in hex form, using the remainder as hex column, read upward: $1CA3BB_{16}$ (Table 3.27).

In the case of decimal fraction numbers, we follow the algorithm we developed in the binary case, multiplying repeatedly by 16 and keeping track of the integral part of the products.

TABLE 3.27 Converting a decimal number to hex

Division	Remainder as Decimal Number	Remainder as Hex	Quotient
1876923 ÷ 16	11	B	117307
117307 ÷ 16	11	B	7331
7331 ÷ 16	3	3	458
458 ÷ 16	10	A	28
28 ÷ 16	12	C	1
1 ÷ 16	1	1	0

As was the case for binary representations, we list the steps as follows:

1. Multiply the decimal fraction by 16.
2. Store the integral part of the product and convert it to hex form, if needed.
3. Repeatedly multiply the remaining decimal fraction by 16 at each stage of the process. Continue until the fractional part becomes 0.
4. The integral parts (in hex form) of the products, when read in order and placed after the hex separator, yield the equivalent hex number.

As a practical matter, we should remark that the conversion rarely results in terminating hex representations, but in the example that we will explore, the outcome will be nice, by design.

Suppose we wanted to convert .31640625 to hex. The process would be as shown in Table 3.28. The equivalent hex number would be $.51_{16}$.

TABLE 3.28 Converting a decimal number to hex

Multiplication	Integral Part as Decimal Number	Integral Part as Hex	Decimal Fraction
.31640625 × 16	5	5	.0625
.0625 × 16	1	1	0

EXAMPLE 3.18 Convert the decimal number .000202026367188 to hex form.

SOLUTION

The recurrence of the decimal fraction .24 indicates that we have reached the beginning of a repetition of the hex number, and thus our representation is (reading the integral part as hex column, downward) $.000D3D70A_{16}$ (Table 3.29).

EXAMPLE 3.19 Convert the decimal number 1876923.000202026367188 to hex form.

SOLUTION

Combining the results of Examples 3.15 and 3.16, we obtain the equivalent hex number: $1CA3BB.000D3D70A_{16}$.

TABLE 3.29 Converting a decimal number to hex

Multiplication	Integral Part as Decimal Number	Integral Part as Hex	Decimal Fraction
.000202026367188 × 16	0	0	.003232421875
.003232421875 × 16	0	0	.05171875
.05171875 × 16	0	0	.8275
.8275 × 16	13	D	.24
.24 × 16	3	3	.84
.84 × 16	13	D	.44
.44 × 16	7	7	.04
.04 × 16	0	0	.64
.64 × 16	10	A	.24

It is amusing to consider the translation of the result of the last example into binary form. Recalling that we need only translate the hex number into equivalent nibbles, we find

$$0001\ 1100\ 1010\ 0011\ 1011\ 1011.0000\ 0000\ 0000\ 1101\ \overline{0011\ 1101\ 0111\ 0000\ 1010}_2$$

You are probably aware that your computer monitor uses the primary colors of light (red, green, and blue) to display all other possible colors. The device uses different intensities of each primary color, combines them, and concentrates them onto a pixel (picture element) of the screen. The colors range from white (in which 100% of red, green, and blue are displayed in the pixel) to black (in which 0% of red, green, and blue are displayed).

HTML manages the **RGB** combinations by using six-digit hex numbers, called **color codes**. The color code designation begins with a pound symbol (#) followed by the hex number, which is actually a combination of three two-digit hex numbers denoting RGB intensities. These allow you to set the red, green, and blue values of a color, and the combination of the relative strengths of the individual colors produces the color desired. The greatest intensity of any particular one of the individual primary colors is FF, which has decimal equivalent 255.

For instance, the color orange can be produced by an intense application of red and a mildly strong green, combined with a relatively weak application of blue. The color code for orange is #FF8040, which denotes the strongest (FF) possible red, a strong but not overly powerful (80) green, and a fairly weak blue (40).

Exercises

Answer the following questions in complete sentences, using correct spelling and grammar.

1. In your own words, explain the process of converting binary numbers to hexadecimal form.
2. In your own words, explain the process of converting hexadecimal numbers to binary form.

Convert the following binary numbers to hexadecimal form.

3. $1101\ 1100_2$
4. $101\ 1111_2$
5. $1\ 1011\ 1101_2$
6. $101\ 1111\ 0110_2$
7. 1101.0111_2
8. $0.0101\ 11_2$

Convert the following hexadecimal numbers to binary form.

9. 33_{16}
10. $9F7_{16}$
11. $2AEB_{16}$
12. $937.2A_{16}$
13. $AF.CAD_{16}$

Convert the following decimal numbers to hexadecimal form.

14. 171
15. 2013
16. 4097
17. 501.1

Write a brief explanation for the following question.

18. You may recall from your experience that certain colors "combine" to form other colors. For instance, "Yellow and blue make green." That is, if you were to use a mixture of yellow pigment and blue pigment and blend them, the result would be green pigment. Are the RGB codes constructed in such a way that addition of RGB codes yields the "blended" code? Investigate this supposition and explain your conclusion.

3.4 The Octal Number System

The octal number system, or base-8 system, is constructed in a manner similar to the previously discussed systems. In this case, the radix is 8, and all numbers are represented using only the digits 0, 1, 2, 3, 4, 5, 6, and 7. We can define the **octal separator** (or octal point) just as we did for the other systems and again define the place values using the powers of the base.

As was the case with hex, our primary interest in octal numbers lies in the ease of conversion between numbers represented in that form and binary. Just as hex numbers were significantly shorter than their binary equivalents, octal numbers typically use one-third the number of digits as the associated binary number. A natural question might be why we would need octal numbers, since we have worked so hard to develop and understand hex, and the rationale lies in the observation that octal numbers use familiar digits, while hex requires the additional symbols A through F, and this can be confusing at first sight. Because, in practice, both hex and octal serve the same purpose, we can probably make do with one or the other. Nonetheless, we are well served by developing fluency with both numeration systems in case a particular application is dealt with more conveniently in one form than the other.

If you understood the method for converting a new base system to and from the decimal system, then you will be comforted to know that the method is identical for the octal system. To translate an octal number into decimal form, produce the expanded form of the number (using the digits multiplied by the appropriate powers of 8) and simplify. This process is well known to us by now and will be left for the exercises.

As was the case for binary and hexadecimal numbers, conversion from decimal form into the new base is trickier, so we will demonstrate the procedure to ensure clarity. For an integral number, we divide by 8 repeatedly, maintaining track of the remainders and proceeding until we reach a quotient of 0. Reading the remainders (converted to octal form) in reverse order and identifying that the number is octal by appending a subscript of 8, we obtain the base-8 representation of the number. Let's consider an example.

EXAMPLE 3.20

Convert the decimal number 1762 to octal form.

SOLUTION

As was done before, we use a tabular approach in Table 3.30 to make the process both systematic and well organized. Reading the remainders (as octal numbers) in reverse order, we obtain the octal equivalent to 1762 as 3342_8.

TABLE 3.30 Converting a decimal number to octal

Division	Remainder as Decimal Number	Remainder as Octal Number	Quotient
1762 ÷ 8	2	2	220
220 ÷ 8	4	4	27
27 ÷ 8	3	3	3
3 ÷ 8	3	3	0

EXAMPLE 3.21

Convert the decimal number 276315 to octal form.

SOLUTION

Having reached a quotient of 0, the process terminates, and we can read our base-8 representation as 1033533_8 (Table 3.31).

TABLE 3.31 Converting a decimal number to octal

Division	Remainder as Decimal Number	Remainder as Octal Number	Quotient
276315 ÷ 8	3	3	34539
34539 ÷ 8	3	3	4317
4317 ÷ 8	5	5	539
539 ÷ 8	3	3	67
67 ÷ 8	3	3	8
8 ÷ 8	0	0	1
1 ÷ 8	1	1	0

If we turn to the case where the decimal number is a decimal fraction, we utilize the same process as was employed in base-2 and base-16, multiplying repeatedly by the radix and keeping track of the integral part of the product (in octal form) until the fractional part of the number is 0. The octal form of the number is then obtained by reading the integral part of the product in advancing order and appending a base-8 subscript.

EXAMPLE 3.22 Convert the decimal number .00625 to octal form.

SOLUTION

Note that the recurrence of the decimal fraction .4 indicates to us that we have reached a point of repetition, and thus the octal form of the .00625 is $.003\overline{146}_8$ (Table 3.32).

TABLE 3.32 Converting a decimal number to octal

Multiplication	Integral Part as Decimal Number	Integral Part as Octal Number	Decimal Fraction
.00625 × 8	0	0	.05
.05 × 8	0	0	.4
.4 × 8	3	3	.2
.2 × 8	1	1	.6
.6 × 8	4	4	.8
.8 × 8	6	6	.4

As we saw with the other bases, if the decimal number were to be a combination of integral and decimal fraction parts, we would obtain the base-8 equivalences independently and then combine the results. Such an illustration will be left for you to attempt in the exercises.

You might be wondering why the base-10 decimal fraction numbers seem to often generate repeating decimal representations in base-2, base-8, and base-16, and this is a question worthy of investigation. The answer has to do with the fact that all the digits to the right of the binary, hex, and octal separators are powers of 2 and have no other prime factors. A decimal number such as .1, which we saw in the base-2 section has a nonterminating binary equivalent, has fraction form 1/10, the denominator of which is not a power of 2, and has a prime factor of 5. Such numbers will always have nonterminating decimal equivalents in the bases we have examined because of this relationship between the denominators.

Conversion between Binary Form and Octal Form

We saw that one of the virtues of hexadecimal numbers was the ease with which we could translate between that base and binary form. A similar relationship exists between octal numbers and binary numbers.

Recall that, when converting from binary to hex form and vice versa, the key step was the recognition that each hex digit had an equivalent four-digit base-2 representation and that we could translate from one form to the other through direct substitution of the equivalent hex digit for the four binary digits and vice versa.

For base-8, each of the eight octal digits is equivalent to a three-digit binary number, and the translation between octal and binary forms merely requires that (for translation from binary to octal) the three binary digits be replaced by their octal equivalent. Conversion from octal to binary is the precise inversion of that procedure.

Before viewing some examples, let's produce the equivalencies for our reference in Table 3.33.

TABLE 3.33 Equivalencies between decimal, binary, and octal numbers

Decimal Number	Binary Equivalent	Octal Equivalent
0	000	0
1	001	1
2	010	2
3	011	3
4	100	4
5	101	5
6	110	6
7	111	7

EXAMPLE 3.23

Convert the binary number $11\ 0101\ 1101\ 1001\ 1111_2$ to octal form.

SOLUTION

In order to make the conversion more obvious, we will begin by reorganizing our binary number, divided into nibbles (as is the common practice) so that its digits are in clusters of three: $110\ 101\ 110\ 110\ 011\ 111_2$.

Next, referencing the table of equivalencies, we will replace each grouping of three binary digits into its octal equivalent. That the number is octal will be indicated, as usual, by the appended subscript of 8 at the end of the representation: 656637_8.

EXAMPLE 3.24

Convert the binary number $1\ 0110\ 1011\ 1001_2$ into octal form.

SOLUTION

In this case, note that the binary number contains 13 digits, and thus reorganization into groups of three cannot be done with the number in its present form. You may recall that this issue was encountered in our consideration of hex numbers, and it was decided that we would incorporate enough preceding 0s to allow for the subdivision into groups required. The smallest multiple of 3 that is greater than 13 is 15, and hence we must include two preceding 0s to our number and then perform the reallocation of digits into sets of three:

$$001\ 011\ 010\ 111\ 001_2$$

Again, we consult the table of binary-octal equivalencies, and the translation to octal form is immediate:

$$13271_8$$

EXAMPLE 3.25 Convert the octal number 7321_8 to binary form.

SOLUTION

Each of the octal digits is replaced by its binary equivalent, yielding

$$111\ 011\ 010\ 001_2$$

In order to maintain our practice of exhibiting binary digits broken into nibbles, we reorganize this expression in groups of four digits, obtaining $1110\ 1101\ 0001_2$.

Exercises

Answer the following questions using complete sentences, with correct spelling and grammar.

1. What is the difference between binary numbers and octal numbers?
2. What is the process for converting binary numbers to octal numbers?
3. What is the process for converting octal numbers to binary numbers?

Convert the following decimal numbers to octal form.

4. 133
5. 2984
6. 618.5
7. 9377.125

Convert the following octal numbers to decimal form.

8. 33_8
9. 6452_8
10. 255.43_8
11. 0.12_8

Convert the following binary numbers to octal form.

12. $1101\ 0110_2$
13. 100.111_2
14. $0.1011\ 111_2$
15. 1111.111_2

Convert the following octal numbers to binary form.

16. 173_8
17. 3722_8
18. 4625.553_8
19. 0.165_8

3.5 BINARY AND 8421 CODES

A **BCD (binary-coded decimal) code** is a device used to express one of the 10 decimal digits (0 through 9) using a nibble of binary digits. Such codes are used extensively in various applications of digital logic, such as encryption devices, arithmetic circuits, and so forth, as well as in computer applications that provide error detection and correction. We will briefly explore several of these codes and indicate why one might be preferable over another.

Perhaps the most straightforward method for encoding the decimal digits in binary form is the **8421 BCD code**. This is actually the direct representation of the decimal digits in their binary equivalent form and hence is a code for which we know the basis. Note that this is not the same as representing the complete decimal number in its equivalent binary form—far from it! The code translates the individual decimal digits into their binary equivalents, displaying the number in a string of nibbles. The representations of the digits are shown in Table 3.34 as a reminder of our past work: because we are using nibbles to represent the 10 decimal digits and there are $2^4 = 16$ actual nibbles possible, there are six nibbles that are considered invalid in 8421 BCD code.

TABLE 3.34 8421 BCD equivalents for decimal numbers

Decimal Number	8421 BCD Representation
0	0000
1	0001
2	0010
3	0011
4	0100
5	0101
6	0110
7	0111
8	1000
9	1001

Note that when a decimal number is represented in BCD code, the subscript of 2, denoting the base-2 representation of a decimal number, is not used. That is, if we wanted to represent the decimal number 368 in BCD, we would have the representation as 0011 0110 1000, and there would be no subscript of 2 at the end of the string of nibbles.

A code such as 8421 BCD is referred to as a **weighted code**, which means that the representation of the base-10 digits is performed using a four-bit combination for which each position within the nibble carries a particular weight of 8, 4, 2, or 1. That is, when we give a nibble of BCD code, such as 1001, the first digit carries a place weight of 8, the second a place weight of 4, the third a place weight of 2, and the final a place weight of 1. Because this is precisely the manner in which the first four places of binary numbers were defined, this terminology may seem a bit superfluous, but there are other BCD codes, such as 2421 BCD, for which the place weights are not the same as binary representation of decimal numbers. In that code, the nibble 1101 would be the representation of the decimal digit 7 because (in expanded form) it is equivalent to $1 \times 2 + 1 \times 4 + 0 \times 2 + 1 \times 1 = 7$.

In some texts, there is an effort to make explicit the point that a string of nibbles is representative of a decimal number in BCD code by appending a subscript of BCD where one would typically see the base identifier. That is, the BCD representation of the decimal number 368 could be given as $0011\ 0110\ 1000_{BCD}$, but we will (in this text) ensure that the use of BCD representation is clear by context and will not employ that notation.

EXAMPLE 3.26

Compare the binary representation of the decimal number 843 to the 8421 BCD code for that number.

SOLUTION

This is a good time to review our method for conversion of decimal numbers to binary form: divide the number repeatedly by 2, keeping track of the remainders and proceeding until the quotient of the division process becomes 0. The remainders, read in reverse order, form the binary equivalent to the decimal number (Table 3.35). The binary representation of the decimal number 843 is $11\,0100\,1011_2$.

TABLE 3.35 Converting 843 to binary

Division	Remainder	Quotient
843 ÷ 2	1	421
421 ÷ 2	1	210
210 ÷ 2	0	105
105 ÷ 2	1	52
52 ÷ 2	0	26
26 ÷ 2	0	13
13 ÷ 2	1	6
6 ÷ 2	0	3
3 ÷ 2	1	1
1 ÷ 2	1	0

The BCD code representation of the decimal number 843 will consist of the coded representations of the individual digits 8, 4, and 3. Referring to our chart or recalling the base-2 representation for each digit, we obtain 1000 0100 0011. Note that the two results are completely unrelated to one another, and this reinforces our admonition that the two processes are extremely different.

A decimal number encrypted using 8421 BCD code can be decoded into its decimal equivalent by replacing every nibble in the code by the appropriate digit, which we can either obtain from the table given earlier or by translating each nibble individually back to base-10.

EXAMPLE 3.27

Find the decimal number represented by the BCD code 0010 1001 0111 0101.

SOLUTION

We can recognize the nibbles as being the binary equivalents to 2, 9, 7, and 6, respectively, and therefore the equivalent decimal number would be 2976.

EXAMPLE 3.28

Find the decimal number represented by the BCD code 0010 1101 0001 0101.

SOLUTION

Proceeding as in the previous example, we find that the decimal equivalents of the individual nibbles would be 2, 13, 1, and 5. Observe that the second of these,

1101, does not represent a single digit but rather is equivalent to the decimal number 13; thus, this BCD code is not a possible representation for a decimal number. Recall that, earlier, we had indicated that there were six nibbles considered to be invalid in 8421 BCD code, and this is one of them.

The examples given to this point have all been integral valued, but BCD code can be used for decimal fraction numbers as well. We simply convert the digits before and after the decimal point to their BCD equivalents and place a separator point in the decimal point's position.

EXAMPLE 3.29

Give the BCD form of the decimal number 193.482.

SOLUTION

Referring once more to our table of BCD equivalents for the decimal digits, we obtain the following representation:

0001 1001 0011.0100 1000 0010

All the codes discussed so far have been weighted codes, but there certainly are **nonweighted codes** as well. Two examples of these are **Excess-3 (XS-3) code** and **Gray codes**. The former can be used for arithmetic operations, the latter for mechanical switching systems.

Full exploration of these codes would take us yet farther afield, but Gray codes are sufficiently interesting that we will spend a brief time considering them. The code type is named in honor of the creator of its first iteration, Frank Gray of Bell Labs. Note that the source of the code type is the name of the inventor, and therefore it would be grammatically incorrect to use any of the other possible spellings of the word Gray (such as "gray" or "grey," for instance).

The difficulty with using binary codes applied to switching systems is this: if a device were to indicate position by opening and closing switches, where open and closed are represented by 1 and 0, respectively, then two adjacent positions (011 and 100, for example) are such that the transition between one position and the next would require the transition of three switches synchronously. In the moment of transition, all three switches will be changing at a slightly different rate, and this could spawn misleading readings for an observer. It would be highly desirable for successive states to be represented by strings of binary numbers that have minimal differences, such as (in an ideal case) variation of only a single bit. A code capable of representing a succession of integer values possessing this quality is referred to as a Gray code.

There is no single Gray code; rather, the name applies to any code that represents each number in a sequence of integers (from 0 through $2^n - 1$, inclusive) as a string of binary digits of length n, such that strings that differ only in one bit represent adjacent integers in the sequence. The rationale for this construction is that advancing through the list of integers requires changing, or flipping, the value of one bit at a time.

One method of constructing a Gray code to represent, for example, the integers from 0 through 15, inclusive, offers an interesting exercise, which we shall leave to you. One starts with a string of four 0s (we use four digits because we know that there are 16 possible strings of four binary digits) and then successively flip the rightmost bit that produces a completely new string of digits. When we're through, we will (hopefully!) have 16 distinct sets of four binary digits to which we can assign the numbers 0 through 15. The set will, by construction, satisfy the definition of what it means to be a Gray code.

A second method is a process of reflecting and prepending; we will demonstrate the process here, since it is less intuitive than the method already described. The point is that there are many ways of constructing Gray codes rather than a single canonical process.

Our procedure will be to start with the digits 0 and 1, "mirror" those digits ("prepend" 0 to the first half and 1 to the last half of digits on the list), and then iterate the process until the desired number of digits (in this case, 16) is obtained. That may sound a bit complicated, but you'll find that it is, in practice, fairly easy.

Step One: **Begin with the digits 0 and 1 and mirror that set:**

- 0, 1, 1, 0 (note that "mirroring" merely means to repeat the digits in reverse order)

Step Two: **To the first half of strings on that list, prepend 0; to the second half of strings on the list, prepend 1:**

- 00, 01, 11, 10

Step Three: **Mirror the set once more:**

- 00, 01, 11, 10, 10, 11, 01, 00

Step Four: **Prepend 0s to the first half of strings and 1s to the second half of strings:**

- 000, 001, 011, 010, 110, 111, 101, 100

Step Five: **Mirror the set once more (this will produce a set of 16 strings):**

- 000, 001, 011, 010, 110, 111, 101, 100, 100, 101, 111, 110, 010, 011, 001, 000

Step Six: **Prepend 0s to the first half of strings and 1s to the second half of strings:**

- 0000, 0001, 0011, 0010, 0110, 0111, 0101, 0100, 1100, 1101, 1111, 1110, 1010, 1011, 1001, 1000

Observe there are no repetitions on the list and that each string differs from its predecessor in a single bit. We could thus make a unique assignment of the decimal numbers from 0 through 15, inclusive, to these binary strings, in order.

Exercises

Answer the following questions using complete sentences, with correct spelling and grammar.

1. What is meant by a BCD code?
2. What is meant by a weighted code?
3. What is a Gray code?

In the following problems, convert the decimal numbers to binary form and then give the 8421 BCD representation of the number.

4. 54
5. 6666
6. 347.625
7. 23.875

In the following problems, convert the 8421 BCD representations to the equivalent decimal number.

8. 1001 0111 0101
9. 0111 0101 1001 0011 0001
10. 1001.0111 0011
11. 0111 0011.1001

In this section, we examined the 8421 BCD weighted code, but there exist other weighted codes, such as the 2421 BCD code. The development of the code is similar to that of 8421 BCD, but the weights are 2, 4, 2, and 1, respectively. The following are 2421 BCD representations of decimal numbers. Find the decimal numbers with the shown representations.

12. 0001 1011 1110 0011
13. 0101 0001 1111.1100 1101
14. 1111.1110 1011 0010
15. 0000.0011 0100 1110

Write an explanation for the the solution of the following.

16. We have introduced one method for generating a Gray code and suggested an alternative approach to represent the integers from 0 through 15, inclusive. Begin with a string of four 0s (we use four digits because we know that there are 16 possible strings of four binary digits) and then successively flip the rightmost bit that produces a completely new string of digits. When we're through, you will have 16 distinct sets of four binary digits to which we can assign the numbers 0 through 15. Demonstrate this procedure generates a Gray code.

Summary

In this chapter, you learned about:

- The decimal, binary, and hexadecimal number systems.
- The arithmetic operations and applications of the binary number system.
- The relationship between binary and hex, decimal and hex, and binary and octal.
- The binary and 8421 codes.

Glossary

ASCII (American Standard Code for Information Interchange) An encryption device for the English alphabet using binary numbers.

base The number of distinct characters employed for numeration within a particular number system. Also called the radix.

base-16 number system A system of numeration using 16 distinct symbols: 0, 1, 2, 3, 4, 5, 6, 7, 8, 9, A, B, C, D, E, and F. Also called the hexadecimal number system.

base-10 number system The decimal number system.

BCD (binary-coded decimal) code Any of the codes used to express the decimal digits (0 through 9) using a nibble.

binary number system Also called base-2, a numeration system in which all numbers are represented using the digits 0 and 1 only.

binary point The analog to the decimal point in base-10; the point of demarcation between the whole number and fractional parts of a binary number. Also called the binary separator.

binary separator The analog to the decimal separator in the binary system. Also called the binary point.

bit A two-digit pairing in the binary number system.

byte An eight-bit grouping in the binary number system.

color codes Six-digit hex number HTML designations that identify the text color in a document.

decimal number system A system of numeration using 10 distinct symbols: 0, 1, 2, 3, 4, 5, 6, 7, 8, and 9.

decimal separator A dot used to indicate the separation from whole number to fraction number places within a number's base-10 representation.

8421 BCD code The particular binary decimal code using the binary equivalents of the numbers 0 through 9 as the encryption code.

Excess-3 (XS-3) codes A BCD code and numeral system using a prespecified value N as a bias value.

expanded form The decomposition of a number as a sum where each term displays a product of a digit of the number's representation with its place value.

exploded form Another term for the expanded form of a number.

Gray codes A binary numeral system where two successive values differ only in one bit.

hexadecimal number system Also called base-16, a numeration system in which all numbers are represented using the symbols 0, 1, 2, 3, 4, 5, 6, 7, 8, 9, A, B, C, D, E, and F only.

hexadecimal point The analog to the decimal point in base-16; the point of demarcation between the whole number and fractional parts of a base-16 number. Also called the hexadecimal separator.

hexadecimal separator The analog to the decimal point in the hexadecimal number system.

least significant binary digit (LSB) The base-2 instance of the least significant digit. Also called the least significant bit.

least significant digit (LSD) The nonzero digit farthest right within a number's representation.

most significant binary digit (MSB) The base-2 instance of the most significant digit. Also called the most significant bit.

most significant digit (MSD) The nonzero digit farthest left within a number's representation.

nibble A four-bit grouping in the binary number system.

nonweighted codes: Codes that employ a weighted positioning system.

octal number system Also called base-8, a numeration system in which all numbers are represented using the symbols 0, 1, 2, 3, 4, 5, 6, and 7 only.

octal separator The analog to the decimal point in the octal number system.

place value The value associated with a particular digit location in a numerical representation.

radix The base of a number system.

RGB The red/green/blue ratios used to construct colors in HTML; also called color codes.

symmetric property of equality The property stating that A = B is equivalent to B = A.

weighted code a code in which representation of the base-10 digits is performed using a four-bit combination for which each position within the nibble carries a particular weight.

End-of-Chapter Problems

In each of the following, identify the place name and place value of every digit.

1. 4527.89
2. 346.567
3. 1274.3

In each of the following, give the expanded form of the decimal number.

4. 1954.37
5. 88.238
6. 883.5678

For the following, give the decimal number whose expanded form is shown.

7. $4 \times 10^3 + 2 \times 10^2 + 6 \times 10^1 + 0 \times 10^0 + 8 \times 10^{-1} + 6 \times 10^{-2}$
8. $5 \times 10^0 + 3 \times 10^{-1} + 9 \times 10^{-2} + 4 \times 10^{-3}$
9. $3 \times 10^5 + 8 \times 10^4 + 0 \times 10^3 + 0 \times 10^2 + 5 \times 10^1 + 4 \times 10^0$

For each of the following, identify the LSD and MSD.

10. 4527.89
11. 536200
12. $5 \times 10^0 + 3 \times 10^{-1} + 9 \times 10^{-2} + 4 \times 10^{-3}$

Give examples of decimal numbers satisfying the following conditions.

13. A whole number having an MSD of 4 and an LSD of 7.
14. A number having 9s in the tenths, hundreds, and thousandths places.

For each of the following decimal numbers, convert to binary form.

15. 5249
16. 127

For each of the following binary numbers, convert to decimal form.

17. $1010\ 1110_2$
18. 101.1110_2

Perform the following binary additions. Check your work by converting the addends and sum to decimal form and adding these to show that the same result is obtained.

19. $1111_2 + 1\ 1000_2$
20. $101.101_2 + 1001.0001_2$

Perform the following binary subtractions. Check your work by converting minuend, subtrahend, and difference to decimal form and subtracting these to show that the same result is obtained.

21. $1\ 1000_2 - 1111_2$
22. $1010.011_2 - 1.1_2$

Perform the following binary multiplications. Check your work by converting the factors and the product to decimal form and multiplying these to show that the same result is obtained.

23. $(1\ 1000_2)(110_2)$
24. $(111_2)(1000_2)$

Perform the following binary divisions. Check your work by converting the dividend, divisor, and quotient to decimal form and dividing these to show that the same result is obtained.

25. $1000\ 0001_2 \div 111_2$
26. $1000\ 0100\ 1100_2 \div 1\ 0010_2$

Convert the following phrases to ASCII representation.

27. This conversion is tedious.
28. Reading books is fun.

Convert the following binary representations into English, using the ASCII conversion chart.

29. 01010100 01101000 01101001 01110011 00100000 01101001 01110011 00100000 01101000 01100001 01110010 01100100 00100000 01110111 01101111 01110010 01101011 00101110
30. 01001000 01100001 01110000 01110000 01111001 00100000 01101110 01101111 01110111 00111111

Convert the following binary numbers to hexadecimal form.

31. $1001\ 1110\ 1010_2$
32. $11\ 0010.0100\ 0110_2$

Convert the following hexadecimal numbers to binary form.

33. BA_{16}
34. $9347FC_{16}$

Convert the following decimal numbers to hexadecimal form.

35. 510
36. 374.25

Convert the following decimal numbers to octal form.

37. 128
38. 4577.0625

Convert the following octal numbers to decimal form.

39. 235_8
40. 4156.25_8

Convert the following binary numbers to octal form.

41. $1011\ 1100\ 0010_2$
42. 101.1_2

Convert the following octal numbers to binary form.

43. 377_8
44. 44125.66_8

In the following problems, convert the decimal numbers to binary form and then give the 8421 BCD representation of the number.

45. 255
46. 97.625

In the following problems, convert the 8421 BCD representations to the equivalent decimal number.

47. 0001 1001 0101 0100
48. 0011 0111.0101 0110

In the following problems, convert the 2421 BCD representations to the equivalent decimal number.

49. 1111 0111.1110 0100 0101
50. 1000 0111 1100 1111 1110 0001

Chapter 4 Straight-Line Equations and Graphs

Consider the following rather common (and distressing!) situation encountered by many consumers: a new wide-screen television is purchased and installed, a favorite television show is tuned in, and all the actors appear to have suddenly gained an inordinate amount of weight. Actresses who had always seemed to be rail thin now look like candidates for the current fad diet, and automobiles are longer than the longest of land yachts from the deepest depths of the early 1970s.

What has happened is that the aspect ratio of the image is different from the proportions of your screen. Most television programs are presented in what is called a 4:3 aspect ratio, which means that the width of the image is 33% more than its height. Wide-screen television screens have monitors that are constructed with a 16:9 aspect ratio, and it is in this format that most high-definition television programs are presented.

The extent to which a 4:3 aspect ratio image has been distorted when "stretched" to fill a wide-screen television can be determined through the use of linear equations. We will introduce such equations in this chapter and consider some applications of them.

Objectives

When you have successfully completed the materials of this chapter, you will be able to:

- Graph ordered pairs with rectangular coordinates.
- Graph a linear equation with two variables.
- Write the slope-intercept form of the equation for a straight line.
- Write the two-point form of the equation for a straight line.
- Write the point-slope form of the equation for a straight line.
- Write the two-intercept form of the equation for a straight line.
- Graphically solve a two-variable linear equation.

4.1 The Basics of the Cartesian Plane

When we speak of graphs, we are discussing sets of points (possibly forming a straight line, possibly forming a curve, possibly discrete and separated from one another by some positive distance) in the Cartesian plane. In this chapter, we will be considering the particular case where the points are aligned in a straight-line path, but before we can fully examine that, we need to set forth the fundamentals of the Cartesian plane, also referred to more simply as the "*xy* plane" or even more generally as "the plane."

The Basic Features of the Plane

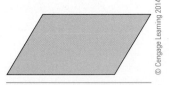

Figure 4.1 Portion of a plane, showing no frame of reference

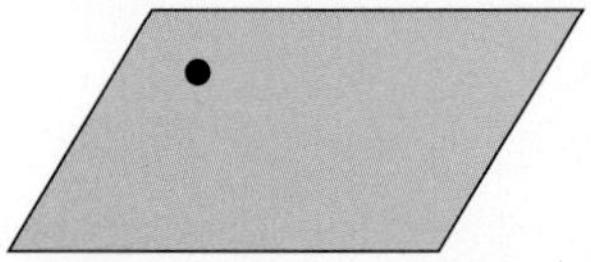

Figure 4.2 Portion of a plane, showing no frame of reference, with a single point identified

The **Cartesian plane** (named in honor of its creator, the French mathematician René Descartes) is a construct that allows us to systematize our investigation of position in a two-dimensional mathematical plane. A general plane, of course, is simply a flat surface that has infinite extension in two dimensions, which we may refer to as length and width, and absolutely no curvature. The Cartesian plane imposes on such a surface a position-identifying grid. Through reference to this, we can specify location of a particular point within the plane.

That explanation may sound mysterious, so let's illustrate the confusion we are attempting to overcome. Consider a representation of an arbitrary plane. We will show a portion of this plane by presenting a rectangle, as done in Figure 4.1. On the plane, we will identify a particular point by placing a dot, as shown in Figure 4.2.

Note that neither a mathematical plane (having no thickness whatsoever) nor a mathematical point (having neither length nor width) can have a real, physical existence. These things are mere abstractions that we represent visually as we have done here.

Now, suppose you are asked to describe where the point is located within the plane (keeping in mind that, although we have shown a fragment of the plane, it continues forever). You have no "reference points" through which you could guide someone to the location of the particular point! It is necessary, for us to be able to speak of the location of the point, that we impose a "frame of reference" grid, referred to as the Cartesian coordinate system.

The grid consists of a pair of perpendicular number lines, crossing at their mutual 0 location. One number line, positioned so that our view of it is horizontal, is referred to as the **x-axis**, and the other (vertical, since it is perpendicular to the first number line) is referred to as the **y-axis**. When we speak of the two number lines together, we call them "axes" (pronounced "akseez"), which is the plural of "axis" (Figure 4.3).

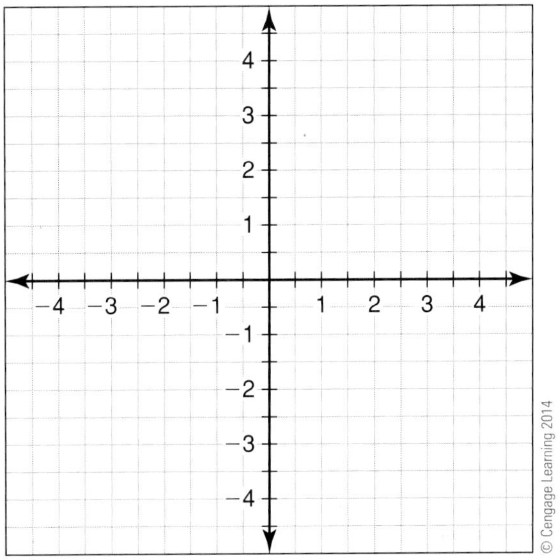

Figure 4.3 The Cartesian plane

The convention is that the positive direction on the horizontal axis is to the right, and the positive direction on the vertical axis is upward. This is sometimes referred to as a "right-handed system," the name of which is derived from an application one sees in physics.

Where the two number lines cross is said to be the **origin** of the Cartesian plane, and the four vast regions into which the plane is subdivided by the crossing axes are called the **quadrants**, numbered I, II, III, and IV, counterclockwise, as shown in Figure 4.4.

The utility of imposing this pair of number lines on an abstract plane is that location within the plane can now be specified through the use of **ordered pairs**. Any point within the plane can be aligned vertically with a particular value "a" on the x-axis and horizontally with a particular value "b" on the y-axis. The point can then be specified with the ordered pair (a, b). The term "ordered" refers to the fact that the order of the values a and b is significant; that is, (a, b) is typically not the same as (b, a). In the ordered pair (a, b), the value of a is called the **x-coordinate**, or the **abscissa**, and the value of b is called the **y-coordinate**, or the **ordinate** of the point.

Note

The quadrants are identified using Roman numerals, beginning with I in the portion of the upper right corner of the plane and advancing through II, III, and IV in a counterclockwise direction!

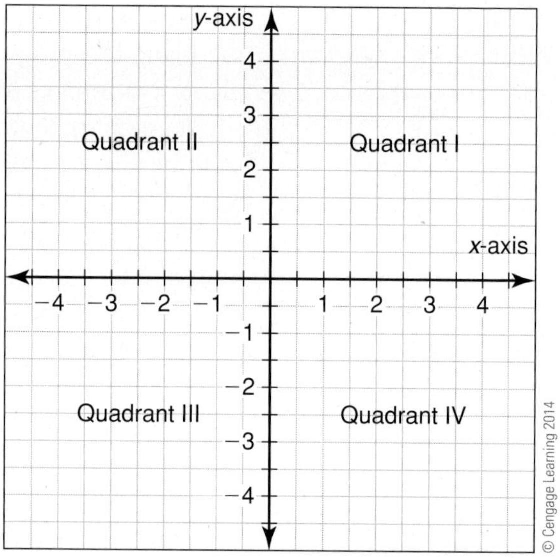

Figure 4.4 The Cartesian plane with quadrants numbered

When we place a "dot" at a location in the plane identified by an ordered pair, we say that we are plotting the point. A Cartesian plane on which a collection of isolated points have been plotted is referred to as a **scatter plot**. A collection of ordered pairs is called a **relation**, and a relation in which all the points have distinct x-coordinates (that is, there are no two points in the collection having the same x-coordinate) is called a **function**.

Let's put the ideas of plotting points together with the concept of the quadrants. In the first quadrant, what can we say about the values of x and y? Due to the convention of choice for the positive direction on the number lines, points in quadrant I are such that both the x- and the y-coordinates of the point must be positive (Figure 4.5).

In the second quadrant, we observe that points have negative x-coordinates and positive y-coordinates. Similar observations can be made for points lying in the other two quadrants, and this nearly puts to rest the interrelationship of coordinate signs and the quadrants. The only remaining question is about those

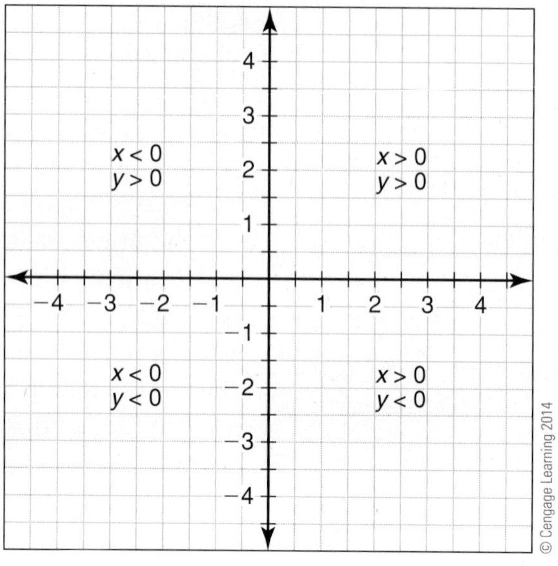

Figure 4.5 Note that, according to the numeration of the axes, in quadrant I both x and y are positive

points that happen to not fall within any of the particular quadrants. These are the points lying on one or both of the coordinate axes, and they are slightly more cumbersome to describe.

Points lying on one or both of the axes are referred to as **quadrantal points** and are not considered to fall within any particular quadrant. All quadrantal points have at least one coordinate equal to 0, with the sign of the other coordinate determined by the part of the axis on which it lies. The only exception to this is the origin, which is quadrantal with both coordinates being 0.

EXAMPLE 4.1

Plot the points (4, −2), (−3, 5), (0, 3), and (4, 6) on a single Cartesian plane.

SOLUTION

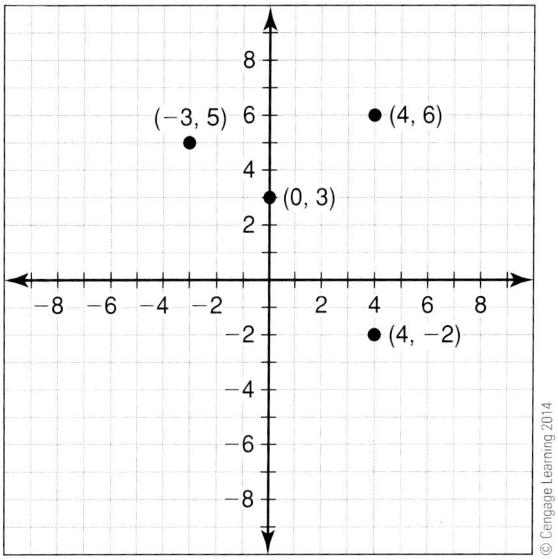

Figure 4.6 Solution to Example 4.1

EXAMPLE 4.2

Name the quadrant in which each of the points named in Example 4.1 lies.

SOLUTION

If we consider the point (4, −2), we notice that the x-coordinate is positive and that the y-coordinate negative. According to the labeling conventions of the Cartesian plane, this places the point in quadrant IV. The point (−3, 5), on the other hand, has a negative x-coordinate and a positive y-coordinate, and this places the point in quadrant II. By similar reasoning, the (4, 6) lies in quadrant I.

The only point whose status is ambiguous is the point (0, 3), which lies on the positive y-axis. Points on the axes are called quadrantal and are not considered to lie in any particular quadrant; therefore, (0, 3) lies in none of the four quadrants.

In a relation (or function), the set consisting of all the x-coordinates of the points of the relation (or function) is called the **domain**, and the set consisting of all the y-coordinates of the points of the relation (or function) is called the range.

Note

In any relation or function, the set of all numbers occurring as x-coordinates of points on the graph of the relation or function is called the domain, while the set of all numbers occurring as y-coordinates of points on the graph of the relation or function is called the range!

EXAMPLE 4.3

Give the domain and the range of the relation given in Example 4.1. Determine if the points form a function or if they are merely a relation.

SOLUTION

Recalling the domain of a relation is the set of all x-coordinates of ordered pairs in the relation, we can see that since the relation consists of the points $(4, -2)$, $(-3, 5)$, $(0, 3)$, and $(4, 6)$, the domain is the set $\{4, -3, 0\}$. Since the order of elements of a set presented in roster form can be given in any order we choose, we can give this answer with the values rearranged to suit our taste. Similarly, since the range of a relation is the set of all y-coordinates of ordered pairs in the relation, the range is $\{-2, 5, 3, 6\}$.

The Distance Formula and the Midpoint Formula

Once we have established the notion of position in the plane, it seems natural to raise the question regarding the distance separating two positions. Positions are specified through the use of ordered pairs, so we will define distance through the use of ordered pairs.

We will be using two distinct points in the plane, and therefore we need a method for symbolically distinguishing between point 1 and point 2. The general practice is to refer to point 1 as having coordinates (x_1, y_1). This is read as "x one, y one" or "x sub one, y sub one," where "sub" is the diminutive of "subscript." All this is intended to indicate is that x_1 is the x-coordinate of point 1 and that y_1 is the y-coordinate of point 1. The second point will be identified in a similar manner as (x_2, y_2).

When we plot two points in the plane, (x_1, y_1) and (x_2, y_2), we can form a right triangle, as in Figure 4.7, with the straight-line path from the first point to the second being the shortest distance between the points. The length of this path is defined to be the distance between the points, and its value can be found using the Pythagorean Theorem (see Figure 4.7).

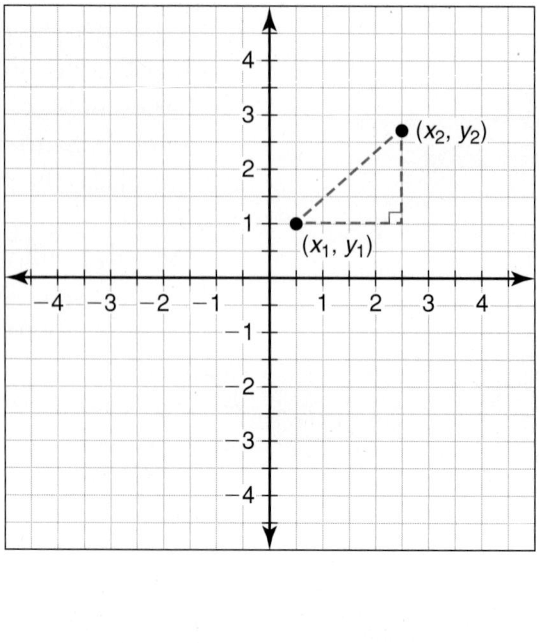

Figure 4.7 A right triangle relating the positions of two points in the plane

whenever we do an application, we will always be careful to make sure that we express our interpretation of the points involved.

The Attributes of a Straight Line in the Plane

There are several features of straight lines of which we should take notice because they have high relevance in applications. In particular, we'll be taking a look at the "intercepts" of the line and the "slope" of the line, and we will consider the importance of those concepts in situations represented by linear graphs.

Intercepts

Some lines intersect the x-axis at a single point (a horizontal line that coincides with the x-axis intersects the x-axis at every point on the line). The point at which a nonhorizontal line intersects the x-axis is called the **x-intercept** of the line. The x-intercept will always have the form $(a, 0)$, because the y-coordinate of such a point must be 0. In casual conversation, we often say "the x-intercept of the line is a," with the recognition that this phrase is meant to indicate that the line crosses the x-axis at the point $(a, 0)$. That is a bit informal, but it is common practice, and we will not swim against that tide.

EXAMPLE 4.7 Identify the x-intercept for the line shown in Figure 4.9.

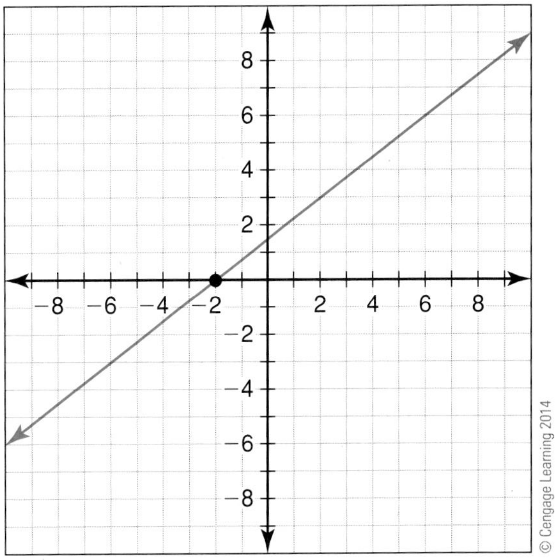

Figure 4.9 The line of Example 4.7

Note

Always remember to give your intercepts in ordered-pair form, since they are actually points in the plane. In general conversation, we often do not give the ordered pair when describing the intercept, but when giving your answers, always use ordered-pair form!

SOLUTION

By inspection, we can see that the line crosses the x-axis at the mark labeled -2. This is, in ordered pair form, the point $(-2, 0)$, and this is the x-intercept of the line.

Some lines intersect the y-axis at a single point (a vertical line that coincides with the y-axis intersects the y-axis at every point on the line). The point at which a nonvertical line intersects the y-axis is called the **y-intercept** of the line. The y-intercept will always have the form $(0, b)$ because the x-coordinate of such a point must be 0. As with the x-intercept, in casual conversation we often say that "the y-intercept of the line is b," again recognizing that this phrase is meant to indicate that the line crosses the y-axis at the point $(0, b)$.

EXAMPLE 4.8

For the shown line, find the y-intercept.

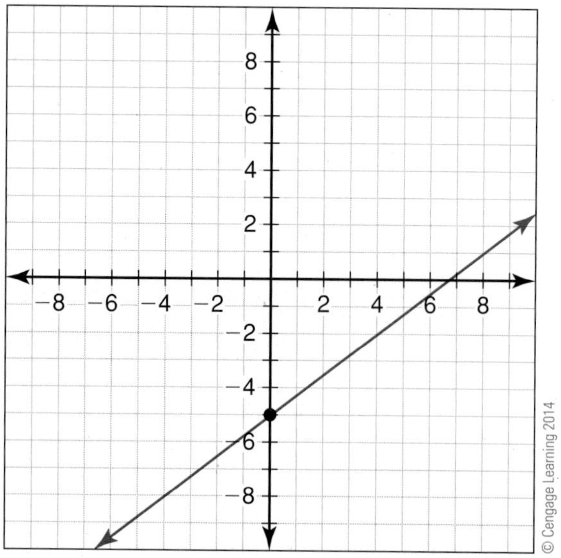

Figure 4.10 The line of Example 4.8

SOLUTION

By inspection, we can see that the graph crosses the y-axis at the mark labeled -5. This is, in ordered-pair form, the point $(0, -5)$, and this is the y-intercept of the line.

EXAMPLE 4.9

In what circumstance would the y-intercept of a line be the same as the x-intercept of the line?

SOLUTION

Recalling that the y-intercept is that point where the line crosses the y-axis and that the x-intercept is that point where the line crosses the x-axis, there is only one situation in which the y- and x-intercepts could coincide, and that would be the case where the line happened to be nonvertical, nonhorizontal, and passing through the origin. The point $(0, 0)$ would be the mutual y- and x-intercept point.

In the explanations of the intercept types, we made careful use of the terms "nonhorizontal" and "nonvertical." Let's take a moment to consider the situations where the lines are horizontal and where the lines are vertical, ignoring those instances where the lines coincide with the axes, which we have already mentioned.

In Figure 4.11, we present a horizontal line. Observe that such a line does not intersect the x-axis at all and therefore has no x-intercept. You do not want to use the term "undefined" to describe the x-intercept because this is incorrect. The x-intercept for such a line does not exist but is not referred to with the technical term "undefined."

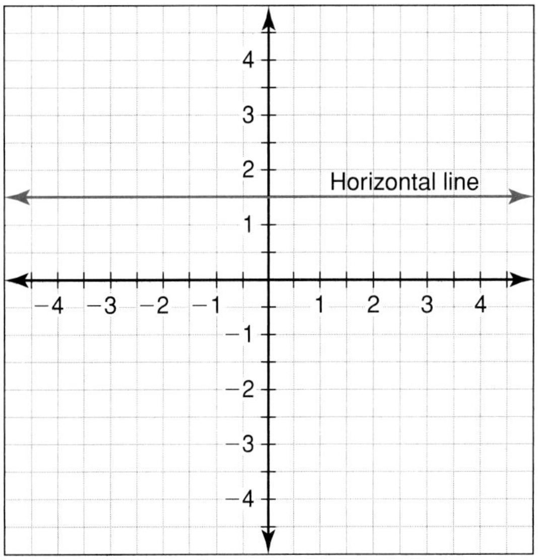

Figure 4.11 Horizontal line in the plane

In Figure 4.12, we present a vertical line. Similar to what was observed with the horizontal line, we note that this line does not intersect the y-axis and consequently has no y-intercept.

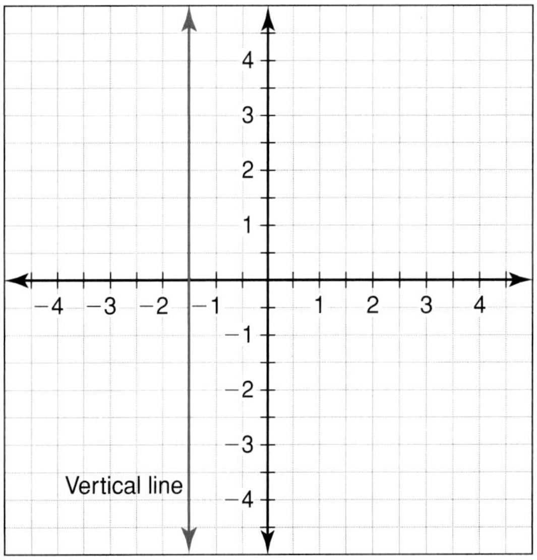

Figure 4.12 Vertical line in the plane

EXAMPLE 4.10

Is it possible for a line, not coincident with one of the coordinate axes, to have neither an x-intercept nor a y-intercept?

SOLUTION

No, this is not possible. Such a line must cross one or both of the coordinate axes at some point, and therefore at least one of the intercept types must occur.

Slope of a Line: The Rise-over-Run Method

The third feature of a line that has significant importance is the **slope** of the line. The slope of a line, which we usually represent with the lowercase letter m, gives us a way to quantify how far the line is tilted from horizontal. From this

Note

The slope of a horizontal line is 0. The slope of a vertical line is undefined.

perspective, a line with no such tilt (a horizontal line, that is) has slope $m = 0$. A vertical line is said to have "undefined" slope for reasons we shall make clear shortly. Some people mistakenly refer to a vertical line as having "no slope," but this is vague and imprecise and should *never* be used. The only correct term used to describe the slope of a vertical line is "undefined."

The slope of a line can be calculated in two ways. One technique is visual and (while useful) can be fraught with peril if a high-quality graph is not available. We will examine that method first, with the clear understanding in mind that we are running a bit of a risk in doing so.

The technique is sometimes referred to as the **rise-over-run method** and hinges on your ability to precisely determine the coordinates of two distinct points on the line. If you cannot absolutely determine the coordinates of two points on the line with precision, you should not attempt to compute the slope of the line in this manner! It is also inadvisable to attempt to use this technique if the points whose coordinates are known have fractional coordinates, as it is difficult to perform the computations for such points.

Suppose we are confronted with the line shown in Figure 4.13. Because the graph is high quality, we can detect the coordinates of at least two points on the line. We will choose, arbitrarily, to work with the points $(-2, 5)$ and $(3, -1)$, which are clearly on the line. Whenever we look at points in the plane, we adopt the convention of referring to objects in relative position to one another, so it is unambiguous to describe the point $(-2, 5)$ as being the leftmost point with which we have chosen to work.

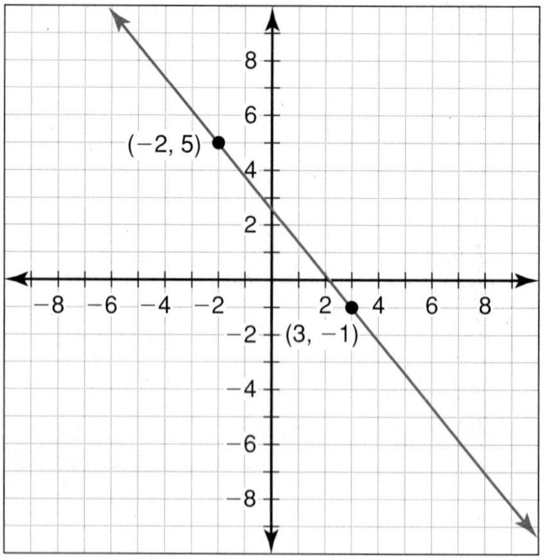

Figure 4.13 Line passing through the points $(3, -1)$ and $(-2, 5)$

Beginning at the leftmost point we've chosen, we will now move to the rightmost point we've chosen. The motion must be done in the following manner: Beginning at the leftmost point, move vertically until we are level with the rightmost point and then move horizontally until the rightmost point is reached. The vertical motion is signed motion, meaning that we consider movement upward to be positive and movement downward to be negative. The horizontal motion is always considered positive.

The distance moved vertically (including the sign, as indicated) is referred to as the **rise**, while the distance moved horizontally is referred to as the **run**. The slope, represented by m, is defined to be the simplified fraction $m = \frac{rise}{run}$. Keep in mind that, according to the convention we've chosen to adopt, the rise can be either positive or negative (depending on the situation), but the run is always positive. Unless told specifically to do otherwise, we'll simplify the fraction, but we will not convert it to an equivalent decimal or decimal approximation or change improper fractions to mixed numbers. The value of the slope should be left as a fraction, possibly improper, except in the singular case where the denominator of the simplified fraction should turn out to be 1, in which case we can express the slope as an integer.

EXAMPLE 4.11 Find the slope of the line shown in Figure 4.14 using the rise-over-run method.

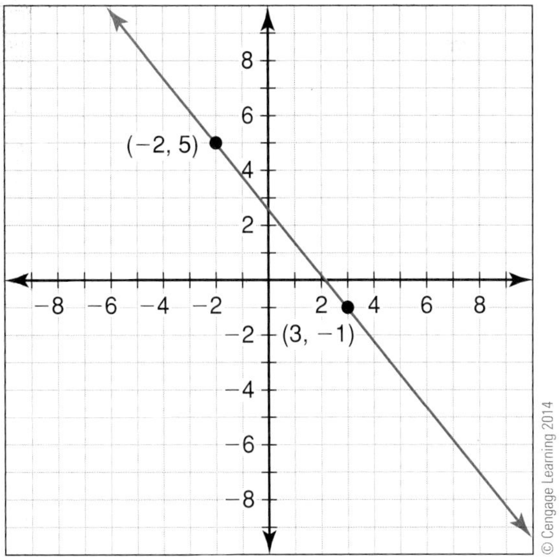

FIGURE 4.14 Image showing rise and run from one point to another on line

SOLUTION

In order to calculate the slope, we will have to choose two points whose coordinates can be determined precisely. Here, we will select the points $(-2, 5)$ and $(3, -1)$. Note that, in this case, to move from the leftmost point to the rightmost point in the manner described earlier, we must move downward 6 units and then to the right 5 units. This indicates that the quantity we have referred to as the rise (being a downward and hence negative movement) is -6, while that which we have referred to as the run is 5, and therefore $m = \frac{-6}{5}$.

EXAMPLE 4.12 Find the slope of the line shown in Figure 4.15 using the rise-over-run method.

SOLUTION

To find the slope, we must choose two points on the line whose coordinates can be determined with certainty. A pair of such points would be $(0, -1)$ and $(3, 1)$. You should note that, if you were to pick any other pair of points on the same

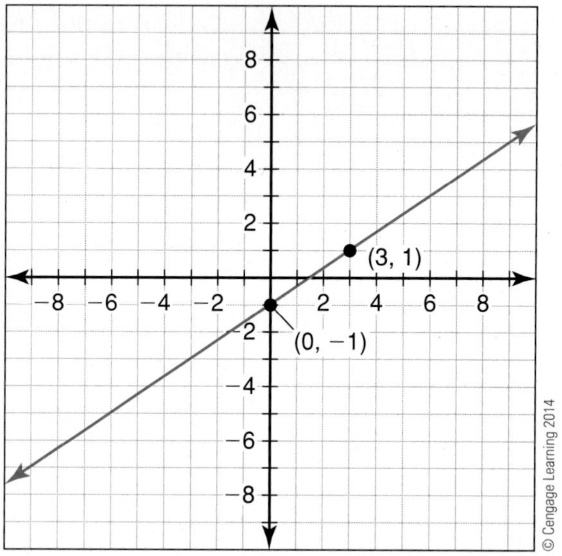

FIGURE 4.15 Line in the plane passing through the points (0, −1) and (3, 1)

line, your computations will, ultimately, lead you to the same value for the slope of the line. That is, the computation of slope is independent of the choice of the points used for the calculations. Beginning at the point on the left, (0, −1), we move vertically until we are horizontally aligned with the point on the right. To accomplish this, our vertical movement is 2 units upward, and hence the rise is 2. Moving right to reach the second point, we must move 3 units, and thus the run is 3. Consequently, the slope of the line is $\frac{2}{3}$.

EXAMPLE 4.13

Find the slope of the line shown in Figure 4.16 using the rise-over-run method.

SOLUTION

Once more, we need to find a pair of points on the graph of the line whose coordinates are clearly visible. In this case, we can see the intercepts fairly clearly, and they are the points (0, 6) and (14, 0). Those are suitable for our purpose, and we'll use them to find the slope. Beginning at the *y*-intercept, we move vertically

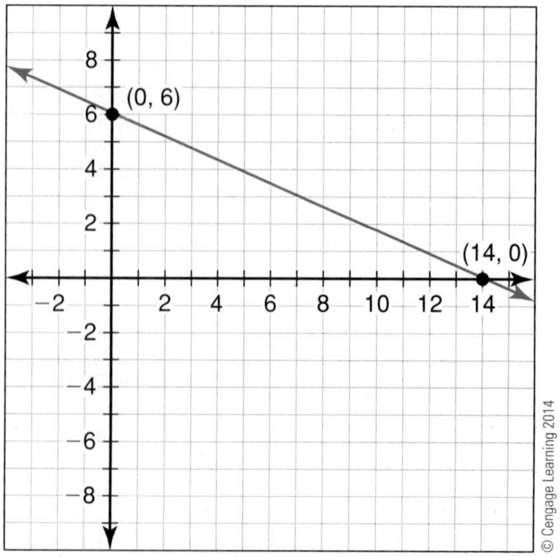

FIGURE 4.16 Line passing through the points (0, 6) and (14, 0)

downward 6 units to reach a point horizontally aligned with the point to the right. This indicates, since the movement is downward, that the run is −6. From that location, we move 14 units to the right, and therefore the run is 14. Inserting these values into the ratio of rise over run, we obtain a value of $\frac{-16}{14}$, which reduces to $\frac{-3}{7}$, and this is the slope of the line.

Earlier, we remarked that horizontal lines had slope zero and that vertical lines had undefined slope. We should consider how those notions relate to the rise-over-run method of calculating slope.

Suppose we have a line that is horizontal, such as shown in Figure 4.17. We can identify two particular points on the line, such as (3, 5) and (7, 5). Following our agreed-on algorithm, we will calculate the slope by determining the rise and the run and then forming a fraction with the rise as the numerator and the run as the denominator.

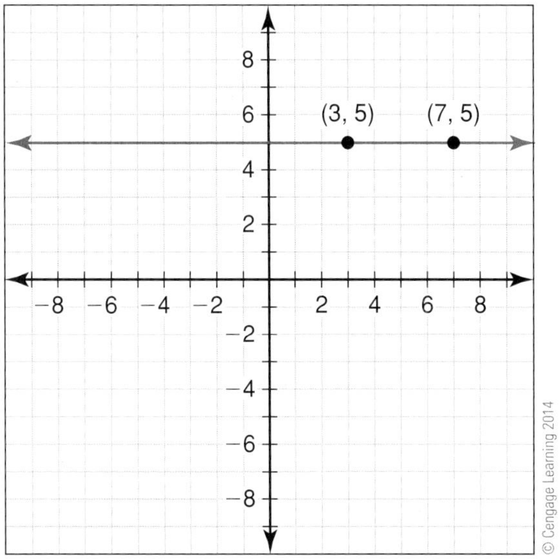

FIGURE 4.17 Horizontal line through (3, 5) and (7, 5)

In this case, the leftmost point is already aligned horizontally with the rightmost point, and therefore no vertical motion is necessary. This tells us that the rise is 0. For the particular example, our run is a direct horizontal movement to the right of 4 units, so the run is 0. Using our definition of slope, $m = \frac{4}{0}$, and any fraction whose numerator is 0 and whose denominator is nonzero is equal to 0. Thus, the rise-over-run computation of slope supports our earlier contention that the slope of a horizontal line was 0.

For the vertical line case, consider the graph shown in Figure 4.18. Again choosing a pair of points on the line, such as (−2, 1) and (−2, 5), we compute our rise and our run and then use them to form the ratio we have defined to be equal to the slope. There is no leftmost point, since the two points are equally far to the left in the plane, so we will arbitrarily choose the first point, (−2, 1), as being the leftmost for the sake of computation of rise and run.

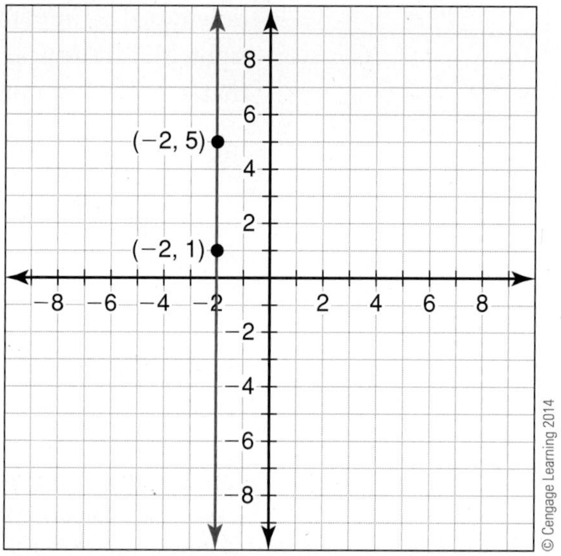

FIGURE 4.18 Vertical line through (−2, 1) and (−2, 5)

To move from the first point to the second point, a rise of 4 units is needed, and a run of 0 is required. Combining these into our definition of slope, we find $m = \frac{4}{0}$. A fraction whose numerator is nonzero and whose denominator is 0 is equivalent to division by 0, and this is a concept we refer to as undefined. This is the motivation for the earlier claim that the slope of a vertical line is undefined.

Note

Lines with positive slope rise as you follow them from left to right through the Cartesian plane. Lines with negative slope fall as you follow them from left to right through the Cartesian plane.

Before leaving this rise-over-run concept behind, we want to present an observation that follows as a consequence of the "signed direction" determined from the rise of a line. A line that moves upward as we follow it from left to right must have a positive rise. Since the run is always a positive quantity, this indicates that a line that rises from left to right has positive slope (Figure 4.19).

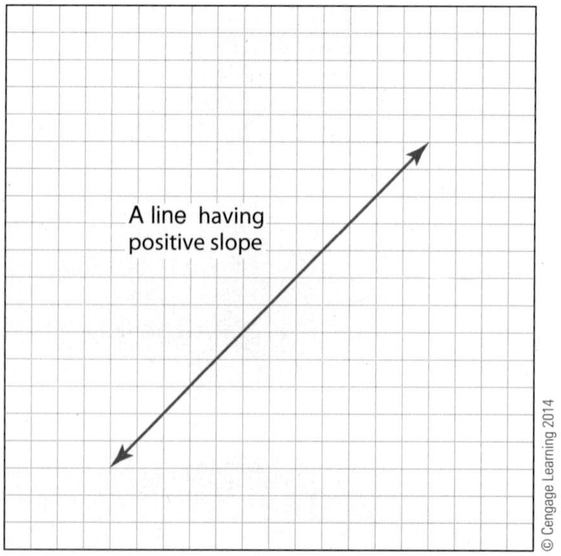

FIGURE 4.19 Line with positive slope, rising from left to right

Similarly, a line that descends as we follow it from left to right must have a negative rise value. Since the run is always positive, this forces the slope of such a line to be negative (Figure 4.20).

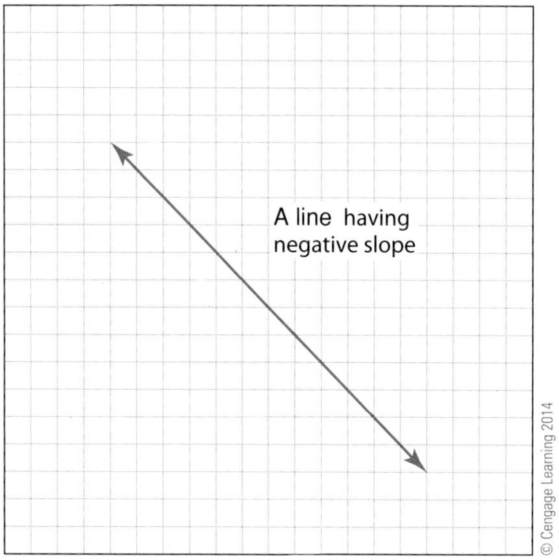

FIGURE 4.20 Line with negative slope, falling from left to right

Slope of a Line: The Slope Formula

The rise-over-run method for finding the slope, which relies on a visual counting of movement from one point on a line to another, is a fine method for calculating the slope and is extremely efficient when a high-quality graph is available and relatively close points with integer-valued coordinates are easily determined. Unfortunately, this is not always the case, since we can often find ourselves in situations where finding points with integer-valued coordinates is not easy or the points that do have such values are inconveniently far apart.

In these situations, it is desirable to have a computational method for calculating slope that is impervious to these potential pitfalls. The formula involves subscript notation, as well—do not confuse subscripts with exponents, for doing so is guaranteed to lead you to disaster! Taking the two points on the line to have coordinates (x_1, y_1) and (x_2, y_2), we now present the defining formula for slope:

$$m = \frac{y_2 - y_1}{x_2 - x_1}$$

The formula presented is also valid if both numerator and denominator are reversed, since that yields the same simplified result. That is, an alternative formula is $m = \frac{y_1 - y_2}{x_1 - x_2}$. Note that the only difference between the two forms of the defining formula is the order of the subscripts. Further, note that in each case, the y-coordinates are placed in the numerator, the x-coordinates are placed in the denominator, and the subscripts in the denominator appear in the same order as the subscripts in the numerator.

EXAMPLE 4.14 Find the slope of the line passing through the points $(-2, 5)$ and $(3, -1)$.

SOLUTION

We actually considered this line in an earlier example, and we are intentionally using it a second time as a demonstration that the rise-over-run method and the defining formula for slope will yield the same results.

In the computation of slope, it is irrelevant which point we consider as the first point and which we consider as the second, so we will choose to assign $(x_1, y_1) = (-2, 5)$ and $(x_2, y_2) = (3, -1)$. Under this assignment, we will insert the coordinate values into the defining formula and simplify the resulting expression:

$$m = \frac{y_2 - y_1}{x_2 - x_1} = \frac{-1-5}{3-(-2)} = \frac{-6}{5}$$

This is precisely the same result we found in our earlier computation using rise over run.

You might feel that the rise-over-run method is superior to the computational approach, and for the previously mentioned example, this may well be the case. Let's look at two examples that establish the utility of the defining formula for slope.

EXAMPLE 4.15 Find the slope of the line passing through the points $(-23, 15)$ and $(31, -28)$.

SOLUTION

The complication we would face if we were to attempt to use the rise-over-run method would be that these points are separated by significant distance in the plane. This would make the sketch of the line rather tedious, so it becomes useful to find the slope with the defining formula.

We assign $(x_1, y_1) = (-23, 15)$ and $(x_2, y_2) = (31, -28)$ and compute as follows:

$$m = \frac{-28-15}{31-(-23)} = \frac{-43}{54}$$

As the numerator is prime, we cannot reduce this fraction, so that rather unattractive value is the slope of the given line.

EXAMPLE 4.16 Find the slope of the line through (3/4, 7) and (1/3, 5/6).

SOLUTION

In this case, we face yet another complication that would make the application of the rise-over-run technique challenging: the appearance of coordinates that are fractions. In a case like this, we are well advised to use the defining formula for slope. Again, we will use the order of occurrence in the problem as our motivation for choice of first point and second point, so we have and $(x_1, y_1) = \left(\frac{3}{4}, 7\right)$ and $(x_2, y_2) = \left(\frac{1}{3}, \frac{5}{6}\right)$. Substitution into the formula for slope yields

$$m = \frac{\frac{5}{6} - 7}{\frac{1}{3} - \frac{3}{4}}$$

We pause here to note that the expression here is what we call a **complex fraction**, where we have a fraction whose numerator, denominator, or both contain fractions. While we could work with the expression as presented, we can clear the complex fraction by multiplying numerator and denominator by the least common denominator of the individual fractions within the complex fraction. The least common denominator of those interior fractions is seen to be 12, so we multiply all terms by that value:

$$m = \frac{12\left(\frac{5}{6}\right) - 12(7)}{12\left(\frac{1}{3}\right) - 12\left(\frac{3}{4}\right)}$$

Simplifying, we obtain

$$m = \frac{10 - 84}{4 - 9} = \frac{-74}{-5} = \frac{74}{5}$$

Note that, although this result is an improper fraction, we had said that the convention was to leave such fractions alone rather than convert to mixed-number form, and consequently this is our result for the slope.

EXAMPLE 4.17 Find the slope of the line through (9, 5) and $(-2, 5)$.

SOLUTION

Identifying $(x_1, y_1) = (9, 5)$ and $(x_2, y_2) = (-2, 5)$, we find

$$m = \frac{5 - 5}{-2 - 9} = \frac{0}{-11}$$

A fraction whose numerator is 0 is equal to 0, and therefore the slope of this line is 0. If we were to sketch the line (plot the two points and draw a straight line through them), we would see that the line is horizontal, and our earlier description of a horizontal line as having slope 0 is further supported.

How might the concept of a line's slope find its way into our applications? Possibly the best way to appreciate this is to recall the "ratio" interpretation of a fraction. A fraction containing measurement of units in its numerator and denominator typically uses the notion of "per" that we derive from such a ratio. For instance, we can think of the quantity "60 miles per hour" as the ratio $\frac{60\ miles}{1\ hour}$. The fraction bar, in this context, is viewed as being synonymous with the word "per."

Consider the following illustration. It is found, through a study of homes in the United States, that the number of satellite dish receivers in the year 2001 was 5.9 million and that that total had increased linearly to 22.3 million in 2007. If we were to graph the growth of the number of satellite dishes in the United States over time, we could put the year on the horizontal axis and the number of dishes on the vertical axis. The two data values described would then correspond to two points in the plane: (2001, 5.9) and (2007, 22.3).

What is the slope of that line, and what does it mean? Using our definition of slope, we can calculate

$$m = \frac{22.3 - 5.9}{2007 - 2001} = \frac{16.4}{6}$$

For applications, we often allow ourselves to obtain a decimal approximation, so (dividing numerator by denominator) we find the approximate value of m to be 2.73.

Now, what is the meaning of that value? The numerator, recall, is the number of satellite dishes in the United States, in millions, and the denominator is the year. The slope was found to be positive, so the line is rising from left to right, and its value tells us that the number of satellite dishes in the United States is rising at the rate of 2.73 million satellite dishes *per* year.

EXAMPLE 4.18

According to a particular government survey, in 2001 there were 3.54 million persons employed in the information technology field. While that number fell briefly, it has essentially followed a linear pattern of growth and rose to 3.84 million in 2006. What is the slope of the line depicting the growth of employment in the information technology field, and what is the interpretation of that value of the slope?

SOLUTION

In this case, we have a situation similar to the previous example, and we can again use the horizontal axis to represent time, while in this case the vertical axis represents the number of folks employed in the information technology field. With this identification of axes, our ordered pairs will take the form (year, number of workers in millions). The data indicate that two known points on the graph would be (2001, 3.54) and (2006, 3.84).

Using our defining formula for slope, we find

$$m = \frac{y_2 - y_1}{x_2 - x_1} = \frac{3.84 - 3.54}{2006 - 2001} = \frac{0.3}{5} = 0.06$$

This suggests that the number of information technology professionals was rising (note the slope is positive!) at the rate of 0.06 million workers per year between 2001 and 2006, or at the rate of 60,000 workers per year.

The slope of a line has a particular use we should mention prior to concluding our discussion. When two lines are being compared, it is sometimes of interest to us to determine if the lines are either parallel or perpendicular. When considering the lines visually, we must be careful to avoid a hasty conclusion about lines being parallel or perpendicular because lines might appear to have one of those relationships when they are not quite parallel or not quite perpendicular. Use of the slope will help us to avoid a risky conjecture based on appearance.

RULE: Two lines are parallel precisely when their slopes are identical. Two lines are perpendicular precisely when one of two situations exists: first, one line is vertical and the other horizontal, or, second, the product of the slopes of the lines is -1.

EXAMPLE 4.19

Show that the line passing through the points $(-3, 1)$ and $(0, 2)$ is perpendicular to the line passing through the points $(2, 3)$ and $(3, 0)$.

SOLUTION

The slope of the first line, found using the slope formula, is $= \dfrac{2-1}{0-(-3)} = \dfrac{1}{3}$. The slope of the second line, found in the identical way, is $m = \dfrac{0-3}{3-2} = -\dfrac{3}{1} = -3$.

Computation of the product of these slopes yields $(1/3)(-3) = -1$, and thus the two lines are perpendicular.

Exercises

Answer each of the following, using complete sentences, proper grammar, and spelling.

1. When we use the phrase "rise over run" as a means of determining slope, what is meant by the "rise," and what is meant by the "run?"
2. What term is used to describe the slope of a vertical line, and why do we use that term?
3. How would you be able to visually determine the sign of the slope of a nonvertical, nonhorizontal line?

Sketch the line passing through the given pair of points and then use the rise-over-run method to find the slope of the line.

4. $(1, 4)$ and $(3, -2)$
5. $(-4, 6)$ and $(5, 1)$
6. $(2, 7)$ and $(-3, 0)$
7. $(-2, 1)$ and $(4, 3)$
8. $(-1, 9)$ and $(5, 4)$
9. $(7, -3)$ and $(-5, 4)$
10. $(4, 6)$ and $(4, -3)$
11. $(2, 1/2)$ and $(2, -3)$

Each of the following problems refers back to an earlier problem. Using the formula for slope, calculate the slope and show that your results are the same as those computed using the rise-over-run method.

12. Exercise 4
13. Exercise 5
14. Exercise 6
15. Exercise 7
16. Exercise 8
17. Exercise 9
18. Exercise 10
19. Exercise 11

In the following exercises, find the slope of the line through the given points, using the method of your choice.

20. $(1/4, 5)$ and $(2/5, 9)$
21. $(0.3, 1.8)$ and $(1, 2.6)$
22. $(0, 5)$ and $(2.1, 4)$ (round your answer to the nearest thousandth)
23. $(9, 6)$ and $(4, -1)$

Answer the following, using complete sentences and offering supporting evidence.

24. Suppose it is found that there were 550 workers in a manufacturing company in 1997 and that there were 610 workers at the same company in 2009. If we assume that the company grows at a steady (linear) pace, what is the rate of change of the level of employment at this company? Can you use that result to extrapolate the projected employment at the company in the year 2015?

25. A study is made on the relationship between hours of study and performance on a particular mathematics examination. It is found that a student studying 0 hours will achieve a score of 43 points on the examination, while a student studying 4 hours will achieve a score of 62 on the exam. Placing the time studied on the horizontal axis and the score on the vertical axis, plot points corresponding to the two known data values. Compute the slope of the line passing through those two points and interpret its value in this context. Can you use the slope found to predict the score of a person who studies 8 hours for the examination? If so, make the prediction. If not, explain why not.

In the following problems, two lines are described using a pair of points on each line. Compute the slope of each line and use the value of the slopes to determine if the lines are parallel, perpendicular, or neither.

26. Line 1 passes through $(1, 2)$ and $(5, -2)$, line 2 through $(3, 4)$ and $(7, 8)$.
27. Line 1 passes through $(0, 3)$ and $(4, 0)$, line 2 through $(0, 1)$ and $(2, 3)$.
28. Line 1 passes through $(0, -3)$ and $(3, 1)$, line 2 through $(-3, -4)$ and $(0, 0)$.
29. Line 1 passes through $(-2, 0)$ and $(0, 2)$, line 2 through $(2, 3)$ and $(5, 0)$.

4.3 The Equation of a Straight Line

Undoubtedly, by this time we have developed a fairly good understanding of the notion of the slope of a straight line. We've also seen that we can interpret the slope in applications and that the interpretation, essentially, tells us the rate at which a quantity is changing (recall the concept of "per").

Every line in the plane can be represented by a particular type of equation, referred to as a **linear equation**. The equation will have the property that it will contain *x*, *y*, or both, with each letter (when occurring) appearing to the power 1 only, and be such that it can be rewritten in the **general form** $Ax + By + C = 0$. At first glance, it may not be clear why such equations can be associated with straight-line graphs in the plane, but this relationship will be made clear shortly.

Prior to exploring the various forms a linear equation can take, we should take a moment to clarify the need for the previously mentioned general form. It will turn out that the same line can be represented by infinitely many expressions, each of which is equivalent to the other. When we say "equivalent," we mean that the equation can be rewritten using the rules of algebra into another form that may appear different from the form with which we started.

With this distressing quantity of all possible equations representing the same line, it is necessary to arbitrarily assign one form to serve as the general form, that form on which we can all agree shall serve as a reference point in our calculations. If two individuals have obtained distinct equations purporting to represent the same line, the two equations can be rewritten, using algebra, in general form and compared. If each equation has the same general form, the expressions were equivalent and truly do represent the same line.

Note

Do not confuse the term "nonnegative" with the similar term "positive." The use of the term "nonnegative" in reference to the value of *A* indicates that *A* could be positive but could also be 0!

RULE: The general form of the equation of a line is $Ax + By + C = 0$, where A, B, and C are integers having no common factors, with A nonnegative and at least one of A or B nonzero.

Note

The definition of "general form" is not universal from textbook to textbook! There are situations in which $Ax + By = C$ is designated as the general form, and this is fine as long as uniqueness within a particular context is assured!

This particular form of the equation of a line is fairly restrictive and serves as a point of agreement where equations in different forms can be compared for equivalence without risk of ambiguity.

Now that we've introduced a general form, it is natural to wonder what other named forms of linear equations exist. In turn, we will examine various types, including the slope-intercept form, the point-slope form, the two-point form, and the two-intercept form. You might suspect that one form would be *the* preferred form, but you would be mistaken. Each of the various forms has a particular usefulness, and having knowledge of all of them will make our examination of particular types of problems easier. Remember that it is always good to have a variety of tools at our disposal because some methods might be more effective than others for specific exercises. Were we to acknowledge the existence of only one form of the equation of a line, we would be severely limiting the efficiency of our work!

Now, let's return to the general form we defined and pose the following question. Why do equations of the form $Ax + By + C = 0$ represent straight lines? The graph of any equation consists of the set of all points (x, y) in the plane whose coordinates satisfy the equation. That is, substitution of the coordinate values into the equation creates a true equality. We can produce the graph of the equation by constructing a table of values, plotting the points obtained from that table, and sketching a graph that follows the perceived pattern demonstrated by those points.

Let's do that now, for the example $3x - 5y = 25$. Choosing some values for x, such as 0, 5, 10, and 15, we can produce Table 4.1.

TABLE 4.1 Table of values

x	y
0	−5
5	−2
10	1
15	4

The values of y are found by inserting our arbitrarily chosen values of x into the equation and solving for y. That is, when $x = 0$, we insert that value into the equation and perform the following computation:

$$\begin{aligned} 3(0) - 5y &= 25 \\ 0 - 5y &= 25 \\ -5y &= 25 \\ y &= -5 \end{aligned}$$

A similar computation is done to find every entry in the y-column of the table.

Now, suppose that those table entries were used to create a set of points, $(0, -5)$, $(5, -2)$, $(10, 1)$, and $(15, 4)$. If those are plotted on the plane, we notice that the points are aligned in a straight path, and the implication is that the set of all points satisfying the equation would fall on that straight line (Figure 4.21). We could gain further evidence of this if we were to increase the number of entries in the table.

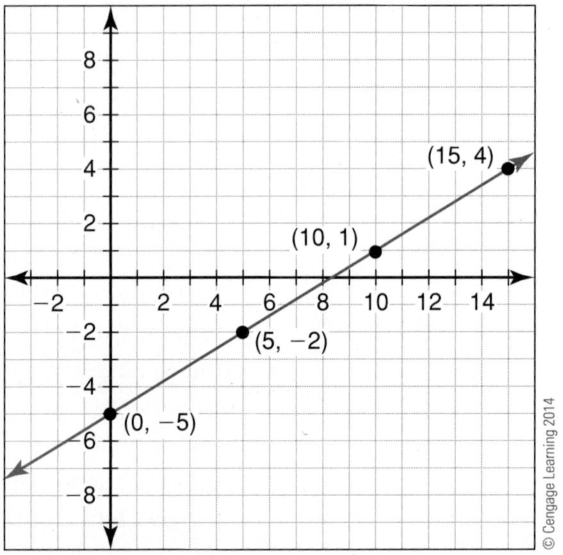

FIGURE 4.21 Graph of equation $3x - 5y = 25$

EXAMPLE 4.20 Create a table of values for and sketch the graph of the equation $2x + 3y - 6 = 0$.

SOLUTION

We will arbitrarily choose the following values for x: 0, 3, 6, 9, and 12. Then we will construct a table of values (Table 4.2).

TABLE 4.2 Table of values

x	y
0	2
3	0
6	−2
9	−4
12	−6

As before, the values of y are found by inserting values of x into the equation and solving for y. For instance, when $x = 0$, we obtain

$$2(0) + 3y - 6 = 0$$
$$3y = 6$$
$$y = 2$$

A similar calculation can be done for each value of x, producing the shown table values. If the entries of the table are converted to the points (0, 2), (3, 0), (6, −2), (9, −4), and (12, −6) and those points are plotted on the plane, we see that the points follow a linear path and draw a straight line passing through the points (Figure 4.22).

While the general form of a line's equation provides a useful reference point at which various forms of a line's equation can be compared, it turns out to not be a particularly useful form for the equation of a line in other situations. We

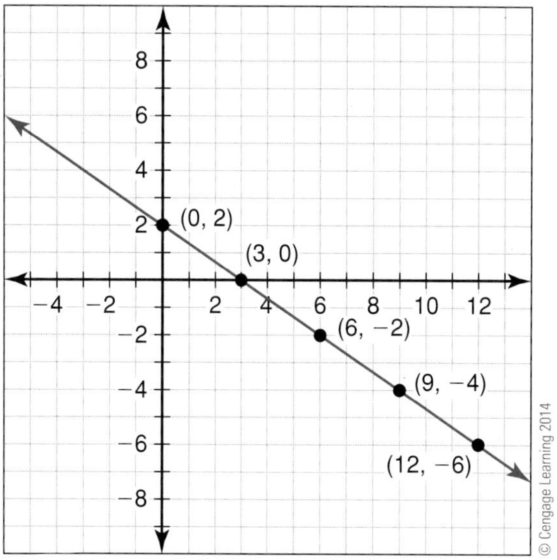

FIGURE 4.22 Graph of $2x + 3y - 6 = 0$

will now examine some of the other forms the equation of a line can take and describe when they might prove useful to have in hand.

Slope-Intercept Form of a Linear Equation (also called $y = mx + b$ form)

When a linear equation is arranged in such a manner that the variable y is isolated, we say that the equation is in **slope-intercept form**, or $y = mx + b$ **form**. Examples of equations in this form would be

$$y = 3x + 2$$
$$y = \frac{2}{3}x - 4$$
$$y = 5$$

When the equation is given in this form, two useful pieces of information are immediately available to us: the slope of the line and the y-intercept of the line. The coefficient of x is the slope of the line, while the constant term identifies the location of the y-intercept.

This is the motivation for referring to the form as $y = mx + b$ because the coefficient of x is the value of m (recall that m was the letter chosen to always represent slope!) and the constant term, b, tells us that the y-intercept point is $(0, b)$. In the case where the x-term is missing, we can visualize it as being a suppressed $0x$, and in the case where the plus sign is a minus sign, we can associate the minus sign with the value of b.

EXAMPLE 4.21

Find the slope and the y-intercept of the lines represented by the following equations:

$$y = 3x + 2$$
$$y = \frac{2}{3}x - 4$$
$$y = 5$$

SOLUTION

In the first equation, we can clearly see the form $y = mx + b$ present and thus can read off the value of $m = 3$ and $b = 2$. This tells us the slope is 3 and the y-intercept the point (0, 2). Remember, although it is common to speak informally about the y-intercept being 2, the precise answer is that the y-intercept is the point (0, 2).

For the second equation,

$$y = \frac{2}{3}x - 4$$

the coefficient of x is 2/3, so that is the slope, and the constant term is −4, which tells us that the y-intercept is the point (0, −4). Note that the minus sign is "attached" to the 4.

In the final example, $y = 5$, the x-term is missing, so we picture the equation as being equivalent to $y = 0x + 5$ and conclude that the slope is 0 and the y-intercept the point (0, 5). Note that we remarked earlier that lines with slope 0 are horizontal, and therefore the graph of this particular equation would be a horizontal line through the point (0, 5).

Note

The slope-intercept form of the equation of a line is generally the preferred form from which to graph the line!

Slope-intercept form is a very convenient form of the equation of a line to have in our possession. It is, as was demonstrated in the example, a form from which the slope and the y-intercept can be readily obtained. It is also an efficient form to use when graphing the line represented by the equation.

To graph a line from slope-intercept form, one point (the y-intercept) is automatically known to us. To find a second point (which is all we truly need in order to graph a line), we have the option of using rise over run to find a second point or of choosing a convenient value of x to substitute in the equation to find a second point.

Note

The use of the term "convenient" is, admittedly, vague. You can actually use any values of x you wish, but thoughtless choice of x-values can lead to complicated calculations. Good choices for x that typically work well are 0 and the denominator of the value of m (recalling that whole numbers can be considered fractions with denominator 1!).

EXAMPLE 4.22

Graph the line given by $y = \frac{2}{5}x - 4$, using the rise-over-run technique.

SOLUTION

By inspection, we can see the y-intercept of the line is the point (0, −4). We plot that point, and then note the slope is 2/5, and hence our rise is 2 and our run 5. Starting at the point (0, −4), we rise 2 units upward and then move five points to the right to reach the points (5, −2). Drawing a line through those points, we have the graph of the line, as shown in Figure 4.23.

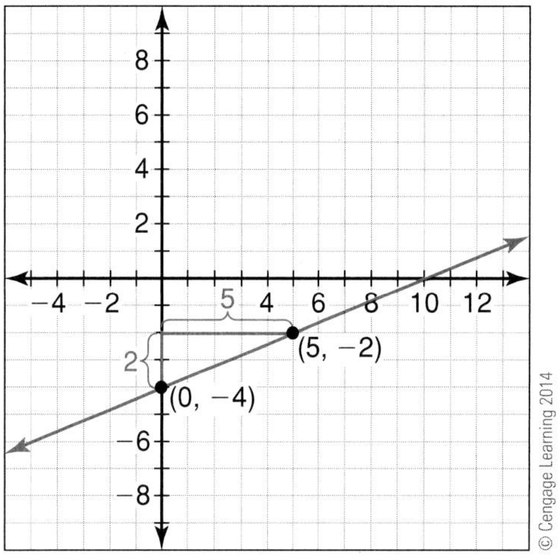

FIGURE 4.23 Graph of $y = (2/5)x - 4$

EXAMPLE 4.23

Graph the line given by $y = \frac{-1}{3}x + 6$ by finding the y-intercept and one additional point.

SOLUTION

Again, we easily see the y-intercept is the point (0, 6). To find a second point, we will choose a convenient value of x. One could choose any value of x, in principle, but we want to choose a value that will make our computations as straightforward as possible. Noting that our inserted value of x will have to be multiplied by $-1/3$ when we attempt to simplify, we would be wise to choose a value of x that will simplify nicely with that fraction. A good choice would be to choose the value of x to be the denominator of the slope. Thus, choosing $x = 3$, we find the value of y by computing

$$y = \frac{-1}{3}(3) + 6$$
$$y = -1 + 6$$
$$y = 5$$

The second point on the line is then (3, 5), which we can plot on the same plane with (0, 6) and then draw the line passing through the two points, as in Figure 4.24.

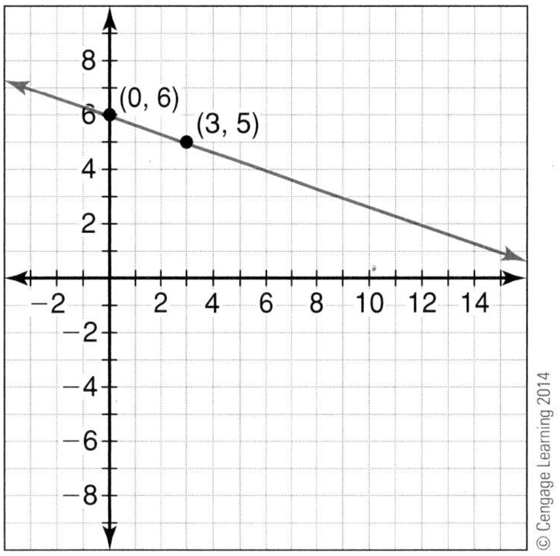

FIGURE 4.24 Graph of $y = -(1/3)\ x + 6$

EXAMPLE 4.24

What is the general form of the equation of the line given by $y = \frac{-1}{3}x + 6$?

SOLUTION

Recall that the general form of the equation of a line is $Ax + By + C = 0$, where A, B, and C are integers with A nonnegative and at least one of A or B being nonzero. Noticing that we have a noninteger coefficient of x in this equation, our first step is to clear that fraction by multiplying all terms by 3. This produces the equivalent equation $3y = -x + 18$.

We can reach general form now by moving all terms to the left of the equals sign, obtaining

$$x + 3y - 18 = 0$$

This is the general form of the equation of the line.

We have now shown, in the last example that the equation $y = \frac{-1}{3}x + 6$ is equivalent to the equation $x + 3y - 18 = 0$. We stated earlier that every line has infinitely many equivalent forms, and we can demonstrate that here. If the general form is multiplied by any nonzero constant, we obtain an equivalent equation. Since we have an infinite number of choices for that constant, we can produce infinitely many equivalent equations.

EXAMPLE 4.25

Produce the slope-intercept form of the equation of the graph shown in Figure 4.25.

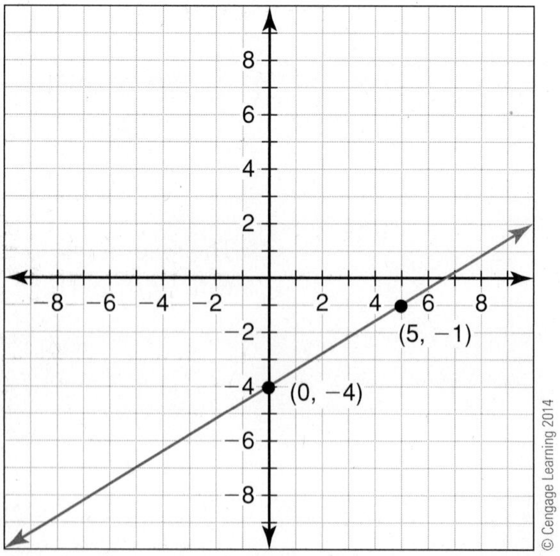

FIGURE 4.25

SOLUTION

To find the equation of the line, we need two bits of information: the y-intercept of the line and the slope of the line. By inspection, we can see the y-intercept of the line appears to be $(0, -4)$. Observing that the point $(5, -1)$ is a second point on the line and using the rise-over-run interpretation of slope, we can determine that to move from $(0, -4)$ to $(5, -1)$ we must rise 3 units vertically and run 5 units horizontally. Thus, the slope is 3/5, and the equation of the line is $y = \frac{3}{5}x - 4$.

The slope-intercept form of the equation of a line is, as we mentioned, a very convenient form to have if you intend to graph a line. The general form is useful as a method for comparing two disparate forms of a line's equation, to confirm that the equations do, in fact, represent the same line. There are other forms of the equation of a line, which prove useful in other circumstances.

The Point-Slope Form of the Equation of a Line

Suppose we want to know the equation of a line given that we know information about the line. If we were extraordinarily fortunate and happened to know the slope and the y-intercept point, then it would be a triviality to produce the slope-intercept form of the equation, from which we could use algebra to obtain any other form of the equation. Naturally, we are rarely that fortunate, so we must prepare ourselves for other situations.

If we knew two points on the line, we could always use them and either the rise-over-run interpretation or the defining formula for the slope to calculate m, and therefore we may as well assume that we know the slope and some arbitrary point on the line, which we can call (x_1, y_1). Is it possible to find the equation of the line, given that information?

The answer to that question is yes, and it leads us to what we call the **point-slope form of the equation of a line.** The equation form is derived from the defining formula for slope, as we'll show. You should be aware that the derivation of the equation form is, in itself, not terribly important, but you never want to accept a rule without supporting evidence, unless you have no alternative!

Recall that the formula for slope is $m = \frac{y_2 - y_1}{x_2 - x_1}$. Suppose that (for mysterious reasons to be made clear in a moment) we decided to clear the fraction in that formula. It would be easy enough to accomplish: multiply both sides of the equality by $(x_2 - x_1)$, and we would obtain

$$m(x_2 - x_1) = y_2 - y_1$$

By the symmetric property of equality, we can reverse the order of the equation and obtain

$$y_2 - y_1 = m(x_2 - x_1)$$

Now, suppose we make the second point, (x_2, y_2), even more generic, and simply call it (x, y). Then the equation becomes

$$y - y_1 = m(x - x_1)$$

This is the point-slope form of the equation of a line. If we know the value of m, the slope, and the coordinates of a particular point on the line, (x_1, y_1), we can substitute into this equation (allowing x and y to remain variables) and obtain an equation for the line, which we can then rewrite in whatever form we wish.

Note

The point-slope form of the equation of a line is $y - y_1 = m(x - x_1)$, and the nonsubscripted x and y *remain variables!* Also, this form is sometimes referred to as the "point-slope formula."

EXAMPLE 4.26

A line passes through the point (7, 5), and has slope 4. Find its equation using the point-slope form of the equation of a line and then rewrite that expression in general form.

SOLUTION

Substituting 7 for x_1, 5 for y_1, and 4 for m, we obtain

$$y - 5 = 4(x - 7)$$

This is the point-slope form of the line's equation. To get to the general form, we simplify and move all terms to the left side of the equation:

$y - 5 = 4x - 28$ (by the distributive law)

$-4x + y + 23 = 0$ (moving all terms to the left of the equals sign)

$4x - y - 23 = 0$ (multiplying all terms by -1 to satisfy the criterion from the definition of general form that the coefficient of x be nonnegative)

The Two-Point Form of the Equation of a Line

In our presentation of the point-slope form of the equation of a line, we began by stating that if two points on the line were known, they could be used to find the slope, and then one of the points could be disregarded to obtain the equation of the line in point-slope form. The slope, recall, would be found using the defining formula for slope, which was

$$m = \frac{y_2 - y_1}{x_2 - x_1}$$

Suppose that, in the point-slope formula, we explicitly showed the computation of the slope using this formula? That is, if we replaced the m in the form

$$y - y_1 = m(x - x_1)$$

by the formula for slope, we would obtain

$$y - y_1 = \left(\frac{y_2 - y_1}{x_2 - x_1}\right)(x - x_1)$$

This is called the **two-point form** of the equation of a line. It is really nothing more than the point-slope form, with the formula for the calculation of slope explicitly shown.

EXAMPLE 4.27

Find the two-point form of the equation of the line through (3, 2) and $(-1, 7)$. Simplify the result to the general form.

SOLUTION

We can identify the given points as $(x_1, y_1) = (3, 2)$ and $(x_2, y_2) = (-1, 7)$ and substitute into the formula

$$y - y_1 = \frac{y_2 - y_1}{x_2 - x_1}(x - x_1)$$

$$y-2=\frac{7-2}{-1-3}(x-3)$$
$$y-2=\frac{5}{-4}(x-3)$$
$$y-2=-\frac{5}{4}(x-3)$$

This, observe, is the point-slope form, which we would have obtained had we worked out the slope initially. We were asked to simplify to the general form, so we will continue:

$$y-2=-\frac{5}{4}x+\frac{15}{4}$$
$$4y-8=-5x+15$$
$$5x+4y-23=0$$

This is the equation in general form.

The Two-Intercept Form of the Equation of a Line

We know that there are two types of intercepts a line *can* have: x-intercepts and y-intercepts. We also know that for x-intercepts, the y-coordinate must be 0, while the reverse is true for y-intercepts.

Suppose the x-intercept of line $Ax + By + C = 0$ was $(a, 0)$ for some nonzero constant value a. What would this imply? By substitution of the value $x = a$ and $y = 0$, we see

$$A(a) + B(0) + C = 0$$
$$A(a) + C = 0$$
$$A(a) = -C$$
$$A = \frac{-c}{a}$$

By a similar computation, assuming the y-intercept is $(0, b)$ for some nonzero constant b, we can show that in $Ax + By + C = 0$, it must be that $B = \frac{-c}{b}$.

Therefore, the general form of the equation of the line must be $-\frac{C}{a}x-\frac{C}{b}y+C=0$.

We can divide all terms by $-C$ and move the constant term to the right side of the equation to produce $\frac{x}{a}+\frac{y}{b}=1$.

This may seem to be a lot of effort for not much gain, but notice that in this form (called the **two-intercept form of the equation of the line**), the denominators of the x- and y-terms are precisely the locations of the respective intercepts! An example will illustrate the convenience of this occurrence.

EXAMPLE 4.28 A line is known to have x-intercept (6, 0) and y-intercept (0, 5). Find its two-intercept equation form and simplify the result to the general form.

SOLUTION

Here $a = 6$ and $b = 5$, so direct substitution into the two-intercept form of the equation produces

$$\frac{x}{6} + \frac{y}{5} = 1$$

We clear the fractions by multiplying through by 30, the least common denominator, and find

$$5x + 6y = 30$$

which simplifies to the general form

$$5x + 6y - 30 = 0$$

Parallel and Perpendicular Lines Revisited

We pointed out earlier that two lines are parallel precisely when the slopes of those lines are identical and that two lines are perpendicular when the product of their slopes is -1. With these thoughts in mind and now armed with the knowledge that we can express linear equations in slope-intercept form, we have another means of determining whether lines are parallel, perpendicular, or neither.

EXAMPLE 4.29 What is the relationship (parallel, perpendicular, or neither) of the lines given by the following?

$$y = 3x - 5 \text{ and } y = \frac{1}{3}x + 1$$

SOLUTION

The slope of the first line is 3, while for the second line it is 1/3. The slopes are not identical, and thus the lines are not parallel. The product of the two slopes is 1, and thus the lines are not perpendicular. In this case, the lines are neither parallel nor perpendicular.

EXAMPLE 4.30 What is the relationship (parallel, perpendicular, or neither) of the lines given by the following?

$$5x - 2y = 10 \text{ and } y = -\frac{2}{5}x + 8$$

SOLUTION

The slope of the second line is clearly $-2/5$, but the slope of the first line is not immediately obvious. We rewrite that equation in slope-intercept form so the slope is apparent by isolating y:

$$5x - 2y = 10$$
$$-2y = -5x + 10$$
$$y = \frac{5}{2}x - 5$$

The slope of this line is 5/2, which is clearly not identical to the first line's slope, so the lines are not parallel. The product of the slopes is $\left(\frac{-2}{5}\right)\left(\frac{5}{2}\right)=-1$, and hence the lines are perpendicular.

Note

In many situations, more than one of the forms of a line's equation may prove useful. It is a matter of practice and experience to determine which form would be best to use in a given problem, but keep in mind that *any* of the forms will, ultimately, lead you to your answer, since they are all equivalent!

We've now seen a series of forms of the equation of a straight line. Each of these forms has its own particular usefulness and could be seen as a "preferable" form under the right conditions.

Let's summarize the different forms of the equation of a line and recall the utility of each:

- General form: $Ax + By + C = 0$, where A, B, and C are integers with no common factors and A is nonnegative. This form is useful as a tool for comparison of disparate but equivalent forms of the equation of a line.
- Slope-intercept form: $y = mx + b$. This form is easy from which to graph or to extract the value of the slope or the y-intercept by inspection.
- Point-slope form: $y - y_1 = m(x - x_1)$. This form is useful for obtaining the equation of a line, if one point on the line and the slope are known.
- Two-point form: $y-y_1=\frac{y_2-y_1}{x_2-x_1}(x-x_1)$. This form is useful for obtaining the equation of a line if two points on the line are known, actually the point-slope form of the equation of the line, with the formula for slope explicitly displayed.
- Two-intercept form: $\frac{x}{a}+\frac{y}{b}=1$. This form is useful as a bridge to obtain the equation of a line, if both of the intercepts are known.

Exercises

Answer the following questions, using complete sentences, proper grammar and spelling.

1. What is meant by the "general form" of the equation of a line?
2. What is the usefulness of the general form?
3. What is meant by the "intercepts" of the graph of a linear equation?
4. Is it necessary that a line have both an x-intercept and a y-intercept?

For the following linear equations, construct a five-entry table of values and sketch the graph of the line.

5. $4x + y = -2$
6. $3x - 6y + 18 = 0$
7. $x + 4y - 9 = 3$
8. $y = 3x - 5$
9. $2y - 8x + 6 = 0$
10. $4y + 2x - 10 = 0$

For the following linear equations, convert the equations to slope-intercept form. In each case, identify the slope of the line and the coordinates of the y-intercept.

11. $3x - 2y = 5$
12. $5x - 3y = 12$

13. $\frac{1}{2}x - 4y + 3 = 0$

14. $\frac{2}{7}x - 3y + 2 = 0$

15. $3x + 7y = 4$

16. $5x - 9y = 18$

Graph the following linear equations by plotting the *y*-intercept and finding an additional point using the rise-over-run interpretation of slope.

17. $y = \frac{3}{5}x - 4$

18. $y = 4x - 2$

19. $y = \frac{-2}{3}x + 4$

20. $y = \frac{-4}{5}x + 9$

21. $y = 5 - 3x$

22. $y = 7 - \frac{1}{6}x$

Find the equations (in general form) represented by the graphs shown.

23.

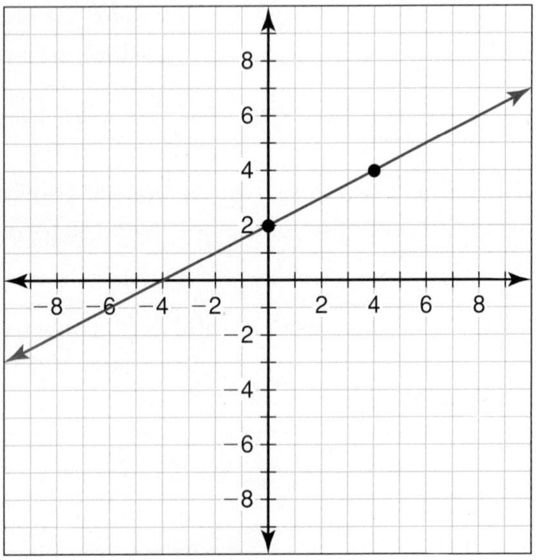

24.

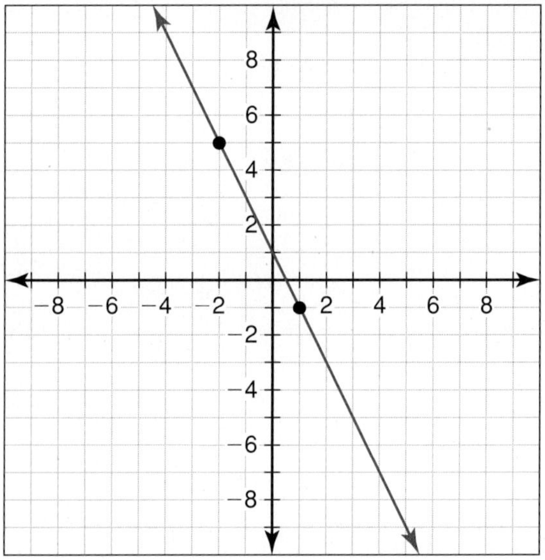

25.

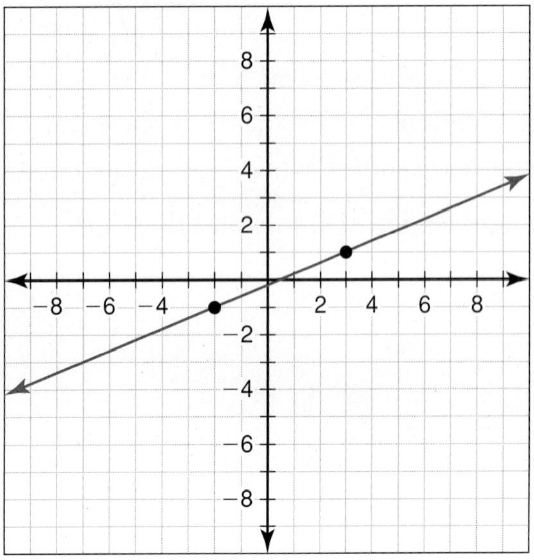

26.

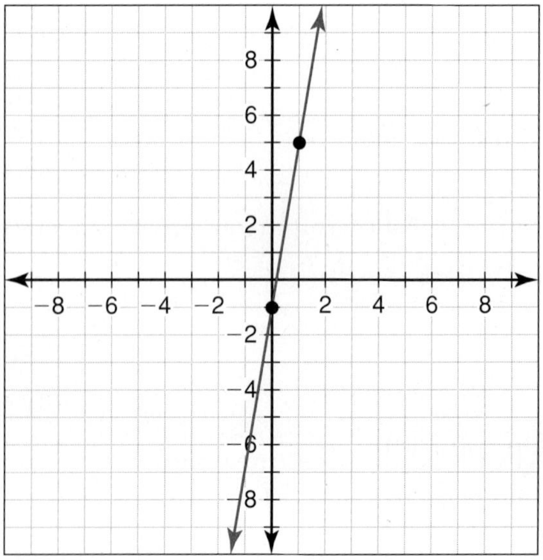

27.

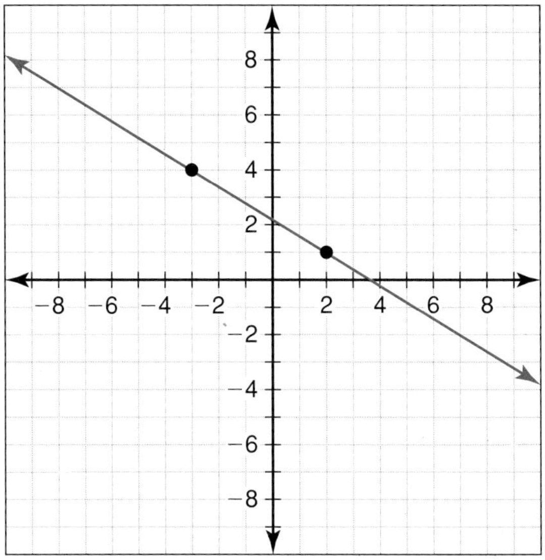

28.

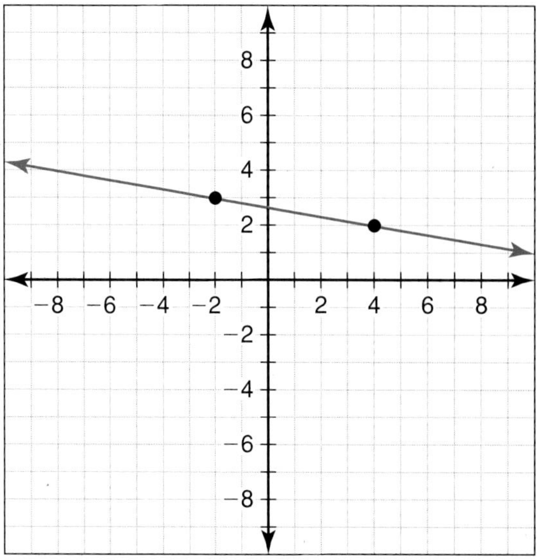

In the following exercises, use the given point and slope to find the equation of the line in general form.

29. Point: (3, 5), $m = 2/3$
30. Point: (−1, 4), $m = 5/7$
31. Point: (−3, 5), $m = 4$
32. Point: (2, 1), $m = 3$
33. Point: (−7, −4), $m = 2/9$
34. Point: (−5, −6), $m = 7/4$

Find the two-point form of the equation of the line passing through the given pair of points. Use the first point given as (x_1, y_1).

35. (4, 5) and (−2, 6)
36. (1, −7) and (−4, 9)
37. (3, 3) and (5, −4)
38. (−2, −2) and (3, 2)
39. (5, −9) and (−3, 4)
40. (0, 1) and (7, −3)

Find the two-intercept form of the equation of the line having the indicated intercepts.

41. (0, 5) and (3, 0)
42. (0, 2) and (−5, 0)
43. (1/2, 0) and (0, 2/3)
44. (3/5, 0) and (0, 2/7)
45. (1, 0) and (0, 7)
46. (11, 0) and (0, −8)

In the following exercises, an earlier exercise is referenced. Give the equation of the line in the specified exercise in general form.

47. Exercise 35
48. Exercise 36
49. Exercise 37
50. Exercise 38
51. Exercise 39
52. Exercise 40
53. Exercise 41
54. Exercise 42
55. Exercise 43
56. Exercise 44
57. Exercise 45
58. Exercise 46

Determine if the lines represented by the following equations are parallel, perpendicular, or neither.

59. $x - 3y = 2$ and $3x - 9y = 15$
60. $4x + 7y = 0$ and $2x + 3y = 1$

61. $2x - y = 5$ and $(-1/2)x - y = 3$
62. $6x - 2y = 10$ and $3y + x = 7$
63. $y - 2x = 0$ and $2y - 4x = 1$
64. $4y - x = 2$ and $(1/4)y - x = 3$

4.4 Solving a System of Linear Equations

We defined a linear equation to be an expression capable of being written in the form $Ax + By + C = 0$, where A, B, and C were integers having no common factors, such that A was nonnegative. Such equations, we saw, had graphs that were straight lines in the Cartesian plane. In the preceding section, we explored the various forms such an equation could take and mentioned the reason one might choose to work with one form over another.

Now we consider the situation where two linear equations are combined to form a **system**. An example of such a construction would be

$$\begin{aligned} 3x + 5y &= 26 \\ 2x - 3y &= -8 \end{aligned}$$

Note

Recall that two lines are parallel precisely when their slopes are *identical*.

If we were to graph the two equations on the same Cartesian plane, there are three possible results: first, the lines could intersect at a single point of the plane; second, the lines might not intersect at all; or third, the lines could coincide, in which case they intersect at every point on their mutual graph. In the first case, we say the system is **consistent** and give the solution in ordered pair form. In the second case, we say the system is **inconsistent** and that the system has no solution. In the final case, the system is called **consistent and dependent**, and the system has infinitely many solutions (which we will express in set-builder notation).

Our objective now is to find the solution(s) to a system, if any exist. There are several ways to accomplish this, the first of which is the method of graphing. There are also computational methods for solving systems, and we will examine these in detail when we reach Chapter 5.

The graphing method is sometimes referred to as a "brute-force" technique. When mathematicians refer to the use of brute force, they are suggesting that we will use the direct approach regardless of the potential for tedious calculations, difficulty, and risk.

The solution to a system of equations has been described as the intersection point or points (if any) of the graphs of the equations. Very well. We shall graph the two equations and make a visual survey to locate the apparent point(s) of intersection. Naturally, it is possible that the graphs may be difficult to obtain or that the point(s) of intersection will be challenging to identify. So be it.

Our system was

$$\begin{aligned} 3x + 5y &= 26 \\ 2x - 3y &= -8 \end{aligned}$$

An efficient method of graphing would be to transform the two equations into two-intercept form, from which we can graph the lines by plotting the intercepts and sketching the line through those points. Two-intercept form can be obtained by dividing all terms by the constant on the right side of the equation, as follows:

$$\frac{3x}{26}+\frac{5y}{26}=1$$

$$\frac{x}{26/3}+\frac{y}{26/5}=1$$

For the first equation, the graph has intercepts of (26/3, 0) and (0, 26/5). Similarly,

$$\frac{2x}{-8}+\frac{-3y}{-8}=1$$

which simplifies to

$$\frac{x}{-4}+\frac{y}{8/3}=1$$

and the intercepts of this line are (−4, 0) and $\left(0, \frac{8}{3}\right)$ The graphs of the two lines on a single Cartesian plane are shown in Figure 4.26, and by inspection it seems that (2, 4) is the solution to the system.

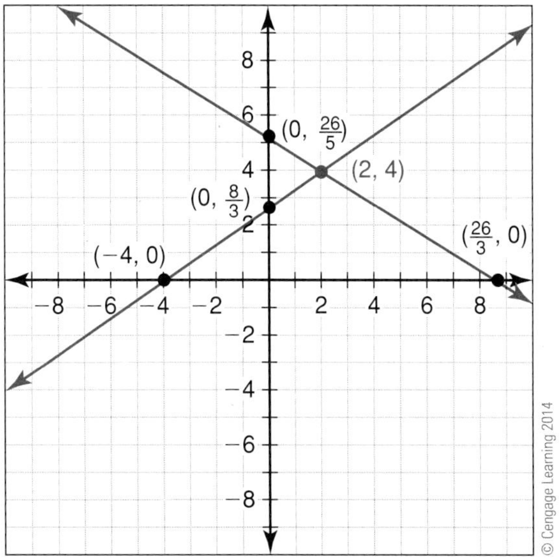

FIGURE 4.26 Graph of the System ________ and show the system.

We can check our solution by inserting the values $x = 2$ and $y = 4$ into both of the equations of the system and confirming that a true equality results for both substitutions:

$$3(2) + 5(4) = 26$$

$$6 + 20 = 26 \surd$$

$$2(2)-3(4) = -8$$

$$4-12 = -8 \surd$$

Since both substitutions produced true equalities, the point (2, 4) has been confirmed as the solution to the system.

EXAMPLE 4.31

Use the method of graphing to solve the following system of equations:

$$4x + 2y = 18$$
$$3x + 9y = 6$$

SOLUTION

An alternative to using the two-intercept method for graphing the equations is to convert to slope-intercept form and graph from there. To demonstrate, we'll do that in this example. The first equation is equivalent to $y = -2x + 9$ and the second to $y = -\frac{1}{3}x + \frac{2}{3}$. We can graph each of these equations by using the y-intercept as one point on the line and then choosing a convenient x-value to generate a second point.

For $y = -2x + 9$, the y-intercept is (0, 9), and a convenient x-value to use to find a second point would be $x = 1$, for which $y = -2(1) + 9 = 7$. The second point on the line is (1, 7).

For $y = -\frac{1}{3}(1) + \frac{2}{3}$, the y-intercept is $\left(0, \frac{2}{3}\right)$ and a convenient x-value to use to find a second point would again be $x = 1$, for which $y = -\frac{1}{3}(1) + \frac{2}{3} = \frac{1}{3}$. The second point on the line is (1, 1/3).

Graphing the two equations on a single Cartesian plane (Figure 4.27), we determine that the point (5, −1) is the solution to the system.

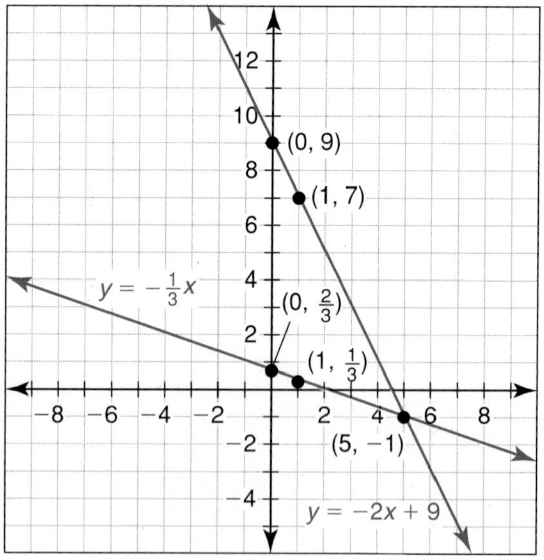

FIGURE 4.27 Graph of the system

EXAMPLE 4.32

Use the method of graphing to solve the system of equations:

$$4x - 5y = 20$$
$$8x - 10y = 30$$

SOLUTION

In this example, we will convert to the two-intercept forms of the equations by dividing the terms of each equation by the constant on the right side of the equation. For the first equation, we divide all terms by 20 and obtain $\frac{x}{5}+\frac{y}{-4}=1$. Note that we have incorporated the minus sign into the second term, since the two-intercept form specified that the terms on the left would be combined by addition. By inspection, we see that the intercepts of the graph are (5, 0) and (0, −4).

Dividing the terms of the second equation by 30, we find $\frac{4x}{15}+\frac{y}{-3}=1$, which is equivalent to $\frac{x}{15/4}+\frac{y}{-3}=1$, and therefore the intercepts of the graph of this equation are $\left(15/4, 0\right)$ and (0, −3).

If we graph the two lines on the Cartesian plane (Figure 4.28), we see that they appear to be parallel to one another, and consequently we suspect that the system does not have a solution. Recall our earlier admonition, however: it is difficult to know with certainty that two lines are parallel based on appearance! *We should compare the slopes of the two lines to ensure that our conclusion is correct.*

We rewrite the equations in slope-intercept form to test our conclusion:

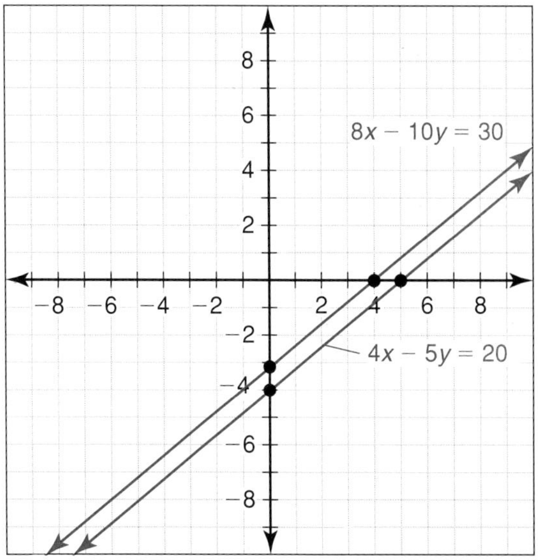

FIGURE 4.28 Graph of the system

$4x - 5y = 20$ is equivalent to $y=\frac{4}{5}-4$, while $8x - 10y = 30$ is equivalent to $y=\frac{4}{5}x-3$

The lines each have slope 4/5 and consequently are parallel. We can now safely conclude that the system is inconsistent and has no solution.

EXAMPLE 4.33 Use the method of graphing to solve the system of equations:

$$6x + 3y = 12$$
$$8x + 4y = 16$$

Note

Do not confuse the concept of having "infinitely many solutions" with having *all* points in the plane as solutions! Those are vastly different statements! The infinitely many points that *are* solutions happen to lie on the graph of the equations of the system but no others.

SOLUTION

By converting the equations to two-intercept form, we find that each can be simplified to $\frac{x}{2}+\frac{y}{4}=1$, and thus both lines have the same intercepts, (2, 0) and (0, 4), and the same graph (Figure 4.29). This indicates that the system is consistent and dependent and has infinitely many solutions.

We can express this in set-builder notation as $\{(x, y)|\ 6x + 3y = 12\}$, which expresses that the solution is the set of all points (x, y) such that the values of x and y satisfy the equation shown. We could have used the other equation in the system, if we wished, or any equivalent form of the equation, since all would contain the same set of points.

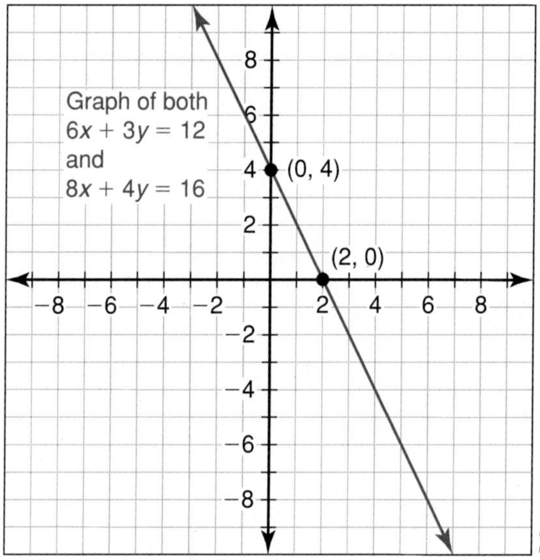

FIGURE 4.29 Graph of the inconsistent system

We mentioned that the graphing method of solution had inherent weaknesses. Primary among those was the tedium of producing the graphs and the potential ambiguity involved in the identification of the solution. The computational methods we will examine in the next chapter will be immune to those weaknesses but have the drawback of requiring that some choices be made and algebraic manipulations performed. The hazard created by having to make choices is that you may wind up choosing poorly, and this could lead to computations that are more involved than they might have been had other choices been made. Fear not, however: things will never become so complicated that even the most careless of choices will lead us to a situation from which we cannot algebraically extricate ourselves with the mathematical weapons we have at hand.

Exercises

Use the method of graphing to find the solution to the system of equations, if a solution exists.

1. $3x-4y=18$
   $5x-y=13$

2. $x-4y=14$
   $2x+3y=6$

3. $4x+2y=-2$
   $x-3y=-18$
4. $4x-y=16$
   $2x+5y=8$
5. $3x-6y=-15$
   $4x+y=-2$
6. $5x+3y=-18$
   $2x-6y=1$
7. $3x+2y=3$
   $6x+4y=12$
8. $x+y=0$
   $3x+2y=1$
9. $4x+2y=10$
   $x-y=1$
10. $2x-3y=5$
    $-4x+6y=1$

Summary

In this chapter, you learned about:

- The Cartesian plane: ordered pairs, domains, and ranges and the distance and midpoint formulas.
- Straight lines in the plane: slope and intercepts.
- Equations of a straight line: slope-intercept form, point-slope form, two-point form, and two-intercept form.
- Solving a system of linear equations graphically.

Glossary

abscissa: That element of an ordered pair corresponding to position relative to the horizontal axis of the plane.

Cartesian plane A two-dimensional construct allowing us to systematize location within a mathematical plane.

complex fraction: A fraction whose numerator, denominator, or both contains one or more fractions.

consistent: A system of equations having a finite number of solutions.

consistent and dependent: A system of equations having infinitely many solutions.

distance formula (distance between two points): Formula to calculate the length of the line segment connecting the point (x_1, y_1) to the point (x_2, y_2).

$$D=\sqrt{(y_2-y_1)^2+(x_2-x_1)^2}=\sqrt{|y_2-y_1|^2+|x_2-x_1|^2}$$

domain: The set of input values of a function for which the function is defined.

function: A rule of assignment that associates with each element of one set (the domain) exactly one element of a second set (the range).

general form of the equation of a line: $Ax + By + C = 0$, where A, B, and C are integers having no common factors, with A nonnegative and at least one of A or B nonzero.

inconsistent: A system of equations having no solution.

line segment: A connected portion of a straight line, having a specific starting and ending point.

linear equation: An equation whose variables appear to the power of 1 only, capable of being rewritten in the general form $Ax + By + C = 0$.

midpoint: That point lying in the exact center of a line segment connecting two specified points in the plane.

midpoint formula: The formula used to calculate the midpoint of the points (x_1, y_1) and (x_2, y_2) $\left(\frac{x_1+x_2}{2}, \frac{y_1+y_2}{2}\right)$.

ordered pairs: A coupled pair of numbers having the form (a, b), where a and b are real numbers. The ordered pair specifies the point in the plane at the intersection of the vertical line $x = a$ and the horizontal line $y = b$.

ordinate: That element of an ordered pair corresponding to position relative to the vertical axis of the plane.

origin: The intersection of the horizontal and vertical axes of the plane.

point-slope form of the equation of a line: The form of a line's equation given by $y - y_1 = m(x - x_1)$, where m is the line's slope and (x_1, y_1) is a point lying on the line.

quadrantal points: Points lying on either of the coordinate axes in the plane.

quadrants: The four regions into which the plane is subdivided by the horizontal and vertical axes.

ray: A half line having an initial point and infinite extension in one direction only.

relation: A rule of assignment that associates to each element of one set one or more members of a second set.

rise: The vertical change in position within the plane when moving along a straight-line path connecting one point to another.

rise-over-run method: A counting technique used to calculate the value of the slope of a line.

run: The horizontal change in position within the plane when moving along a straight-line path connecting one point to another.

scatter plot: A graph depicting with distinct dots the locations of distinct points within the plane.

slope: A numerical value quantifying how far a line is tilted from horizontal.

slope intercept form of the equation of a line: The form of a line's equation $y = mx + b$, where m is the slope of the line, and the point $(0, b)$ is the y-intercept of the line. Also called the "$y = mx + b$" form.

straight line: A path in the plane having no curvature, defined as a geometric primitive.

system: Two or more linear equations, considered as a group.

two-intercept form of the equation of a line: The form of a line's equation $\frac{x}{a} + \frac{y}{b} = 1$, where $(a, 0)$ and $(0, b)$ are the x- and y-intercepts of the line, respectively.

two-point form of the equation of a line: The form of a line's equation $y - y_1 = \left(\frac{y_2 - y_1}{x_2 - x_1}\right)(x - x_1)$, where (x_1, y_1) and (x_2, y_2) are points on the line.

x-axis: The horizontal axis in the Cartesian plane.

x-coordinate: The abscissa in an ordered pair in the Cartesian plane.

x-intercept: The point on the x-axis intersected by a nonhorizontal line.

y-axis: The vertical axis in the Cartesian plane.

y-coordinate: The ordinate in an ordered pair in the Cartesian plane.

y-intercept: The point on the y-axis intersected by a nonvertical line.

$y = mx + b$ form of the equation of a line: The form of a line's equation $y = mx + b$, where m is the slope of the line, and the point $(0, b)$ is the y-intercept of the line. Also called "slope-intercept" form.

List of Equations

Eqn 4.1 $D = \sqrt{(y_2 - y_1)^2 + (x_2 - x_1)^2}$ (distance formula)

Eqn 4.2 $Midpoint = \left(\frac{x_1 + x_2}{2}, \frac{y_1 + y_2}{2}\right)$ (midpoint formula)

Eqn 4.3 $m = \frac{y_2 - y_1}{x_2 - x_1}$ (formula for slope)

Eqn 4.4 $Ax + By + C = 0$ (general form of the equation of a line)

Eqn 4.5 $y = mx + b$ [(slope-intercept form of the equation of a line having slope m and y-intercept $(0, b)$)]

Eqn 4.6 $y - y_1 = m(x - x_1)$ (point-slope form of the equation of a line)

Eqn 4.7 $y - y_1 = \left(\dfrac{y_2 - y_1}{x_2 - x_1}\right)(x - x_1)$ (two-point form of the equation of a line)

Eqn 4.8 $\dfrac{x}{a} + \dfrac{y}{b} = 1$ [(two-intercept form of the equation of a line, where $(a, 0)$ and $(0, b)$ are the intercepts of the line)]

End-of-Chapter Problems

For each of the following, sketch a Cartesian plane and indicate the following features on it.

1. Quadrants II and III
2. The points $(3, 4)$, $(-3, 4)$, $(5, -4)$, and $(-2, -4)$
3. The negative x-axis

For each of the following, determine if the set of ordered pairs represents a function. Briefly explain your answer. Whether or not it represents a function, give the domain and range of the set of ordered pairs.

4. $\{(1, 5), (3, -2), (-2, 8), (4, 3)\}$
5. $\{(1, 4), (4, 1), (1, 5), (-4, 3)\}$
6. $\{(2, 5), (3, 1/2), (5, 7), (-4, 2)\}$

For each of the following, conditions describing a set of points are given. Identify the quadrant(s) in which the points are located so that the conditions are satisfied.

7. $x > 0$ and $y < 0$
8. $y = -5$
9. $x < -2$ and $y > 0$

Find the distance separating the given pairs of points. Give the exact value and a decimal approximation accurate to three decimal places, as needed.

10. $(4, 3)$ and $(-4, 2)$
11. $(13, 4)$ and $(15, 8)$
12. $(1, 8)$ and $(7, 13)$

Find the midpoint of the line segment connecting the two given points.

13. $(3, 5)$ and $(-3, 2)$
14. $(4, 11)$ and $(7, 11)$
15. $(-2, 4)$ and $(3, 8)$
16. Use the midpoint formula twice to find the point that lies three-quarters of the distance along the straight line connecting the point $(3, 4)$ to the point $(8, 12)$.

Sketch the line passing through the given pair of points and then use the rise-over-run method to find the slope of the line.

17. (5, 4) and (−1, −1)
18. (7, 8) and (3, 2)
19. (−2, 2) and (4, −4)

Using the formula for slope, calculate the slope of the line passing through the given pair of points.

20. (3, 8) and (12, −4)
21. (0, 0) and (8, 4)
22. (2, 5) and (8, 3)

In the following exercises, find the slope of the line through the given points, using the method of your choice.

23. (4.2, 8) and (−2.1, 6) (round your answer to the nearest thousandth)
24. (1/8, 1) and (3/8, 9/4)
25. (−3, 5) and (−3, 8)
26. Suppose that the Virtual University was found to have 2030 students in 2001 and 3590 students in 2011. If we assume that the number of students grew linearly, what is the rate of change of the number of students enrolled? Extrapolate to find the expected enrollment in 2016.

In the following problems, two lines are described using a pair of points on each line. Compute the slope of each line and use the values of the slopes to determine if the lines are parallel, perpendicular, or neither.

27. Line 1 passes through (0, 0) and (8, 4) and line 2 through (−2, 4) and (5, −10).
28. Line 1 passes through (2, 4) and (8, 10) and line 2 through (4, 5) and (8, 9).

For the following linear equations, construct a five-entry table of values and sketch the graph of the line.

29. $2x + 3y = 12$
30. $3x - 5y - 12 = 0$
31. $x - y = -3$

For the following linear equations, convert the equations to slope-intercept form. In each case, identify the slope of the line and the coordinates of the y-intercept.

32. $14x + 7y = 7$
33. $-3x + 4y - 8 = 0$
34. $2x - 5y + 7 = 0$

Graph the following linear equations by plotting the y-intercept and finding an additional point using the rise-over-run interpretation of slope.

35. $y = 4x - 7$
36. $y = 9 - 0.5x$

Find the equations (in general form) represented by the shown graphs.

37.

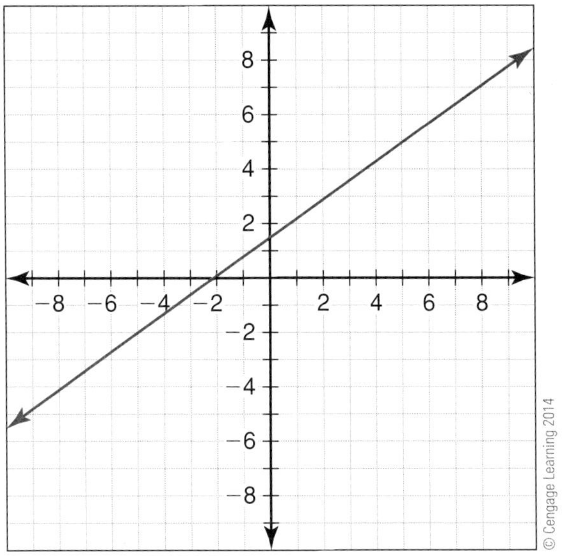

38.

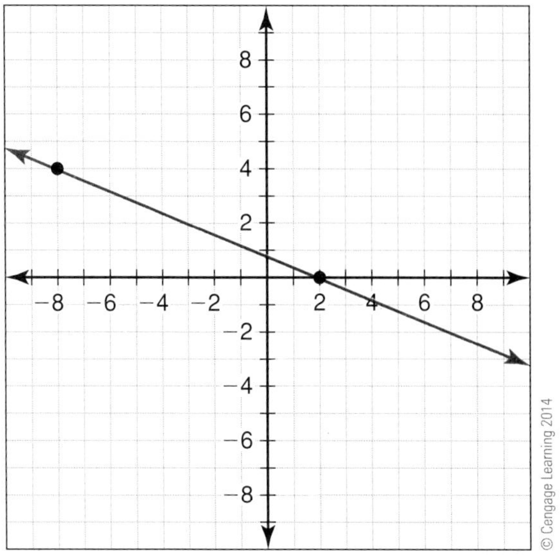

In the following exercises, use the given point and slope to find the equation of the line in general form.

39. Point: $(-1, -5)$, $m = ½$

40. Point: $(4, 3)$, $m = -3$

Find the two-point form of the equation of the line passing through the given pair of points. Use the first point given as (x_1, y_1).

41. $(3, 5)$ and $(4, 6)$

42. $(8, -4)$ and $(-8, 4)$

Find the two-intercept form of the equation of the line having the indicated intercepts.

43. $(0, 8)$ and $(4, 0)$

44. $(1/4, 0)$ and $(0, 13)$

Give the equation of the line in general form.

45. $y-8=\dfrac{12-8}{6-4}(x-4)$

46. $y-3=\dfrac{6-3}{12-2}(x-2)$

Determine if the lines represented by the following equations are parallel, perpendicular, or neither.

47. $x - 4y = 2$ and $y - 4x = 3$

48. $3x + 2y = 7$ and $4x - 5y = 3$

Use the method of graphing to find the solution to the system of equations, if a solution exists.

49. $5x - 3y = 2$ and $4x + y = 5$

50. $3x + 5y = 17$ and $x + 7y = 11$

Chapter 5 Solving Systems of Linear Equations Algebraically and with Matrices

At the end of Chapter 4, we encountered systems of linear equations and saw that they could be solved by determining the point(s) of intersection of the graphs of the equations. In the event that a single point of intersection occurred, that point was identified as the solution, and the system was called consistent. If the lines were parallel (and hence no intersection would occur), the system was called inconsistent, and we said there was no solution. In the rather rare instance where the two lines had the same graph, the system was called consistent and dependent, and the solution set consisted of all points lying on their mutual graph.

That visual method of solving systems possesses inherent weaknesses. First, a high-quality graph must be in your possession, with lines drawn that are extremely precise. Even the slightest imperfection in the graph of one or both of the lines could lead you to an incorrect determination of the solution. Second, even with a high-quality graph in hand, if the solution point were such that its coordinates were not integer valued, exact identification of the solution could prove to be difficult or impossible.

These issues lead us to consideration of methods for solution that are immune from the ambiguity of graphical analysis, methods that are wholly algebraic and computational. Such methods are free from the requirement that a high-quality graph be in our possession and are also immune from the problem of identifying noninteger coordinates.

We will consider several computational techniques, including the methods of substitution and elimination and the matrix method. We will see that these approaches offer us protection from the vagaries of the graphing method and that they employ mathematical techniques already well known to us.

Objectives

When you have successfully completed the materials of this chapter, you will be able to:

- **Solve a system of linear equations by the method of substitution.**
- **Solve a system of linear equations by the method of elimination.**
- **Use substitutions to convert systems of equations into linear systems and solve them.**
- **Understand the basic definitions and operations of matrices.**
- **Solve a system of linear equations using matrices.**

5.1 Solving Systems of Linear Equations by the Substitution Method

Let's consider a system of linear equations whose solution would be difficult to determine using the method of graphing. You will note that there is nothing in the appearance of the equations that suggests that we are headed for trouble by attempting to solve by graphing; the obstacles arise only when the graph is in our possession:

$$2x + 3y = 3$$
$$6x + 12y = 11$$

Suppose we attempt to solve the system by graphing and see what happens. You are already familiar with the procedure for producing the graphs, so we will suppress that work and present the result for analysis (Figure 5.1). You will note that the point of intersection is somewhat difficult to determine. The coordinates are clearly not integer valued, so we could start to make conjectures about their exact values and check our guesses in the original equations of the system, but that is rather risky and inefficient. If we suppose that the coordinates of

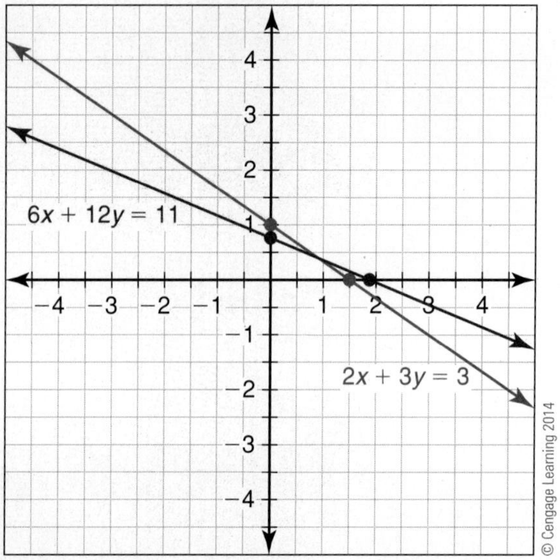

Figure 5.1 Graph of a System of Linear Equations

the intersection are $\left(\frac{1}{2}, \frac{3}{4}\right)$, for instance, how do we know that they are not $\frac{11}{23}, \frac{5}{8}$? Or some other fractions that are close to $\frac{1}{2}$ and $\frac{3}{4}$? We could make guesses all day long and never find the actual answers, and that situation is clearly unsatisfactory.

Let's now approach the problem from a purely computational angle. There are, as we've mentioned, several techniques from which to choose, and none is universally better than any of the others, and all will work, so it's really a matter of our choosing one to work with. The first of the methods we will call the method of **substitution**.

Substitution proceeds in the following manner:

1. Choose one equation to work with and in that equation isolate one of the variables.
2. Go to the equation not chosen in step 1 and replace the variable you isolated in step 1 by the expression to which you found it equal when the variable was isolated.
3. Solve the resulting equation.
4. Recalling the result of step 1, where one variable was isolated, substitute the solution found in step 3 and compute the value of the missing variable.
5. Combine your results into an ordered pair and check to see if it verifies each of the original equations of the system. If it does not, then a computational error must exist somewhere in your work—examine your steps to find your error.

Those steps may sound challenging, but we'll be up to it. Let's recall the system given earlier and apply the process.

EXAMPLE 5.1

Solve the system by the method of substitution:

$$2x + 3y = 3$$
$$6x + 12y = 11$$

SOLUTION

There is no reason to pick any particular equation with which to work, so we'll choose the first one, and in that equation we'll choose to isolate x. You'll see, as you go along, that there are occasionally situations where it is to your advantage to choose a particular equation to work with and in that equation a particular variable to isolate, but typically it is irrelevant which one you choose.

> **Note**
>
> It is imperative that you go to the *other* equation in step 2! If you reuse the same equation you chose in step 1, you will find that you are led to a "dead end" in your computations!

Step One: Solve the first equation for x:

$$2x + 3y = 3$$
$$2x = 3 - 3y$$
$$x = (3 - 3y)/2$$

The work above is the process described in step 1 of the substitution process. We selected an equation and a variable within it to isolate and accomplished that goal. Now we advance to step 2.

Step Two: In the other equation, $6x + 12y = 11$, replace the x with the expression $(3 - 3y)/2$:

$$6[(3 - 3y)/2] + 12y = 11$$

Step Three: Solve this equation. Note that, in the first term, the 6 reduces with the 2, yielding

$$3(3 - 3y) + 12y = 11$$
$$9 - 9y + 12y = 11$$
$$9 + 3y = 11$$
$$3y = 2$$
$$y = \frac{2}{3}$$

Step Four: Return to $x = (3 - 3y)/2$ and insert the value of y that we obtained previously. This gives us

$$x = \left[\frac{3 - 3\left(\frac{2}{3}\right)}{2}\right]$$
$$x = (3 - 2)/2$$
$$x = \frac{1}{2}$$

Combining the results into an ordered pair, we find the solution is (apparently) the point $\left(\frac{1}{2}, \frac{2}{3}\right)$. It remains for us to check that this is correct, but it appears to agree with the result suggested by the graph considered earlier.

Step Five: Remember that any check must be performed in the original equations we were given. Once we modified those equations, it was no longer clear whether anything was correct because we may have made computational errors. Therefore, we must check our answer in

$$2x + 3y = 3$$
$$6x + 12y = 11$$

Inserting $x = \frac{1}{2}$ and $y = \frac{2}{3}$, we find

$$2\left(\frac{1}{2}\right) + 3\left(\frac{2}{3}\right) = 1 + 2 = 3\checkmark$$

and

$$6\left(\frac{1}{2}\right) + 12\left(\frac{2}{3}\right) = 3 + 8 = 11\checkmark$$

and thus we have confirmed that our solution is correct.

EXAMPLE 5.2

Solve the following system, using the method of substitution:

$$\begin{aligned} 4x + 2y &= 6 \\ 5x - 5y &= 18 \end{aligned}$$

SOLUTION

As was the case in Example 5.1, choose one equation and isolate one variable within that equation. Here, consideration of the alternatives leads us to suspect that isolating y in the first equation would be the easiest approach (do you see why?).

Step One:

$$\begin{aligned} 4x + 2y &= 6 \\ 2y &= 6 - 4x \\ y &= 3 - 2x \end{aligned}$$

Step Two: Turning to the other equation of the system, we make a substitution for y:

$$5x - 5(3 - 2x) = 18$$

Step Three: Solve the equation for x:

$$\begin{aligned} 5x - 15 + 10x &= 18 \\ 15x - 15 &= 18 \\ 15x &= 18 + 15 \\ 15x &= 33 \\ x &= \frac{33}{15} \\ x &= \frac{11}{5} \end{aligned}$$

Step Four: Find the value of the other variable (y) by back-substitution:

$$y = 3 - 2\left(\frac{11}{5}\right)$$

$$y = 3 - \left(\frac{22}{5}\right)$$

$$y = \frac{15}{5} - \frac{22}{5}$$

$$y = -\frac{7}{5}$$

Combining the values found in steps 3 and 4, we conclude that the solution to the system is the point $\left(\frac{11}{5}, -\frac{7}{5}\right)$.

Step Five: Check the solution. Inserting the values of x and y into the original equations of the system, we find

$$4\left(\frac{11}{5}\right) + 2\left(\frac{-7}{5}\right) = 6$$

$$\frac{44}{5} - \frac{14}{5} = 6$$

$$\frac{30}{5} = 6\surd$$

$$5\left(\frac{11}{5}\right) - 5\left(\frac{-7}{5}\right) = 18$$

$$\frac{55}{5} + \frac{35}{5} = 18$$

$$11 + 7 = 18\surd$$

Since the solution checks in both original equations, we conclude that our answer is correct. Note that we could gain further support for the correctness of this answer by graphing the two equations of the system on the same plane, and observing the point of intersection agrees with the solution point we found computationally.

What would it look like, you might wonder, if a system were to have one of the other possible outcomes, that is, if it were an inconsistent system or consistent and dependent? We want to be sure that we can recognize those occurrences, and a few examples will help make the situation clear.

EXAMPLE 5.3

Solve the system of equations, using the method of substitution:

$$2x - 3y = 5$$
$$4x - 6y = 9$$

SOLUTION

Again, without any particular reason, we will choose to isolate x in the first equation. Keep in mind that it generally does not matter too much which equation we choose to work with and which letter in that equation we choose to isolate.

Step One: Solving the first equation for x, we find

$$2x - 3y = 5$$
$$2x = 3y + 5$$
$$x = (3y + 5)/2$$

Step Two: Next, we substitute that expression in place of x in the second equation. This gives

$$4[(3y + 5)/2) - 6y = 9$$

Step Three: Again, we can reduce the 4 with the 2 and obtain

$$\begin{aligned} 2(3y + 5) - 6y &= 9 \\ 6y + 10 - 6y &= 9 \\ 10 &= 9 \end{aligned}$$

Notice that we have reached a rather unsettling point in our computations. The variable y has completely disappeared from the equation, and we are left with a false statement of equality (10 equal to 9 is just about as false as you can get!). This is what we refer to as a contradiction, and it indicates to us that the equation we are attempting to solve has no solution. The no-solution case is also referred to as the **inconsistent** case, and hence we can conclude that the system we are attempting to solve is inconsistent and has no solution.

> **Note**
> If, in the process of solving by substitution, you reach a point where your equation has become a contradiction, then stop. Your system is inconsistent and has no solution!

Graphing the equations of the system, we find that the two lines do not intersect. Remember, the solution to a system of linear equations is equivalent to the point of intersection of the graphs of the equations of the system, and thus a system with no solutions would generate a pair of equations such as this, which are parallel to one another.

EXAMPLE 5.4

Solve the following system of equations, using the method of substitution:

$$\begin{aligned} 5x + 8y &= 10 \\ 15x + 24y &= 30 \end{aligned}$$

SOLUTION

To shake things up a bit, we will choose to work with the second equation and to isolate y in that equation. There is no particular reason for this; we have just chosen to do something different.

Step One:

$$\begin{aligned} 15x + 24y &= 30 \\ 24y &= 30 - 15x \\ y &= (30 - 15x)/24 \\ y &= (10 - 5x)/8 \end{aligned}$$

Step Two: Turning to the other equation, we substitute this expression for y and solve:

$$5x + 8[(10 - 5x)/8] = 10$$

Step Three: The 8 reduces with the 8, giving us

$$\begin{aligned} 5x + 10 - 5x &= 10 \\ 10 &= 10 \end{aligned}$$

> **Note**
> If, in the process of solving by substitution, you reach a point where your equation has become an identity, then stop. Your system is consistent and dependent and has infinitely many solutions! Express the solution set using set-builder notation.

As in Example 5.2, we see the variable has subtracted out, but in this case, we are left with a mathematically true equality. This is what we refer to as a mathematical **identity**, and its appearance in the substitution process tells us that our system is consistent and dependent and has infinitely many solutions. As was done in the graphing method, we express this solution using set-builder notation and one of the original equations: $\{(x, y) \mid 5x + 8y = 10\}$.

EXAMPLE 5.5

Solve the system of equations, using the method of substitution:

$$\begin{aligned} 9x + 4y &= 13 \\ 5x + y &= 7 \end{aligned}$$

SOLUTION

In this case, we will make our first attempt to be shrewd in our choices. We observe that, in the second equation of the system, it would be very easy to isolate the variable y, so we will make the choice to work with the second equation and to solve that equation for y. We do not have to make that choice, but it certainly appears that this would lead to the least amount of work for us.

Step One:

$$\begin{aligned} 5x + y &= 7 \\ y &= 7 - 5x \end{aligned}$$

Step Two: Turning to the other equation in the system, we replace the variable y by this expression and solve:

$$9x + 4(7 - 5x) = 13$$

Step Three:

$$\begin{aligned} 9x + 28 - 20x &= 13 \\ -11x + 28 &= 13 \\ -11x &= -15 \\ x &= \frac{15}{11} \end{aligned}$$

This is not the most attractive solution we could imagine, but it is the one we are stuck with, so we must live with the rather unappealing fraction. We substitute this value into the expression we obtained when isolating y and find the following.

Step Four:

$$\begin{aligned} y &= 7 - 5\left(\frac{15}{11}\right) \\ y &= 7 - \frac{75}{11} \\ y &= \frac{77}{11} - \frac{75}{11} \\ y &= \frac{2}{11} \end{aligned}$$

Step Five: Our conclusion is that the solution to the system appears to be $\left(\frac{15}{11}, \frac{2}{11}\right)$. It remains to check this in the original equations of the system to verify that our work is correct:

$$\begin{aligned} 9x + 4y &= 13 \\ 9\left(\frac{15}{11}\right) + 4\left(\frac{2}{11}\right) &= 13 \end{aligned}$$

$$\frac{135}{11}+\frac{8}{11}=13$$
$$\frac{143}{11}=13\surd$$
$$5x+y=7$$
$$5\left(\frac{15}{11}\right)+\frac{2}{11}=7$$
$$\frac{75}{11}+\frac{2}{11}=7$$
$$\frac{77}{11}=7\surd$$

The solution satisfies both of the original equations of the system, and we conclude that we have verified that our solution is correct.

Before moving on to the next computational method of solution, we should address the question of when it would be appropriate to choose to use this method over the others we will see. We have already noted that all the methods will work, but when would it be to our advantage to choose substitution?

Generally, substitution is best used either when it is easy to isolate one variable in one of the equations (as was the case in Example 5.4) or in a problem where one of the equations is given to us with a variable already isolated. In other situations, it generally requires a bit more work to use substitution than it would to use other methods. The amount of extra work will never be onerous, so you should not shun the method out of fear that it will be unnecessarily complicated, but do be aware that in other situations, the alternative methods we will present are likely to be more efficient.

Note

If it appears that one variable is computationally easy to isolate in one of the equations in the system, consider attempting to solve by substitution!

Exercises

Solve the following systems using the method of substitution.

1. $x-y=-7$
   $8x+10y=16$
2. $3x+3y=6$
   $x-y=0$
3. $2x-y=2$
   $6x+8y=39$
4. $2x-y=10$
   $x+5y=-6$
5. $-3x+y=-7$
   $9x-3y=21$
6. $4x+3y=13$
   $5x-y=2$
7. $4x-3y=20$
   $8x-6y=40$
8. $0.5x+3.2y=9$
   $0.2x-1.6y=-3.6$
9. $2x-2y=-2$
   $4x+5y=3$
10. $2x=7-4y$
    $y=3-2x$

11. $\frac{1}{2}x - \frac{3}{2}y = -1$

$5x + \frac{2}{3}y = 10$

12. $2x - 5y = 11$

$3x + 5y = 4$

13. $y = 3x$

$x - 2y + 6 = 0$

14. $2x + 5y = 6$

$3x - y + 6 = 0$

15. $x - 2y - 4 = 0$

$5x - 3y = 10$

The following problems can be solved by constructing and solving a suitable system of linear equations. Construct the appropriate system, solve using the method of substitution, and then interpret the solution in the context of the problem.

16. A hot dog stand entrepreneur invests \$8000 to open his business. The owner knows that a hot dog and bun will cost him \$1.10, and he intends to sell his hot dogs for \$1.85. How many hot dogs must he make and sell in order to break even (obtain a profit of \$0)? How many hot dogs must he make and sell in order to amass a profit of \$10,000?
17. You are offered two jobs as a computer salesperson. The first job offers you a salary of \$600 each week, plus 3% commission on sales. The second job offers you a straight salary of 7% commission on sales. How much would your sales have to be in order to make the straight commission salary the better offer?
18. A chemistry student wants to create a mixture that contains a 16% alcohol solution. She has two premixed ingredients, one of which is a 10% alcohol solution and the other a 70% alcohol solution. If she wants to obtain 50 liters of the 16% solution, how much of each of her ingredients should she use?
19. Suppose the wheat supply in a particular country is approximated by the model $p = 7x - 400$ (where p is in U.S. dollars), while the wheat demand in that country is modeled by $p = 510 - 3.5x$. Here, p is the price of wheat per metric ton, and x is the quantity of wheat in millions of metric tons. Find the price at which the supply and demand are equal.
20. A widget salesman has been offered two possible compensation plans. The first plan offers a monthly salary of \$2000 plus a royalty of 10% of the total of his monthly sales. The second plan offers a monthly salary of \$1000 plus a royalty of 20% of the total of his monthly sales. What must his widget sales be in a month in order for the two compensation plans to yield the same monthly salary?
21. A construction company needs to rent a dump truck and is comparing two potential rental companies. Big Truck Rental offers the truck for \$75 per day plus \$0.40 per mile, while Tiny Truck Rental offers the same type of truck for \$105 per day plus \$0.25 per mile. How far would the truck have to be driven in order to have the costs of rental for both companies be equal, and what would be the cost involved in such a rental?

5.2 Solving Systems of Equations Using the Method of Elimination

An alternative algebraic technique for solving systems of linear equations is sometimes referred to as "annihilation" but is more commonly called the method of **elimination**. It is particularly useful in the case where the equations are given to you in the form $Ax + By = C$, but since any linear equation can be rewritten in that form, the method applies to all systems of linear equations.

We will assume that the coefficients of the equations of the system have integer values. If they are not integers, then for each equation having fractional coefficients, multiplication of all terms in the particular equation by the least common denominator of the fractions will clear the fractions to produce integer coefficients.

EXAMPLE 5.6

Rewrite the equation $\frac{2}{3}x + \frac{4}{5}y = 8$ with integer coefficients.

SOLUTION

The least common denominator of the fractions in the equation is 15, so we multiply all the terms by that value and simplify:

$$15\left(\frac{2}{3}x\right) + 15\left(\frac{4}{5}y\right) = 15(8)$$

$$10x + 12y = 120$$

The method of elimination bears a superficial resemblance to the method of substitution in that choices are being made, but the procedure is completely different. For elimination, the series of steps we must follow are these:

1. Pick one variable to eliminate from the system.
2. Multiply one or both of the equations, as needed, by constants chosen in such a way that the coefficients of the variable you chose to eliminate will be equal but opposite in sign.
3. Add the equations of the system together, eliminating the variable you chose in step 1.
4. Solve the resulting equation, producing one coordinate of the solution point.
5. If the solution you found in step 4 is an integer, substitute it into one of the original equations of the system and solve to find the value of the other variable. This is the second coordinate of the solution point. If the solution you found in step 4 is not an integer, you may choose to repeat the elimination process, this time opting to eliminate the other letter. This, again, leads you to the second coordinate of the solution point.
6. Combine your results in an ordered pair and check to see that it satisfies each of the original equations of the system. If it does not, then a computational error must exist somewhere in your work—examine your steps to find your error.

EXAMPLE 5.7

Solve the following system, using the method of elimination:

$$\begin{aligned} 8x - 3y &= -25 \\ 5x + 6y &= 8 \end{aligned}$$

SOLUTION

Step One: We will want to try to make our choice in step 1 shrewdly, but we could make an arbitrary choice. The system was

$$\begin{aligned} 8x - 3y &= -25 \\ 5x + 6y &= 8 \end{aligned}$$

The objective in step 2 will be to make the coefficients of one variable in the system be equal but opposite in sign. We will, essentially, want to change the coefficients of the variable into their least common multiple, with opposite signs. We could do this with either variable, but consideration of our options reveals that it would be slightly easier to work with y than with x.

Step Two: The least common multiple of the coefficients of y is 6, and one of the equations already has that coefficient for its y term.

In order to match the coefficients of y (with opposite signs), we need to multiply the first equation by 2:

$$\begin{aligned} 2(8x - 3y &= -25) \\ 5x + 6y &= 8 \end{aligned}$$

which produces

$$\begin{aligned} 16x - 6y &= -50 \\ 5x + 6y &= 8 \end{aligned}$$

Step Three: If the equations are now added together, we obtain

$$21x = -42$$

Step Four: Solving the equation obtained in step 3, we find

$$x = -2$$

Step Five: This value (being an integer) is substituted into one of the original equations, and we solve for the other variable, y:

$$\begin{aligned} 8(-2) - 3y &= -25 \\ -16 - 3y &= -25 \\ -3y &= -9 \\ y &= 3 \end{aligned}$$

Step Six: Our solution appears to be the point $(-2, 3)$. We confirm this by substitution into the original equations of the system:

$$\begin{aligned} 8(-2) - 3(3) &= -25 \\ -16 - 9 &= -25\checkmark \\ 5(-2) + 6(3) &= 8 \\ -10 + 18 &= 8\checkmark \end{aligned}$$

Since both of the equations are satisfied by the coordinates of the point $(-2, 3)$, we conclude that our solution is correct.

As was the case for substitution, we should be aware of how we would recognize the inconsistent or the dependent cases. These will be determined in step 3 when the equations are added together to annihilate the variable chosen in step 1. The inconsistent case is identified by the appearance of a contradiction at this stage and the dependent case by the appearance of an identity.

Remember that a contradiction would be a false equality. Both of the variables will be eliminated from the equation, and you will be confronted with an equality of the form $0 = k$, where k is some nonzero number. An identity is a universally true equality, which in this situation will be $0 = 0$. In either case, the process will terminate after step 3 and the solution identified.

EXAMPLE 5.8

Solve the system of equations, using elimination:

$$\begin{aligned} 2x - 5y &= 10 \\ 4x - 10y &= 15 \end{aligned}$$

SOLUTION

Step One: We choose to eliminate the letter x, since the least common multiple of the coefficients of the x terms is seen to be 4.

Step Two: To create a situation where the coefficients of x are equal but opposite in sign, the first equation is multiplied by (-2):

$$\begin{aligned} (-2)(2x - 5y &= 10) \\ 4x - 10y &= 15 \end{aligned}$$

which yields

$$\begin{aligned} -4x + 10y &= -20 \\ 4x - 10y &= 15 \end{aligned}$$

Step Three: When we add the equations together, note that we obtain

$$0 = -5$$

This, being a contradiction, tells us that the system is inconsistent and has no solution.

EXAMPLE 5.9

Solve the system of equations by the elimination method:

$$\begin{aligned} 7x - 3y &= 2 \\ 21x - 9y &= 6 \end{aligned}$$

SOLUTION

Step One: Again, we see that it would not be too difficult to eliminate x, since the least common multiple of the x-coefficients is 21.

Step Two: Since we want to create a situation where the coefficients of x are equal but opposite in sign, we multiply the first equation by (-3):

$$\begin{aligned} (-3)(7x - 3y &= 2) \\ -21x + 9y &= -6 \end{aligned}$$

which yields

$$-21x + 9y = -6$$
$$21x - 9y = 6$$

Step Three: Adding the equations together, we obtain $0 = 0$, an identity. The conclusion is that the equations are consistent and dependent, and the system has infinitely many solutions. The set of solutions, in set-builder notation, is $\{(x, y) \mid; 7x - 3y = 2\}$.

Note

During the elimination process, if (after adding the equations together) you obtain a contradiction of the form $0 = k$ (for some nonzero number k), *stop*—the system has no solution and is *inconsistent*. If (after adding the equations together) you obtain the identity $0 = 0$, *stop*—the system is *consistent and dependent* and has infinitely many solutions (which you should express in set-builder notation).

Exercises

Answer each of the following, using complete sentences and proper grammar and spelling.

1. If the process of elimination should lead to an identity during the process of solving by elimination, what conclusion can be drawn?
2. If the process of elimination should lead to a contradiction during the process of solving by elimination, what conclusion can be drawn?

Solve the following systems of linear equations, using the method of elimination.

3. $x - 4y = 3$
   $3x + 8y = 5$
4. $2x - y = 1$
   $4x - 2y = 2$
5. $2x - 3y = 3$
   $x + 2y = 5$
6. $2x - 3y = 3$
   $-4x + 6y = 6$
7. $7x - 3y = 10$
   $14x - 6y = 5$
8. $0.3x - 0.2y = 0.7$
   $0.8x + 0.4y = 0$
9. $2x + y = 9$
   $3x - y = 16$
10. $x - 2y = -9$
    $x + 3y = 16$
11. $4x - 3y = 25$
    $-3x + 8y = 10$
12. $12x - 13y = 2$
    $-6x + 6.5y = -2$
13. $12x - 3y = 6$
    $4x - y = 2$
14. $2x + 3y = 5$
    $x - 5y = -17$

The following application problems can be solved through the use of a system of linear equations. Construct a system appropriate to each application example, solve the system using the method of elimination, and interpret the solution in the context of the problem.

15. An airplane flies into a headwind and travels the 1000-mile distance between two cities in 2 hours and 10 minutes. On the return flight, the

same distance is covered in exactly 2 hours. Find the airspeed of the plane and the speed of the wind, assuming that both have a constant magnitude.

16. A school is presenting a play where tickets for the event cost \$5 for adults and \$2.50 for children; 548 total tickets are sold, generating revenue of \$2460. How many children's tickets were sold, and how many adult tickets were sold?
17. How many liters of a 20% alcohol solution must be mixed with 40 liters of a 50% alcohol solution in order to create a 30% alcohol solution?
18. Sterling silver is 92.5% pure silver. How many grams of sterling silver must be mixed with a 90% silver alloy to obtain 500 grams of a 91% silver alloy?
19. The demand and supply functions for a new type of cell phone are

$$p = 200 - 0.00001x$$
$$p = 80 + 0.00002x$$

where p is the price in dollars for the phone and x is the number of units sold. Find the equilibrium point of the system, which is the value of p and the value of x that will satisfy both of the equations in the system simultaneously.

20. If a total of \$18,000 is invested in a pair of bonds that pay 9% and 6%, respectively, how should you divide your investment if you want to earn \$1410 in interest?
21. Suppose peanuts cost \$2.50 per pound, and cashews cost \$4.00 per pound. If the two types of nuts are to be mixed together and sold as a blend, how much of each type should be used to produce 25 pounds of a mix that will cost \$3.00 per pound?
22. Suppose Juan has \$8.05 in nickels and quarters. If he has 29 more nickels than he does quarters, how many of each type of coin does he have?
23. If two cities are 500 miles apart and Joan leaves city A at noon and travels toward city B at 60 miles per hour while Sonya leaves city B at 1:00 p.m. and heads toward city A at 50 miles per hour, when will they pass one another?
24. Two runners start from the same spot on the same quarter-mile oval track. One runner runs at 6 miles per hour and the other at 8 miles per hour. If they start at the same moment and run in the same direction, when will the faster runner be one lap ahead of the slower runner?

5.3 Substitutions That Lead to Systems of Linear Equations

Suppose an equation had the following form:

$$\frac{h}{x} + \frac{k}{y} = C$$

where h, k, and C are constants. In order to avoid a situation where the expression becomes undefined, it's clear that we can have neither $x = 0$ nor $y = 0$, and hence neither the origin nor any quadrantal points can lie on the graph of the equation.

Under the assumption that neither variable can be 0, we can rewrite the equation using the following substitutions: $u = 1/x$, $v = 1/y$. With these substitutions, the equation becomes

$$hu + kv = C$$

which is a linear equation in the unknowns u and v.

Although not a common situation, it is possible that we could encounter a system in which each equation has the form $\frac{h}{x} + \frac{k}{y} = C$, and in that event, the previously mentioned substitutions would lead us to a system of linear equations that we could solve by one of our established methods. Naturally, since the solutions to that new system will be in terms of u and v, we will have to translate those in terms of x and y and maintain vigilance that the original expression is not undefined for the solutions we obtain.

Let's consider an example and see how the process would work.

EXAMPLE 5.10 Solve the following system of equations:

$$\frac{3}{x} + \frac{5}{y} = 2$$

$$\frac{7}{x} - \frac{3}{y} = 10$$

SOLUTION

Using the substitutions $u = 1/x$, $v = 1/y$, the system becomes

$$3u + 5v = -2$$
$$7u - 3v = 10$$

We are now free to choose a method with which to solve this system. Inspection of the equations does not reveal a particular variable that would be easy to isolate, and consequently we steer clear of the method of substitution. Attempting to solve by graphing might be possible, but if the point of intersection in the u, v plane is not apparent, this could be wasted effort. Consequently, we choose to use the method of elimination.

The coefficients of v are already opposite in sign, so the variable we choose to eliminate in this situation will be v. The least common multiple of the v-coefficients is 15, so the first equation will be multiplied by 3 and the second by 5:

$$9u + 15v = -6$$
$$35u - 15v = 50$$

We add the equations together, producing

$$44u = 44$$

or

$$u = 1$$

Since this result is an integer, we can easily find the value of v by inserting this value of u into one of the original equations. Choosing the first equation,

$$\begin{aligned} 3(1) + 5v &= -2 \\ 3 + 5v &= -2 \\ 5v &= -5 \\ v &= -1 \end{aligned}$$

Note

It is a complete coincidence that the coordinates of the solution in the x, y plane are identical to the coordinates found in the u, v plane. This is not typically the case, as we will see in coming examples!

Substitution of the values of $u = 1$ and $v = -1$ into each of the original equations confirms this answer to be correct.

We must now recall that our original system involved x and y rather than u and v, so this cannot be the solution to our original problem. We must now recall our substitution expressions and use them to determine the values of x and y.

Since $u = 1/x$, we have $1 = 1/x$, or (by taking the reciprocal of each side of the equation) $1 = x$. Similarly, since $v = 1/y$, we have $-1 = 1/y$, or $-1 = y$, and hence the solution in the x, y plane is the point $(1, -1)$.

EXAMPLE 5.11

Solve the following system:

$$\frac{4}{x} + \frac{7}{y} = 5$$

$$\frac{2}{x} + \frac{8}{y} = 3$$

SOLUTION

Again we make the substitutions $u = 1/x$ and $v = 1/y$, which transforms the given system into

$$\begin{aligned} 4u + 7v &= 5 \\ 2u + 8v &= 3 \end{aligned}$$

In this case, there again is no clear choice for a variable to isolate, so substitution appears to be unwise. Choosing to opt for the method of elimination and considering the coefficients of u and v, it appears that it would be easier to eliminate the variable u from the system. We multiply the second equation by -2, which makes the coefficients of u equal but opposite in sign:

$$\begin{aligned} 4u + 7v &= 5 \\ -4u - 16v &= -6 \end{aligned}$$

Adding the equations together produces an equation involving v only:

$$-9v = -1$$

which we solve to find

$$v = \frac{1}{9}$$

Although this is not a terribly attractive value of v, it is (as fractions go) not horrible, so we will substitute into one of the system's equations and solve for u. Arbitrarily choosing the second equation with which to work, we find

$$2u+8\left(\frac{1}{9}\right)=3$$

$$2u+\frac{8}{9}=3$$

$$18u + 8 = 27$$
$$18u = 19$$
$$u=\frac{19}{18}$$

Substitution of these values into the equations of the system indicates that our work is correct, and now it merely remains to convert these solutions into x, y terms.

Note

When the substitutions involved are $u = 1/x$ and $v = 1/y$, the value of x is the reciprocal of the value of u, and the value of y is the reciprocal of the value of v!

At first, we may be hesitant to tackle the computations because the values of u and v are fractions, and this may create the concern that difficult fractions could arise in our computations, but this is actually never the case. Since our substitutions were $u = 1/x$ and $v = 1/y$, what is actually the case is that u is the reciprocal of x and v the reciprocal of y. Therefore, to find the solutions in terms of x and y, it is merely necessary to take the reciprocals of our u and v solutions. Hence, $x=\frac{18}{19}$ and $y=\frac{9}{1}$, and our solution in the x, y plane is $\left(\frac{18}{19}, 9\right)$. Substitution of these values into the original equations of our system confirms the correctness of the solution.

Note that this is not the only substitution that could lead to a linear system of equations in u and v. Consider this case:

$$\frac{3}{2-x}+\frac{5}{1+y}=11$$
$$\frac{7}{2-x}-\frac{4}{1+y}=10$$

In this case, the substitutions involved would be $u = 1/(2 - x)$ and $v = 1/(1 + y)$. This would lead us to a system of linear equations in u and v, but when we translate those answers into x, y coordinates, the process will not be as simple as merely taking the reciprocal of u and v! In such a case, when the substitution is more involved, we must insert the values of u and v into the substitution expressions and compute the values of x and y directly.

In this case, the system in terms of u and v would be

$$3u + 5v = 11$$
$$7u - 4v = 10$$

There is no obvious choice for a variable to eliminate, so we will choose to eliminate v (since the coefficients of the v terms already have opposite signs). We multiply the first equation by 4 and the second by 5, yielding

$$12u + 20v = 44$$
$$35u - 20v = 50$$

When we add the equations together, we find

$$47u = 94$$
$$u = 2$$

Substitution into the first equation of the system then shows us that $v = 1$, and a check in both of the equations confirms the accuracy of this result.

To find the values of x and y, it is necessary to actually insert the values of u and v into the substitution expressions and compute rather than simply take the reciprocal of what we have already found:

$$2 = \frac{1}{2-x}$$
$$2(2-x) = 1$$
$$4-2x = 1$$
$$-2x = -3$$
$$x = \frac{3}{2}$$

and

$$1 = \frac{1}{1+y}$$
$$1(1+y) = 1$$
$$1+y = 1$$
$$y = 0$$

Substitution of these values into the equations of the original system verifies their correctness, and we conclude the solution to the original system was the point $\left(\frac{3}{2}, 0\right)$.

Exercises

Solve the following systems of linear equations using an appropriate substitution.

1. $\frac{4}{x} - \frac{3}{y} = 28$

   $\frac{9}{x} - \frac{1}{y} = -6$

2. $\frac{1}{x} + \frac{2}{y} = 7$

   $\frac{3}{x} + \frac{12}{y} = 2$

3. $\frac{-1}{10x} + \frac{1}{2y} = 4$

   $\frac{2}{x} - \frac{10}{y} = -80$

4. $\frac{3}{x} + \frac{2}{y} = 0$

   $\frac{-1}{x} - \frac{2}{y} = 8$

5. $\frac{6}{x}+\frac{4}{y}=12$

$\frac{9}{x}+\frac{6}{y}=18$

6. $\frac{8}{x-1}-\frac{3}{y}=-3$

$\frac{5}{x-1}-\frac{2}{y}=-1$

7. $\frac{3}{x}-\frac{6}{y}=12$

$\frac{4}{x}-\frac{8}{y}=16$

8. $\frac{-5}{3-x}-\frac{7}{y+2}=10$

$\frac{8}{3-x}+\frac{3}{y+2}=-3$

9. $\frac{11}{x+7}+\frac{5}{y}=\frac{1}{3}$

$\frac{2}{x+7}-\frac{5}{y}=\frac{25}{3}$

10. $7x-\frac{8}{y}=15$

$3x+\frac{2}{y}=9$

11. $\frac{3}{2x}-\frac{2}{3y}=0$

$\frac{3}{4x}+\frac{4}{3y}=\frac{5}{2}$

12. $\frac{4}{x}+\frac{4}{y}=\frac{10}{3}$

$\frac{5}{x}-\frac{5}{y}=\frac{5}{6}$

13. $\frac{3}{x}-\frac{2}{y}=-30$

$\frac{2}{x}-\frac{3}{y}=-30$

14. $\frac{6}{x}+\frac{12}{y}=-6$

$\frac{2}{x}-\frac{1}{y}=-7$

15. $\frac{3}{2+x}-\frac{4}{5+y}=9$

$\frac{1}{2+x}+\frac{2}{5+y}=8$

16. $\frac{1}{x-3}+\frac{1}{y+1}=\frac{9}{20}$

$\frac{1}{x-3}-\frac{1}{y+1}=\frac{1}{20}$

5.4 Introduction to Matrices

The methods shown thus far as tools for the solution of systems of linear equations work well in the case where two equations in two unknowns are present. When more variables and more equations are involved, it becomes more efficient to use matrices to solve systems.

The word "matrices" is the plural of **matrix**, and applications of matrices are both far-flung and exotic. They are used extensively in computer programming, particularly in the rendering of images numerically (that is, in the digital representation of images rather than preservation on film) and in the creation of virtual reality three-dimensional simulations. It is prudent to introduce matrices now, when work with them will remain relatively elementary, so that we will have familiarity with them when the need arises to use them in more elaborate situations.

We should mention that we are, in this text, going to only introduce the topic of matrices and not present an exhaustive development. A full examination of matrices and their properties is better suited to a course in linear algebra or matrix theory, and we will leave quite a bit of the theory and examination of matrix properties to those courses.

The Fundamentals of Matrices

The technical definition of a matrix is that it is an array of numbers. This is not particularly helpful unless we are familiar with the notion of an array, so we should include the definition of that term in our introduction to matrices. An array is an arrangement of objects into horizontal **rows** and vertical **columns**.

Note

Always keep in mind that *rows* go across, and *columns* go up and down! Also, try to accustom yourself to referring to rows first and columns second!

The array of numbers that make up the matrix is typically enclosed within oversized brackets or parentheses to indicate the limitations of the array and is named with some capital letter, such as *A*, *B*, or *C*. You can use any letter you wish to name matrices as long as you remember that the common practice is to use capital letters.

An example of a matrix would be

$$A = \begin{bmatrix} 3 & 0 \\ 1 & -2 \\ 4 & 5 \end{bmatrix}$$

Observe that this matrix has three rows and two columns. The numbers within the array are referred to as the **entries** of the matrix and are referred to using their row and column location (note that, once again, the rows are referenced first and the columns second). The rows are numbered top to bottom and the columns left to right. Thus, in the example, the entry 5 is said to be in row 3, column 2, and is described as the 3, 2 entry.

Note

In a matrix, the entry located in row *i* and column *j* is referred to as the *i*, *j* entry. Always mention the row first and the column second!

When a matrix has been named with a capital letter, it is customary to refer to the entries of the matrix using subscripted notation, incorporating the lowercase form of the letter used to name the matrix. That is, when we said the 5 was the 3, 2 entry, the typical practice is to express this as $a_{3,2} = 5$, and one often sees the notation $A = [a_{i,j}]$ to emphasize that individual entries will be identified in that manner.

The size, or **order**, of a matrix is a statement of the number of rows and columns existing in the array. The notation $m \times n$ is used to specify the order of a matrix, where m is the number of horizontal rows and n the number of vertical columns. The example

Note

A matrix having m horizontal rows and n vertical columns is said to be of order $m \times n$, or to have size $m \times n$.

$$A = \begin{bmatrix} 3 & 0 \\ 1 & -2 \\ 4 & 5 \end{bmatrix}$$

is then a 3×2 matrix, read as "three by two."

EXAMPLE 5.12

Specify the order of the matrices shown below and identify the particular entry indicated, if it exists:

a.
$$C = \begin{bmatrix} 2 & -5 & 3 \\ 4 & 1/2 & 0 \end{bmatrix}, \text{ entry } c_{2,2} \text{ and } c_{3,1}$$

b.
$$D = \begin{bmatrix} 4 & -2 \\ 7 & -1 \\ 1/4 & 10 \end{bmatrix}, \text{ entry } d_{1,2} \text{ and } d_{2,1}$$

SOLUTION

a. The matrix C has two rows and three columns, and hence its order is 2×3. The 2, 2 entry is that value in the second row, second column, and here that would be ½. The 3, 1 entry does not exist, since that would be in the third row, but matrix C does not have a third row.

b. The matrix D has three rows and two columns, and hence its order is 3×2. The 1, 2 entry of matrix D is the entry lying in the first row, second column, and that would be the entry -2. The 2, 1 entry, on the other hand, is in the second row, first column, and that would be the entry 7.

In the particular case where a matrix happens to have exactly as many rows as it has columns, the matrix is referred to as a **square matrix**. The source of this term is the square shape taken on by the array when the number of rows and columns are equal.

Two matrices are said to be **equal** if they have precisely the same order and entries in corresponding locations are equal. In technical form, we can say $A = B$ if $a_{i,j} = b_{i,j}$ for all possible i, j entries.

EXAMPLE 5.13

Matrix A is known to be equal to matrix B. Give the values of the missing entries, identifying them using proper notation:

$$A = \begin{bmatrix} 2 & \_\_ \\ 3 & 5 \end{bmatrix}, \quad B = \begin{bmatrix} \_\_ & 4 \\ 3 & \_\_ \end{bmatrix}$$

SOLUTION

The missing entry in matrix A is $a_{1,2} = 4$, since the 1, 2 entries of the two matrices must agree. Similarly, the missing entries in matrix B are $b_{1,1} = 2$ and $b_{2,2} = 5$.

There are some special matrices we should mention at this point. Any matrix (of whatever size) having all entries being 0 is referred to as a **zero matrix**. When you are interested in specifying a particular zero matrix, you might identify it (for instance) as the "3×4 zero matrix," possibly using the notation $0_{3,4}$, which would tell the reader that you were talking about a zero matrix of that specific size.

A square matrix having all entries being 0 except for 1 in the entries in the 1, 1 position, the 2, 2 position, the 3, 3 position, and so on is referred to as an **identity**

matrix of order n, where n is the number of rows in the matrix. Since the matrix is square, this also happens to be the number of columns in the matrix. An identity matrix is always identified as I_n, where n is the number of rows in the matrix.

EXAMPLE 5.14 Write the matrix we would refer to as I_3.

SOLUTION

The matrix is the 3 × 3 identity matrix, so all its entries are 0 except for the 1, 1 entry, the 2, 2 entry, and the 3, 3 entry, all of which are 1. The matrix would be

$$\begin{bmatrix} 1 & 0 & 0 \\ 0 & 1 & 0 \\ 0 & 0 & 1 \end{bmatrix}$$

The entries of a square matrix that are in the 1, 1 position, the 2, 2 position, the 3, 3 position, and so on form what we call the **main diagonal** entries of the square matrix. We will examine the main diagonal of a square matrix in greater detail shortly. Any matrix having all entries not lying on the main diagonal being 0, with main diagonal entries being either 0 or nonzero (of whatever value), is called a **diagonal matrix**.

EXAMPLE 5.15 Give an illustration of a diagonal matrix having three rows.

SOLUTION

There are infinitely many such matrices, one of which is

$$\begin{bmatrix} 3 & 0 & 0 \\ 0 & 2 & 0 \\ 0 & 0 & 0 \end{bmatrix}$$

Observe that the main diagonal entries need not all be 0, as the 3, 3 entry in this matrix illustrates. An identity matrix is a particular instance of a diagonal matrix, where all the main diagonal entries happen to be 1.

Addition and Subtraction of Matrices

When two matrices are added, which we denote by $A + B$, the entries in corresponding locations are added to find the entry in the matrix sum. Note that this requires that the two matrices A and B must have the same order.

Formally, we define $A+B=\left[a_{i,j}\right]+\left[b_{i,j}\right]=\left[a_{i,j}+b_{i,j}\right]$, which is the way to express the content of the preceding paragraph in slightly more formal notation.

EXAMPLE 5.16 Add the two matrices.

$$\begin{bmatrix} 3 & 5 & -1 \\ 0 & 1/2 & 7 \end{bmatrix} + \begin{bmatrix} -6 & 3 & 11 \\ 4 & 2/3 & 3 \end{bmatrix}$$

SOLUTION

Both the matrices are 2 × 3, and therefore we can add them together to form a new 2 × 3 matrix sum. All we need to do is to add the corresponding entries and simplify the results:

$$\begin{bmatrix} 3+(-6) & 5+3 & -1+11 \\ 0+4 & \frac{1}{2}+\frac{2}{3} & 7+3 \end{bmatrix}$$

$$\begin{bmatrix} -3 & 8 & 10 \\ 4 & \frac{7}{6} & 10 \end{bmatrix}$$

If we were to take a matrix A and add to it a zero matrix (having the same order as matrix A), then the sum would simply be the original matrix A. For this reason, a zero matrix is referred to as an **additive identity matrix.**

Because the addition of matrices amounts to addition of the individual entries of the array, the familiar properties of addition of real numbers carries over to addition of matrices. Matrix addition is **commutative** and **associative.** That is, $A + B = B + A$, which means that the order of addition can be reversed, and also $(A + B) + C = A + (B + C)$, which means that matrices can be added in whatever grouping we wish.

Matrix subtraction acts, as was the case with matrix addition, entry by entry. In order to perform $A - B$, we subtract entries in corresponding locations, noting once more that the matrices must have precisely the same order. It may be excessive to do so, but we will admonish you to be careful about your signs when performing subtraction!

EXAMPLE 5.17 Perform the matrix subtraction.

$$\begin{bmatrix} 4 & -5 \\ 3 & 1/2 \end{bmatrix} - \begin{bmatrix} -6 & 2 \\ 7 & 4 \end{bmatrix}$$

SOLUTION

Both the matrices are 2 × 2, so subtraction is defined. We subtract entry by entry as follows:

$$\begin{bmatrix} 4-(-6) & -5-2 \\ 3-7 & \frac{1}{2}-4 \end{bmatrix}$$

Note that we are being very careful about our signs in the shown work:

$$\begin{bmatrix} 10 & -7 \\ -4 & -\frac{7}{2} \end{bmatrix}$$

Note

While matrix addition possesses the commutative and associative properties, matrix subtraction does not!

As was the case with subtraction of real numbers, subtraction of matrices is neither commutative nor associative. It is always the case that we must be more careful when subtracting than when adding, and particular caution should be applied when working with signed numbers.

Multiplication of a Matrix by a Scalar

When we talk about numbers occurring within the terms of polynomials, we refer to those numbers as "coefficients." Numbers occurring within the array of a matrix are referred to as the "entries" of the matrix. We say that such numbers appear in a "context." Numbers not appearing in any particular context are referred to as **scalars**.

Given a matrix A, we may want to consider multiplication of that matrix by a number such as 3, which we would describe as "three times matrix A," or $3A$. The 3, in this case, is what we would consider a scalar.

To multiply a matrix by a scalar, each entry of the matrix is multiplied by the scalar. In a sense, the scalar "distributes" through all the entries of the matrix.

EXAMPLE 5.18 For the matrix $A = \begin{bmatrix} 2 & 4 & -3 \\ 5 & 1/3 & 7 \end{bmatrix}$, find $6A$.

SOLUTION

We multiply each of the entries of the matrix A by the scalar 6 and obtain $\begin{bmatrix} 12 & 24 & -18 \\ 30 & 2 & 42 \end{bmatrix}$.

One can reverse the process of scalar multiplication of a matrix and factor a common scalar factor from the entries of a matrix. There are occasions when this proves useful in computations, as we'll see later.

EXAMPLE 5.19 Factor the largest possible scalar from matrix $B = \begin{bmatrix} 81 & 27 \\ 99 & -36 \\ 45 & 108 \end{bmatrix}$ while maintaining integer-valued entries in the matrix.

SOLUTION

The entries of the matrix have greatest common factor 9, so we will factor the 9 from each of the entries. This produces $9\begin{bmatrix} 9 & 3 \\ 11 & -4 \\ 5 & 12 \end{bmatrix}$. Noting that the entries of the matrix have no other common factors, we see that we have factored the largest possible scalar from the entries.

EXAMPLE 5.20 For $A = \begin{bmatrix} 3 & -2 \\ 7 & 5 \end{bmatrix}$ and $B = \begin{bmatrix} -2 & 4 \\ 12 & 3 \end{bmatrix}$, find and completely simplify $5A - 3B$.

Note

When performing matrix operations, continue to follow the order of operations introduced in algebra!

SOLUTION

In this case, the matrix A will be multiplied by the scalar 5 and the matrix B by the scalar 3, and then we will subtract. Observe that, even now, we are following the well-known order of operations as we simplify our expressions:

$$5A - 3B = \begin{bmatrix} 15 & -10 \\ 35 & 25 \end{bmatrix} - \begin{bmatrix} -6 & 12 \\ 36 & 9 \end{bmatrix}$$

$$= \begin{bmatrix} 15-(-6) & -10-12 \\ 35-36 & 25-9 \end{bmatrix} = \begin{bmatrix} 21 & -22 \\ -1 & 16 \end{bmatrix}$$

Multiplication of Matrices

To this point, the operations with matrices have been somewhat intuitive in the sense that the operations are performed in precisely the manner you would expect them to be done. Matrix multiplication is not quite so straightforward, though it is not particularly difficult to master.

Before introducing the method, we need to specify when the operation of matrix multiplication is defined and the notation we will use. In general, matrix multiplication is denoted by juxtaposition of the matrices, or their names. That is, if one matrix is written directly alongside another, the implication is that the matrices are to be multiplied. Alternatively, if our matrices are named A and B, then if we write AB, we mean that we want to multiply matrix A by matrix B.

Matrix multiplication is defined in the particular case only when the number of columns of the first matrix matches precisely the number of rows in the second matrix. Note that this is not necessarily going to happen for matrices of precisely the same order, except in the case where the matrices are square and of the same order!

This concept can be a bit slippery to grasp when first encountered, so let's illustrate. Suppose that matrix A is of order $m \times n$ and matrix B of order $p \times q$. How do we determine if multiplication is even possible? It all depends on the orders of the matrices involved and their "inner dimensions." The term "inner dimensions" is not universally used, but it offers an important visual reference. When we write the matrix names alongside one another and their respective sizes below them, as in

$$\begin{array}{c} AB \\ (m \times n)(p \times q) \end{array}$$

Note

The matrix product AB, where A is an $m \times n$ matrix and B is a $p \times q$ matrix, is possible only if $n = p$, and in that case the product matrix will have size $m \times q$!

the values of n and p are what we are referring to as the inner dimensions because of their relative positioning when writing the sizes below the matrix names. The multiplication is possibly *only* if the inner dimensions are equal. That is, the matrix multiplication shown for AB is possible only if $n = p$.

If those values match precisely, the *outer* dimensions (the number of rows in A and the number of columns in B) tell us the size of the product matrix. That is, if it turns out that we have $n = p$, then the product AB will be of order $m \times q$!

This definition presents a surprising result, unique in our experience to this point: noncommutativity of multiplication. Consider the following example as an illustration.

EXAMPLE 5.21

Suppose that A is of order 3×5 and B of order 5×3. Determine if AB is defined and if BA is defined. If either or both multiplications are defined, state the order of the product matrix.

SOLUTION

For AB, we note that (written in order) the sizes are $(3 \times 5)(5 \times 3)$. The inner dimensions match, and thus the multiplication is defined, and the outer dimensions tell us that the product matrix will be of order 3×3.

For BA, we again note that the orders are $(5 \times 3)(3 \times 5)$, and, as before, the inner dimensions match. This time, however, the outer dimensions tell us that the size of the product matrix is 5×5!

This is truly unusual for us. It is possible to calculate (by a method not yet discussed) each of the products AB and BA, but they will not have the same sizes as one another and therefore will not be equal! In some cases, it may even be possible to compute the product AB, while the product BA will not be defined.

Note

Even when both orders of multiplication are defined, it is not necessarily the case that $AB = BA$! That is, matrix multiplication is not guaranteed to possess the quality of commutativity!

So, how does matrix multiplication proceed? When computing the i, j entry of the product matrix AB, we use row i of matrix A and column j of matrix B. The entries of row i are multiplied with the entries of column j, one at a time, and the products added together. The sum obtained will be entry i, j of the product matrix. That sounds mysterious, perhaps, and we could formalize it with what is called "double-subscript notation," but let's see if an example will make the procedure clear.

EXAMPLE 5.22

Suppose $A = \begin{bmatrix} 1 & 2 \\ 3 & 4 \\ 5 & 6 \end{bmatrix}$ and $B = \begin{bmatrix} 7 & 8 \\ 9 & 10 \end{bmatrix}$. Find the product AB and the product BA, if possible. If either or both is not possible, explain.

SOLUTION

Since A is a 3×2 matrix and B is a 2×2 matrix, the product AB is defined and will be a 3×2 matrix. Let's calculate that matrix, moving slowly and taking care to explain the procedure cautiously. We'll call the product matrix P, so its entries can be referred to as $p_{i,j}$:

$$AB = \begin{bmatrix} 1 & 2 \\ 3 & 4 \\ 5 & 6 \end{bmatrix} \begin{bmatrix} 7 & 8 \\ 9 & 10 \end{bmatrix} = \begin{bmatrix} p_{1,1} & p_{1,2} \\ p_{2,1} & p_{2,2} \\ p_{3,1} & p_{3,2} \end{bmatrix}$$

Entry $p_{1,1}$ is found by multiplying the entries of row 1 of matrix A against the entries of column 1 of matrix B and adding those products. That would be $(1)(7) + (2)(9)$. Note that we move across the row of matrix A and down the column of matrix B, in order. This simplifies to 25, and that is the 1, 1 entry of the product matrix.

Entry $p_{1,2}$ is found by multiplying the entries of row 1 of matrix A against the column 2 entries of matrix B and adding those products. That would be $(1)(8) + (2)(10) = 28$, which is the 1, 2 entry of the product matrix.

Continuing in the same manner for the other entries of the product, we obtain the following result:

$$AB = \begin{bmatrix} 1 & 2 \\ 3 & 4 \\ 5 & 6 \end{bmatrix} \begin{bmatrix} 7 & 8 \\ 9 & 10 \end{bmatrix} = \begin{bmatrix} 25 & 28 \\ (3)(7)+(4)(9) & (3)(8)+(4)(10) \\ (5)(7)+(6)(9) & (5)(8)+(6)(10) \end{bmatrix}$$

$$= \begin{bmatrix} 25 & 28 \\ 57 & 64 \\ 89 & 100 \end{bmatrix}$$

For the product BA, recalling that B was a 2×2 matrix and that A was a 3×2 matrix, we see that the product is not defined, since the inner dimensions for BA are not equal.

EXAMPLE 5.23 Perform the matrix multiplication shown, if possible:

$$\begin{bmatrix} 2 & -3 & 1 \end{bmatrix} \begin{bmatrix} 2 & 4 \\ -1 & 5 \\ -2 & -3 \end{bmatrix}$$

SOLUTION

Since the first matrix is 1×3 and the second is 3×2, the multiplication is defined, and we will obtain a product that is a 1×2 matrix. For each i, j entry of the product matrix, we will use row i of the first matrix and column j of the second matrix:

$$[(2)(2) + (-3)(-1) + (1)\,(-2)\ (2)(4) + (-3)(5) + (1)(-3)] = [5 - 10]$$

The Multiplicative Inverse of a Square Matrix

Recalling that a square matrix is a matrix having as many rows as it has columns, we define the **multiplicative inverse of a square matrix A** (often simply referred to as the "inverse of A") to be the square matrix A^{-1}, having the same size as matrix A, such that $AA^{-1} = A^{-1}A = I_n$, where n is the number of rows/columns of matrix A.

EXAMPLE 5.24 Show that the matrices $A = \begin{bmatrix} -1 & 3 \\ 2 & -5 \end{bmatrix}$ and $A^{-1} = \begin{bmatrix} 5 & 3 \\ 2 & 1 \end{bmatrix}$ are inverses of one another.

SOLUTION

To show the matrices are inverses, we must compute the products (remember, commutativity is not assured, so we must compute both AA^{-1} and $A^{-1}A$) and show that both are I_2.

$$\begin{bmatrix} -1 & 3 \\ 2 & -5 \end{bmatrix} \begin{bmatrix} 5 & 3 \\ 2 & 1 \end{bmatrix} = \begin{bmatrix} -5+6 & -3+3 \\ 10+(-10) & 6+(-5) \end{bmatrix} = \begin{bmatrix} 1 & 0 \\ 0 & 1 \end{bmatrix} = I_2$$

$$\begin{bmatrix} 5 & 3 \\ 2 & 1 \end{bmatrix}\begin{bmatrix} -1 & 3 \\ 2 & -5 \end{bmatrix} = \begin{bmatrix} -5+6 & 15+(-15) \\ -2+2 & 6+(-5) \end{bmatrix} = \begin{bmatrix} 1 & 0 \\ 0 & 1 \end{bmatrix} = I_2$$

Since both products are I_2, we conclude that the matrices are inverses of one another.

It turns out, as we will see in the exercises, that if a matrix has an inverse, then that inverse is unique. Thus, it makes sense to talk about "the" inverse of matrix A. When a matrix has an inverse (not all square matrices do!), we say that it is **invertible** or that it is **nonsingular**. If a square matrix has no inverse, then we say that it is **noninvertible**, or **singular**. That terminology is a bit difficult to master, since "invertible" correlates to "nonsingular," while "noninvertible" correlates to "singular." This is unusual, but the terminology involving the word "singular" is inherited from a feature of matrices we will explore shortly.

The inverse of a matrix can be challenging to find. There is a shortcut technique one can apply to 2×2 matrices that we will ask you to uncover in the exercises, but in general the procedure is to follow the Gauss-Jordan reduction method. The procedure can be tedious but is generally not difficult. It is highly susceptible to computational error, however, and thus great care should be taken with all your computations. Even a slight mistake will lead to a result that is tragically far from the truth and is likely to spawn numerical values that are, to be blunt, horrific.

Conceptually, what we will do is take our matrix and "adjoin" to it an identity matrix. In this context, by "adjoin" we mean that we will double the width of the matrix, setting up the right half of the matrix so that it is an identity matrix of appropriate size. Then we will perform **elementary row operations** on this widened matrix until the left half of the expanded matrix is the identity matrix, at which point the right half of the matrix will be the inverse of our original matrix.

The elementary row operations one may perform on matrices are as follows:

- Interchange any two rows.
- Multiply each element of a row by a nonzero number.
- Add any constant multiple of every element of a row to the corresponding element of another row.

There are obvious problems here that we are stalling for the moment, prime among which is how we would determine if a given matrix had an inverse at all! We have said that finding an inverse can be a challenging procedure, so it is natural to ask if we should discuss methods for determining existence before embarking on the path to locating the inverse. Perhaps so, but we'll address that question after introducing the method.

Let's start with the matrix $A = \begin{bmatrix} 5 & 0 & 2 \\ 2 & 2 & 1 \\ -3 & 1 & -1 \end{bmatrix}$ and compute its inverse. The traditional method for computation of an inverse matrix can be tricky business. In some cases, when matrices are set up very nicely, the steps can work out with delightful ease. Most of the time, however, a good deal of ingenuity is involved, and we do not want to explore cases that are limited to the situation where things work out in an ideal manner (this would leave you with an unrealistic notion of the complexity of the problem!). Consequently, we are going to take a slightly different approach than is customary, one that will give you a systematic method that will work no matter how hideous the inverse matrix may become.

The first step is to "adjoin" the identity matrix having the same size as A to the right of matrix A:

$$\begin{bmatrix} 5 & 0 & 2 & 1 & 0 & 0 \\ 2 & 2 & 1 & 0 & 1 & 0 \\ -3 & 1 & -1 & 0 & 0 & 1 \end{bmatrix}$$

The objective now is to use the elementary row operations to transform this matrix into a form where the left half of the array is the identity matrix. Assuming that this can be done, the right half of the resulting matrix will be the inverse matrix we seek.

Our process, which we will not pretend is the most elegant or traditional possible, will be a methodical approach that will not rely on clever observations or trickery. The procedure may generate somewhat large numbers along the way, but things will never be so bad that we cannot handle them with relative ease.

Step One: **Multiply each row by a constant such that all the nonzero first column entries are the same, with the top nonzero entry positive and the others negative.**
Observe that the first column entries are 5, 2, and -3. The least common multiple of these numbers is 30, so we will multiply the top row by 6, the middle row by -15, and the bottom row by 10. This produces

$$\begin{bmatrix} 30 & 0 & 12 & 6 & 0 & 0 \\ -30 & -30 & -15 & 0 & -15 & 0 \\ -30 & 10 & -10 & 0 & 0 & 10 \end{bmatrix}$$

Step Two: **Add the first row to each of the rows having nonzero entries in column 1; only the rows being added to are changing in this process, and the first row is fixed:**

$$\begin{bmatrix} 30 & 0 & 12 & 6 & 0 & 0 \\ 0 & -30 & -3 & 6 & -15 & 0 \\ 0 & 10 & 2 & 6 & 0 & 10 \end{bmatrix}$$

Observe that the entries in the matrix are growing in size. If we wish, we can multiply each row by a suitable entry at any stage in order to remove common factors and thereby make the entries as small as possible as we proceed. This is not necessary, but you may wish to do so. In this case, we can multiply the top row by 1/6, the middle row by 1/3, and the third row by 1/2. This produces

$$\begin{bmatrix} 5 & 0 & 2 & 1 & 0 & 0 \\ 0 & -10 & -1 & 2 & -5 & 0 \\ 0 & 5 & 1 & 3 & 0 & 5 \end{bmatrix}$$

Step Three: **Now we focus on the second column. We want the nonzero entries to be the equal, with the middle entry this time being positive and the other nonzero entries negative.**

Since the 1, 2 entry is already 0, we ignore that entry and consider the other rows. The least common multiple of the existing entries is 10, so we multiply the second row by -1 and the bottom row by -2. This produces

$$\begin{bmatrix} 5 & 0 & 2 & 1 & 0 & 0 \\ 0 & 10 & 1 & -2 & 5 & 0 \\ 0 & -10 & -2 & -6 & 0 & -10 \end{bmatrix}$$

Step Four: **Add the second row to each of the rows having a nonzero entry in column 2:**

$$\begin{bmatrix} 5 & 0 & 2 & 1 & 0 & 0 \\ 0 & 10 & 1 & -2 & 5 & 0 \\ 0 & 0 & -1 & -8 & 5 & -10 \end{bmatrix}$$

Note that the matrix is gradually assuming the form we desire. None of the rows have a common factor for their entries, so we are forced to leave the entries as they are for the moment.

Step Five: **We now wish the nonzero third column entries to be equal, with the third row entry positive and the other entries negative.**

Here, the least common multiple of 2, 1, and 1 is 2, so we multiply the first row by -1, the second by -2, and the third by -2 to yield

$$\begin{bmatrix} -5 & 0 & -2 & -1 & 0 & 0 \\ 0 & -20 & -2 & 4 & -10 & 0 \\ 0 & 0 & 2 & 16 & -10 & 20 \end{bmatrix}$$

Step Six: **Add the third row to the rows having nonzero entries in the third column:**

$$\begin{bmatrix} -5 & 0 & 0 & 15 & -10 & 20 \\ 0 & -20 & 0 & 20 & -20 & 20 \\ 0 & 0 & 2 & 16 & -10 & 20 \end{bmatrix}$$

Step Seven: **We are nearly to the end. All that remains is to multiply each row by the reciprocal of the first nonzero entry in that row, and this will complete the process.**

Note that this could create fractions, but we have stalled off that possibility as long as possible. The first row must be multiplied by $\frac{-1}{5}$, the second by $\frac{-1}{20}$, and the third by $\left(\frac{1}{2}\right)$ This yields

$$\begin{bmatrix} 1 & 0 & 0 & -3 & 2 & -4 \\ 0 & 1 & 0 & -1 & 1 & -1 \\ 0 & 0 & 1 & 8 & -5 & 10 \end{bmatrix}$$

We conclude the inverse is $A^{-1} = \begin{bmatrix} -3 & 2 & -4 \\ -1 & 1 & -1 \\ 8 & -5 & 10 \end{bmatrix}$, which we can confirm by computing the product (in both orders) of A with its alleged inverse. We will suppress that work here and ask you to trust that we have checked its correctness!

EXAMPLE 5.25 Find the inverse of the matrix $C = \begin{bmatrix} 4 & 7 \\ -2 & 3 \end{bmatrix}$, given that the inverse truly does exist.

SOLUTION

As our first step, we adjoin the identity matrix to the matrix C:

$$\begin{bmatrix} 4 & 7 & 1 & 0 \\ -2 & 3 & 0 & 1 \end{bmatrix}$$

Next, we arrange things so that the first column entries are equal, with the top entry in the column being positive and the lower entry negative. The existing entries are 4 and -2, and they possess common multiple 4, so we merely need to multiply the entire second row by 2 in order to achieve our objective:

$$\begin{bmatrix} 4 & 7 & 1 & 0 \\ -4 & 6 & 0 & 2 \end{bmatrix}$$

The top row is now added to the bottom row, producing

$$\begin{bmatrix} 4 & 7 & 1 & 0 \\ 0 & 13 & 1 & 2 \end{bmatrix}$$

We now turn our attention to the second column. The existing entries are 7 and 13, which have common multiple 91. We wish the second row to have a positive 2, 2 entry and the top row to have a negative 1, 2 entry, so the top row is multiplied by -13 and the bottom by 7. This does, as we forewarned, produce rather large numbers, but they are integer valued and are not beyond our ability to handle:

$$\begin{bmatrix} -52 & -91 & -13 & 0 \\ 0 & 91 & 7 & 14 \end{bmatrix}$$

We add the second row to the first row, which gives us a diagonal matrix in the left half of our expanded array:

$$\begin{bmatrix} -52 & 0 & -6 & 14 \\ 0 & 91 & 7 & 14 \end{bmatrix}$$

In order to complete the computation of the inverse matrix, we now multiply each row by the reciprocal of its first nonzero entry. The first row is multiplied by $-\dfrac{1}{52}$ and the second row by $\dfrac{1}{91}$. This produces rather frightening fractions, but, as we never have to perform calculations with them, it is of little concern:

$$\begin{bmatrix} 1 & 0 & 6/52 & -14/52 \\ 0 & 1 & 7/91 & 14/91 \end{bmatrix}$$

Reducing our entries in the right half of the array, we find

$$\begin{bmatrix} 1 & 0 & 3/26 & -7/26 \\ 0 & 1 & 1/13 & 2/13 \end{bmatrix}$$

Hence, our inverse matrix is

$$C^{-1} = \begin{bmatrix} 3/26 & -7/26 \\ 1/13 & 2/13 \end{bmatrix}$$

A straightforward calculation reveals that both CC^{-1} and $C^{-1}C$ simplify to I_2, confirming the correctness of our work.

Testing for Invertibility

As we have seen, finding the inverse of a matrix can be a lengthy procedure. We had indicated prior to presenting the method for computation of the inverse that it would be desirable to have a means for determining the existence of an inverse prior to undertaking the process of calculating the matrix, and it is to that issue we now turn.

Of course, there is already a means at hand for us to determine that an inverse would fail to exist, though we have not spoken of it. It could well be that the process we were attempting in the examples would be impossible, and that would demonstrate the nonexistence of an inverse.

EXAMPLE 5.26 Show that the matrix $A = \begin{bmatrix} 3 & 4 \\ 12 & 16 \end{bmatrix}$ does not have an inverse.

SOLUTION

Proceeding as before, we attempt to calculate the inverse:

$$\begin{bmatrix} 3 & 4 & 1 & 0 \\ 12 & 16 & 0 & 1 \end{bmatrix}$$

The least common multiple of the first column entries is 12, so we multiply the first row by 4 and the second by -1, yielding

$$\begin{bmatrix} 12 & 16 & 4 & 0 \\ -12 & -16 & 0 & -1 \end{bmatrix}$$

Now we add the first row to the second, which gives the following rather unusual result:

$$\begin{bmatrix} 12 & 16 & 4 & 0 \\ 0 & 0 & 4 & -1 \end{bmatrix}$$

Observe that the first half of the second row has all 0 entries. It will be impossible to generate an entry of 1 in the 2, 2 entry, which is needed for us to transform the left half of the matrix into the identity matrix. Consequently, computation of the inverse is, in fact, impossible.

The previous example illustrates that it is possible to determine directly the failure for an inverse to exist. We can, if we wish, always proceed as in that example, knowing that (if our matrix is not invertible) we will ultimately reach a stage where the left half of one of the rows in our augmented matrix is all 0s and that this establishes noninvertibility.

There is an alternative to this approach that uses the **determinant of the matrix**. We will introduce computation of determinants using **cofactor expansion** across the first row of the matrix. We point out here that this is, in practice, not the most efficient manner of calculating a determinant, but the other methods rely on further exploration of matrices, and this would take us beyond the scope of our interest here. Parenthetically, we point out that in applications it is not uncommon to use matrices whose size exceeds 25×25 and that cofactor expansion of a square matrix with n rows typically requires at least $n!$ multiplications. For a 25×25 matrix, this would call for 25! computations, or approximately 1.5×10^{25}. If you had a computer capable of performing 1 billion multiplications per second, it would require more than 500,000,000 years to complete the determinant calculation! In our work, we will not see matrices this large, however.

For a 2×2 matrix, $A = [a_{i,j}]$, the determinant, which we symbolize as det A, or $|A|$, is defined by the formula $a_{11}a_{22} - a_{1,2}a_{2,1}$. This can be thought of as the difference of the products of the entries lying on the crossing diagonals of the matrix.

EXAMPLE 5.27 Find the determinant of the matrix $A = \begin{bmatrix} 3 & 5 \\ -7 & 2 \end{bmatrix}$.

SOLUTION

We compute the determinant by taking the product of the main diagonal entries 3 and 2 and subtracting the product of the entries on the other diagonal, -7 and 5. That is, $\det A = (3)(2) - (-7)(5) = 6 - (-35) = 6 + 35 = 41$.

Before proceeding, we should mention that the value of the determinant, among other things, provides us with the answer to our question: does A have an inverse? The answer is yes if the determinant of A is nonzero and no if the determinant of A is 0. Consequently, the matrix A in the example is invertible.

RULE: The matrix A has an inverse if and only if its determinant is nonzero.

We will not prove this result in this text, but it is a fundamental result established in a course on linear algebra.

For matrices larger than 2×2, we use a recursive definition, where each larger square matrix determinant is defined on the basis of the determinant of the next-smaller square matrix.

For $n \geq 2$, the determinant of an $n \times n$ matrix $A = [a_{i,j}]$ is computed according to the following rule:

$$\det A = \sum_{j=1}^{n} (-1)^{1+j} a_{1,j} \left[\det A_{1j} \right]$$

where A_{1j} is the $(n-1) \times (n-1)$ matrix obtained from matrix A by deleting row 1 and column j. In the formula, note that $a_{1,j}$ is the entry in row 1, column j, of the original matrix A. The formula looks intimidating to use but in practice is not particularly difficult, as is shown in the next example.

EXAMPLE 5.28 Compute the determinant of matrix $A = \begin{bmatrix} 2 & 5 & 0 \\ 3 & -4 & 1 \\ 2 & 0 & 5 \end{bmatrix}$.

SOLUTION

We will be working across the first row and will walk you through the computations cautiously. You will note that the factor in the summation $(-1)^{1+j}$ will cause an alternation in sign from term to term, beginning with a positive factor:

$$\det A = (-1)^{1+1}(2)\det \begin{bmatrix} -4 & 1 \\ 0 & 5 \end{bmatrix} + \ldots$$

We pause here to remark on where that matrix "came from." The value of j in the first term of the summation is 1, and this accounts for the form of the exponent on (-1). Also, the 2 is the 1, 1 entry of matrix A. The 2×2 matrix can be found if we examine matrix A and think of what would remain if the first row and first column of A were deleted. That is precisely the 2×2 matrix shown.

Continuing, we find

$$\det A = (-1)^{1+1}(2)\det \begin{bmatrix} -4 & 1 \\ 0 & 5 \end{bmatrix} + (-1)^{1+2}(5)\det \begin{bmatrix} 3 & 1 \\ 2 & 5 \end{bmatrix}$$
$$+ (-1)^{1+3}(0)\det \begin{bmatrix} 3 & -4 \\ 2 & 0 \end{bmatrix}$$

Notice that each of these computations relies on the ability to compute the determinant of a 2×2 matrix, but that has already been established. This is what is meant by a "recursive" definition: computation of determinants of a particular size relies on the determinants of matrices of the "next-smaller size."

Consequently, we simplify to obtain

$$\begin{aligned}\det A &= (-1)^2(2)\det\begin{bmatrix} -4 & 1 \\ 0 & 5 \end{bmatrix} + (-1)^3(5)\det\begin{bmatrix} 3 & 1 \\ 2 & 5 \end{bmatrix} \\ &\quad + (-1)^4(0)\det\begin{bmatrix} 3 & -4 \\ 2 & 0 \end{bmatrix} \\ &= (1)(2)[-20 - 0] + (-1)(5)[15 - 2] + (1)(0)[0 - (-8)] \\ &= -40 - 65 + 0 \\ &= -105\end{aligned}$$

The determinant of A has been shown to be nonzero, and hence we can conclude (though this was not part of the question at hand!) that A is invertible.

Notice that the computation of the determinant was made simpler by the appearance of 0s within the matrix. This fact underlies the rationale for other, faster techniques for computation of determinants that are presented in texts on linear algebra.

Exercises

For the following questions, answer in complete sentences using proper grammar and spelling.

1. What is meant by the order of a matrix, and how is it determined?
2. Explain the method for determining when two matrices can be multiplied.
3. How does one determine if two matrices are equal?
4. What is meant by the term "diagonal matrix"?
5. When two matrices have the same order, under what conditions can the matrices be added, subtracted, and multiplied?

Perform the following matrix operations or explain why the operations are impossible.

6. $\begin{bmatrix} 2 & 5 \\ -1 & 8 \\ 3 & -4 \end{bmatrix} - \begin{bmatrix} 4 & -7 \\ 4 & 3 \\ 5 & -2 \end{bmatrix}$

7. $\begin{bmatrix} 4 & -5 \\ 3 & 6 \end{bmatrix}\begin{bmatrix} 1 & -3 & 2 \\ 7 & -10 & 4 \end{bmatrix}$

8. $\begin{bmatrix} 5 & 0 & 1 \\ 2 & -3 & 7 \\ 2 & -1 & -3 \end{bmatrix}\begin{bmatrix} 5 \\ 7 \\ -1 \end{bmatrix}$

9. $\begin{bmatrix} 6 \\ 5 \end{bmatrix}\begin{bmatrix} 4 & -5 \\ 5 & 7 \end{bmatrix}$

10. $\begin{bmatrix} -2 \\ 5 \\ 6 \end{bmatrix}\begin{bmatrix} -5 & 4 & -6 \\ 3 & 10 & 2 \\ 1 & -2 & 4 \end{bmatrix}$

11. $\begin{bmatrix} 8 & -1 & 4 \end{bmatrix} + \begin{bmatrix} 3 \\ 1 \\ -4 \end{bmatrix}$

12. $\begin{bmatrix} 3 & -2 \end{bmatrix} + \begin{bmatrix} -4 & 7 \end{bmatrix}$

13. $3\begin{bmatrix} 4 & -5 \\ 8 & -1 \\ -3 & 7 \end{bmatrix} - 8\begin{bmatrix} -1 & 2 \\ 3 & -2 \\ 5 & 2 \end{bmatrix}$

14. $\frac{1}{3}\begin{bmatrix} 6 & 7 \\ 5 & 9 \end{bmatrix} - \frac{3}{4}\begin{bmatrix} 8 & 4 \\ 5 & 3 \end{bmatrix}$

In the following exercises, let the matrices A through F be given as follows. Perform the indicated operations or explain why the operations cannot be performed.

$$A = \begin{bmatrix} 1 & -2 \\ -4 & 5 \end{bmatrix},\ B = \begin{bmatrix} 5 & \frac{2}{3} & 4 \\ -1 & 2 & 5 \end{bmatrix},\ C = \begin{bmatrix} 1 & -2 & 0 \\ \frac{1}{3} & 5 & -2 \end{bmatrix},$$

$$D = \begin{bmatrix} 8 & -2 \end{bmatrix},\ E = \begin{bmatrix} -2 \\ -5 \\ 7 \end{bmatrix},\ F = \begin{bmatrix} 3 & 4 & -5 \\ 2 & -6 & 1 \\ 0 & -2 & 5 \end{bmatrix}$$

15. $B - C$
16. $B + E$
17. ED
18. DE
19. FE
20. AD
21. BC
22. A^2
23. CB
24. F^2
25. $C + 7B$
26. $B - C$
27. $\frac{2}{3}B - \frac{1}{4}C$
28. AD
29. $C - 8A$

For the following matrices, find their determinants, if possible. If it is not possible to find their determinants, explain why.

30. $\begin{bmatrix} 3 & 4 \\ -7 & 2 \end{bmatrix}$

31. $\begin{bmatrix} -2 & 5 \\ 3 & -1 \end{bmatrix}$

32. $\begin{bmatrix} 1 & -2 & 1 \\ 4 & 2 & 3 \end{bmatrix}$

33. $\begin{bmatrix} 4 \\ -2 \\ 5 \end{bmatrix}$

34. $\begin{bmatrix} 1 & -1 & 9 \\ 0 & 0 & -4 \\ 3 & -1 & 1 \end{bmatrix}$

35. $\begin{bmatrix} 2 & 0 & -3 \\ 1 & 5 & -1 \\ 2 & -4 & 3 \end{bmatrix}$

36. $\begin{bmatrix} 3 & 2 & 6 \\ 1 & 1 & 2 \\ 2 & 2 & 5 \end{bmatrix}$

37. $\begin{bmatrix} 1 & 1 & 1 \\ 2 & -3 & 1 \\ 3 & -1 & 4 \end{bmatrix}$

For the following pairs of matrices A and B, compute the products AB and BA and use the results to determine if the matrices are inverses of one another.

38. $A = \begin{bmatrix} 7 & -6 \\ -5 & 4 \end{bmatrix},\ B = \begin{bmatrix} 1 & 2 \\ 1 & -3 \end{bmatrix}$

39. $A = \begin{bmatrix} -2 & -1 \\ -1 & 1 \end{bmatrix},\ B = \begin{bmatrix} 1 & 1 \\ 1 & 2 \end{bmatrix}$

40. $A = \begin{bmatrix} 2 & 5 \\ -1 & 4 \end{bmatrix}, B = \begin{bmatrix} 4/13 & -5/13 \\ 1/13 & 2/13 \end{bmatrix}$

41. $A = \begin{bmatrix} 3 & -7 \\ -4 & 4 \end{bmatrix}, B = \begin{bmatrix} -1/4 & -7/16 \\ -1/4 & -3/16 \end{bmatrix}$

42. $A = \begin{bmatrix} -2 & 1 & -1 \\ -5 & 2 & -1 \\ 3 & -1 & 1 \end{bmatrix}, B = \begin{bmatrix} 1 & 0 & 1 \\ 2 & 1 & 3 \\ -1 & 1 & 1 \end{bmatrix}$

43. $A = \begin{bmatrix} -1/19 & -5/19 & 2/19 \\ -4/19 & -1/19 & 8/19 \\ 9/19 & 7/19 & 1/19 \end{bmatrix}, B = \begin{bmatrix} 3 & -1 & 2 \\ -4 & 1 & 0 \\ 1 & 2 & 1 \end{bmatrix}$

For the following matrices, find the inverse matrix (if it exists).

44. $\begin{bmatrix} 3 & 5 \\ -1 & 3 \end{bmatrix}$

45. $\begin{bmatrix} 2 & 3 \\ -3 & 4 \end{bmatrix}$

46. $\begin{bmatrix} 1 & 8 \\ -1 & 6 \end{bmatrix}$

47. $\begin{bmatrix} 2 & -3 \\ 3 & 5 \end{bmatrix}$

48. $\begin{bmatrix} 4 & -1 & 0 \\ 3 & 1 & 1 \\ 1 & 2 & 1 \end{bmatrix}$

49. $\begin{bmatrix} 1 & 5 & -1 \\ 0 & 3 & -1 \\ 2 & 4 & -1 \end{bmatrix}$

50. $\begin{bmatrix} 7 & -2 & 3 \\ 1 & 0 & 2 \\ 4 & 1 & 1 \end{bmatrix}$

51. $\begin{bmatrix} 3 & 2 & 3 \\ 5 & 10 & 2 \\ 8 & 3 & -4 \end{bmatrix}$

52. $\begin{bmatrix} -2 & 5 & 1 \\ 7 & -2 & 6 \\ 4 & -1 & -3 \end{bmatrix}$

53. $\begin{bmatrix} 1 & 1 & 1 \\ 2 & -3 & 1 \\ 3 & -1 & 4 \end{bmatrix}$

5.5 Using Matrices to Solve Systems of Linear Equations

With the basics of matrices in our hands and well under control, it is time to see how they can be used to solve systems of linear equations. Because the known methods of substitution and elimination work well and efficiently for systems of two linear equations, we typically apply the matrix method only to systems having at least three variables, but the method works for two-equation systems as well. Equations involving three or more variables are considered linear in the case where each of the variables is raised to the power one only and no term contains more than a single variable.

An example of such a system would be

$$\begin{aligned} x+y+z&=10 \\ 2x-3y+z&=4 \\ 3x-y+4z&=0 \end{aligned}$$

Systems with more than two variables also have "points" as their solution, but they are no longer ordered pairs. The points will have one coordinate for each variable occurring within the system, and thus the shown system would have an **ordered triple** as its solution point. When three coordinates are involved, the standard ordering of coordinates is alphabetical, as (x, y, z), but if more than three variables are present in a system, we typically dispense with distinct letters and use subscripted variables such as x_1, x_2, x_3, x_4, and so on. This indicates to us the order of occurrence of the variables as coordinates within the point.

It is not uncommon to find systems of extremely large size occurring within applications. When aircraft construction companies create computer models to examine airflow around experimental aircraft, the systems can have as many as 2 million equations and variables! Such systems are, of course, impractical to solve by hand and require the use of some of the world's fastest supercomputers for their solution.

When we attempt to use matrices to solve systems, we express the left side of the equation as a matrix product and the right side as a one-column matrix. The result will be a matrix equation of the form $AX = B$.

If we view the equation in this way, it is natural to suppose that the equation can be solved by "dividing" each side of the equation by the matrix A, but we have never seen an exhibition of matrix division (for very good reason—such a process is undefined!). However, if we were to multiply each side of the equation on the left by A^{-1}, note that we can solve the equation for X:

$$\begin{aligned} A^{-1}AX &= A^{-1}B \\ X &= A^{-1}B \end{aligned}$$

Naturally, this process relies on the existence of A^{-1}, but if the inverse does not exist, then the system of linear equations does not have a unique solution. Let's see how the method would work by examining the system once more and producing the matrices involved.

Recall that we were considering

$$\begin{aligned} x+y+z&=10 \\ 2x-3y+z&=4 \\ 3x-y+4z&=0 \end{aligned}$$

To express this as a matrix equation, we will use two matrices on the left: a **coefficient matrix** (whose entries are the coefficients of the equations on the left side of the equations) and a **variable matrix** (a single-column matrix whose entries are the variables of the system). The right side of the matrix equation will be a single-column matrix whose entries are the constants to the right of the equals signs.

Thus, in this case, $A = \begin{bmatrix} 1 & 1 & 1 \\ 2 & -3 & 1 \\ 3 & -1 & 4 \end{bmatrix}$, $X = \begin{bmatrix} x \\ y \\ z \end{bmatrix}$, and $B = \begin{bmatrix} 10 \\ 4 \\ 0 \end{bmatrix}$.

With these matrices, the system of equations is equivalent to the matrix equation $AX = B$, which is easily seen if one multiplies the matrices A and X, and we recall the definition of equality of matrices.

The determinant of matrix A is -9 (see Exercise 37 in Section 5.4), and hence A is invertible, and the system has a solution. The inverse of matrix A is $A^{-1} = \begin{bmatrix} 11/9 & 5/9 & -4/9 \\ 5/9 & -1/9 & -1/9 \\ -7/9 & -4/9 & 5/9 \end{bmatrix}$ (see Exercise 53 in Section 5.4), and hence we can find the solution to the system as

$$X = \begin{bmatrix} 11/9 & 5/9 & -4/9 \\ 5/9 & -1/9 & -1/9 \\ -7/9 & -4/9 & 5/9 \end{bmatrix} \begin{bmatrix} 10 \\ 4 \\ 0 \end{bmatrix}$$

$$X = \begin{bmatrix} 130/9 \\ 46/9 \\ -86/9 \end{bmatrix}$$

The solution to the system is the ordered, triple $\left(\frac{130}{9}, \frac{46}{9}, \frac{86}{9} \right)$.

EXAMPLE 5.29 Solve the system of linear equations

$$\begin{aligned} x + 3y + 4z &= -3 \\ x + 2y + 3z &= -2 \\ x + 4y + 3z &= -6 \end{aligned}$$

, using the inverse matrix method.

SOLUTION

The system of equations can be expressed as the matrix product $AX = B$, where we have the matrices $A = \begin{bmatrix} 1 & 3 & 4 \\ 1 & 2 & 3 \\ 1 & 4 & 3 \end{bmatrix}$, $X = \begin{bmatrix} x \\ y \\ z \end{bmatrix}$, and $B = \begin{bmatrix} -3 \\ -2 \\ 6 \end{bmatrix}$. A straightforward computation reveals that the determinant of the matrix A is $\left(\frac{1}{2} \right)$, and therefore the inverse matrix exists, and the system has a unique solution.

Using our technique of Section 5.4, we find the inverse matrix to be $A^{-1} = \begin{bmatrix} -3 & 7/2 & 1/2 \\ 0 & -1/2 & 1/2 \\ 1 & -1/2 & -1/2 \end{bmatrix}$, and thus the solution to the system can be found by computation of $A^{-1}B$. The computation yields the product matrix $\begin{bmatrix} -1 \\ -2 \\ 1 \end{bmatrix}$,

and hence the solution to the system of equations is the ordered triple $(-1, -2, 1)$.

We can confirm this result by substitution into the equations of the original system, demonstrating the coordinates satisfy the original system equations simultaneously:

$$(-1) + 3(-2) + 4(1) = -1 - 6 + 4 = -3 \surd$$
$$(-1) + 2(-2) + 3(1) = -1 - 4 + 3 = -2 \surd$$
$$(-1) + 4(-2) + 3(1) = -1 - 8 + 3 = -6 \surd$$

EXAMPLE 5.30

Solve the system of linear equations $\begin{aligned} 2x+4y+10z&=3 \\ 4x+6y+16z&=10 \\ -2x+2y+4z&=5 \end{aligned}$ using the inverse matrix method.

SOLUTION

The coefficient matrix of the system is $A = \begin{bmatrix} 2 & 4 & 10 \\ 4 & 6 & 16 \\ -2 & 2 & 4 \end{bmatrix}$, which has determinant -8, and hence is invertible. The inverse matrix is found using our method of Section 5.4 and is equal to the matrix $A^{-1} = \begin{bmatrix} 1 & -1/2 & -1/2 \\ 6 & -7/2 & -1 \\ -5/2 & 3/2 & 1/2 \end{bmatrix}$.

The solution to the system is found by computation of the matrix product $X = A^{-1}B = \begin{bmatrix} 1 & -1/2 & -1/2 \\ 6 & -7/2 & -1 \\ -5/2 & 3/2 & 1/2 \end{bmatrix} \begin{bmatrix} 3 \\ 10 \\ 5 \end{bmatrix} = \begin{bmatrix} -9/2 \\ -22 \\ 10 \end{bmatrix}$. As in the previous example, this indicates that the solution to the system of equations is the ordered triple $\left(\frac{-9}{2}, -22, 10 \right)$.

Exercises

Solve each of the following systems of linear equations, using the matrix method.

1. $4x - 3y = 1$
   $3x + y = 9$

2. $8x - 14y = -46$
   $2x + 5y = -3$

3. $-8x + 6y = 6$
   $4x - 6y = -5$

4. $6x - 5y = 1$
   $8x + 3y = 11$

5. $2x+9y=5$
$3x+2y=-4$

6. $2x-y=-2$
$6x+4y=22$

7. $\frac{3}{4}x+y=\frac{7}{2}$
$\frac{1}{3}x+2y=4$

8. $x-\frac{1}{2}y=7$
$\frac{1}{3}x+\frac{1}{2}y=5$

9. $2x-3y+z=11$
$3x-y+z=12$
$x-2y-5z=-5$

10. $5x-6y-4z=-47$
$7x+8y-2z=27$
$3x+y+4z=14$

11. $3x-2y+z=-1$
$5x+3y=13$
$x+y-2z=13$

12. $4x-y+2z=9$
$3x-3y+z=19$
$2x-y+3z=3$

13. $x+7y-2z=30$
$3x+2y-4z=25$
$2x+4x-z=18$

14. $x-2y+z=-8$
$3x+y-z=5$
$2x-4y+2z=-16$

Summary

In this chapter, you learned about:

- Solving systems of linear equations by the substitution method and the method of elimination.
- Conversions into linear equations by substitution.
- Matrices, their attributes, and their operations; equality, addition, subtraction, multiplication, and multiplicative inverse; and solving systems of linear equations.

Glossary

additive identity matrix A matrix all of whose entries are 0s. Also called a zero matrix.

associative A reorganization of mathematical operations using symbols of grouping, which does not change its outcome.

coefficient matrix A matrix created using the coefficients of the variables of a linear system of equations.

cofactor expansion A technique for calculating the determinant of a matrix.

column of a matrix Any of the vertical alignments of entries of a matrix.

commutative An operation whose order can be reversed without changing its outcome.

determinant of a matrix A numerical value corresponding to a particular square matrix, used to determine invertibility.

diagonal matrix A square matrix whose entries not on the main diagonal are all 0.

elementary row operations Adding a multiple of any row of a matrix to another, multiplication of any row of a matrix by a nonzero scalar, and interchanging of rows of a matrix.

elimination One of the techniques of solving systems of equations.

entries The numbers within a matrix.

equal matrices Two matrices having the same order and identical entries in corresponding locations.

identity A true statement of equality.

identity matrix A square matrix having 1s in all main diagonal entries and 0s elsewhere.

inconsistent system A system of equations having no solution.

invertible matrix A square matrix for which a multiplicative inverse exists.

main diagonal In a square matrix, those entries occurring in row k, column k.

matrix An array of real numbers.

multiplicative inverse of a square matrix A A second square matrix, named A^{-1}, having the same size as matrix A, such that $AA^{-1} = A^{-1}A = I$.

noninvertible matrix A matrix for which no multiplicative inverse exists.

nonsingular matrix An invertible matrix.

order The size of a matrix, given as $m \times n$, where m is the number of rows and n the number of columns within the matrix.

ordered triple A three-coordinate expression identifying a location in three-dimensional real space.

row of a matrix Any of the horizontal alignments of entries within a matrix.

scalar A number not appearing within any particular mathematical context.

singular matrix A square matrix having no inverse.

square matrix A matrix having as many rows as columns.

substitution One of the techniques of solving systems of equations.

variable matrix A column matrix whose entries are the variables of a linear system of equations.

zero matrix A matrix all of whose entries are 0s. Also called an additive identity matrix.

End-of-Chapter Problems

Solve the following systems using the method of substitution.

1. $x - y = 3$
   $8x + y = 7$
2. $x + y = 1$
   $7x + 7y = 7$
3. $x + 3y = 9$
   $2x + 5y = 14$
4. $2x - 3y = 7$
   $3x + 4y = 11$

The following problems can be solved by constructing and solving a suitable system of linear equations. Construct the appropriate system, solve using the method of substitution, and then interpret the solution in the context of the problem.

5. Toys of type A cost \$0.20, and toys of type B cost \$0.30. You want to spend \$7.00 and buy 40 toys. How many toys of type A do you need to buy and how many of type B? Does your answer make sense?
6. The demand for computers on a college campus is given by the equation $2342 - 0.01p$, where p is the price in dollars. The supply is given by the equation $253 + 0.01p$. At what price are the supply and demand equal?

7. One train leaves a point 37 miles east of Chicago, heading east at 60 miles per hour. At the same time, another train leaves a point 3 miles east of Chicago, heading east at 90 miles per hour. When will the second train overtake the first, and how far east of Chicago will they be?

8. George has \$1.05 in pennies and nickels. He has 15 more pennies than nickels. How many of each kind of coin does he have?

9. You want to eat 250 grams of food, and consume a total of 1500 calories. Carbohydrates run 4 calories/gram, while fat runs 9 calories/gram. How many grams of carbohydrates should you eat and how many grams of fat?

Solve the following systems of linear equations using the method of elimination.

10. $3x-4y=7$
    $9x-3y=37$

11. $x-y=5$
    $3x-2y=7$

12. $\frac{3}{2}x+7y=19$
    $4x-5y=13$

13. $4x+2y=\frac{1}{3}$
    $3x+3y=2$

The following application problems can be solved through the use of a system of linear equations. Construct a system appropriate to each application example, solve the system using the method of elimination, and interpret the solution in the context of the problem.

14. The County Fair admission fee is \$8 for adults and \$6 for seniors; 5432 people were admitted, and \$38,024 was collected. How many adults and how many seniors were admitted?

15. Suppose you mix x liters of one fluid with y liters of another and end up with 30 liters of fluid weighing 40 kg. Furthermore, suppose that the first fluid weighs 1 kg/liter and that the second fluid weighs 1.5 kg/liter. What are the values of x and y?

16. You can buy shingles at store A for \$175 per square, but the store charges a flat rate of \$25 for delivery. At store B, shingles are only \$160 per square, but the charge for delivery is \$85. What is the break-even point, where the total cost is the same from the two stores?

17. 14-karat gold is $\frac{14}{24}$ gold, while 24-karat gold is pure gold. How many grams of 14-karat gold must be mixed with 24-karat gold to obtain 20 grams of 18-karat $\left(\frac{18}{24}\right)$ gold?

18. A cell phone company has two different plans. In the first plan, you pay \$10 per month plus \$0.25/minute of calling time. In the second plan, you pay \$20 per month but only \$0.10/minute of calling time. How many minutes do you need to talk per month to make the two plans cost the same?

Solve the following systems of linear equations using an appropriate substitution.

19. $\frac{3}{x+2} - \frac{4}{2y-3} = 7$

$\frac{5}{x+2} + \frac{15}{2y-3} = 12$

20. $\frac{8}{3-x} + \frac{7}{3y+1} = 17$

$\frac{12}{3-x} - \frac{2}{3y+1} = 33$

21. $\frac{3}{x-2} + \frac{12}{3-y} = 12$

$\frac{5}{x-2} - \frac{17}{3-y} = 5$

22. $\frac{1}{2x-3} + \frac{3}{5y+4} = 7$

$\frac{3}{2x-3} - \frac{5}{5y+4} = 5$

Perform the following matrix operations or explain why the operations are impossible.

23. $\begin{bmatrix} 1 & 2 \\ 3 & 5 \end{bmatrix}\begin{bmatrix} 5 & 6 \\ 2 & 3 \end{bmatrix}$

24. $\begin{bmatrix} 5 & 3 \\ 2 & 4 \end{bmatrix} + \begin{bmatrix} 8 & 12 & 3 \\ 5 & 7 & 8 \end{bmatrix}$

25. $\begin{bmatrix} 8 & 7 & 12 \\ 3 & 6 & 9 \\ 8 & 0 & 1 \end{bmatrix} - \begin{bmatrix} 5 & 8 & 6 \\ 0 & 10 & 3 \\ 4 & 5 & 9 \end{bmatrix}$

26. $\begin{bmatrix} 2 & 8 & 3 \\ 5 & 0 & 4 \end{bmatrix}\begin{bmatrix} 1 & 8 & 4 \\ 2 & 0 & 1 \\ 3 & 5 & 7 \end{bmatrix}$

In the following exercises, let the matrices *A* through *D* be given as follows. Perform the indicated operations or explain why the operations cannot be performed.

$$A = \begin{bmatrix} 1 & 5 \\ 2 & 3 \end{bmatrix},\ B = \begin{bmatrix} 8 & 8 \\ 3 & 9 \end{bmatrix},\ C = \begin{bmatrix} 1 & 4 & 5 \\ 2 & -3 & 4 \end{bmatrix},\ D = \begin{bmatrix} 4 & 1/2 \\ 2 & 0 \\ 3 & 4 \end{bmatrix}$$

27. $A + B$
28. $A + C$
29. $A - B$
30. AC
31. DC
32. A^2
33. $2A - 3D$
34. $3B - A$

For the following matrices, find their determinants, if possible. If it is not possible to find their determinants, explain why.

35. $\begin{bmatrix} 3 & 8 \\ 4 & 2 \end{bmatrix}$

36. $\begin{bmatrix} 1 & 0 & 2 \\ 3 & 3 & 4 \\ 2 & 8 & 3 \end{bmatrix}$

37. $\begin{bmatrix} 2 & 5 \\ 6 & 15 \end{bmatrix}$

38. $\begin{bmatrix} 2 & 8 \\ 3 & 1 \\ 5 & 2 \end{bmatrix}$

For the following pairs of matrices A and B, compute the products AB and BA and use the results to determine if the matrices are inverses of one another.

39. $A=\begin{bmatrix} 2 & -1/4 \\ -1/2 & 1 \end{bmatrix}, B=\begin{bmatrix} 1/2 & -4 \\ -2 & 1 \end{bmatrix}$

40. $A=\begin{bmatrix} 1 & 1/2 \\ 1/3 & 2 \end{bmatrix}, B=\begin{bmatrix} 12/11 & -3/11 \\ -2/11 & 6/11 \end{bmatrix}$

41. $A=\begin{bmatrix} 1 & -1/2 & -1/4 \\ 3/2 & -1/2 & 1/3 \\ 3 & -1 & 1/4 \end{bmatrix}, B=\begin{bmatrix} 1 & -2 & -4 \\ 1/2 & 3/5 & 1 \\ -1/4 & 1 & 1/2 \end{bmatrix}$

42. $A=\begin{bmatrix} 3 & -1 & 0 \\ -2 & 4 & 1 \\ 1/2 & -3/2 & -1 \end{bmatrix}, B=\begin{bmatrix} 5/12 & 1/6 & 1/6 \\ 1/4 & 1/2 & 1/2 \\ -1/6 & -2/3 & -5/3 \end{bmatrix}$

For the following matrices, find the inverse matrix (if it exists).

43. $\begin{bmatrix} 1 & 3 \\ 2 & 4 \end{bmatrix}$

44. $\begin{bmatrix} 4 & 2 & 3 \\ 5 & 1 & 2 \\ 8 & 7 & 3 \end{bmatrix}$

45. $\begin{bmatrix} 2 & 8 \\ 3 & 12 \end{bmatrix}$

46. $\begin{bmatrix} 3 & 7 & 5 \\ 4 & 6 & 2 \\ 15 & 35 & 25 \end{bmatrix}$

Solve each of the following systems of linear equations using the matrix method.

47. $2x+3y=7$
$\frac{x}{2}-7y=3$

48. $5x+7y=35$
$x+y=2$

49. $5x+3y+4z=12$
$3x+7y-2z=14$
$7x-4y+3z=35$

50. $-2x+3y-4z=17$
$3x-2y+3z=19$
$x-4y+2z=21$

Chapter 6 Sequences and Series

The concept of sequences and series is widely applicable to events in everyday life. The feature common to such events is often the existence of a pattern repeated with successively increasing or decreasing values, but it is also possible that the values would either be or would eventually become constant. An illustration of such events are a bouncing ball whose height decreases in a predictable manner over time or compound interest, where the values of an investment increase over time because of the accumulation of interest.

If we can develop an understanding of sequences, we stand a very good chance of being able to assess and interpret situations where these types of events occur and to construct models that describe and predict the future conditions of those situations.

Objectives

When you have successfully completed the materials of this chapter, you will be able to:

- Differentiate between sequences and series.
- Understand the concepts of arithmetic and geometric progressions.
- Understand the principles of mathematical induction.
- Understand and apply the binomial theorem.

6.1 Sequences and Summation Notation

An Introduction to Sequences

There are situations in which we encounter a progression of real numbers, which may be randomly generated or which may follow some predetermined pattern. If the order of the numbers within the progression is significant (that is, if we consider a reordering of the numbers to be a different progression from the original), then we say that the progression is a **sequence** of real numbers. Keep this idea of the importance of order in mind: if you change the order of the numbers in the progression, you change the sequence! In advanced mathematics classes, when sequences are studied in greater detail, such modifications are referred to as a "rearrangement" of the sequence.

Much like the sets we considered earlier, the numbers within the sequence are referred to as the **members of a sequence**, or elements of the sequence (sometimes called the **terms of a sequence**). The number of terms within the progression is referred to as the **length of a sequence**, and that length may be either finite or infinite. As mentioned at the outset and unlike what we saw when we examined sets, the ordering of the members is relevant, and elements may appear more than once within the progression.

The terms of a sequence are generally identified by their order of occurrence, where we speak of the first term, second term, third term, and so forth. This is typically accomplished by using a lowercase subscripted letter, where the subscript identifies the number's position within the progression. For instance, the first term of the sequence might be identified as a_1, the second by a_2, the third by a_3, and so on. There are other notations in common use, such as $a(1)$, $a(2)$, $a(3)$, and so on, but we'll use the subscripted notation in this book.

Formally, a sequence is a function that associates with each positive number n another value, a_n. That is, if our progression of numbers is given by the list 2, 4, 6, 8, 10, …, then we can say $f(1) = 2, f(2) = 4, f(3) = 6, f(4) = 8, f(5) = 10$, and so forth. Typically, the function notation is not used in the context of sequences;

rather, we use the subscripted lowercase letter a_n. We would say $a_1 = 2$, $a_2 = 4$, $a_3 = 6$, $a_4 = 8$, $a_5 = 10$, and so forth.

Depending on the nature of the sequence, we have a variety of possible "presentations" for the terms. If a sequence is finite and of relatively short length, the terms can be given as an ordered listing of the terms. If it is finite but rather lengthy and if the elements follow a particular pattern that is easily detected by the reader, then enough terms can be given to establish the pattern with clarity, followed by ellipsis and the final term of the sequence. Naturally, the perception of what constitutes "lengthy" could vary from person to person, so it is a matter of personal judgment when determining whether an ellipsis should be used, but it is mandatory that enough terms be shown in order to ensure the clarity of the pattern.

EXAMPLE 6.1

1. 2, 4, 6, 8, 10, 12 This is an illustration of a finite sequence. Since the list is relatively brief, we have simply listed the elements in order of occurrence.
2. 1, 3, 5, 7, ..., 99 This finite sequence consists of the positive odd integers up to and including 99. Notice that the pattern is clearly indicated through the list of the first few terms in the progression, and hence there is no ambiguity in the determination of the nature of the suppressed terms of the sequence.
3. 1, 1, 2, 3, 5, ..., 233 This is also a finite sequence whose members are following a particular pattern. Observe that there is a repeated element, since 1 occurs as both the first and the second term of the sequence. The implication is that the terms of the sequence are following a definite pattern, one that we should be able to detect. A bit of contemplation reveals that the elements of the sequence are produced in the following manner: two successive terms are added to find the next term in the progression. That is, $1 + 1 = 2$, $1 + 2 = 3$, $2 + 3 = 5$, and so on. This is the beginning of a sequence that, if allowed to continue, forms the "Fibonacci sequence," which has application in perspective geometry in the construction of the so-called Golden Spiral.
4. $1, -1, 2, -2, \ldots, -100$. This progression contains alternately positive and negative integers, whose absolute values are increasing in value to 100.
5. $\frac{1}{2}, \frac{2}{3}, \frac{3}{4}, \ldots$ This progression contains fractions in which the numerator and denominator increase by 1 with each successive term. It seems evident that the following terms would be $\frac{4}{5}, \frac{5}{6}, \frac{6}{7}$, and so forth.

It is essential, when using ellipsis in giving the list of terms in the sequence, that the pattern of numbers in the progression be self-evident. It would not be appropriate, for instance, to give a sequence as $\{1, -3, 4, 7, -2, \ldots, 100\}$, since there is no readily apparent pattern that would allow a reader to determine the terms lying between -2 and 100 in the progression or how many terms would occur within the progression. Consequently, we should always take care

Note

When giving the terms of a sequence using a list of terms, be sure that the pattern followed by subsequent terms is absolutely clear! Readers should never be in doubt regarding the values of subsequent terms!

to ensure that our progression pattern makes obvious the intervening values of the terms of the sequence.

The General Term of a Sequence

Infinite sequences, which are somewhat more common in applications, are typically given by a formula with which any particular term of the sequence can be computed. One formulaic method for describing the terms of a sequence is by presentation of a **general term for a sequence** (an expression into which a value of n can be inserted to compute the nth term of the progression). One format commonly used is the notation $\{a_n\}$, which indicates we shall use a_n as a genaeral descriptor for the terms of the sequence and then give an expression for a_n, where n is taken to be the independent variable. The formula should be defined for all natural numbers; otherwise, one or more of the terms in the progression would be impossible to calculate.

EXAMPLE 6.2

Give the first five terms of the sequence whose general term is shown:

a. $\{a_n\}$, where $a_n = 4^n$

b. $\{a_n\}$, where $a_n = \dfrac{3+n}{n}, n>1$

SOLUTION

a. In this case, the nth term of the sequence is calculated by inserting particular values for n in the expression 4^n. That is, the first term is $4^1 = 4$, the second term is $4^2 = 16$, the third term is $4^3 = 64$, and the fourth term is $4^4 = 256$.

b. The second example also uses a formula to generate each term of the sequence. In this case, the first term a_1 is found by inserting 1 for n in the expression, producing $a_1 = \dfrac{3+1}{1} = \dfrac{4}{1} = 4$. The second term is found by inserting 2 for n in the expression, yielding $a_2 = \dfrac{3+2}{2} = \dfrac{5}{2}$. Continuing, we find the third and fourth terms to be $a_3 = \dfrac{6}{3} = 2$ and $a_4 = \dfrac{7}{4}$.

Note that the use of lowercase a is not mandatory, though it is common. You might just as easily choose to express the general term of a sequence as $\{b_n\} = \{3n^2 - 5\}$, in which case we would identify the first four terms of the sequence as $b_1 = -2$, $b_2 = 7$, $b_3 = 22$, and $b_4 = 43$. In other words, choice of the lowercase letter is essentially arbitrary.

The use of a general term to present the terms of a sequence is particularly useful if we wish to determine a specific term within the sequence. For instance, in Example 6.2(b), we might wonder what the two-hundredth term in the sequence might be. It is found easily enough by substitution of 200 into the expression in place of n. Here, that would yield $a_{200} = 203/200$.

EXAMPLE 6.3

For each of the following sequences, calculate the value of the two-hundredth term:

a. $a_n = (-1)^n \left(\frac{4n-5}{2n} \right)$

b. $b_n = n - n^2$

SOLUTION

a. Substituting 200 for n, we find $a_{200} = (-1)^{200} \left[\frac{4(200)-5}{2(200)} \right] = \frac{795}{400} = \frac{159}{80}$.

b. Again, we substitute 200 for n and obtain $b_{200} = 200 - 200^2 = 200 - 40{,}000 = -39{,}800$.

In the preceding examples, we were given a formula for the terms of the sequence, but this question can be reversed. That is, we can be given the terms of the sequence and asked to develop a general term formula for the progression. Typically, this is a much more challenging problem to solve and is complicated by the possibility that more than one suitable formula could exist. Although we will not prove it, if we are given a finite set of terms of a sequence, there will be infinitely many possible formulas that will generate the shown terms. Consequently, when you are asked to produce such a formula, you should not be overly concerned if your answer does not match the one we present in the solutions section of the textbook. Provided that your formula produces the desired terms in the indicated order, it is and should be considered correct.

EXAMPLE 6.4

Produce the general term for the sequences shown:

a. 1, 4, 7, 10, ...

b. 2, 5, 10, 17, ...

SOLUTION

a. The sequence appears to be generated by starting with 1 and adding 3 repeatedly to obtain successive terms. One possible formula we can use would be $a_n = 3n - 2$. Note that substitutions of $n = 1, 2, 3, \ldots$ will yield the terms of the sequence. While it is conceivable that other general term formulas may exist, production of any particular one suffices, and we can accept this as our answer.

b. This sequence is a bit more challenging. Observe that the successive terms are not, as was the case in (a), obtained by repeatedly adding a particular constant to the preceding term. We happen to recognize (after some consideration and experimentation!) that the terms of the sequence are always precisely 1 more than the value of a perfect square. That is, the first term is $1^2 + 1$, the second term is $2^2 + 1$, the third term is $3^2 + 1$, and so on. Consequently, one form for the general term of the sequence is $a_n = n^2 + 1$.

Let's make a remark about the preceding example, tying it to the immediately preceding statement. We had said that we could find other formulas that would produce the same progression, and we should demonstrate this to convince you that it can be done. The formula we mentioned in the solution of the example may be the most obvious, but consider the general term $b_n = \frac{6n+7}{2} - \frac{11}{2}$. Note that the first four terms of the sequence are $b_1 = 1$, $b_2 = 4$, $b_3 = 7$, and $b_4 = 10$. This list is identical with the list given in the example, and thus b_n should be accepted as a general term expression for the sequence.

There is no specific technique we can suggest for obtaining the general term of a sequence. In practice, the creation of such formulas is often the outcome of trial and error, experimentation, and experience. While construction of these expressions is challenging, we encourage you to persevere! Think of the derivation of such a formula as a logical puzzle that (we guarantee!) will be within the scope of your ability, at least for all the examples we encounter in this text.

Sequences Defined Using a Recursive Relationship

In some cases, the terms of a sequence are defined using a **recursive relationship**. By this, we mean that the initial term (or possibly initial few terms) of the sequence is stated explicitly and forms what we call the basis of the recursion relation. This is followed by a recursive relation, where subsequent terms of the sequence are defined in terms of one or more of their predecessors. Such relationships are common in problems of finance, biology, and physics, where future values of variables depend on the value of the variable at some earlier point in time.

We'll examine some applications shortly, but let's start with some numerical examples.

EXAMPLE 6.5

The given sequence is defined recursively, using the following basis and recursion relation:

- $a_1 = 4$
- $a_n = 2a_{n-1}$ for $n \exists 2$

Give the first five terms of the sequence.

SOLUTION

In this situation, a_1 is known to us, so it remains to find the next four terms. Observe that a_2 is defined in terms of a_1, since the recursion relation tells us that $a_2 = 2a_1 = 2(4) = 8$. Each subsequent term of the sequence depends, in this case, on the value of its immediate predecessor in the progression.

Thus, we find $a_3 = 2a_2 = 2(8) = 16$, $a_4 = 2a_3 = 2(16) = 32$, and $a_5 = 2a_4 = 2(32) = 64$.

EXAMPLE 6.6

The given sequence is defined recursively, using the following basis and recursion relation:

- $a_1 = 1$, $a_2 = 1$
- $a_n = a_{n-1} + a_{n-2}$, for $n \exists 3$

Give the first eight terms of the sequence.

SOLUTION

The first two terms are provided, so we begin our calculations with a_3. The recursive relation tells us that $a_3 = a_2 + a_1 = 1 + 1 = 2$.

Continuing, we find that $a_4 = a_3 + a_2 = 1 + 2 = 3$ and that $a_5 = a_4 + a_3 = 3 + 2 = 5$. Do you recognize the sequence yet?

$$a_6 = a_5 + a_4 = 5 + 3 = 8$$
$$a_7 = a_6 + a_5 = 8 + 5 = 13$$
$$a_8 = a_7 + a_6 = 13 + 8 = 21$$

The first eight terms of the sequence are 1, 1, 2, 3, 5, 8, 13, and 21. This is the list of the first eight terms of the Fibonacci sequence given in Example 6.1.

Now that we've examined some strictly numerical illustrations of recursively defined sequences, it's time to turn our attention to examples from everyday life. At first glance, you might not realize that these examples lead to recursively defined sequences, but on consideration, you'll see that the value of the variable at any given point depends entirely on its value at some earlier time.

EXAMPLE 6.7

A ball is dropped from a platform 10 feet above a level concrete surface. Each time it bounces, the ball rebounds to a height half that from which it had fallen. What are the heights of the first five bounces of the ball?

SOLUTION

When the ball is released, it falls from a 10-foot height, bounces, and rebounds to a height of 5 feet. From the apex of the first rebound, the ball falls 5 feet and rebounds to a height of 2.5 feet. Each time it bounces, the height of the rebound is halved, and thus the heights of the successive bounces are given by the terms of the sequence 5, 2.5, 1.25, 0.625, and 0.3125, each of which is measured in feet.

EXAMPLE 6.8

Suppose an investor deposits \$2000 into an account earning 8% interest compounded annually. Find the balance of the account at the end of each of the first 4 years of the investment.

SOLUTION

To say that the interest is compounded annually is to say that at the end of each year, the interest earned is computed and added to the preexisting value of the account, and the interest cycle begun anew.

At the end of the first year, the value of the account has increased by 8% over its initial level. In other words, at the end of the first year, the amount of money in the account is 108% of that with which we started. Computationally, the method for calculating this new balance is (\$2000)(1.08) = \$2160.

Similarly, at the end of the second year, the balance of the account is (\$2160)(1.08) = \$2332.80, while at the end of the third year, the value of the account is (\$2332.80)(1.08) = \$2519.42, and at the end of the fourth year it is \$2720.98.

Thus, the first four terms of the sequence of account values are (in dollars) {2160, 2332.80, 2519.42, 2720.98}.

It is interesting to observe that, while it may not be apparent, one can develop a general term formula for the terms of the sequence if we deconstruct the individual computations. The value for each year is found by multiplying the preceding year's account balance by 1.08. Note that, as an example, the third year's balance is computed as (2160)(1.08), but the value of 2160 was found by computing (2000)(1.08), and therefore the value of the account at the end of the third year is actually $(2000)(1.08)(1.08) = (2000)(1.08^2)$. In a similar manner, the balance at the end of the fourth year can be expressed as $(2000)(1.08^3)$. By taking this view, we can see the general term formula is $a_n = (2000)(1.08^{n-1})$.

EXAMPLE 6.9

Suppose an international company opened a manufacturing plant in 2007 with an initial employee count of 5000. The company plans for rapid and continuing growth, with projections of an increase in workforce of 10% per year. Find the workforce for the company's new facility in each of the first 5 years of its existence and devise a recursive formula describing the growth of the facility.

SOLUTION

In each year of its existence, the plan is expected to increase its workforce by 10%, suggesting that the staffing in a particular year can be found by multiplying the employee count in the preceding year by 1.10. Taking $a_1 = 5000$, we find $a_2 = 5000(1.10) = 5500$, $a_3 = 5500(1.10) = 5000(1.10)(1.10) = 5000(1.10)^2$, and so on. This suggests the recursion relation to be as follows: $a_n = 5000(1.10)^{n-1}$ for $n \exists 2$.

We note that the terms of the sequence may not be integer valued, and this is problematic in the "real-life" interpretation of the values of the sequence. We can agree that (since there are no fractional people employed at the plant) the actual level of employment will be found by truncating the decimal of the calculated sequence term value as necessary.

In the remarks following Example 6.2, we calculated the two-hundredth term of a particular sequence given by a general term formula. This led to a straightforward evaluation of the general term using $n = 200$ and an easy simplification. For the sequence given in Example 6.4, were we to require the value of the two-hundredth term, we would be required to calculate the values of the first 199 terms of the sequence and then use the value of a_{199} in order to determine the value of a_{200}.

This is a significant problem but one for which there is no solution. If a sequence is defined recursively, this is a consequence with which we must live.

Factorial Notation

A common occurrence in the construction of sequences is the product of successive whole numbers. Products of successive whole numbers can be expressed using a mathematical notation referred to as **factorial** notation. The notation for the factorial product is an exclamation point and appears as $n!$, where n is any whole number. The definition of the factorial product (more generally referred to as a factorial) is as follows:

DEFINITION: For any natural number n, the product "n factorial," denoted $n!$, is defined as

$$n! = n(n-1)(n-2)(n-3)\ldots(3)(2)(1).$$

As a special case of factorials, we further define $0! = 1$.

Informally, we can define factorials in this way: setting aside the special case of $0! = 1$, for any positive integer n, we define $n!$ to be the product of n with all smaller natural numbers. That is, as an illustration, $7! = (7)(6)(5)(4)(3)(2)(1) = 5040$.

The computation of 7! illustrates a significant fact: the value of $n!$ increases very rapidly as the values of n increase. If we compute the first 11 factorials, we find the results shown in Table 6.1.

TABLE 6.1 List of Early Factorials

n	*n!*
0	1
1	1
2	2
3	6
4	24
5	120
6	720
7	5040
8	40,320
9	362,880
10	3,628,800

Factorial functions grow extremely fast, outpacing polynomial and even exponential functions in their rate of growth. Most calculators, in fact, can handle factorials only as large as 69!, since 70! exceeds 10^{100}, which is the ceiling on most calculators' ability to accept scientific notation. The value of 69! is literally almost beyond comprehension and is in written form

> 171,122,452,428,141,311,372,468,338,881,272,839,092,270,544,893,520,
> 369,393,648,040,923,257,279,754,140,647,424,000,000,000,000,000.
> We should remark that, though the representation of the number extends through several lines of print, this is all a single number!

If this were counted in seconds, it would be far, far longer than the age of the universe at the time of this writing.

Factorials, should they appear within fractions, will typically spawn interesting reductions. As an example, if we had the fraction 7!/5!, we could compute each separately to obtain 5040/120, which reduces to 42. Alternatively, we could set the expression up as $(7)(6)(5!)/5! = (7)(6) = 42$. This form of reduction will play a role when we consider infinite series shortly.

EXAMPLE 6.10 Reduce the following fractions to lowest terms:

a. $\dfrac{13!}{15!}$

b. $\dfrac{(2n+3)!}{(2n-1)!}$

SOLUTION

a. If we rewrite the denominator using the definition of factorial notation, we find the fraction equal to $\dfrac{13!}{(15)(14)(13!)} = \dfrac{1}{(15)(14)} = \dfrac{1}{210}$.

b. This example is a little harder to understand because of the presence of the variable n. First, we observe that the numerator is larger, since $2n + 3 > 2n - 1$. If we take the product of $(2n + 3)$ and all smaller whole numbers, we find that the product will be

$$(2n + 3)(2n + 2)(2n + 1)(2n)(2n - 1)(2n - 2) \ldots (3)(2)(1).$$

If we truncate this expression at $(2n - 1)$, we find our rational expression to be equal to

$$\frac{(2n+3)(2n+2)(2n+1)(2n)/[(2n-1)!]}{(2n-1)!} = (2n+3)(2n+2)(2n+1)(2n)$$

EXAMPLE 6.11

Give the first six terms of the sequence $\{a_n\}$, where $a_n = (-1)^{n+1}n!$

SOLUTION

This sequence presents us with an interesting factor in the general term, -1 raised to a power. Typically, occurrence of such a factor will produce an alternation of sign from term to term within the sequence.

Referring to the table of factorial values presented earlier, we find

$$\begin{aligned} a_1 &= (-1)^2 1! = 1 \\ a_2 &= (-1)^3 2! = -2 \\ a_3 &= (-1)^4 3! = 6 \\ a_4 &= (-1)^5 4! = -24 \\ a_5 &= (-1)^6 5! = 120 \\ a_6 &= (-1)^7 6! = -720 \end{aligned}$$

Note that because the power on -1 increases by 1 with each passing term, the sign of that factor will be alternately positive and negative, and it is this that generates the alternation in sign.

EXAMPLE 6.12

Factorials play a major role in the field of combinatorics, which is an area of discrete mathematics wherein finite or countable discrete structures are examined. One illustration is an arrangement problem in which we consider the number of distinct orderings of n different objects. It turns out that the number of such orderings is exactly $n!$, which we can establish by a relatively straightforward argument.

Imagine the n objects were going to be lined up before you. For the first item in the line, you could choose from all n objects in the group. Once that item is chosen, there are $(n - 1)$ objects from which you can choose the second object in line. Then, for the third position, you choose from the $(n - 2)$ remaining objects. Continuing in this manner until the arrangement is complete, we see that we can form $n(n - 1)(n - 2) \ldots (3)(2)(1) = n!$ different orderings.

Summation Notation

As mentioned earlier, sequences are progressions of numbers occurring with an unchangeable order. A more technical description of a sequence is that it is a function whose domain is the natural numbers. The sum of the terms of a sequence (whether the sequence is finite or infinite) is called a **series**. If the associated sequence is infinite, we generally refer to the series as an infinite series, and those types of series have broad applications in contemporary mathematics, where complicated functions are approximated to an arbitrarily fine degree using a polynomial of suitable degree.

A series, or the sum of the terms of a sequence, can be expressed compactly using summation notation, or **sigma notation**. The latter name is derived from the fact that the notation incorporates the capital Greek letter Σ, as an integral component, and the choice for sigma is motivated by the first letter in the word "summation."

The notation appears in the following format: $\Sigma_{n=1}^{k} a_n = a_1 + a_2 + \cdots + a_k$. The expression to the left of the equals sign is read as "the sum of a sub n, from $n = 1$ to $n =$ k." The variable n is called the **index of summation**, and value of k is called the **upper limit of summation**. To calculate the sum, we replace the variable n in the general term of the sequence by 1, 2, 3, . . ., k and then add each of the resulting values. This is the meaning of the expression to the right of the equals sign. Since we are adding a collection of numbers together, that process is sometimes referred to as "calculating a sum."

EXAMPLE 6.13

Calculate each of the following sums:

a. $\sum_{n=1}^{5} (n^2)$

b. $\sum_{n=1}^{10} (3n-5)$

c. $\sum_{n=1}^{4} \left(\frac{2}{n}\right)$

d. $\sum_{n=1}^{5} 4$

SOLUTION

a. In this case, the upper limit of summation is 5, so we calculate the value of $a_n = n^2$ for $n = 1, 2, 3, 4$, and 5 and add the results. In this case, we find $1^2 + 2^2 + 3^2 + 4^2 + 5^2 = 1 + 4 + 9 + 16 + 25 = 55$.

b. This is a longer summation, since the upper limit of summation is 10, but the general term is rather nice to work with, so we shouldn't run into any real trouble. The sum will be $[3(1) - 5] + [3(2) - 5] + [3(3) - 5] + [3(4) - 5] + [3(5) - 5] + [3(6) - 5] + [3(7) - 5] + [3(8) - 5] + [3(9) - 5] + [3(10) - 5] = -2 + 1 + 4 + 7 + 10 + 13 + 16 + 19 + 22 + 25 = 105$. This example, by the way, will be revisited shortly. There is something about this computation that suggests that there may be properties of sums that could be of use to us, so it bears some further consideration.

c. Each of the terms involved in this summation will be fractions because of the nature of a_n, and you will note that (as we insert the successive values of n) we will simplify those fractions immediately. The reason for this is that we will need to find a common denominator in order to add the terms, so we may as well reduce the fractions to lowest terms in order to obtain the smallest possible common denominator. The summation will be $\frac{2}{1}+\frac{2}{2}+\frac{2}{3}+\frac{1}{2}=\frac{12}{6}+\frac{6}{6}+\frac{4}{6}+\frac{3}{6}=\frac{25}{6}$.

d. In this example, notice that the general term contains no instance of the variable n, the index of summation. This may be a bit confusing at first, but recall the instructions for calculating one of these sums: we are to insert successive values of n (in this case 1, 2, 3, 4, and 5) into the general term wherever we see an n and then add the resulting expressions. The general term here is 4, and no occurrence of n is to be found. Thus, when $n = 1$, the general term's value is 4. When $n = 2$, the general term's value is 4. For every value of n, the general term's value is 4. Thus, the sum is the rather suspicious-looking expression $\sum_{(n=1)}^{5} 4 = 4 + 4 + 4 + 4 + 4 = 20$.

There are many times when the sum of a sequence of numbers is of vital interest to particular individuals. The U.S. government and politicians of both major political parties use such information for political advantage all the time.

EXAMPLE 6.14

According to the Bureau of Labor Statistics (as reported in the *Monthly Labor Review*, March 2010), in 2009 the U.S. unemployment rate grew significantly. Table 6.2 shows the changes in the level of employment per month (through November) in 2009, where the number indicated depicts the jobs lost, in thousands, and $n = 1$ corresponds to January 2009. The sign of the number being negative indicates that the value represents jobs lost.

TABLE 6.2 Job Losses per Month in the United States, 2009

n	a_n
1	−779
2	−726
3	−753
4	−582
5	−347
6	−504
7	−344
8	−211
9	−225
10	−224
11	74
12	−85

Write a series whose sum represents the total job losses in the United States during 2009 and calculate its sum. Interpret the sum of the series (does the total represent an overall gain or a loss for the year?).

SOLUTION

The sum of the finite series is found by adding the values in the a_n column. Here, that would be

$$(-779) + (-726) + (-753) + (-582) + (-347) + (-504) + (-344) + (-211) + (-225) + (-224) + (74) + (-85) = -4706$$

Since the change in the number of jobs was, in this table, represented in thousands, the sum indicates that 4706 thousand jobs were lost during 2009 or (more directly put) that a total of 4,706,000 jobs were lost in the United States during that time.

Exercises

Answer each of the following, using complete sentences and correct grammar and spelling.

1. Define a sequence.
2. What is meant by the term "recursive sequence"?
3. What is the "general term" of a sequence?
4. What is meant by the term "factorial"?

For the following sequences whose general term is given, find the first five terms of the progression.

5. $a_n = 3n - 7$
6. $b_n = \dfrac{7+n}{2n-1}$
7. $c_n = \dfrac{(-1)^n}{3^n}$
8. $d_n = n^2 + 2^n$

The following problems refer back to Problems 5 to 8. For each problem, find the specified term of the sequences in Problems 5 to 8.

9. Find the tenth term of the sequences.
10. Find the fifteenth term of the sequences.
11. Find the twentieth term of the sequences.
12. Find the twenty-fifth term of the sequences.

The following problems give the first four terms of a sequence. Find a general term for the sequences (answers are not unique!).

13. $8, 4, 2, 1, \ldots$
14. $\dfrac{1}{2}, -\dfrac{5}{2}, \dfrac{25}{2}, -\dfrac{125}{2}, \ldots$
15. $6, 18, 54, 162, \ldots$
16. $50, 100, 200, 400, \ldots$
17. $1, 5, 9, 13, \ldots$
18. $6, -3, -12, -21, \ldots$
19. $3, 6, 18, 72, \ldots$
20. $2, 4, 10, 30, \ldots$

The following problems refer back to Problems 13 to 20. For each, find the indicated term of the sequence.

21. In Problem 13, find the tenth term of the sequence.
22. In Problem 14, find the tenth term of the sequence.
23. In Problem 15, find the fifteenth term of the sequence.
24. In Problem 16, find the fifteenth term of the sequence.
25. In Problem 17, find the twentieth term of the sequence.
26. In Problem 18, find the twentieth term of the sequence.
27. In Problem 19, find the seventh term of the sequence.
28. In Problem 20, find the ninth term of the sequence.

The following sequences are recursively defined, with the indicated basis and recursive relation defined form ∃ 2. Give the first six terms of the sequence.

29. $a_1 = 3,\ a_n = 3a_{n-1} + 5$
30. $a_1 = 1028,\ a_n = \frac{1}{2}a_{n-1}$
31. $a_1 = 2,\ a_n = (a_{n-1})^2 + 3$
32. $a_1 = 5,\ a_n = (-1)^2(5a_{n-1})$
33. $a_1 = \frac{1}{2},\ a_n = 3(a_{n-1}) + n$
34. $a_1 = 1,\ a_n = 4 - \frac{3}{a_{n-1}}$

Simplify the following expressions, which involve factorials.

35. $\frac{4!}{2!}$
36. $\frac{18!}{5!}$
37. $\frac{17!}{5!12!}$
38. $\frac{28!}{23!5!}$
39. 2! 3! 4!
40. 5!7! − 3! 4!
41. $\frac{203!}{200!}$
42. $\frac{9!}{2!0!}$
43. 3! 10! − 3! 12!
44. $\frac{49! - 51!}{48!}$

Calculate the following sums.

45. $\sum_{n=1}^{7}(3n)$
46. $\sum_{n=1}^{6}(-n^2)$
47. $\sum_{n=3}^{8}(2n-7)$
48. $\sum_{n=2}^{7}n^3$
49. $\sum_{n=1}^{7}(n^2 - n^3)$
50. $\sum_{n=10}^{11}(n^2 - 2n - 3)$
51. $\sum_{n=2}^{10}(4 - 7n)$
52. $\sum_{n=-1}^{4}(4 + 3n^2)$

Each of the following problems can be solved using a finite summation. Construct the summation involved in the problem and then use it to solve the problem.

53. Ali puts \$55 into a safety deposit box and makes a personal resolution to put \$15 per week into the box to create an emergency reserve fund. Give a summation that gives the amount of money in the box after n weeks and use the formula to find Ali's savings box total after 184 weeks.

54. Suppose that Hector borrows \$15,000, interest free, from his father-in-law. He agrees to pay back the loan in monthly installments of \$450. Find a summation that gives the balance owed by Hector after n months and use the summation to find his balance owed after 11 months.

55. A pendulum is constructed in such a manner that the length of each swing is 0.99 that of the previous swing. If the initial swing is 20 feet in distance (from the point of initial release to the point where the pendulum ends its swing, pauses, and swings in the opposite direction), find a summation formula to express the total distance swung by the pendulum after n swings. Then use the formula to find the total distance traveled by the pendulum during its first 10 swings (disregard the initial swing of 20 feet). Approximate your answer to the nearest hundredth.

56. A golf ball is dropped from a height of 50 feet and on each bounce rebounds to a height that is 80% of the distance it fell. Construct a summation to express the total distance traveled (up and down) by the golf ball as it bounces n times (disregard the initial fall). Use the formula to find the total distance traveled by the golf ball during its first 8 bounces.

6.2 Arithmetic Sequences

The Definition of an Arithmetic Sequence

We've already mentioned that a sequence is an ordered progression of real numbers that may or may not follow a prescribed pattern. If the order is changed, the sequence is changed, so the order of those numbers is fixed once it has been established.

Let's consider a particular type of sequence whose terms follow a specific sort of pattern. The pattern will probably be evident, but we'll point it out to ensure clarity.

$$\{a_n\} = \{4, 9, 14, 19, \ldots\}$$

The terms of the sequence happen to satisfy the condition that they increase by a particular constant value (in this case, 5) from term to term.

Such a sequence is called an **arithmetic sequence**. The emphasis in the word "arithmetic" is on the "met" fragment rather than the "rith" you may be used to using, so the pronunciation is "arith*met*ic sequence."

The constant value that is the difference between the terms of an arithmetic sequence is, appropriately, referred to as the **common difference** for the terms of the sequence, and we say that the terms follow an **arithmetic progression**. A useful fact about such sequences is that they are completely determined by the value of the first term, a_1, and the common difference, which we can refer to with the variable d.

Let's examine the terms of an arithmetic sequence a bit more closely by considering the example already presented. The value of the first term was $a_1 = 4$. To find succeeding terms in the sequence, we repeatedly add the common difference, which was (for that sequence) the constant value 5. That is, $a_2 = 4 + 5 = 9$, $a_3 = 9 + 5 = 14$, $a_4 = 14 + 5 = 19$, and so on. If we dissect the construction of the terms a bit more, we can observe the common tie that binds the terms together and allows for the formation of the all-important general term of the sequence a_n.

Consider $a_3 = 9 + 5$ once more, and let's express it using only the first term of the sequence and the common difference. The term a_3 can be broken down into the following form: $a_3 = 4 + 5 + 5 = 4 + 2(5)$. This, though it might not seem to be hugely important at the moment, underlies the construction of the general term. Now look at a_4 and repeat the analysis. We have $a_4 = 4 + 5 + 5 + 5 = 4 + 3(5)$.

Let's list the terms we've considered so far, all in this breakdown using the initial term and the common difference, and see if we can detect the pattern:

$$\begin{aligned} a_1 &= 4 \\ a_2 &= 4 + 5 \\ a_3 &= 4 + 2(5) \\ a_4 &= 4 + 3(5) \end{aligned}$$

If we try to express a_1 and a_2 in a form identical to the latter two terms, we find

$$\begin{aligned} a_1 &= 4 + 0(5) \\ a_2 &= 4 + 1(5) \\ a_3 &= 4 + 2(5) \\ a_4 &= 4 + 3(5) \end{aligned}$$

Now the situation appears much more clear. The term a_n can be expressed in a form reliant only on the initial term and the common difference.

Note

A sequence is arithmetic if the terms of the sequence always increase (or decrease) by a constant value as the sequence progresses! To find the common difference, subtract any term from its successor in the sequence.

RULE: The general term of an arithmetic sequence is $a_n = a_1 + (n - 1)d$, where d is the common difference in the arithmetic sequence.

One of the important observations you want to make when considering a sequence is how you can recognize that it happens to be a sequence of a particular type, in this case arithmetic. The key observation here is that the terms of the sequence always change by the same amount from term to term. The common difference can be determined by finding that difference between the terms.

EXAMPLE 6.15 The following sequences may or may not be arithmetic. Identify the common difference in each series that is arithmetic or establish that the sequence is not arithmetic:

a. $13, 21, 29, 38, \ldots$

b. $5, 2, -1, -4, \ldots$

c. $\frac{1}{2}, \frac{5}{8}, \frac{3}{4}, \frac{7}{8}, \ldots$

d. $2, 11, 20, 29, \ldots$

SOLUTION

a. We notice that the difference between the first two terms is 8 and that this difference is the same as that between the second and third terms. At first glance, this suggests that the sequence is arithmetic, but we must be sure the progression continues as we move forward. Observe that the difference between the third and fourth terms is not 8 but actually 9. This means that the difference between successive terms is not constant as we move through the sequence, and consequently the sequence is not arithmetic.

b. The terms of the sequence are decreasing, so we need to be cautious as we perform our investigation. Subtracting pairs of successive terms, we find $(2 - 5) = -3$, $(-1 - 2) = -3$, and $[-4 - (-1)] = -3$. The successive terms have a common difference of -3, and consequently the sequence is arithmetic.

c. The terms of the sequence are fractions, and this presents the problem of having to find the difference between the terms using common denominators. In fact, we might be wise to rewrite the terms with the common denominator to enhance the clarity of our analysis. Here, the common denominator is 8, so the terms of the sequence can be rewritten with that denominator: $\frac{4}{8}, \frac{5}{8}, \frac{6}{8}, \frac{7}{8}, \ldots$ In this format, it is evident that the sequence is arithmetic, with a common difference of $\frac{1}{8}$.

d. This sequence has integer-valued terms, so we are on somewhat safer ground than we were in Example 6.15c. Noting that $11 - 2 = 20 - 11 = 29 - 20 = 9$, we see that there is a common difference of 9 for the sequence and that it is arithmetic.

We had said earlier that an arithmetic sequence is entirely determined by its first term and the common difference for the succeeding terms. Consequently, if we are given that information, we should be able to produce the terms in the form of a list.

EXAMPLE 6.16 For the following arithmetic sequences having the given initial term and common difference, give the first four terms of the sequence as a list, ending with ellipsis:

a. Initial term 3, common difference 7

b. Initial term 2, common difference $\frac{2}{3}$

c. Initial term $\frac{1}{2}$, common difference -2

d. Initial term 0, common difference -4

SOLUTION

a. The first four terms of the sequence are 3, 10, 17, 24, ...

b. The first four terms of the sequence are 2, $\frac{8}{3}$, $\frac{10}{3}$, 4, ...

c. Here we will consider the common difference to be $-\frac{4}{2}$ (to simplify the calculation of the sequence terms), and this means that the terms of the sequence will be $\frac{1}{2}, -\frac{3}{2}, -\frac{7}{2}, -\frac{11}{2}, \ldots$.

d. This example is one that is often a source of confusion, since many times we forget to include 0 as the initial term. The first four terms of the sequence are $0, -4, -8, -12, \ldots$.

Finding and Using the General Term of an Arithmetic Sequence

Recall that the general term took the form $a_n = a_1 + (n - 1)d$, where d was the common difference. Consequently, we can produce the general term for all such sequences with comparative ease.

EXAMPLE 6.17

Find the general term for the following arithmetic sequences:

a. 3, 8, 13, 18, ...

b. $-5, -1, 3, 7, \ldots$

c. $4, \frac{11}{2}, 7, \frac{17}{2},$

SOLUTION

a. Note that in this example, the initial term is 3, and the common difference is 5. Using our formula for the general term of an arithmetic sequence, we find $a_n = 3 + (n - 1)(5)$.

b. In this sequence, the initial term is -5, and the common difference is 4. Therefore, the general term of the sequence is $a_n = -5 + (n - 1)(4)$.

c. As before, when fractions are involved, we need to be particularly careful. Thinking of the terms as $\frac{8}{2}$, $\frac{11}{2}$, $\frac{14}{2}$, and $\frac{17}{2}$, we see the common difference is $\frac{3}{2}$, and thus the general term of the sequence is $a_n = 4 + (n - 1)(\frac{3}{2})$.

When the general term of an arithmetic sequence is known, we can use it to find an arbitrary term of the sequence by evaluating the general term for the value of n corresponding to the relevant term.

EXAMPLE 6.18 Find the one-hundredth term of the given arithmetic sequence:

a. $a_n = -4 + (n - 1)(7)$
b. $b_n = 12 + (n - 1)(-3)$
c. $c_n = \frac{1}{2} + (n - 1)\left(\frac{1}{3}\right)$

SOLUTION

a. To find the one-hundredth term of any sequence, including an arithmetic sequence, all that is required is to evaluate the general term expression for $n = 100$. In this case, $a_{100} = -4 + (99)(7) = 689$.
b. Again evaluating the expression for $n = 100$, we find $b_{100} = 12 + (99)(-3) = -285$.
c. Once more, using the value of $n = 100$, we find $c_{100} = \frac{1}{2} + (99)\left(\frac{1}{3}\right) = \frac{1}{2} + 33 = \frac{67}{2}$.

Calculating the Sum of the First k Terms of an Arithmetic Sequence

As we saw when we introduced series in Section 6.1, we can find the sum of the terms of a finite sequence and express that sum in sigma notation. Let's turn our attention to the arithmetic sequences and see if we can deduce a pattern that will permit us to find the sum of the first k terms of such a sequence. The technique for finding the formula for the sum involves a bit of cleverness, but it's within our grasp, as we'll see when we consider an illustrative example.

Suppose we consider the first eight terms of the following arithmetic sequence and attempt to determine the sum of the terms 2, 5, 8, 11, 14, 17, 20, and 23. If we denote the sum by S_8 (the subscript of 8 referring to the fact that we are adding the first eight terms of the sequence together), we find

$$S_8 = 2 + 5 + 8 + 11 + 14 + 17 + 20 + 23$$

An elementary computation reveals the sum to be 100, but let's set that result aside for the moment. Here is where the cleverness enters the procedure. It is true, though not obviously important, that addition is commutative, so we can reverse the order of the terms in the expression, and thus,

$$S_8 = 23 + 20 + 17 + 14 + 11 + 8 + 5 + 2$$

Adding the two sums together term by term, we find

$$2S_8 = 25 + 25 + 25 + 25 + 25 + 25 + 25 + 25$$

That is an unexpected and interesting result. Each of the pairings added to the same value, 25, which happens (not coincidentally) to be the sum of the first and last terms in the finite subset of the arithmetic sequence. Writing each of those 25s as the sum of the first and last terms of the list, we find

Note

To find the sum of the first k terms of an arithmetic sequence, add the first and kth terms, multiply by k, and divide the result by 2.

$$\begin{aligned}2S_8 = {} & (2 + 23) + (2 + 23) + (2 + 23) + (2 + 23) \\ & + (2 + 23) + (2 + 23) + (2 + 23) + (2 + 23)\end{aligned}$$

or

$$2S_8 = 8(2 + 23)$$

We can solve for S_8 by dividing both sides of the equation by 2 and find

$$S_8 = 8(2 + 23)/2$$

Note

Recall that the nth term of an arithmetic sequence can be found using the general term expression $a_n = a_1 + (n - 1)d$.

There is a reason we have not simplified the result, and that is because we want to analyze the simplified expression. The 8 in the numerator is precisely the number of terms of the arithmetic sequence we were adding, and the 2 and 23 within the parentheses are a_1 and a_8, respectively. This suggests the following formula for finding the sum of the first k terms of an arithmetic sequence:

$$S_k = \frac{k(a_1 + a_2)}{2}$$

EXAMPLE 6.19

Find the sum of the first 20 terms of the following arithmetic sequences:

a. Initial term 3, common difference 7

b. Initial term 2, common difference $\frac{2}{3}$

c. Initial term 0, common difference -4

SOLUTION

a. Here, $a_1 = 3$, and we find $a_{20} = 3 + (19)(7) = 3 + 133 = 136$. Consequently, the sum of the first 20 terms is $S_{20} = \frac{(20)(3+136)}{2} = (10)(139) = 1390$.

b. For this sequence, we find $a_1 = 2$ and $a_{20} = 2 + (19)(2/3) = 2 + 38/3 = 44/3$. Therefore, the sum of the first 20 terms is $S_{20} = \frac{(20)(2 + 44/3)}{2} = (10)(50/3) = \frac{500}{3}$.

c. With initial term $a_1 = 0$ and $a_{20} = 0 + (19)(-4) = -76$, the sum of the first 20 terms of the sequence is $S_{20} = \frac{(20)[0+(-76)]}{2} = (10)(-76) = -760$.

There are a wide variety of applications to which we can apply arithmetic sequences, and we'll close this section by examining a few problems where such sequences appear.

EXAMPLE 6.20

Suppose a college plans to design a field house in which a particular section of a balcony is constructed in such a way that the first row has 22 seats, the second 25 seats, the third 28 seats, and so on. If the balcony has enough space to permit the construction of 17 rows, how many individuals can be seated in this balcony section?

SOLUTION

Observing that the provided information does not tell us the number of seats available in the last row, we need to begin our calculations by finding that final term in the progression. The seventeenth term in the sequence is $a_{17} = 22 + (17 - 1)(3) = 22 + 48 = 70$. Therefore, the sequence of available seats per row is $\{22, 25, 28, \ldots, 70\}$ and is arithmetic with common difference 3. Using our formula for the sum of a finite arithmetic sequence, we can calculate the total number of seats available in the balcony section:

$$S_{17} = \frac{(17)(22+70)}{2} = \frac{17(92)}{2} = 782$$

Thus, a total of 782 individuals can be seated in this particular balcony section of the field house.

EXAMPLE 6.21

We know that a grandfather clock that strikes only on the hour will strike once at 1:00 p.m., twice at 2:00 p.m., and so forth and then once at each intervening quarter-hour mark. How many times, in total, will the clock strike in a particular 12-hour period from 12:01 p.m. until midnight?

SOLUTION

The sequence that is involved, describing the number of times the clock has struck in each hour (and including the three strikes during the quarter hours between hours), has term values equal to 3 more than the actual hour on the clock face. Thus, the sequence is $\{4, 5, 6, \ldots, 15\}$. The sequence is arithmetic, with common difference 1, and the number of strikes performed during the 12-hour period is $S_{12} = \frac{(12)(4+15)}{2} = \frac{(12)(19)}{2} = (6)(19) = 114$. Thus, the clock strikes 114 times during the 12-hour span of time.

EXAMPLE 6.22

An employer offers a new hire a starting salary of \$47,500 per year, with annual raises of \$2700 for the time of employment. If the employee works for the company for 16 years, what is the total amount of compensation she will earn during this time?

SOLUTION

Since the annual increase in salary is constant, the sequence whose terms are the annual salary amounts will be arithmetic with common difference 2700. The first term of the sequence is 47,500, and the sixteenth term is $a_{16} = 47{,}500 + (15)(2700) = 88{,}000$.

Therefore, the total compensation can be found using the sum of the arithmetic series formula, and we find $S_{16} = \frac{(16)(47{,}500+88{,}000)}{2} = \frac{(16)(135{,}500)}{2} = (8)(135{,}500) = 1{,}084{,}000$, and therefore the total comes to \$1,084,000 over the course of her time on the job.

Exercises

For each of the following, answer using complete sentences and proper grammar and spelling.

1. Explain what is meant by the term "arithmetic sequence"?
2. When we refer to the "common difference" of an arithmetic sequence, what do we mean?
3. What is meant by "ellipsis," and what do ellipses signify?

For the following sequences, determine whether they are arithmetic. If they are arithmetic, determine the common difference. If they are not arithmetic, explain.

4. $3, 7, 11, 15, \ldots$
5. $\frac{1}{2}, \frac{7}{8}, \frac{5}{4}, \frac{13}{8},$
6. $4, 13, 22, 30, \ldots$
7. $9, 4, -1, -6, \ldots$
8. $1, 1.1, 2.1, 3.1, \ldots$
9. $-1, 2, -3, 4, \ldots$

The following problems give an initial term and common difference for arithmetic sequences. Give the first five terms of the sequence, ending the list with ellipsis.

10. $a_1 = 7$, common difference -3
11. $a_1 = \frac{1}{2}$, common difference $\frac{5}{7}$
12. $a_1 = 3.2$, common difference 0.5
13. $a_1 = 1$, common difference π
14. $a_1 = \frac{2}{3}$, common difference $\frac{1}{3}$
15. $a_1 = \sqrt{7}$, common difference $2\sqrt{7}$

The following problems give the first few terms of an arithmetic sequence. Find the general term of the sequence.

16. $3, 7, 11, 15, \ldots$
17. $-4, -11, -18, -25 \ldots$
18. $10, 20, 30, 40, \ldots$
19. $3, 6, 9, 12, \ldots$
20. $2, 6, 10, 14, \ldots$
21. $-5, -3, -1, 1, \ldots$
22. $-1, 10, 21, 32, \ldots$
23. $14, 19, 24, 29, \ldots$

Each of the following problems refers to a previous problem in this exercise set. Find the one-hund.edth term of the sequence from the indicated problem.

24. The sequence of Problem 16
25. The sequence of Problem 17
26. The sequence of Problem 18
27. The sequence of Problem 19
28. The sequence of Problem 20
29. The sequence of Problem 21
30. The sequence of Problem 22
31. The sequence of Problem 23

For the following arithmetic sequences given by a general term, find the sum of the indicated terms.

32. For the sequence where $a_n = 3 + (n - 1)(9)$, find the sum of the first 45 terms.
33. For the sequence where $a_n = 5 + (n - 1)(-3)$, find the sum of the first 50 terms.
34. For the sequence where $a_n = 5 + (n - 1)(-2)$, find the sum of the first 100 terms.
35. For the sequence where $a_n = 8 + (n - 1)(4)$, find the sum of the first 100 terms.
36. For the sequence where $a_n = \left(\frac{1}{2}\right) + (n - 1)\left(\frac{3}{2}\right)$, find the sum of the first 40 terms.
37. For the sequence where $a_n = \left(\frac{3}{4}\right) + (n - 1)\left(\frac{5}{4}\right)$, find the sum of the first 30 terms.
38. For the sequence where $a_n = 9 + (n - 1)(0.5)$, find the sum of the first 80 terms.
39. For the sequence where $a_n = -3 + (n - 1)(0.7)$, find the sum of the first 75 terms (round to the nearest tenth).

6.3 Geometric Sequences

The Definition of a Geometric Sequence

When we considered the arithmetic sequences, we observed that each term differed from its predecessor by a constant value, referred to as the common difference. For a **geometric sequence**, the principle is similar, but in the case of a geometric sequence, each term differs from its predecessor by a constant multiple, referred to as the **common ratio**. Just as an arithmetic sequence is also known as an arithmetic progression, a geometric sequence is also said to represent a **geometric progression**.

Geometric sequence terms are determined entirely by the first term in the progression, a_1, also called the initial term, and the common ratio, typically denoted with the variable r. The common ratio can often be determined by inspection, but it can also be computed by dividing any particular term in the sequence by its predecessor. Since every term is the same multiple of its predecessor as any other term, this quotient will always be the same, no matter which two successive terms are chosen.

EXAMPLE 6.23 The following sequences may or may not be geometric sequences. If they are geometric, identify the common ratio. If they are not geometric, explain why they are not geometric:

a. $\{4, 12, 36, 108, \ldots\}$
b. $\{9, 18, 27, 36, \ldots\}$
c. $\{8, 4, 2, 1, \ldots\}$

SOLUTION

a. When we examine the terms of the sequence, we can see that $a_2 = 3a_1$, that $a_3 = 3a_2$, and finally that $a_4 = 3a_3$. Each term is precisely 3 times its predecessor, and therefore the sequence is geometric with common ratio 3. This can also be seen by dividing each term by its predecessor. Observe that $\frac{a_2}{a_1} = \frac{12}{4} = 3$, that $a_3/a_2 = 36/12 = 3$, and so forth. Again, we see that the common ratio is 3 and that the sequence is geometric.
b. At first glance, the second sequence appears that it could be geometric. The terms clearly follow a pattern, but is the progression geometric? We notice that the second term in the sequence is 2 times the first term but that the third term is not 2 times the second term. Thus, the sequence is not geometric, since there is no common ratio. In fact, this is an example of the sequences we examined in Section 6.2 and is arithmetic.
c. To determine if the sequence is geometric, we can proceed in either of the methods shown in Example 6.23(a), but we'll choose the option of dividing one term by its predecessor. Doing so, we observe that $\frac{a_2}{a_1} = \frac{4}{8} = \frac{1}{2}$, that $\frac{a_3}{a_2} = \frac{2}{4} = \frac{1}{2}$, and finally that $\frac{a_4}{a_3} = \frac{1}{2}$. Each term is precisely $\frac{1}{2}$ times its predecessor, and therefore the sequence is geometric, with common ratio $\frac{1}{2}$.

As we did with the arithmetic sequences, we now turn our attention to the problem of finding a general term formula. By definition, we know that each term is a constant multiple of its predecessor and that the progression depends entirely on that multiple and the initial term. Thus, we can present the terms in the following manner:

$$\begin{aligned}
a_1 &= a_1 \\
a_2 &= a_1 r \\
a_3 &= a_2 r = (a_1 r)r = a_1 r^2 \\
a_4 &= a_3 r = (a_1 r^2)r = a_1 r^3
\end{aligned}$$

and so forth. If we recall that $r^0 = 1$, then a_1 can be viewed as $a_1 r^0$, so all the terms can be expressed in a common format.

RULE: The general term of a geometric sequence is $a_n = a_1 r^{n-1}$, where r is the common ratio in the geometric sequence.

EXAMPLE 6.24 If a sequence is geometric with the shown initial term and common ratio, list the first four terms of the sequence:

a. $a_1 = 3, r = 5$

b. $a_1 = 6,\ r = \frac{1}{3}$

c. $a_1 = \frac{1}{2}, r = -\frac{1}{5}$

SOLUTION

a. Beginning with a_1, the terms of the sequence are obtained by repeated multiplication, and therefore we can start with $a_1 = 3$ and multiply by 5 repeatedly in order to obtain the following three terms. Therefore, $a_2 = 15$, $a_3 = 75$, and $a_4 = 375$.

b. Again, we start with the initial term and multiply by $\frac{1}{3}$, finding $a_1 = 6$, $a_2 = 2$, $a_3 = \frac{2}{3}$, and finally $a_4 = \frac{2}{9}$.

c. The final example requires that we pay attention to our operations with fractions. Starting with the initial term, we see $a_1 = \frac{1}{2}$, $a_2 = -\frac{1}{10}$, $a_3 = \frac{1}{50}$, and $a_4 = -\frac{1}{250}$.

As you can see in the previous example, geometric sequence terms can undergo a rather dramatic rate of change as the sequence progresses. Such progressions can have serious consequences in business applications and underlie rather notorious and disreputable schemes.

EXAMPLE 6.25 Suppose a particular home business is constructed in such a way that each person hired as an independent contractor is rewarded when he or she has recruited three new employees. Assuming that each contractor is limited to recruiting three new workers, analyze the growth of the additions to the company payroll. Further assume that each new hire is successful in attaining their recruitment goals.

SOLUTION

On the assumption that a single person originates the business model and that that individual hires three other workers, we see that the first round of hiring brings in three new workers. In the second round of hiring, each of those workers recruits three individuals, and therefore that round pulls in nine new workers. Therefore, the number of new employees in each round of hiring follows the following geometric progression: $\{1, 3, 9, 27, 81, 243, \ldots\}$.

The preceding example suggests the potential hazard of attempting to join such a hiring plan after the first few rounds of the company's existence. We know that the number of hires in any particular round can be found using the general term formula for the geometric sequence, which is $a_n = 1(3^{n-1})$. In the eleventh round of hiring, according to this scheme, 59,049 new employees must be added, and in the fifteenth round of hiring, 4,782,969 employees must be added. Such plans typically base the profits of an individual hired at any particular time on the successful recruiting of those hired afterward, and consequently (since the available workforce is limited) it becomes increasingly difficult to achieve profitability the later one enters the plan.

Such organizations are sometimes referred to as "pyramid schemes" because the base needed to support the growth of the organization increases in size at such an alarming (and uncontrollable) rate. After 21 rounds of hiring, the scheme described here would require nearly the entire population of the world to advance to the next level.

EXAMPLE 6.26

For a geometric sequence with the given initial term and common ratio, find the value of the tenth term in the progression:

a. $a_1 = 3, r = 2$

b. $a_1 = ½, r = 3$

SOLUTION

a. In this case and in the subsequent examples, the value of n is 10, so we insert our given values into the expression derived for the general term of the geometric sequence: $a_n = a_1 r^{n-1}$. Here, that gives $a_{10} = 3(2^9) = 1536$.

b. As in the first illustration, we use the general term expression and find $a_{10} = (½)(3^9) = 9841.5$ to be the tenth term of the progression.

Calculating the Sum of the First k Terms of a Geometric Sequence

As was the case with the sum of the first k terms of an arithmetic sequence, we can find a formula for the sum of the first k terms of a geometric sequence. The derivation of such formulas is a source of boundless amusement for mathematicians, since they invariably involve some subtle trickery.

Consider the sum $a_1 + a_1 r + a_1 r^2 + \cdots + a_1 r^{k-1} = S_k$. The subscript k, once again, refers to the fact that we are adding the first k terms of the sequence. If the common ratio were 1, of course, the sum would be easy to find and equal to $a_1 k$. Assuming that this is not the case, if we were to multiply both sides of this equation by r, the common ratio of the sequence, we would find the following:

$$a_1 + a_1 r + a_1 r^2 + \cdots + a_1 r^{k-1} = S_k$$
$$a_1 r + a_1 r^2 + \cdots + a_1 r^{k-1} + a_1 r^k = rS_k$$

Now we subtract the lower line from the upper. This produces an annihilation of many of the terms in the two expressions and yields

$$a_1 - a_1 r^k = S_k - rS_k$$

or

$$a_1(1 - r^k) = S_k(1 - r)$$

Since r is not 1, we can divide both sides by $(1 - r)$ to isolate S_k, and we find

$$S_k = a_1(1 - r^k)/(1 - r)$$

This formula gives us the sum of the first k terms of a geometric sequence with common ratio r.

RULE: The sum of the first k terms of a geometric sequence, having initial term a_1 and common ratio $r \neq 1$, is given by $S_k = \dfrac{a_1(1-r^k)}{(1-r)}$.

EXAMPLE 6.27

For geometric sequences having the given common ratio and initial term, find the sum of the first 20 terms of the sequence:

a. $a_1 = 4, r = 5$
b. $a_1 = 10, r = -2$
c. $a_1 = 2, r = \dfrac{1}{3}$

SOLUTION

a. Using our known formula for the sum of the first k terms of a geometric series, with $k = 20$, we find that the sum will be $S_{20} = \dfrac{(4)(1-5^{20})}{1-5} = 95,367,431,640,625$. In case you're curious, that's 95 trillion, 367 billion, 431 million, 640 thousand, six hundred and twenty-five. We told you that geometric progressions grew rapidly!
b. Again, inserting our initial term and common ration into the formula for the sum of the terms, we find $S_{20} = \dfrac{(10)[1-(-2)^{20}]}{1-(-2)} = \dfrac{(10)(1-1{,}048{,}576)}{3}$ $= -3{,}495{,}250$. In this case, observe that the sum is negative. Is this surprising to you? The common ratio is -2, and therefore every other term a_n (the even values of n) is going to be negative and twice as large in absolute value as its predecessor. If we were to compute S_{21}, we would find that the result would be positive!
c. This example is a challenging one, since it will require us to work with complex fractions. Substitution into the sum formula yields

$$S_{20} = \frac{(2)\left[1-(1/3)^{20}\right]}{1-1/3} = \frac{(2)\left(\frac{3{,}486{,}784{,}400}{3{,}486{,}784{,}401}\right)}{2/3}$$

$$= (2)\left(\frac{3{,}486{,}784{,}400}{3{,}486{,}784{,}401}\right)\left(\frac{3}{2}\right) = \frac{3{,}486{,}784{,}400}{1{,}162{,}261{,}467}$$

a truly hideous result and one that can be shown to be irreducible. This may be somewhat surprising to you, but we assure you it is the case (we checked it!).

The examples shown illustrate that geometric progressions grow in magnitude rather quickly for values of r such that $|r| > 1$. This feature of rapid geometric growth is evident if we consider a famous story from antiquity.

EXAMPLE 6.28

Legend has it that a king who loved chess was visited by a sage who challenged the king to a game of chess. The sage requested that, if he were to defeat the king, his reward would be in rice, according to the following plan: on a chessboard, in the first square, a single grain of rice would be placed. In each successive square of the board, the number of rice grains would be doubled until every square of the board was accounted for. The king agreed and (naturally, for what else would you expect in such a legend?) lost the game. When settling accounts was begun, the true calamity of the price of defeat came home to him. What is the quantity of rice committed to by the king as payment?

SOLUTION

Since the number of grains of rice on each successive square is precisely twice the quantity applied to the preceding square, the number of grains on the successive squares followed a geometric progression: 1 grain, 2 grains, 4 grains, 8 grains, and so forth. The initial term of the sequence is $a_1 = 1$, and the common ratio is $r = 2$. Since a chessboard has 64 squares, the total number of grains of rice earned by the sage can be calculated using the sum of terms of a geometric sequence: $S_{64} = \dfrac{1\left(1-2^{64}\right)}{1-2} = 18{,}446{,}744{,}073{,}709{,}551{,}615$ grains of rice. That's more than 18 quintillion grains of rice, and if we assume that one grain of rice weighs 25 mg, the total amount of rice would weigh over 460 billion metric tons and would fill a space having a volume of almost 37 cubic miles!

We should point out that there are many variations on this story, but the principle underlying the legend is consistent: the amount of rice involved is at least 80 times the amount of wheat humanity could produce in one year if *all* arable land on the planet was devoted to wheat harvest!

The Value of an Annuity and Its Relationship to Geometric Sequences

Although the focus of this text is not on the particulars of specific applications, there are times it pays for us to delve into a portion of the wealth of topics to which our ideas apply. In this case, the idea of a geometric progression applies particularly well to an important financial concept: the annuity. An annuity, simply put, is a series of yearly payments made to an individual (usually a recipient of an insurance policy payment or a retiree). If the annuity payment is structured in such a way that it will increase as the years go by at a predetermined, regulated rate, then the progression of the payments involved is geometric.

Let's illustrate with an example where a certain Mr. Rodriguez decides to set up a trust fund for his daughter, Rhiannon, that will pay her a lump sum on the first day of January every year. The first payment will be made on January 1 of the current year and will be \$35,000. On every subsequent New Year's Day, the amount paid to the daughter will increase by 6% over that of the preceding year. We can calculate the payments due the daughter for the first few years, and in so doing we will demonstrate the geometric nature of the progression.

On the first day of the second year, the payment will be 6% over the amount paid in the first year, an increase of (\$35,000)(0.06) = \$2100, so Rhiannon will be paid \$37,100 at the start of the second year. On the first day of the third year, she will receive an increase of 6% over the payment of the second year, an increase of (\$37,100)(0.06) = \$2226, and thus Rhiannon receives a third-year payment of \$39,326. The sequence of payments, {35,000, 37,100, 39,326, ...}, may not appear to be geometric at first glance, but let's consider in more detail how the payments were computed.

To find the value of the second payment, we calculated 6% of the original payment and then added that quantity to the original payment value. That is, we found

$$\$35{,}000 + (\$35{,}000)(0.06) = \$37{,}100$$

However, the left side of this equation can be factored to yield

$$(\$35{,}000)(1 + 0.06) = (\$35{,}000)(1.06)$$

In a similar fashion, we can demonstrate the third payment (through a process of factoring and simplification) to be $(\$35{,}000)(1.06)^2$. The geometric nature of the growth of the payments is now clear: in year n, the payment to Rhiannon will be $(\$35{,}000)(1.06)^{n-1}$, and hence the growth of her payments (that is, the value of the annuity paid to Rhiannon) is an illustration of a geometric progression.

We should mention that we are taking a naive view of annuities because we don't want to venture far afield from our intended purpose here. Annuities are a vast and deep topic, and entire courses are devoted to their study. Our intention is merely to introduce the concept and offer a glimmering of the potential one can uncover if the topic is pursued with greater vigor.

Here, we'll be content with recognizing that the total amount paid into an annuity in the nth year of its existence, where the initial payment is P dollars and each subsequent payment reflects an increase of a set percentage (where i is the decimal form of that percentage) and is given by $A = P(1 + i)^{n-1}$. The values of the payments form a geometric sequence, and the total paid into the annuity can be calculated using the formula for the sum of the first k terms of a geometric sequence, with small modifications (as we'll see after the following example).

EXAMPLE 6.29 Suppose that Alphonse wants to make a series of payments into a retirement account on the final day of every calendar year for the next 27 years (the remainder of his working life) and that he wants to increase the contribution by 7% per year to account for his anticipated increase in salary every year. Ignoring the interest earned by the account and on the assumption that Alphonse's first payment will be \$5000, find the total amount deposited into the account by him in the tenth, twentieth, and twenty-seventh years.

SOLUTION

This is a situation very much like we saw with Mr. Rodriquez and his sequence of payments to his daughter, Rhiannon. There is a superficial difference in that the deposits are being made on the last day of the year rather than the first, but in this context that makes no difference. In some problems, where interest is being calculated, that difference can be significant, but it plays no role here.

In the tenth year of the annuity, the value of n is 10, and therefore the amount deposited into the account will be $A = P(1 + i)^{n-1} = 5000(1 + 0.07)^9 = 9192.30$. Note that, this being a dollar-and-cents problem, the value deposited to the account should be rounded to two decimal places to reflect dollars and cents, and thus the amount deposited by Alphonse in the tenth year would be \$9192.30.

In the twentieth year, the value of n is 20, and again we compute the value of the deposit using our formula for the value of an annuity: $A = P(1 + i)^{n-1} = 5000(1.07)^{19} = 18{,}082.64$, which corresponds (in the context of the problem) to \$18,082.64. Similarly, in year 27, using the value $n = 26$, we find that the amount deposited by Alphonse to the account would be \$29,036.76, at which time he plans to retire.

An interesting problem to consider is the total amount paid into an annuity, such as we examined in Example 6.29 and the paragraphs immediately preceding it. That is, how much did Rhiannon actually receive from her father, and how much did Alphonse deposit into the account?

Since the values of the annuity followed a geometric progression, we know that we can compute the totals involved using the sum of the first k terms of the geometric sequence formula. Recall that this formula was $S_k = \dfrac{a_1\left(1-r^k\right)}{1-r}$, and in this context, a_1 corresponds to the value of the initial payment and r to the multiple needed in order to find the increased value of the payments of the annuity from year to year, which we had determined to be $(1 + i)$, where i is the decimal form of the percentage rate corresponding to the rate of increase of the annual payments.

EXAMPLE 6.30

Referring to the situation of Mr. Rodriguez and his daughter, Rhiannon, how much will Rhiannon receive during the first 15 years of the annuity?

SOLUTION

In this case, we know that Rhiannon's first payment was \$5000 and that the annual rate of increase of her payments was $i = 0.06$, so the general term of the geometric sequence describing her annuity payments was $(35{,}000)(1.06)^{n-1}$. Using our formula for the sum of the first 15 terms of a geometric sequence, we find that the total paid to Rhiannon is $S_{15} = \dfrac{35{,}000\left(1-1.06^{15}\right)}{1-1.06} = 814{,}658.95$, and thus she received \$814,658.95 in payments from her father over the first 15 years of the annuity.

EXAMPLE 6.31

Referring to the situation of Alphonse in Example 6.29, what is the total he deposits to his account over the period of 27 years?

SOLUTION

Recall that Alphonse had originally deposited \$5000 to his account and intended to increase that amount by 7% per year for the period of 27 years. The pattern of payments he made followed a geometric progression, and the general term of the sequence was $5000(1.07)^{n-1}$.

Using the formula for the sum of the first 27 terms of the sequence, we find that the total amount deposited by Alphonse is $\frac{5000\left(1-1.07^{27}\right)}{1-1.07}=372{,}419.12$, which indicates that a total of \$372,419.12 was deposited.

This may seem surprising when compared to the result for Rhiannon, since Alphonse was depositing so much less than did Rhiannon's father, but keep in mind that the length of time Alphonse is working with is significantly longer!

Let's turn the problem around a bit and see if we can attack the question from another angle. Suppose that we know how much, in total, was paid into an annuity, that we were aware of how much the value paid into the annuity was increasing year to year, *and* that we knew the length of time involved. Could we work backward to find the value of the initial payment into the annuity, knowing what we know to this point?

Consider the equation for the sum of the first k terms of a geometric sequence:

$$S_k=\frac{a_1\left(1-r^k\right)}{1-r}$$

Can this equation be solved for a_1? If so, then the answer to our question is an emphatic yes! Multiplying both sides of the equality by $(1-r)$ to clear the fraction on the right, we find

$$S_k(1-r)=a_1(1-r^k)$$

Now, divide each side by $(1-r^k)$, and the value of the first term of the progression (which is equivalent to the first payment of the annuity) is isolated:

$$\frac{S_k(1-r)}{\left(1-r^k\right)}=a_1$$

EXAMPLE 6.32

Phillippe had been paying into an annuity over a period of 10 years, increasing his payments by 13% per year. At the end of this time, he had paid a total of \$138,148.12 into the annuity. Calculate the amount of his initial payment level.

SOLUTION

Based on the information at hand, we can employ the formula derived just prior to the example to compute the initial payment of the annuity:

$$\frac{S_k(1-r)}{(1-r^k)} = a_1$$

In this case, the value of $r = 1.13$, and the value of $S_{10} = 138{,}148.12$. By substitution, we find

$$\frac{S_k(1-r)}{(1-r^k)} = a_1$$

$$\frac{(138{,}148.12)(-0.13)}{(1-1.13^{10})} = a_1$$

$$7500 \approx a_1$$

Thus, the initial amount Phillippe paid into the annuity was \$7500 (allowing for round-off error).

Exercises

Answer the following questions using complete sentences, proper grammar, and correct terminology.

1. What is meant by a "geometric sequence"?
2. What is the "initial term" of a geometric sequence?
3. What is meant by the "common ratio" of a geometric sequence, and how do you find it?

The following sequences may or may not be geometric. If they are, identify the common ratio. If they are not, explain why they are not geometric.

4. $3, 9, 27, 81, \ldots$
5. $5, 10, 20, 40, \ldots$
6. $9, -3, 1, -\frac{1}{3}, \ldots$
7. $1, \pi, 2\pi, 3\pi, \ldots$
8. $4, 8, 12, 16, \ldots$
9. $1, 1, 1, 1, \ldots$
10. $2, -2, 2, -2, \ldots$
11. $8, 6, \frac{9}{2}, \frac{27}{10}, \ldots$
12. $10, 2, \frac{2}{5}, \frac{2}{25}, \ldots$

In the following problems, the first term and the common ratio of a geometric sequence are given. Find the first four terms of the sequence.

13. $a_1 = 5, r = 3$
14. $a_1 = -2, r = \frac{1}{4}$
15. $a_1 = 10, r = \pi$
16. $a_1 = \frac{1}{4}, r = 6$
17. $a_1 = -5, r = 5$
18. $a_1 = 14, r = 2$
19. $a_1 = \frac{1}{3}, r = 9$
20. $a_1 = 100, r = \frac{1}{2}$
21. $a_1 = \frac{2}{3}, r = \frac{3}{5}$
22. $a_1 = \frac{5}{6}, r = \frac{2}{3}$

The following problems refer back to earlier exercises and their geometric sequences. Find the tenth term of the sequence referenced in the given problem.

23. Problem 13
24. Problem 14
25. Problem 15
26. Problem 16
27. Problem 17
28. Problem 18
29. Problem 19
30. Problem 20
31. Problem 21
32. Problem 22

In the following problems, the first few terms of a geometric sequence are given. Find the common ratio and then find the indicated term of the sequence.

33. 7, 14, 28, ... Find the eleventh term
34. 9, −18, 36, ... Find the seventh term
35. 3, 12, 48, ... Find the tenth term
36. $\frac{1}{2}, \frac{1}{3}, \frac{2}{9}, \ldots$ Find the twentieth term
37. $\frac{3}{7}, \frac{6}{7}, \frac{12}{7}, \ldots$ Find the fifteenth term
38. $3, \sqrt{3}, \ldots$ Find the fifth term
39. 4, 24, 144, ... Find the eighth term
40. $5, -5, \sqrt{5}, 25, \ldots$ Find the ninth term
41. 3, 36, 432, ... Find the tenth term
42. $18, -\frac{2}{3}, \ldots$ Find the sixth term

The following problems reference a preceding problem. Find the sum of the first *n* terms of the sequence, where *n* is given.

43. Problem 13, $n = 8$
44. Problem 14, $n = 10$
45. Problem 15, $n = 6$ (round to the nearest tenth)
46. Problem 16, $n = 7$
47. Problem 17, $n = 15$
48. Problem 18, $n = 16$
49. Problem 19, $n = 12$
50. Problem 20, $n = 11$
51. Problem 21, $n = 10$
52. Problem 22, $n = 30$

Find the sum of the given finite geometric sequence.

53. $\sum_{n=1}^{12} 3(2/3)^n$
54. $\sum_{n=1}^{15} 5(3/4)^n$
55. $\sum_{n=1}^{25} 5(2)^n$
56. $\sum_{n=1}^{14} 4(3)^n$
57. $\sum_{n=1}^{8} \left(\frac{-1}{4}\right)^n$
58. $\sum_{n=1}^{10} 3\left(\frac{2}{3}\right)^n$

59. $\sum_{n=1}^{20} 250(1.08)^n$ (round to the nearest tenth)

60. $\sum_{n=1}^{25} 400(1.15)^n$ (round to the nearest tenth)

61. $\sum_{n=1}^{15} 5\left(-\frac{1}{3}\right)^n$ (round to the nearest tenth)

62. $\sum_{n=1}^{30} 2\left(-\frac{4}{5}\right)^n$

In the following problems, use the techniques of geometric sequences to solve the applications.

63. Suppose a principal of \$8000 is invested at 5% interest. Find the value of the investment after 10 years if the interest is compounded (a) monthly and (b) quarterly.

64. Suppose a principal of \$10,000 is invested at 7% interest. Find the value of the investment after 10 years if the interest is compounded (a) monthly and (b) quarterly.

65. Suppose a deposit of \$125 is made at the beginning of each month into an account that earns 5% interest compounded monthly. The balance of the account at the end of 10 years is $125\left(1+\frac{0.05}{12}\right)^1 + 125\left(1+\frac{0.05}{12}\right)^2 + \cdots + 125\left(1+\frac{0.05}{12}\right)^{120}$. Find the balance at the end of 10 years.

66. Suppose a deposit of \$75 is made at the beginning of each month into an account that earns 6% interest compounded monthly. The balance of the account at the end of 7 years is $75\left(1+\frac{0.06}{12}\right)^1 + 75\left(1+\frac{0.06}{12}\right)^2 + \cdots + 75\left(1+\frac{0.06}{12}\right)^{84}$. Find the balance at the end of 7 years.

67. What initial deposit must be made into an account with interest rate of 7% compounded monthly for 5 years if the final balance of the account is to be \$10,000?

68. What initial deposit must be made into an account with interest rate of 4% compounded monthly for 8 years if the final balance of the account is to be \$6000?

69. Anthony wants to open an account by depositing \$4000 on the first day of the year. Suppose he intends to make subsequent deposits on the first day of each subsequent year for the next 20 years, increasing his deposit by 10% each year. Ignoring interest, how much will Anthony's deposit be in years 5, 10, and 15, and how much will he deposit in total over the course of the 20 years?

70. Serena opens an account on the first day of a year by depositing \$5500 and plans to increase her deposits by 15% on the first day of the year of each subsequent year for the next 12 years. Ignoring interest, find Serena's deposit in years 4, 8, and 12 and find the total she deposits over the 12 years.

6.4 The Principles of Mathematical Induction

There are many formulas in mathematics that hold or that are valid for all positive integers. For instance, the great German mathematician Carl Friedrich Gauss established the following formula to be true:

$$1+2+3+\cdots+n=\frac{n(n+1)}{2}$$

This formula tells us that the sum of the first n positive integers can be found by multiplying n and $(n + 1)$ and dividing that sum by 2.

Proving such formulas can be tricky business. It is not enough to show by experimentation that the formula is valid for lots and lots of numbers. It could be that the formula holds for all the numbers that we take the time to check but that there are other numbers beyond the list with which we experimented for which the formula fails. A mathematical proof establishes with certainty that a formula such as this one holds for all positive integer values.

There are many different methods of mathematical proof, most of which lie beyond the scope of our discussion here. One that bears investigation is the method of **mathematical induction**. The method relies on an axiom (a fact that is not proven but is self-evidently true) called the mathematical induction principle. The principle states the following:

To prove a given mathematical fact about the natural numbers is true, it is necessary to establish two facts: first, that the fact is true for n = 1 (or some other suitable value of n), and second, if the fact holds for a particular unspecified value of n, call that value k, then it also holds for the next integer, (k + 1).

Once these conditions have been established, the axiom tells us that the fact we are attempting to establish must hold for all natural numbers beyond the initial value for which we had shown the fact was true.

For instance, let's establish the result of Gauss mentioned above by induction.

EXAMPLE 6.33 Prove the following by mathematical induction:

$$1 + 2 + 3 + \cdots + n = \frac{n(n + 1)}{2}$$

SOLUTION

To establish this as a proven fact, we will perform the two steps previously mentioned. We begin by showing that the formula holds for $n = 1$. This amounts to showing

$$1 = \frac{1(1 + 1)}{2}$$

We simplify the right side and find

$$1 = \frac{1(2)}{2}$$

$$1 = 1$$

Thus, the formula holds for $n = 1$.

Next, we assume that the formula is valid for $n = k$. That is, we accept as a fact that

$$1 + 2 + 3 + \cdots + k = \frac{k(k + 1)}{2}$$

Based on this accepted fact, we now will attempt to show that the formula holds for $n = k + 1$.

That is, we attempt to show

$$1 + 2 + 3 + \cdots + (k + 1) = \frac{(k + 1)[(k + 1) + 1]}{2}$$

To do this, we will try to employ some cleverness so that we can exploit our accepted fact. Note that the left side can be written as

$$1 + 2 + 3 + \cdots + k + (k + 1)$$

and that the first k terms of this sum is equal to, by assumption, $\frac{k(k+ 1)}{2}$, so we can rewrite the expression as

$$\frac{k(k+1)}{2} + (k+1)$$

Now, if we obtain a common denominator and add these expressions together, we get

$$\frac{k(k+1)}{2} + \frac{2k+2}{2}$$

or

$$\frac{k^2+k+2k+2}{2}$$

which simplifies to

$$\frac{k^2+3k+2}{2}$$

Now, if we factor the numerator, we find

$$\frac{(k+1)(k+2)}{2}$$

or (breaking apart the second factor in the numerator so that it looks the way we had wanted it to look)

$$\frac{(k+1)[(k+1)+1]}{2}$$

This is precisely what we were attempting to show, and thus the result must, by the axiom of mathematical induction, be true for all natural numbers n.

You may have to read through Example 6.33 several times in order to convince yourself that what we've said is true. Mathematical proof is one of the most abstract concepts ever devised, so do not be daunted if, at first glance, you find yourself asking what we just showed there. It's not at all intuitive!

EXAMPLE 6.34 Prove $1 + 4 + 9 + \cdots + n^2 = \dfrac{n(n+1)(2n+1)}{6}$ for all positive integers n.

SOLUTION

We will attempt to prove this by induction and will start by showing that the result holds for $n = 1$. This means that we need to confirm that $1 = \dfrac{1(1+1)(2+1)}{6}$.

A straightforward computation shows that the numerator on the right is (1)(2)(3) = 6, so the formula holds for $n = 1$.

Next, we assume the formula to be valid for $n = k$, so $1 + 4 + \cdots + k^2 = \dfrac{k(k+1)(2k+1)}{6}$, and now we attempt to confirm the result for $1 + 4 + \cdots + k^2 + (k+1)^2$.

The first k terms of the sum are assumed equal to $\dfrac{k(k+1)(2k+1)}{6}$, so we now must confirm that

$$\frac{k(k+1)(2k+1)}{6} + (k+1)^2 = \frac{(k+1)[(k+1)+1][2(k+1)+1]}{6}$$

The right side of that equality may not be obvious. What we've done is to examine $\dfrac{k(k+1)(2k+1)}{6}$ and replace the ks with $(k+1)$s. If we can now show that the two sides are truly equal, the proof will be complete.

Obtaining a common denominator on the left and simplifying on the right, we see

$$\frac{k(k+1)(2k+1)}{6} + \frac{6(k+1)^2}{6} = \frac{(k+1)(k+2)(2k+3)}{6}$$

$$\frac{k(k+1)(2k+1)+6(k+1)^2}{6} = \frac{(k+1)(k+2)(2k+3)}{6}$$

Note

No matter how many examples you can produce that show a result is true, listing examples is *not* acceptable as proof that a statement is true for all positive integers *n*!

Note

You can prove that a result is *not* true by producing a "counterexample." That is, if you can find (somehow) a number *n* for which the result is not true, then that in itself is proof that the result does not hold for all positive integers *n*!

Prepare yourself for some algebraic heavy lifting at this point. We observe that, on the left, a factor of $(k + 1)$ can be removed in the numerator, giving us

$$\frac{(k+1)\left[k(2k+1)+6(k+1)\right]}{6}=\frac{(k+1)(k+2)(2k+3)}{6}$$

As you may have been told at some point in your past, "You're doing a lot of chopping, son, but no chips are flying." At the moment, it may not be clear if we're making progress. We'll press on, simplifying the left side and hoping that something comes along we can use:

$$\frac{(k+1)\left[2k^2+k+6k+6\right]}{6}=\frac{(k+1)(k+2)(2k+3)}{6}$$

$$\frac{(k+1)\left[2k^2+7k+6\right]}{6}=\frac{(k+1)(k+2)(2k+3)}{6}$$

Fortune has smiled on us! The trinomial factor in the left side's numerator happens to factor, and we see

$$\frac{(k+1)\left[(k+2)(2k+3)\right]}{6}=\frac{(k+1)(k+2)(2k+3)}{6}$$

which proves the result. Thus, the result holds for all positive integers *n*.

Before closing this section, we will leave you with two very important notes:

Exercises

Answer the following questions, using complete sentences, proper grammar, and correct terminology.

1. What is meant by "mathematical induction"?
2. In your own words, explain the "induction principle."

Prove the following statements, using mathematical induction.

3. $2^n > n$ for all positive integers $n \exists 4$
4. $2n^2 > (n + 1)^2$ for all positive integers $n > 2$
5. $1 + 3 + 5 + \cdots + (2n - 1) = n^2$
6. $1^3 + 2^3 + 3^3 + \cdots + n^3 = \dfrac{n^2(n+1)^2}{4}$
7. $1^4 + 2^4 + 3^4 + \cdots + n^4 = \dfrac{n(n+1)(2n+1)(3n^2+3n-1)}{30}$
8. $\dfrac{1}{2}+\dfrac{1}{6}+\dfrac{1}{12}+\cdots+\dfrac{1}{n(n+1)}=\dfrac{n}{n+1}$

9. $\dfrac{1}{(1)(3)}+\dfrac{1}{(3)(5)}+\cdots+\dfrac{1}{(2n-1)(2n+1)}=\dfrac{n}{2n+1}$

10. 3 is a factor of $4^n - 1$ for all positive integers n

11. 2 is a factor of $3n^4 - 3n - 12$ for all positive integers n

12. $1^2 + 3^2 + \cdots + (2n-1)^2 = \dfrac{n(2n-1)(2n+1)}{3}$ for all positive integers n

13. $1 + 2 + 4 + 8 + \cdots + 2^n = 2^n - 1$ for all positive integers n

14. $1 + 4 + 7 + 10 + \cdots + (3n-2) = \dfrac{3n^2 - n}{2}$ for all positive integers n

15. $1(1!) + 2(2!) + 3(3!) + \cdots + n(n!) = (n+1)! - 1$ for all positive integers n

16. A wealthy man has only \$2 bills and \$5 in his possession (assume an unlimited supply). You can request any amount of him of \$4 or more, and he can produce the exact quantity of money you request using the bills in his possession.

17. For any integer $n > 23$, there exist nonnegative integers a and b such that we can express $n = 7a + 5b$.

18. For any integer n 1, the number $\sqrt{1+\sqrt{1+\sqrt{1+\cdots}}}$ is irrational, where n is the number of 1s present.

19. This is called the *Towers of Hanoi* problem. You have three pegs and a collection of disks of differing sizes. Initially, all the disks are stacked in one pile on one peg, with the largest peg on the bottom and successively smaller disks atop it (so the smallest disk is on the top). A move in the process consists of moving a disk from one peg to another, subject to the condition that a larger disk may never rest on a smaller one. The objective is to find the number of permissible moves that will transfer all the disks from the first peg to the third peg, making sure the disks are assembled on the third peg in the proper order (according to size). The second peg is an intermediate peg. Prove that it takes $2^n - 1$ moves to move n disks from the first peg to the third.

20. Suppose m is a fixed integer such that m is a factor of $x - y$. Prove the following statement: m is a factor of $x^n - y^n$ for all positive integers n.

21. Prove that n distinct lines, such that no three lines in the set intersect in a single point, in the plane divide the plane into $\dfrac{n^2+n+2}{2}$ regions.

22. Prove that a set with n elements has 2^n subsets.

23. Prove the following special case of Bernoulli's inequality: $4^n > 1 + 4n$ for all integers $n > 1$.

6.5 The Binomial Theorem

A Binomial Expansion

A mathematical expression with two terms is called a binomial expression. For example, $2x - 3$, $1 + x^2$, and $\frac{3}{x} - 2$ and are binomials. Consider the following equations in which a binomial expression $(a + b)$ is raised to successive whole number powers:

$$(a+b)^0 = 1$$
$$(a+b)^1 = a+b$$
$$(a+b)^2 = a^2 + 2ab + b^2$$
$$(a+b)^3 = a^3 + 3a^2b + 3ab^2 + b^3$$
$$(a+b)^4 = a^4 + 4a^3b + 6a^2b^2 + 4ab^3 + b^4$$

We refer to the simplified expressions to the right of the equal signs as **binomial expansions**. Let's make some observations about those binomial expansions:

- Each expansion has one term more than the exponent of the binomial. For example, the expression $(a + b)^1$ has two terms, which is one more than the exponent 1; the expression $(a + b)^2$ has three terms, which is one more than the exponent 2; and so on.
- The exponents of the first and last terms of the expansion are equal to the exponent of the binomial.
- In each term of the expansion, the sum of the exponents of a and b equals the exponent of the binomial.
- As we go from left to right, you will notice that the exponent of a decreases from a value equal to the exponent of the binomial to zero. At the same time, the exponent of b increases from zero to equal the exponent of the binomial.
- The coefficients of the terms in the expansion are symmetrical; for example, in the expansion of $(a + b)^3$, the coefficient of the first and the fifth terms is the same and that of the second and the fourth terms is the same.

The Binomial Theorem

The **binomial theorem** is a formula that expands a binomial that is raised to a power in the form of a series. The following formula helps expand the expression without actually having to multiply:

$$(a+b)^n = a^n + na^{n-1}b + \frac{n(n-1)}{2}a^{n-2}b^2 + \frac{n(n-1)(n-2)}{3(2)}a^{n-3}b^3 + \cdots + b^n$$

That formula, known as the binomial theorem, is useful for expanding powers of binomials. Note that you should not be misled into believing that the theorem applies only to binomials using addition, for $(a - b)$ can be viewed as $[a + (-b)]$, to which the theorem obviously applies.

It turns out that the theorem can also be expressed in an equivalent form using the factorial notation we introduced earlier. The coefficient of each term in the binomial expansion is calculated by what we refer to as the "binomial coefficient formula," ${}_nC_r$, where

$${}_nC_r = \frac{n!}{(n-r)!r!}$$

Note

Recall that $n! = n(n-1)(n-2)(n-3)\ldots(3)(2)(1)$, with the special case $0! = 1$.

The notation ${}_nC_r$ is sometimes read as "n choose r" because of its application in problems of probability, where it allows us to compute the number of distinct ways that r objects can be selected from a collection of n objects. Using this notation, the binomial expansion formula can be written as

$$(a+b)^n = {}_nC_0a^n + {}_nC_1a^{n-1}b + {}_nC_2a^{n-2}b^2 + {}_nC_3a^{n-3}b^3 + \cdots + {}_nC_nb^n$$

Note

The rth term of the binomial expansion is given by ${}_nC_{r-1}\,a^{n-r+1}\,b^{r-1}$.

The binomial expansion of $(a + b)^n$ has $n + 1$ terms. In general, the rth term in the expansion is calculated by

$${}_nC_{r-1}a^{n-r+1}b^{r-1}$$

EXAMPLE 6.35

Expand $(a + b)^6$ using the binomial theorem.

SOLUTION

The coefficients are evaluated using ${}_nC_r = \frac{n!}{(n-r)!r!}$ as follows:

$${}_6C_0 = \frac{6!}{(6-0)!0!} = \frac{6!}{6!(1)} = 1$$

$${}_6C_1 = \frac{6!}{(6-1)!1!} = \frac{6!}{5!(1)} = \frac{6(5!)}{5!} = 6$$

$${}_6C_2 = \frac{6!}{(6-2)!2!} = \frac{6!}{4!(2)(1)} = \frac{6(5)(4!)}{4!(2)} = 15$$

$${}_6C_3 = \frac{6!}{(6-3)!3!} = \frac{6!}{3!(3!)} = \frac{6(5)(4)(3!)}{3!(3!)} = \frac{6(5)(4)}{3(2)(1)} = 20$$

$${}_6C_4 = \frac{6!}{(6-4)!4!} = \frac{6!}{4!(2)(1)} = \frac{6(5)(4!)}{4!(2)} = 15$$

$${}_6C_5 = \frac{6!}{(6-5)!5!} = \frac{6!}{5!(1)} = \frac{6(5!)}{5!} = 6$$

$${}_6C_6 = \frac{6!}{(6-6)!6!} = \frac{6!}{6!(1)} = 1$$

Using the coefficient values, we find

$$(a+b)^6 = a^6 + 6a^5b + 15a^4b^2 + 20a^3b^3 + 15a^2b^4 + 6ab^5 + b^6$$

EXAMPLE 6.36 Expand $(2x - 3y)^5$ using the binomial theorem.

SOLUTION

You can write $(2x - 3y)^5 = [2x + (-3y)]^5$ and apply the binomial expansion formula to obtain the final expression. We begin by computing the coefficients:

$$_5C_0 = \frac{5!}{(5-0)!0!} = \frac{5!}{5!(1)} = 1$$

$$_5C_1 = \frac{5!}{(5-1)!1!} = \frac{5(4!)}{4!(1)} = 5$$

$$_5C_2 = \frac{5!}{(5-2)!2!} = \frac{5!}{3!(2)(1)} = \frac{5(4)(3!)}{3!(2)} = 10$$

$$_5C_3 = \frac{5!}{(5-3)!3!} = \frac{5!}{2!(3!)} = \frac{5(4)(3!)}{2!(3!)} = \frac{5(4)}{2(1)} = 10$$

$$_5C_4 = \frac{5!}{(5-4)!4!} = \frac{5!}{(1)4!} = \frac{5(4!)}{4!} = 5$$

$$_5C_5 = \frac{5!}{(5-5)!5!} = \frac{5!}{(1)5!} = \frac{5!}{5!} = 1$$

Using these coefficients and the appropriate powers of $2x$ and $(-3y)$ in the binomial expansion, we find

$$(2x-3y)^5 = {_5C_0}(2x)^5 + {_5C_1}(2x)^{5-1}(-3y) + {_5C_2}(2x)^{5-2}(-3y)^2 + {_5C_3}(2x)^{5-3}(-3y)^3 + {_5C_4}(2x)^{5-4}(-3y)^4 + {_5C_5}(2x)^{5-5}(-3y)^5$$

$$(2x-3y)^5 = (2x)^5 + 5(2x)^4(-3y) + 10(2x)^3(-3y)^2 + 10(2x)^2(-3y)^3 + 5(2x)^1(-3y)^4 + (2x)^0(-3y)^5$$

$$(2x-3y)^5 = 32x^5 - 240x^4y + 720x^3y^2 - 1080x^2y^3 + 810xy^4 - 243y^5$$

The Pascal Triangle

Without a doubt, the most tedious step in the derivation of the nth power of a binomial expansion is the calculation of the individual coefficients. There is a shortcut method for obtaining the coefficients of the binomial expansion, called the **Pascal triangle** method. Although the result is traditionally named in honor of the French mathematician Blaise Pascal, it actually predates him and was known to the ancient Persians, Indians, Chinese, and Italians before him.

The construction of the Pascal triangle is, as the name implies, triangular, with the rows numbered (for reasons explained shortly) from the top, starting with row 0:

1	**Row 0**
1 1	**Row 1**
1 2 1	**Row 2**
1 3 3 1	**Row 3**
1 4 6 4 1	**Row 4**
1 5 10 10 5 1	**Row 5**
.	
.	
.	

Note how the rows are constructed. Each row begins and ends with 1, and each intermediate entry in the row is obtained by adding the pair of numbers diagonally above it. That is, consider the first 3 in row 4 of the triangle (remember, the top row is numbered row 0!). The 3 is found by looking at the number pair diagonally above it: the 1 and 2 in row 3 of the triangle.

The usefulness of the triangle is evident if we consider the coefficients calculated in Example 6.35. If one examines row 5 of the triangle (actually, keep in mind, the sixth row from the top because of our numbering agreement), the numbers of row 5 are precisely the **binomial coefficients** we computed in the example. This is much easier and faster than finding the individual coefficients by hand.

Let's use the Pascal triangle to assist in the expansion of the following example.

EXAMPLE 6.37 Expand $(5x + y)^4$ using the binomial theorem and Pascal's triangle.

SOLUTION

We consult the triangle first and observe that the numbers in row 4 are 1 4 6 4 1. These will be the coefficients for our expansion.

Recalling the binomial theorem

$$(a+b)^n = {}_nC_0a^n + {}_nC_1a^{n-1}b + {}_nC_2a^{n-2}b^2 + {}_nC_3a^{n-3}b^3 + \cdots + {}_nC_nb^n$$

and using $a = 5x$, $b = y$, and $n = 4$, we obtain

$$\begin{aligned}(5x + y)^4 &= 1(5x)^4 + 4(5x)^3y^1 + 6(5x)^2y^2 + 4(5x)y^3 + 1y^4 \\ &= 625x^4 + 500x^3y + 150x^2y^2 + 20xy^3 + y^4\end{aligned}$$

EXAMPLE 6.38 Expand $\left(3+\frac{1}{x}\right)^5$ using the binomial theorem, obtaining the coefficients from the Pascal triangle.

SOLUTION

The numbers in row 5 of the Pascal triangle are 1 5 10 10 5 1, and these will be the binomial coefficients for our expansion.

Using the binomial theorem, taking $n = 5$, $a = 3$, and $b = \frac{1}{x}$, we find

$$\left(3+\frac{1}{x}\right)^5 = {}_5C_0(3)^5 + {}_5C_1(3)^{5-1}\left(\frac{1}{x}\right) + {}_5C_2(3)^{5-2}\left(\frac{1}{x}\right)^2 + {}_5C_3(3)^{5-3}\left(\frac{1}{x}\right)^3$$

$$+ {}_5C_4(3)^{5-4}\left(\frac{1}{x}\right)^4 + {}_5C_5(3)^{5-5}\left(\frac{1}{x}\right)^5$$

$$= 1(3)^5 + 5(3^4)\left(\frac{1}{x}\right)^1 + 10(3^3)\left(\frac{1}{x}\right)^2 + 10(3^2)\left(\frac{1}{x}\right)^3 + 5(3^1)\left(\frac{1}{x}\right)^4 + 1\left(\frac{1}{x}\right)^5$$

$$= 243 + \frac{405}{x} + \frac{270}{x^2} + \frac{90}{x^3} + \frac{15}{x^4} + \frac{1}{x^5}$$

Although it is not obvious from the examples and explanations provided to this point, the binomial theorem also applies to problems where a binomial is raised to a *negative* power. For such examples, the Pascal triangle is not of use, but we can revert to our original method of computing the binomial coefficients in such an event.

EXAMPLE 6.39 Find the first three terms of the binomial expansion of $\frac{1}{(2+x)^4}$.

SOLUTION

$$\frac{1}{(2+x)^4} = (2+x)^{-4}$$

Using the binomial theorem, we obtain

$$(a+b)^n = a^n + na^{n-1}b + \frac{n(n-1)}{2}a^{n-2}b^2 + \frac{n(n-1)(n-2)}{3(2)}a^{n-3}b^3 + \cdots$$

$$\frac{1}{(2+x)^4} = (2)^{-4} + (-4)(2)^{-4-1}(x) + \frac{(-4)(-4-1)}{2}(2)^{-4-2}(x)^2 + \cdots$$

$$= \frac{1}{16} - \frac{1}{8}x + \frac{5}{32}x^2 + \cdots$$

Exercises

Answer the following questions using complete sentences, proper grammar, and correct terminology.

1. What is the Pascal triangle, how is it constructed, and how is it used?
2. What are the coefficients of a binomial expansion called, and how are they computed?

Calculate the following binomial coefficients.

3. ${}_5C_2$
4. ${}_7C_4$
5. ${}_8C_8$
6. ${}_5C_5$

7. ${}_9C_0$

8. ${}_3C_0$

9. ${}_{12}C_7$

10. ${}_{25}C_{10}$

Use the binomial theorem to expand the following expressions.

11. $(x + 2)^5$

12. $(x + 4)^7$

13. $(x - 3)^6$

14. $(x - 5)^4$

15. $(2x + 3)^5$

16. $(3x + 2)^7$

17. $(4x + 2y)^5$

18. $(7x + 3y)^8$

19. $(x + 4y)^7$

20. $(x + 5t)^7$

21. $(x - 3y)^4$

22. $(x - 8u)^6$

23. $(5x - 7y)^4$

24. $(3a - 5b)^5$

25. $(x^2 + 2y^2)^4$

26. $(x^2 + 4u)^5$

27. $\left(\dfrac{2}{x} + y\right)^4$

28. $(1/u - v)^5$

29. $\left(\dfrac{3}{t} - 4w\right)^5$

Find the expansion of the following expression, using Pascal's triangle as an aid.

30. $(4x - y)^5$

31. $(3 - 2t)^6$

32. $(x + 4y)^7$

33. $(5x - 2y)^5$

Find the first three terms in the expansion of the following expressions.

34. $\dfrac{1}{(x+2y)^3}$

35. $\dfrac{1}{(3x-4y)^5}$

36. $\dfrac{1}{(4x+1)^6}$

37. $\dfrac{1}{(5-x)^7}$

For the following expressions, find the coefficient of x^5.

38. $(x + 2)^{12}$

39. $(x - 3)^8$

40. $(2x - 5)^{10}$

41. $(3x + 4)^9$

For the following expressions, find the fourth term of the expansion.

42. $(2x + 1)^8$

43. $(5x + 3)^9$

44. $(3 - 2x)^5$

45. $(7 - x)^6$

Summary

In this chapter, you learned about:

- Definitions and concepts of sequences and series: the general term and a recursive definition and factorial and summation notation.
- Arithmetic and geometric sequences: finding the general term and calculating the first k terms.
- The principle of mathematical induction.
- Binomial expansion using the binomial theorem and the Pascal triangle.

Glossary

arithmetic progression The pattern followed by the terms of an arithmetic sequence.

arithmetic sequence A sequence in which the difference between successive terms is constant.

binomial coefficients The coefficients of a binomial expansion.

binomial expansion The simplified expression resulting from calculating $(a + b)^n$, for any values a and b, and any whole number n.

binomial theorem A formula with which one can calculate the binomial expansion of an expression.

common difference The difference between successive members of an arithmetic sequence.

common ratio The ratio of one term of a geometric sequence to its predecessor.

factorial An abbreviation for a particular mathematical product; $n!$ represents the product of the natural number n with all lesser natural numbers.

general term of a sequence A formula with which the individual terms of a sequence can be calculated as a function of the order of occurrence.

geometric progression The pattern followed by the terms of a geometric sequence.

geometric sequence A sequence in which each term is a constant multiple of its predecessor.

index of summation The variable used within the formula of sigma notation.

length of a sequence The number of terms in a sequence.

mathematical induction One technique of mathematical proof.

members of a sequence The numbers occurring within a sequence.

Pascal triangle A shortcut method for determining the binomial coefficients of a binomial expansion.

recursive relationship A relationship in which the value of a term in a sequence depends on earlier terms in the sequence.

sequence A progression of real numbers whose ordering cannot be changed.

series The sum of the terms of a sequence.

sigma notation A compact notation representing the sum of the terms of a sequence. Also called summation notation.

terms of a sequence The numbers occurring within a sequence.

upper limit of summation The maximum value of the index of summation in sigma notation.

End-of-Chapter Problems

For the following sequences whose general term is given, find the first five terms of the progression.

1. $a_n = 4n + 6$
2. $a_n = 2^{n-2}$
3. Find the tenth term of the sequence specified in Exercise 1.
4. Find the fifteenth term of the sequence specified in Exercise 2.

The following problems give the first four terms of a sequence. Find a general term for the sequences (answers are not unique!).

5. $-\frac{1}{2}, \frac{1}{4}, -\frac{1}{8}, \frac{1}{16},$
6. 27, 125, 343, 729, ...
7. Find the eleventh term of the sequence specified in Exercise 5.
8. Find the sixth term of the sequence specified in Exercise 6.

The following sequences are recursively defined, with the indicated basis and recursive relation defined for $n \exists 2$. Give the first six terms of the sequence.

9. $a_1 = 3, a_n = 4a_{n-1} + 3$
10. $a_1 = 1, a_n = (2n + 1)a_{n-1}$
11. $a_1 = 7, a_n = \frac{2}{a_{n-1}}$

Simplify the following expressions, which involve factorials.

12. $\dfrac{137!}{134!}$
13. $\dfrac{37! - 35!}{33!}$

Calculate the following sums.

14. $\sum_{n=1}^{7} \frac{1}{2^n}$
15. $\sum_{n=5}^{7} n^3 + 2n^2$

Each of the following problems can be solved using a finite summation. Construct the summation involved in the problem and then use it to solve the problem.

16. A battery in a laptop computer, when it is recharged, does not regain all its original charge. Assume that it recharges to only 99.5% of its previous charge. Construct a summation formula to express the total energy put into the battery after n discharge/recharge cycles. Initially, the battery capacity is 80 Wh. After seven recharge cycles, how much energy has been put into the battery?
17. Hermione spends 17 galleons on books at the beginning of the semester. Each week thereafter, she spends 3 galleons on books. Show a summation that gives the amount of money that Hermione has spent on books n weeks into the semester. Use the formula to determine how much she has spent 13 weeks into the semester.

For the following sequences, determine whether they are arithmetic. If they are arithmetic, determine the common difference. If they are not arithmetic, explain.

18. $-5, -1, 3, 7, \ldots$
19. $17, 20, 24, 29, \ldots$

The following problems give an initial term and common difference for arithmetic sequences. Give the first five terms of the sequence, ending the list with ellipses.

20. $a_1 = 1\frac{1}{2}$, common difference $\frac{3}{8}$
21. $a_1 = -7$, common difference -4

The following problems give the first few terms of an arithmetic sequence. Find the general term of the sequence.

22. $21, 26, 31, 36, \ldots$
23. $-4, -1, 2, 5, \ldots$
24. Find the one-hundredth term of the sequence from Exercise 22.
25. Find the one-hundredth term of the sequence from Exercise 23.

For the following arithmetic sequences given by a general term, find the sum of the indicated terms.

26. For the sequence $a_d = -5 + (n-1)\frac{3}{8}$, find the sum of the first 60 terms.
27. For the sequence $a_d = 45 + (n - 1)(0.1)$, find the sum of the first 75 terms.

The following sequences may or may not be geometric. If they are, identify the common ratio. If they are not, explain why they are not geometric.

28. 1, 5, 25, 125, . . .

29. $1, \pi, \pi^2, \pi^3, \ldots$

30. 0, 3, 15, 255, . . .

In the following problems, the first term and the common ratio of a geometric sequence is given. Find the first four terms of the sequence.

31. $a_1 = 3, r = 4$

32. $a_1 = 8, r = \frac{1}{2}$

33. Find the tenth term of the sequence in Exercise 31.

34. Find the tenth term of the sequence in Exercise 32.

In the following problems, the first few terms of a geometric sequence are given. Find the common ratio and then find the indicated term of the sequence.

35. 5, 15, 45, . . . Find the ninth term.

36. 7, 10.5, 15.75, . . . Find the seventh term.

Find the sum of the given finite geometric sequence.

37. $\sum_{n=1}^{8} 3(2)^{n-1}$

38. $\sum_{n=1}^{7} \frac{1}{3}(\frac{1}{3})^{n-1}$

In the following problems, use the techniques of geometric sequences to solve the applications.

39. Suppose a principal of \$10,000 is invested at 1% interest. Find the value of the investment after 10 years if the interest is compounded (a) annually and (b) quarterly.

40. George is planning to save for retirement. He figures that he'll live 20 years after he retires, and it is 40 years until he retires. He wants \$40,000 the first year of retirement, then 3% more each year after that. He plans on saving a certain amount the first year, then 4% more each year after that. Ignoring interest, how much does George have to save the first year?

Prove the following statements, using mathematical induction.

41. $6+18+54+162+ \quad +2 \bullet 3^n = 3^{n+1} - 3$

42. 5 is a factor of $6^n - 1$ for all positive integers n

Calculate the following binomial coefficients.

43. ${}_8C_3$

44. ${}_{12}C_6$

Use the binomial theorem to expand the following expressions.

45. $(2x + 5)^4$

46. $(3x^2 + \frac{2}{y})^5$

Find the expansion of the following expressions, using Pascal's triangle as an aid.

47. $(3x - y)^4$

48. $(2u + v)^5$

Find the first three terms in the expansion of the following expressions.

49. $\dfrac{1}{(u+v)^5}$

50. $\dfrac{1}{(5t-1)^4}$

Chapter 7 Right-Triangle Geometry and Trigonometry

One of the most frequently applied concepts of geometry is that of a **right triangle**, which is a triangle containing a 90° angle at one vertex. The right triangle holds an important place as a cornerstone of the study of trigonometry, a field whose name (taken literally) means the *study of triangles* and that is the object of our interest in this chapter.

Finding its roots in ancient Greece, trigonometry had its beginnings as the computational aspect of geometry and found application as a tool to measure objects from a distance. What distinguishes trigonometry from geometry is that it is based on the measurements of angles and also on particular quantities determined by those measurements, while geometry is a study of shapes and of measurements of the Earth.

The measures of shapes (side lengths, areas, perimeters, and so on) are determined using any of our well-known scales of measurement, such as inches, feet, meters, and so on. Angles, on the other hand, can be measured using two distinct scales, one being the rather familiar form of degrees, the other the less well-known system of radians. These angle measurements (whatever the form) are related to the lengths of the sides of triangles by the trigonometric functions of sine, cosine, tangent, and others, and we will introduce these in this chapter.

Note that trigonometric functions are not nearly as straightforward to work with as are the algebraic functions with which we have already gained experience.

In fact, historically, mathematical tables were constructed cataloging their values, and, even so, the computations involving the functions proved so unwieldy that they helped to motivate the creation of mathematical tools such as logarithms and slide rules. The challenge of working with the functions is, however, amply rewarded by the wealth of applications in which they serve as mathematical models, and consequently it is well worth our time to examine the functions in some detail.

Objectives

When you have successfully completed the materials of this chapter, you will be able to:

- Convert an angle from degrees to radians and vice versa.
- Find the values of the six trigonometric functions of an angle using a calculator.
- Use a calculator to find an acute angle that has a specific trigonometric function value.
- Find the values of the missing sides and angles of a right triangle.
- Solve practical problems that involve right triangles.

7.1 Measuring an Angle

The Degree System of Angle Measurement

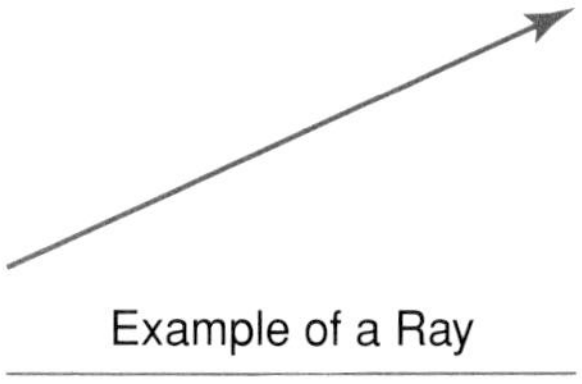

Figure 7.1 A ray

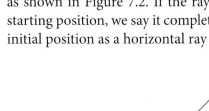

In our study of trigonometry, we will take as primitive the notion of a ray, a half line that begins at a certain point and extends forever in one direction. The point where the ray begins is known as the **end point**, and the direction in which the ray extends is indicated by a half line terminating in an arrowhead (Figure 7.1).

When a ray rotates on its fixed end point from an initial position to a new position, it creates a span referred to as an **angle** between those two positions, as shown in Figure 7.2. If the ray turns around completely and returns to its starting position, we say it completes one revolution. It is common to depict the initial position as a horizontal ray (though this is not necessary).

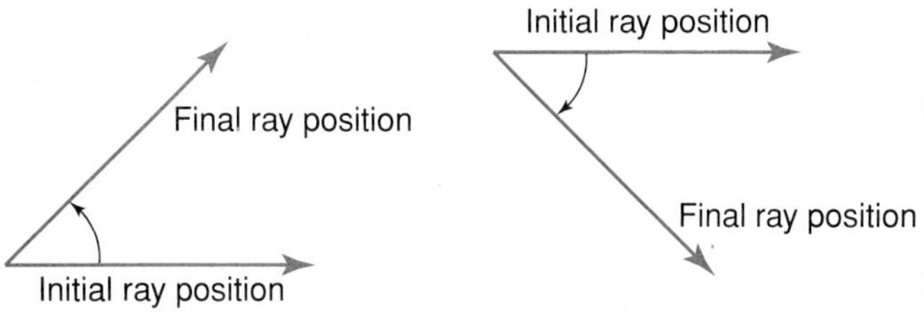

Figure 7.2 Examples of angles

One complete revolution is defined as 360°, also written as 360°, and thus **one degree** is of 1/360 one revolution. One degree can be subdivided into further refined measures, and that subdivision can be written as either a decimal or a fraction, and thus it is not uncommon to encounter such measurements as 83.125°, or 71½°. Measurement of a drawn angle in degrees can be accomplished with a tool called a protractor, which typically has the shape of a half-circle on which 180° is marked off in single-degree units (Figure 7.3).

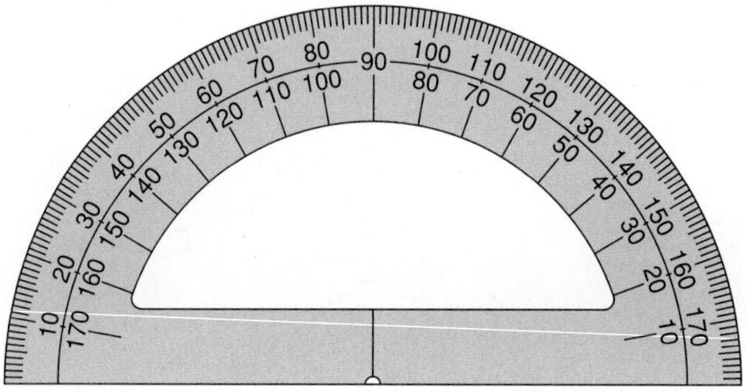

Figure 7.3 A protractor

Another method of subdivision of degree measure is the **sexagesimal system**. The word "sexagesimal" may be unfamiliar, as it is not a word one encounters in everyday conversation, but it refers to a reckoning or subdivision by sixtieths, just as an hour is subdivided into 60 minutes and a minute subdivided into 60 seconds. In the sexagesimal system, one degree is subdivided into 60 minutes, and one minute is further subdivided into 60 seconds. Thus, degree breakdown in this manner resembles the manner in which hours are decomposed into minutes and seconds.

Note

The following equivalences are useful to keep in mind!

1° = 60' (one degree = 60 minutes)

1' = 60" (one minute = 60 seconds)

Symbolically, the use of a single apostrophe denotes minutes, while a double apostrophe indicates seconds. An example of the use of this notation would be 34°15'45", which we would read as "34 degrees, 15 minutes, 45 seconds."

These relationships permit us to construct conversion factors we can use to translate a measurement in one unit of measure to the other unit of measure. In this case, now that we have established the definitions of minutes and seconds in the context of degrees, we obtain the following conversion factors:

$$\left(\frac{1^\circ}{60'}\right), \left(\frac{60'}{1^\circ}\right), \left(\frac{1'}{60''}\right), \text{ and } \left(\frac{60''}{1'}\right)$$

These factors can be used to establish a further fact that $1^\circ = 360''$:

$$\frac{1^\circ}{60'} \times \frac{1'}{60''} = \frac{1^\circ}{3600''}$$

Note

The equivalence between degrees and seconds:

1° = 60' = 3600"

We can convert the fractional or decimal part of an angle into minutes and seconds using these formulas in a manner that we can readily illustrate and that we can use as an opportunity to practice our techniques of unit conversion.

Consider the angle whose measure is 24.6° and find its sexigesimal equivalent. When we do our conversion, the whole number part of the degree measurement will remain as given, but the decimal component must be converted to minutes and seconds. We notice that in order to do the conversion, we can use the equivalences listed above in the following manner:

$$\frac{0.6^\circ}{1} \cdot \frac{60'}{1^\circ} = \frac{36'}{1}$$

Since the conversion is exact, we can now say that 24.6° is exactly equivalent to 24°36'. If we want to show the seconds as well, we would say 24°36'0". This complete representation, showing degrees, minutes, and seconds, is sometimes called **DMS notation**.

EXAMPLE 7.1

Convert the measurement 16.555° to DMS notation.

SOLUTION

We will leave the whole number degree measure as given and focus on the conversion of the decimal degree to minutes and seconds. Using our conversion factors, we will first convert to minutes and then (if needed) proceed to seconds:

$$\frac{0.555^\circ}{1} \times \frac{60'}{1^\circ} = 33.3'$$

Since the conversion does not result in a whole number for minutes, we proceed to transform the decimal part of the minute measurement to seconds:

$$\frac{0.3'}{1} \times \frac{60''}{1'} = 18''$$

This completes the conversion to DMS form, and we conclude that 16.555° is equivalent to the DMS measure 16°33'18".

EXAMPLE 7.2

Convert 95°36'45" to decimal degree form.

SOLUTION

Whichever direction our conversion takes, from decimal degree to DMS or vice versa, the whole number part of the angle measure is preserved, so we can ignore the 95° for the purposes of our computation. We will convert the minutes and seconds individually to decimal degree form and then combine the results to achieve our decimal degree form. Note that, using our conversion factors, $\frac{36'}{1} \times \frac{1^\circ}{60'} = 0.6^\circ$, and $\frac{45''}{1} \times \frac{1^\circ}{3600''} = 0.0125^\circ$. Consequently, if we add our results, we find the equivalent decimal degree form to be 95.6125°.

Degree measurement of angles is probably fairly familiar to you, but (for reasons left unspecified at this time) the use of degrees can ultimately lead to undesirable complications in practical applications of trigonometry. Another system for measuring angles uses a circle of arbitrary radius to define a system of angular measurement, and that system is immune from this problem afflicting the degree system. Because of its use of the radius of a circle in its construction, the system is called the **radian system of measurement**.

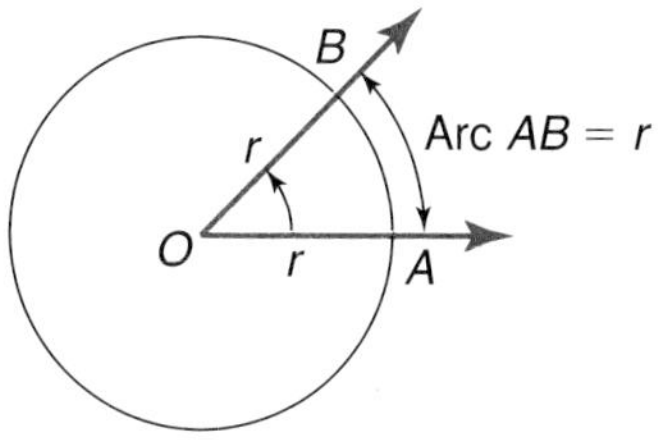

Figure 7.4 The definition of one radian

Construction of the Radian System of Angle Measurement

In a circle of arbitrary radius r with center point O, draw a ray, OA, as shown in Figure 7.4. If the ray rotates on O to trace an arc, AB, which happens to have length exactly that of the radius, then the measure of angle AOB is defined as the unit angle in this system and is said to be **one radian** (rad).

The measure of any angle in the radian system can be found by measuring the arc of the circle between the initial and terminal ray locations and dividing the length of that arc by the radius of the circle. Note that, whatever the units of measurement of arc and radius, creation of this ratio generates an angular measurement that is real number valued only and is otherwise free of units.

Note

1 revolution = 2π radians = 360°, and hence π radians = 180°

Recalling that the circumference of a circle is given by $C = 2\pi r$, it follows immediately that the angle corresponding to a complete revolution of the circle measures 2π radians. The same angle when expressed in degrees measures 360°, 360° and thus we have discovered a crucial equivalence that ties the two angle measurement systems together:

These equalities allow us to expand our list of conversion factors and further our opportunity to convert between different units of measure:

$$\left(\frac{\pi}{180^\circ}\right) \text{and} \left(\frac{180^\circ}{\pi}\right) \text{are our new conversion factors}$$

EXAMPLE 7.3

Convert 2.456 radians to degrees.

SOLUTION

To perform the unit conversion, we place the given measure into a fraction whose denominator is 1, $\frac{2.456 \text{ rad}}{1}$. The notation *rad* is technically not needed, and we will typically suppress it, with the understanding that an angle measure showing no degree symbol is, by default, in radian measure.

The fraction we have created is then multiplied against the conversion factor whose denominator possesses the units we wish to eliminate. In this case, we would employ $\left(\frac{180^\circ}{\pi}\right)$ and perform a multiplication of fractions:

$$\frac{2.456 \text{ rad}}{1} \times \frac{180^\circ}{\pi \text{ rad}} = 140.7^\circ$$

This process is used extensively when working with angle measurements, and you should be sure that you fully understand the simplification before moving forward.

Observe that we are technically abusing notation here. In truth, the value of the degree angle measurement is approximately 140.7° rather than precisely equal to 140.7°. We will, in the examples and computations that follow, allow ourselves the informality of showing decimal approximations without reference.

EXAMPLE 7.4

Convert 60° to radians.

SOLUTION

Here our intention is to convert from degrees to radians, so we must choose the proper conversion factor to suit our purposes. As before, we create a fraction whose numerator is the angle measurement we wish to convert and whose denominator is 1: $\frac{60°}{1}$. In order to eliminate the units of degrees, we must multiply this fraction by a conversion factor whose denominator contains degrees, so the units of degrees will reduce from the expression. That is, the proper conversion factor would be $\left(\frac{\pi}{180°}\right)$. Consequently, we set up and simplify the product of fractions:

$$\left(\frac{60°}{1}\right) \times \left(\frac{\pi}{180°}\right) = \frac{\pi}{3}$$

It would be beneficial at this point to produce a list of some of the more common acute angles and their radian equivalents (Table 7.1). As was done in Example 7.2, the conversion is an immediate application of the conversion factor $\left(\frac{\pi}{180°}\right)$, and we will leave their proof to the exercises.

TABLE 7.1 Common Degree and Radian Equivalences for Acute Angles

Degree Measure	Radian Measure
30°	$\frac{\pi}{6}$
45°	$\frac{\pi}{4}$
60°	$\frac{\pi}{3}$

EXAMPLE 7.5

Convert 20°35'42" to radians.

SOLUTION

We first convert the minutes and seconds into degrees and then add all the parts to obtain the decimal value of the angle in degrees. Then we will convert the degrees to radians, and we will have the desired units.

$$\text{To convert the minutes into degrees, } \frac{35'}{1} \times \frac{1°}{60'} = 0.5833°$$

$$\text{To convert the seconds into degrees, } \frac{42''}{1} \times \frac{1°}{3600''} = 0.0117°$$

Thus, if we combine the degrees, minutes, and seconds in their equivalent degree form, we have 20.5950°. What remains is to convert this measurement to radians:

$$\frac{20.5950°}{1} \times \frac{\pi}{180°} = 0.3595$$

Therefore, our equivalent measure, in radians, is 0.3595 radians.

EXAMPLE 7.6 Convert 2.343782 radians into degrees, minutes, and seconds.

SOLUTION

Our approach here will be to first convert the radians into degrees and then convert the decimal part of the degrees to minutes, the decimal form of which will, in turn, be converted to seconds. That sounds horribly convoluted, but you'll see that it's really rather systematic.

Convert the radians to degrees:

$$\begin{aligned} 2.343782 \text{ rad} &= \frac{180}{\pi} \times 2.343782 \\ &= 134.2885 \text{ degrees} \end{aligned}$$

Convert the decimal part of degrees to minutes:

$$\text{Since } 1^\circ = 60',\ 0.2885^\circ \times 60' = 17.31'$$

Now, convert the decimal part of the minutes to seconds:

$$\text{Since } 1' = 60'',\ 0.31' \times 60'' = 19'$$

Therefore, 2.343782 rad = $134^\circ 33' 19''$

Exercises

Answer each of the following using complete sentences, proper grammar, and correct spelling.

1. What is meant by "one degree"?
2. What is meant by "one radian"?
3. What is the process for converting a degree measure, such as 19.3875°, to DMS notation?
4. What is the process for converting radian measure to degree measure?

Sketch each of the following angles in standard position. Identify the quadrant in which the terminal side of the angle is found.

5. 150°
6. 300°
7. 310°14'
8. 285°
9. −200°
10. 260°
11. −40°
12. 400°

For each of the following angles in degree measure, convert to radian measure.

13. 45°
14. 30°
15. 90°
16. 150°
17. 235°
18. 245°
19. −48.5°
20. 175°

For each of the following angles in radian measure, convert to degree measure.

21. $\frac{\pi}{4}$
22. $\frac{5\pi}{6}$
23. $\frac{13\pi}{12}$
24. $\frac{11\pi}{12}$
25. $\frac{2\pi}{7}$
26. $-\frac{9\pi}{2}$
27. $\frac{15\pi}{8}$
28. 3
29. -4.5
30. 1.35
31. 2.75

Sketch each of the following radian angles in standard position and identify the quadrant in which the terminal side of the angle is found.

32. $\frac{3\pi}{5}$
33. $\frac{7\pi}{6}$
34. $-\frac{5\pi}{4}$
35. $-\frac{11\pi}{6}$
36. 5
37. -4
38. 1.5
39. 3.25

Convert the following decimal degree measurements to DMS notation.

40. $200.25°$
41. $175.375°$
42. $-25.6°$
43. $-130.8°$
44. $0.65°$
45. $-1.875°$

Convert the following measurements to decimal degree form. Round your answer to three decimal places, as needed.

46. $45°15'45''$
47. $-50°18'30''$
48. $-230°50'$
49. $300°10'45''$

Convert the following measurements to radian form. Round your answer to three decimal places, as needed.

50. $95°45'30''$
51. $-120°20'15''$
52. $-300°12'24''$
53. $210°18'36''$

7.2 Trigonometric Functions

In this section, our goal will be to define a particular class of functions, called trigonometric functions, through the use of ratios of sides of a right triangle. As we'll see, trigonometric functions or ratios allow us to find the values of the missing sides and angles of a right triangle, and the ability to do this will permit us to resolve many application problems.

Defining the Trigonometric Functions

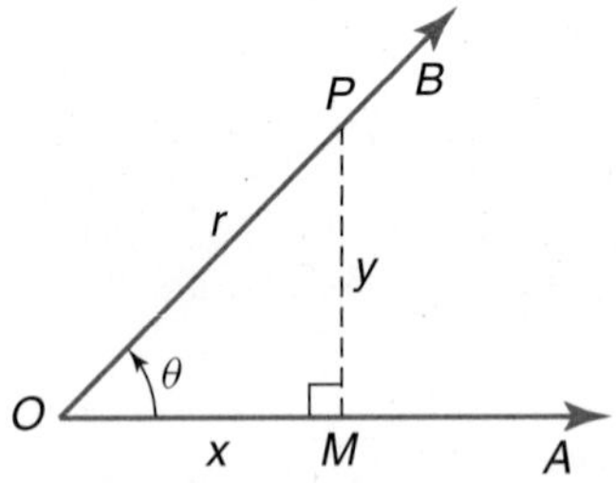

Figure 7.5 The acute angle θ and the sides of a right triangle relative to it.

When a ray, OA (as shown in Figure 7.5), rotates from its starting position to a new position, OB, it traces an acute angle θ. Take any point P on OB and drop a perpendicular line from P to OA at$M$. Let the hypotenuse (the side of the triangle opposite the 90° angle) of the triangle $OP = r$, the perpendicular $PM = y$, and the side adjacent to the angle $OM = x$.

With these right triangle side identifications, the trigonometric functions are defined as

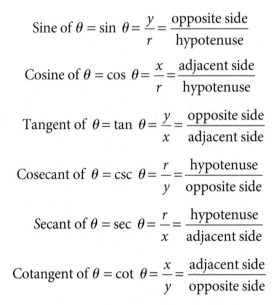

$$\text{Sine of } \theta = \sin\ \theta = \frac{y}{r} = \frac{\text{opposite side}}{\text{hypotenuse}}$$

$$\text{Cosine of } \theta = \cos\ \theta = \frac{x}{r} = \frac{\text{adjacent side}}{\text{hypotenuse}}$$

$$\text{Tangent of } \theta = \tan\ \theta = \frac{y}{x} = \frac{\text{opposite side}}{\text{adjacent side}}$$

$$\text{Cosecant of } \theta = \csc\ \theta = \frac{r}{y} = \frac{\text{hypotenuse}}{\text{opposite side}}$$

$$\text{Secant of } \theta = \sec\ \theta = \frac{r}{x} = \frac{\text{hypotenuse}}{\text{adjacent side}}$$

$$\text{Cotangent of } \theta = \cot\ \theta = \frac{x}{y} = \frac{\text{adjacent side}}{\text{opposite side}}$$

Note

In a right triangle, if y is the side opposite the angle θ, x its adjacent side, and r the hypotenuse, the trigonometric functions for θ are defined as

$\sin\theta = \frac{y}{r}$,

$\cos\theta = \frac{x}{r}$,

$\tan\theta = \frac{y}{x}$,

$\csc\theta = \frac{r}{y}$,

$\sec\theta = \frac{r}{x}$,

$\cot\theta = \frac{x}{y}$.

Some people use the acronym SOHCAHTOA to assist in memorizing this collection of ratios. The acronym can be subdivided into three segments: SOH, CAH, and TOA. These refer to the fact that **sine** is defined as *opposite over hypotenuse* (SOH), **cosine** as *adjacent over hypotenuse* (CAH), and **tangent** as *opposite over adjacent* (TOA). There are other devices one can use to memorize the relationships, but in some manner you should commit the ratios to memory because they will be used extensively in the work to come.

The trigonometric ratios depend only on the value of the angle θ and not on the position of the point P on OB, the terminal location of the rotating ray. This means that if you choose P at another point on OB, the lengths PM, OM, and OP also change, but their ratios remain the same. The reason for this is that a different position of P, say P 1, creates a triangle that is geometrically similar

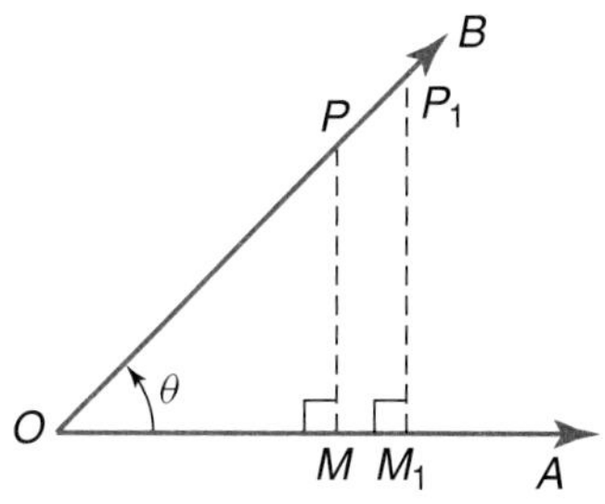

Figure 7.6 Geometrically similar right triangles

to the original one, and in such triangles, the ratios of the corresponding sides are equal, a fact illustrated in Figure 7.6.

The Relationships among the Trigonometric Ratios

The six trigonometric ratios defined previously are interrelated through some rather common algebraic operations. The functions pair up in three reciprocal relationships and also can be used to form two quotient relationships. These relationships are direct consequences of the definitions given for the functions, so they follow naturally once we have set forth those definitions.

The reciprocal relationships are

$$\text{Since } \sin\,\theta = \frac{y}{r} \text{ and, } \csc\,\theta = \frac{r}{y}\text{, we conclude that } \sin\,\theta = \frac{1}{\csc\,\theta}$$

$$\text{Since } \cos\,\theta = \frac{x}{r} \text{ and, } \csc\,\theta = \frac{r}{x}\text{, we conclude that } \cos\,\theta = \frac{1}{\sec\,\theta}$$

$$\text{Since } \tan\,\theta = \frac{y}{x} \text{ and, } \cot\,\theta = \frac{x}{y}\text{, we conclude that } \tan\,\theta = \frac{1}{\cot\,\theta}$$

Note

The primary reciprocal relations among trigonometric functions are

$$\sin\,\theta = \frac{1}{\csc\,\theta}$$

$$\cos\,\theta = \frac{1}{\sec\,\theta}$$

$$\tan\,\theta = \frac{1}{\cot\,\theta}$$

The remaining reciprocal relationships can be found by algebraic manipulation of these.

If we manipulate the ratios a bit, we can find further relationships. Because these connections can allow us to drastically simplify trigonometric expressions, the more relationships we can find, the more likely we'll be to meet with success in the future sections. Consider, as an example, the definitions of sine and cosine and the two fractions we can form using the two functions as the numerator and denominator:

$$\text{Since, } \sin\,\theta = \frac{y}{r}\text{, } \cos\,\theta\ \frac{x}{r}\text{, and } \tan\,\theta = \frac{y}{x}'$$

$$\text{we can form the ratio } \frac{\sin\,\theta}{\cos\,\theta} = \frac{y/r}{x/r} = \frac{y}{x} = \tan\,\theta$$

Note

The ratio (or quotient) relations among trigonometric functions are

$$\frac{\sin\,\theta}{\cos\,\theta} = \tan\,\theta$$

$$\frac{\cos\,\theta}{\sin\,\theta} = \cot\,\theta$$

Additionally,

$$\text{Since, } \sin\,\theta = \frac{y}{r}\text{, } \cos\,\theta\,\frac{x}{r}\text{, and } \cot\,\theta = \frac{x}{y}' \text{ the ratio } \frac{\cos\,\theta}{\sin\,\theta} = \frac{x/r}{y/r} = \frac{x}{y} = \cot\,\theta$$

These relationships form what we call the "ratio" or "quotient" relations of sine and cosine.

Values of the Trigonometric Functions and Expressions

We can find the values of the trigonometric functions of an angle in a right triangle by using their definitions if the magnitudes of the sides of the triangle are known. If the magnitude of the angle is known, you can find its trigonometric functions using a calculator. The function keys sin, cos, and tan are used to find three of the trigonometric ratios, and you can then find their reciprocal or inverse ratios. Expressions that contain trigonometric functions are evaluated just like any other algebraic expression.

A precaution worth mentioning is that, when using a calculator, the "mode" should be set with care. If the measurements are in degrees, be *certain* that your calculator is in "degree mode," and if the measurements are in radians, be *certain* that the calculator is in "radian mode." Failure to ensure that you are in the right mode can lead to catastrophic errors! An interesting exercise is to compare the size of an angle measuring 30° to an angle measuring 30 radians. If we use the more familiar measure of degrees, and convert 30 radians to degrees, we find that 30 radians is equivalent to about 3437°. Thus, one can appreciate that basing calculations on 30° as opposed to 30 radians would result in very different computational outcomes!

Note

Always verify your calculator is set to the proper mode (degree or radian)!!

As we prepare to advance to the computation of values of the trigonometric functions, let's recall a very famous mathematical result from antiquity, the Pythagorean Theorem:

The sum of the squares of the lengths of the legs of a right triangle is equal to the square of the length of the hypotenuse.

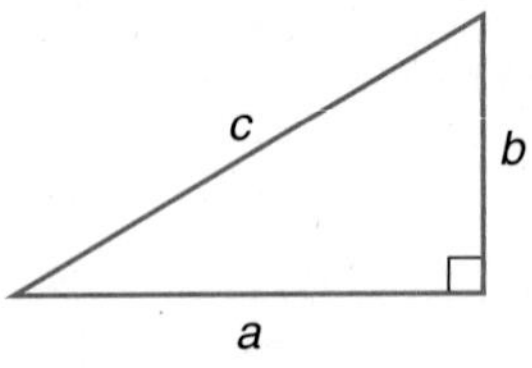

Figure 7.7 A right triangle

If a and b represent the lengths of the two sides (legs) and c represents the length of the hypotenuse (the side opposite the right angle; Figure 7.7), then the statement of the theorem is

$$a^2 + b^2 = c^2$$

EXAMPLE 7.7

Find the values of the six trigonometric functions for the angle θ in the right triangle shown in Figure 7.8.

Figure 7.8 The right triangle from Example 7.7

SOLUTION

To calculate the hypotenuse of the right triangle, use the Pythagorean Theorem:

Hypotenuse $= \sqrt{3.47^2 + 6.35^2} = \sqrt{52.3} = 7.23$

Knowing the three side lengths, the six trigonometric function values can be found using the definitions

$$\sin\,\theta = \frac{opposite}{hypotenuse} = \frac{3.47}{7.23} = 0.480$$

$$\cos\,\theta = \frac{adjacent}{hypotenuse} = \frac{6.35}{7.23} = 0.878$$

$$\tan\,\theta = \frac{opposite}{adjacent} = \frac{3.47}{6.35} = 0.546$$

$$\cot\,\theta = \frac{adjacent}{opposite} = \frac{6.35}{3.47} = 1.83$$

$$\sec\,\theta = \frac{hypotenuse}{adjacent} = \frac{7.23}{6.35} = 1.14$$

$$\csc\,\theta = \frac{hypotenuse}{opposite} = \frac{7.23}{3.47} = 2.08$$

EXAMPLE 7.8

Find the value of $\sin 53°$ accurate to four decimal places.

SOLUTION

Set the calculator in degree mode, enter the value 53, and then press the sin function key. The answer is 0.7986, accurate to four decimal places:

$$\sin 53° = 0.7986$$

Remark

We mention the key-strike order for calculators, but be aware that some calculators use a different key sequence. Some calculators, especially graphing models, would use a key sequence for the previous example that would be "press the sin key, enter the value 53, and then press the ENTER key." You should consult your owner's manual or experiment with your calculator to determine the correct key sequence for your calculator!

EXAMPLE 7.9

Find the value of $\cos(1.298)$ accurate to four decimal places.

SOLUTION

Since there is no degree symbol visible, we conclude that the angle is measured in radians and set the calculator in radian mode. Enter the value 1.298 and then press the cos function key. The answer is 0.2694, accurate to four decimal places.

$$\cos(1.298) = 0.2694$$

EXAMPLE 7.10

Find the value of $\cot 53°$ accurate to four decimal places.

SOLUTION

Set the calculator in degree mode, enter the value 53, and then press the tan function key. The result is the reciprocal of this answer, which is obtained using either the $1/x$ or x^{-1} key. The answer, accurate to four decimal places, is 0.7536:

$$\cot 53° = 0.7536$$

Note

In Example 7.10, we made use of the reciprocal key in conjunction with the tangent key as a means to find a value of cotangent. There is a calculator function $\tan^{-1}$ on your calculator that is not the same as cotangent! The use of this function will be explained shortly.

EXAMPLE 7.11

Find the value of $2.5 \sin 1.3 + 4 \cos 0.6$, accurate to four decimal places.

SOLUTION

Set the calculator in radian mode and compute the values of the trigonometric functions. Then use these values to give

$$2.5 \sin 1.3 + 4 \cos 0.6 = 2.4088 + 3.3012 = 5.71$$

Angle from the Value of the Function

If the value of a trigonometric function of an angle is known, we can find the value of the angle by performing another operation, inversion. This is symbolically written as follows:

$$\text{If } \sin\theta = A, \text{ then } \theta = \arcsin A \text{ or } \theta = \sin^{-1} A$$

We say "the sine of angle theta is A," or "theta is the arcsine of A," or "theta is the inverse sine of A." All three statements are equivalent.

You can perform the same action for the other trigonometric functions. That is, you use operations called arccos A or $\cos^{-1} A$ and arctan A or $\tan^{-1} A$. The approximate value of the angle can be computed using the $\sin^{-1}$, $\cos^{-1}$, and $\tan^{-1}$ keys on the calculator (these functions might have to be accessed through the "2nd" or "INV" key on your keypad).

Note

If $\sin\theta = A$, then $\theta = \arcsin A$ or $\theta = \sin^{-1} A$.

EXAMPLE 7.12

If $\sin\theta = 0.5432$, find θ in degrees and in radians.

SOLUTION

Set the calculator in degree mode and use the function to get $\sin^{-1} 0.5432 \approx 32.9°$. Set the calculator in radian mode and use the $\sin^{-1}$ function to get $\sin^{-1} 0.5432 \approx 0.574$ rad.

EXAMPLE 7.13

If $\csc\theta = 1.841$, find θ in degrees.

SOLUTION

Find $\sin\theta = \dfrac{1}{\csc\theta} = \dfrac{1}{1.841} \approx 0.5432$

Now, set the calculator in degree mode and use the $\sin^{-1}$ function to get

$$\sin^{-1} 0.5432 \approx 32.9°$$

Exercises

Answer the following questions using complete sentences, proper grammar, and proper spelling.

1. How is the notation $y = \csc\theta$ spoken? How is the value of y computed, assuming that the angle is an acute angle in a right triangle?
2. How is the notation $y = \arccos(x)$ spoken? How is the value of y computed, assuming that y is an acute angle in a right triangle?
3. To what does the acronym SOHCAHTOA refer, and how is it used?
4. What is the difference (if any) between the notations $\cos^{-1}x$ and $\sec x$? Do the notations refer to the same concept?

In each of the following, assume that the problem refers to a right triangle with acute angle θ, where x is the leg adjacent to θ and y the leg opposite θ. Find the values of the six trigonometric functions of θ, rounded to three decimal places.

5. $x = 4.2, y = 6$
6. $x = 4.9, y = 2$
7. $x = 3.7, y = 8$
8. $x = 11.3, y = 5$

9. $x = 2.7, y = 3.1$
10. $x = 3.2, y = 9.7$
11. $x = 6, y = 10$
12. $x = 5, y = 11$

For the following angles, find the values of sine, cosine, and tangent, rounded to four decimal places.

13. $35°$
14. $57°$
15. $100°$
16. $120°$
17. $47°$
18. $19°$
19. $45°$
20. $20°$
21. 1.75
22. 2.8
23. 2.5
24. 3

In the following problems, for an acute angle θ in a right triangle, x , y , and r refer to the adjacent, opposite, and hypotenuse sides of the triangle. Find the value of the angle θ in both degrees and radians using an appropriate inverse trigonometric function.

25. $x = 14, y = 11$
26. $x = 9, y = 8$
27. $x = 3.1, y = 2.4$
28. $x = 6.4, y = 5$
29. $x = 3, r = 7$
30. $x = 1.5, r = 5$
31. $y = 3, r = 10$
32. $y = 1.8, r = 2$
33. $y = 4.7, r = 6$
34. $y = 3, r = 7.7$

Evaluate the following trigonometric expressions, rounding your answers to four decimal places.

35. $4 \cos (3.85) + 2 \sin (1.5)$
36. $2 \cos (1.75) - 4 \sin (2)$
37. $3 \cos 45° - 2 \sin 26°$
38. $10 \sin 80° - 5 \cos 30°$
39. $20 \sin 10° + 18 \cos 55°$
40. $\frac{3}{4} \cos 10 - \frac{1}{5} \sin 20$
41. $\frac{2}{7} \sin 40 + \frac{1}{3} \cos 15$
42. $9 \cos 2 + \frac{2}{3} \sin 3$

Given the following information, find the acute angle θ in both radians and degrees. Round your answer to four decimal places as needed.

43. $\sin \theta = 0.295$
44. $\cos \theta = 0.125$
45. $\cos \theta = 0.5$
46. $\sin \theta = 0.777$
47. $\tan \theta = 5$
48. $\tan \theta = 1.5$
49. $\csc \theta = 1.75$
50. $\csc \theta = 3.95$
51. $\sec \theta = 9.9$
52. $\sec \theta = 7.1$
53. $\cot \theta = 0.125$
54. $\cot \theta = 0.375$

7.3 Right Triangles

Calculating the Unknown Measurements of a Right Triangle

One of the angles in a right triangle is always known by virtue of it being a right triangle. You can find the values of all other sides and angles using trigonometric functions and the Pythagorean Theorem if one side and one other angle or two sides are known values. Each trigonometric function for a specific angle is the ratio of two sides, and therefore the strategy is to work with a trigonometric function for which you know the values of two of the three quantities, from which you can then find the unknown quantities. Parenthetically, we recall that the angles of a triangle add to 180°, a fact we shall use freely in the following examples.

EXAMPLE 7.14

Find all the unknown values of the right triangle shown in Figure 7.9.

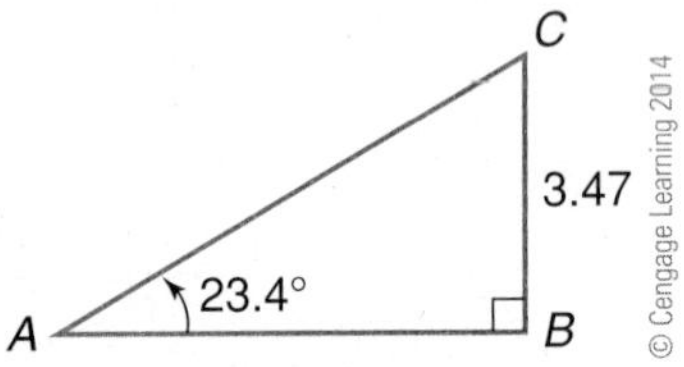

Figure 7.9 The right triangle from Example 7.14

SOLUTION

The sides AB and AC and the angle ACB are unknown. Solving for the sides first, take

$$\sin 23.4° = \frac{3.47}{AC}$$

Solving for AC,

$$AC = \frac{3.47}{\sin 23.4°} \approx \frac{3.47}{0.3971} \approx 8.74$$

Similarly,

$$\tan 23.4° = \frac{3.47}{AB}$$

Solving for AB,

$$AB = \frac{3.47}{\tan 23.4°} \approx \frac{3.47}{0.4327} \approx 8.02$$

To solve the angle ACB, subtract the sum of the two known angles from 180°:

$$\text{Angle } ACB \approx 180° - (90° + 23.4°) = 66.6°$$

EXAMPLE 7.15

Find all the unknown values of the right triangle shown in Figure 7.10.

SOLUTION

The side AB and the angles ACB and CAB are unknown. We can find the unknown side, AB, by using the Pythagorean Theorem:

$$AB = \sqrt{(7.58)^2 - (2.54)^2} \approx 7.14$$

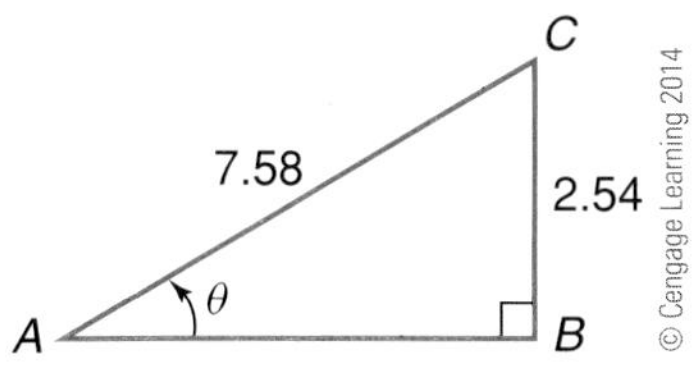

Figure 7.10 The right triangle from Example 7.15

From the figure,

$$\sin\ \theta = \frac{2.54}{7.58} = 0.3351$$

So,

$$\theta = \sin^{-1}(0.3351) \approx 19.6°$$

To find the angle ACB, subtract the sum of the two known angles from 180°:

$$\text{Angle } ACB \approx 180° - (90° + 19.6°) = 70.4°$$

Applications of Right Triangles

Right triangles can be used to find heights and distances of objects. The angle the line of sight makes with the top of an object at a higher level is called its **angle of elevation**. When an object at a lower level is viewed from a vantage point, the angle the line of sight makes with the horizontal is called the **angle of depression**. The angles of elevation and depression for two different objects are shown in Figure 7.11. Trigonometric functions can help solve the right triangle involved.

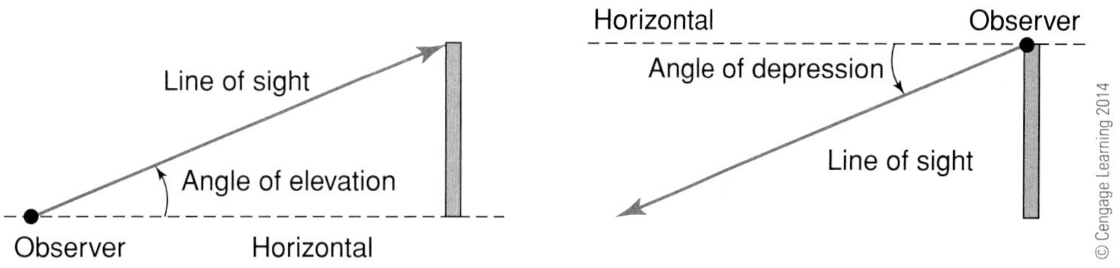

Figure 7.11 Angle of elevation and angle of depression

EXAMPLE 7.16

An observer finds the angle of elevation to the top of a tower to be 30° when seen from a point 600 meters away from the base of the tower. Find the height of the tower.

SOLUTION

Figure 7.12 shows the angles and distances that are involved in the right triangle. If h is the height of the tower, then $\tan 30° = \frac{h}{600}$. Solving for h,

$$h = 600 \tan 30° = 600(0.5774) = 346.4 \text{ m}$$

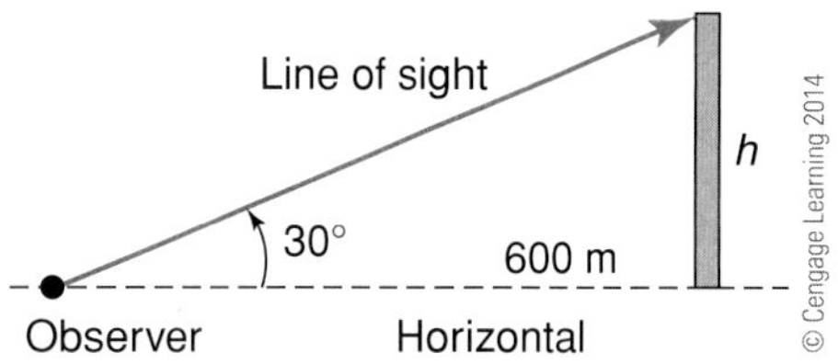

Figure 7.12 The right triangle from Example 7.16

EXAMPLE 7.17

Two observers are on opposite sides of a tower that is 50 meters high. They find the elevations of the tower to be 30° and 60°, respectively. Find the total distance between the two observers, assuming that the tower and the observers stand on level ground.

SOLUTION

Figure 7.13 shows the angles and distances that are involved in the right triangle. If h is the height of the tower, then $\tan 60° = \frac{h}{x}$. Solving for x,

$$x = \frac{50}{\tan 60°} = \frac{50}{1.732} = 28.9 \text{ m}$$

Similarly, $\tan 30° = \frac{h}{y}$. Solving for y,

$$y = \frac{50}{\tan 30°} = \frac{50}{0.5774} = 86.6 \text{ m}$$

Therefore, the distance between the two observers is

$$x + y = 28.9 + 86.6 = 115.5 \text{ m}$$

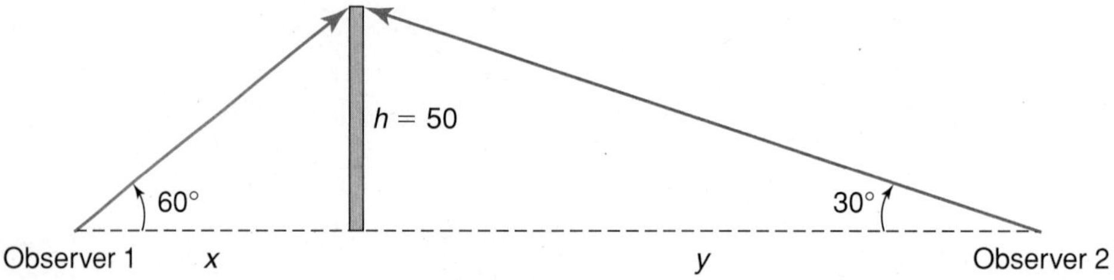

Figure 7.13 The right triangles from Example 7.17

EXAMPLE 7.18

From the top of a hill, the angles of depression of two consecutive milestones on the same side of the hill are 30° and 45°, respectively. Find the height of the hill.

SOLUTION

Figure 7.14 shows the heights and distances that are involved. From the figure,

$$\tan 45° = \frac{h}{x}$$

Cross-multiplying,

$$h = x \tan 45° = x(1) = x$$

Also, from the larger triangle formed,

$$\tan 30° = \frac{h}{x+1}$$

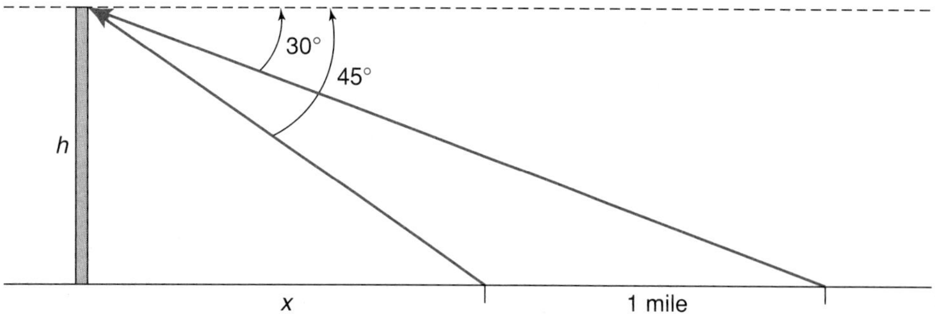

Figure 7.14 The right triangles from Example 7.18

Substituting $h = x$, $\tan 30° = \dfrac{x}{x+1}$. Cross-multiplying,

$$x = (x + 1) \tan 30° = (x + 1)(0.5774)$$

Solving for x,

$$x \approx \frac{0.5774}{0.4226} = 1.4 \text{ miles}$$

EXAMPLE 7.19

Calculate the length of the side of a regular hexagon inscribed in a circle with a radius of 100 centimeters.

SOLUTION

Figure 7.15 shows the heights and distances involved. If you draw the diagonals of the hexagon and consider any one triangle formed by their intersection, it results in an isosceles triangle with

$$OP = OQ = 100 \text{ cm}$$

Note that, since one revolution is 360°, the angle $POQ = 60°$, since the revolution is subdivided into six parts by the creation of the hexagon. Drop a perpendicular OM to the base PQ to bisect the angle POQ. Angle MOQ is easily seen to be 30° in measure:

$$\sin 30° = \frac{QM}{OQ}$$

Solving for $QM = (100) \sin 30° = 100(0.5) = 50$

$$QM = 100 \sin 30° = 100(0.5) = 50$$

Since the perpendicular to the base of an isosceles triangle bisects it,

$$PQ = 2(QM) = 2(50 \text{ cm}) = 100 \text{ cm}$$

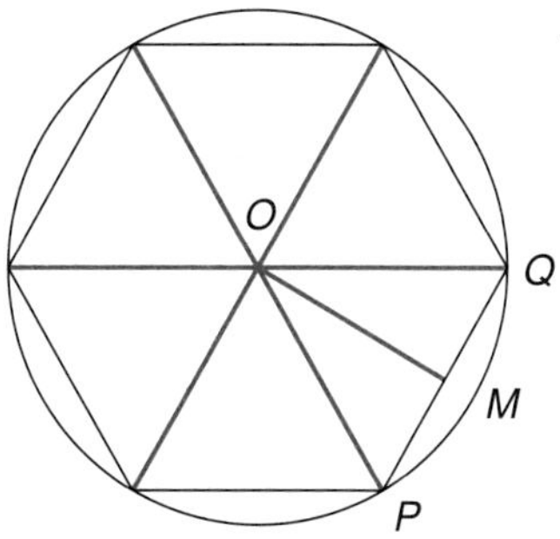

Figure 7.15 A regular hexagon inscribed in a circle.

Exercises

Answer each of the following using proper grammar, complete sentences, and correct spelling and terminology.

1. In a right triangle, what is meant by the side adjacent to one of the acute angles?

2. In a right triangle, what is meant by the side opposite from one of the acute angles?
3. How does the angle of elevation differ from the angle of depression?

For each of the following, find the missing measures of right triangle ABC, where the right angle lies at vertex B. Round your answers to two decimal places, as needed.

4. $A = 35°, a = 10$
5. $A = 50°, a = 8$
6. $a = 3, c = 4$
7. $a = 3, c = 6$
8. $a = 5, b = 8$
9. $a = 13, b = 15$
10. $c = 5, A = 45°$
11. $c = 7, A = 85°$

In each of the following problems, construct a right triangle depicting the situation and use trigonometric functions to solve. Round answers to two decimal places, as needed.

12. A 15-meter-tall lighthouse sits on an islet at a point 5 meters above sea level. A ship at sea sights to the lighthouse and finds that the angle of elevation from the deck (2 meters above sea level) is 20°. What is the distance from the ship to the lighthouse?
13. A man in a window 30 meters above street level sights to a balloon in the distance, hovering over a point 15 kilometers away on level ground. If the angle of elevation to the balloon is 5°, what is the elevation of the balloon?
14. A system designer positions a flat-screen television on a vertical wall 14 meters from the front row of an arrangement of chairs. If the height of the bottom of the screen above the eye level of a person in that chair is 2 meters, what is the angle of elevation from the viewer to the bottom of the screen?
15. Referring to the situation described in Problem 14, if the screen is 1.5 meters high, what is the angle of elevation from the viewer to the top of the screen?
16. A designer wants to create a logo that is a regular octagon (eight equal sides and angles). If the octagon is inscribed within a circle of radius 30 centimeters, what will be the measures of the sides of the octagon?
17. Repeat Problem 16, changing the design to be a nonagon (nine-sided polygon).
18. A computer screen that has a width of 45 centimeters and a height of 40 centimeters has a single pixel illuminated at a point that is 13 centimeters above the bottom of the screen along the line from the lower left corner of the screen to the upper right corner. What is the angle from the lower right corner of the screen to the pixel, and what is the distance from that corner to the pixel?
19. The microphone for a particular manufacturer is ideally positioned 15 centimeters from the mouth of the speaker. If the speaker's mouth, at that ideal distance, is positioned at a point 8 centimeters above the level of the microphone, what is the angle of depression from the speaker's mouth to the microphone?

Summary

In this chapter, you learned about:

- The different methods with which angles can be measured.
- The trigonometric functions and their inverses.
- Trigonometric Identities
- Solving applications using right triangle trigonometry.

Glossary

angle The span between two rays sharing a common initial point.

angle of depression The angle the line of sight makes with the top of an object at a lower level.

angle of elevation The angle the line of sight makes with the top of an object at a higher level.

cosine of an angle The ratio x/r, where x is the x-coordinate of an arbitrary point on the terminal side of the angle in standard position, and r is the distance from the origin to that point.

DMS notation An expression of the measure of an angle using degrees, minutes, and seconds.

end point The starting point of a ray.

one degree $\frac{1}{360}$ of one revolution around a point.

one radian: The measurement of an angle determined by the sweep of an arc of length equal to the radius of an arbitrary circle whose center lies at the common initial point of the two involved rays.

radian system of measurement A means of measuring angles whose basis is a circle of arbitrary radius centered at the origin.

right triangle A triangle containing a 90° angle at one vertex.

sexagesimal system Subdivision of degree measure into minutes and seconds.

sine of an angle The ratio y/r, where y is the y-coordinate of an arbitrary point on the terminal side of the angle in standard position, and r is the distance from the origin to that point.

tangent of an angle The ratio y/x, where x and y are the coordinates of an arbitrary point on the terminal side of the angle in standard position.

End-of-Chapter Problems

Sketch each of the following angles in standard position. Identify the quadrant in which the terminal side of the angle is found.

1. 135°
2. 225°
3. 315°

For each of the following angles in degree measure, convert to radian measure.

4. 225°
5. 38°
6. 75°

For each of the following angles in radian measure, convert to degree measure.

7. $\frac{3\pi}{4}$
8. 1.2
9. 2.7

Sketch each of the following angles in standard position and identify the quadrant in which the terminal side of the angle is found.

10. 3
11. $-\frac{5\pi}{4}$
12. 5

Convert the following decimal degree measurements to DMS notation.

13. 230.87°
14. 90.125°
15. 67.185°

Convert the following measurements to decimal degree form. Round your answer to three decimal places, as needed.

16. 53°12'30"
17. 42°15'37"
18. 73°28'53"

Convert the following measurements to radian form. Round your answer to three decimal places, as needed.

19. 62°12'30"
20. 46.7°
21. 130°12'0"
22. 70°12'6"

In each of the following problems, assume that the problem refers to a right triangle with acute angle θ, where x is the leg adjacent to θ and y the leg opposite θ. Find the values of the six trigonometric functions of θ, rounded to three decimal places.

23. $x = 15, y = 25$
24. $x = 4.9, y = 5.5$
25. $x = 5.3, y = 3.7$
26. $x = 1.5, y = 3.0$

For the following angles, find the values of sine, cosine, and tangent, rounded to four decimal places.

27. 45°
28. $\frac{3\pi}{4}$
29. 27°32'45"
30. $\frac{\pi}{6}$

In the following problems, for an acute angle θ in a right triangle, x, y, and r refer to the adjacent, opposite, and hypotenuse sides of the triangle, respectively. Find the value of the angle θ in both degrees and radians using an appropriate inverse trigonometric function.

31. $x = 1, r = 2$
32. $x = 4.7, y = 5.3$
33. $y = 5, r = 7$
34. $x = 3, y = 2$

Evaluate the following trigonometric expressions, rounding your answers to four decimal places.

35. $5 \sin (4.7) + 3 \tan (1)$
36. $4.7 \cos (1.7) - 3.8 \tan (47)$
37. $5.3 \sin (3.8) - 3.1 \cos (3)$
38. $7.2 \tan (3) + 5.4 \cos (2)$

Given the following information, find the angle θ in both radians and degrees. Round your answer to four decimal places as needed.

39. $\cos \theta = 0.375$
40. $\tan \theta = 1$
41. $\sin \theta = 1$
42. $\sec \theta = 2$

For each of the following, find the missing measures of right triangle ABC, where the right angle lies at vertex B. Round your answers to two decimal places, as needed.

43. $A = 40°, a = 7$
44. $a = 6, c = 8$
45. $a = 6, b = 8$
46. $A = \frac{\pi}{4}, b = 20$

In each of the following problems, construct a right triangle depicting the situation and use trigonometric functions to solve. Round answers to two decimal places, as needed.

47. A telephone pole casts a shadow that is 7 meters long. The sun is 55° above the horizon. How tall is the telephone pole?
48. A man looks out of the window of his apartment building. He looks down and directly across the street at the bottom-most window on the apartment building on the far side of the street. The angle of depression is measured to be 59.8°. He then measures the distance between the two apartment buildings and finds it to be 10 meters. The bottom-most window on the facing apartment building is 1 meter off the ground. How high off the ground is the window the man is looking through?
49. A man looks down from a mountaintop. The mountain is known to be 1237 meters above sea level, and the spot he is looking at is known to be 63 meters above sea level. The angle of depression is 27°. Ignore the curvature of the Earth. How far away is the spot he is looking at?
50. Sarah is using a transit to determine the height of a tree. She measures the angle of elevation to be 30°, and the distance from the transit to the tree is measured to be 70 meters. The transit is 1.5 meters above the ground. How high is the tree?

Chapter 8 Trigonometric Identities

In Chapter 7, we were given an introduction to the six trigonometric functions, including the definitions of those functions and some of their applications. Now we turn our attention to the ways in which the trigonometric functions are related and how expressions involving them can be rewritten in terms of the other trigonometric functions or even as functions of other angles.

Our ultimate goal is to be able to solve trigonometric equations, but in order to accomplish this, we must first take a detour through the rich topic of trigonometric identities. An identity is an equation that is true for all values of the independent variable, and a **trigonometric identity** (in particular) is an equation that is true for all values of an angle.

In this chapter, we'll investigate some of the fundamental trigonometric identities, including those based on the famous Pythagorean Theorem and those based on the reciprocal relationships existing among the trigonometric functions. We will also search for relationships between the trigonometric functions of the sum and difference of two angles as well as functions of a multiple of an angle. In turn, we will then learn to derive new identities based on those that will be introduced to us.

Objectives

When you have successfully completed the materials of this chapter, you will be able to:

- Simplify a trigonometric expression using the fundamental trigonometric identities.
- Derive new trigonometric identities from the fundamental trigonometric identities.
- Simplify expressions using the sum and difference, double-angle, and half-angle formulas.

8.1 Introduction to Trigonometric Identities and Trigonometric Functions

The Basic Trigonometric Identities and Their Derivation

A trigonometric identity is an equation involving trigonometric functions that is true for all values of the involved angle. In Chapter 7, we saw a few trigonometric identities, and we will now investigate some new ones based on the Pythagorean Theorem. So first, let's remind ourselves what the Pythagorean Theorem tells us, then recall the trigonometric relationships we saw in Chapter 7, and then we should be ready to move ahead.

THE PYTHAGOREAN THEOREM: For a right triangle with legs a and b and hypotenuse c, the following relationship exists: $a^2 + b^2 = c^2$.

The Reciprocal Trigonometric Relationships

The reciprocal trigonometric relationships shown in Table 8.1 are actually introductory examples of identities. They are true regardless of the value of the angle theta, and consequently they are often referred to as the **reciprocal identities**.

TABLE 8.1 The Reciprocal Trigonometric Relationships

$\sin\theta = \dfrac{1}{\csc\theta}$	$\csc\theta = \dfrac{1}{\sin\theta}$
$\cos\theta = \dfrac{1}{\sec\theta}$	$\sec\theta = \dfrac{1}{\cos\theta}$
$\tan\theta = \dfrac{1}{\cot\theta}$	$\cot\theta = \dfrac{1}{\tan\theta}$

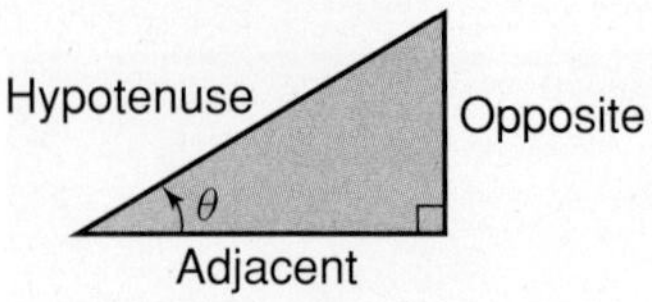

Figure 8.1 The relationships of the sides of a right triangle to the angle

The question now arises as to how the trigonometric functions relate to the Pythagorean Theorem. The relationship can be seen if we consider once more the particular case of an acute angle theta within a right triangle and the description of side lengths as opposite, adjacent, and hypotenuse (Figure 8.1).

We saw that alternative definitions of the trigonometric functions existed that were stated in terms of these side identifications, and although those were valid only for acute angles θ, they will allow us to construct some of the identities that actually apply to all angles.

Table 8.2 shows the alternative definitions of the trigonometric functions of θ.

TABLE 8.2 The Trigonometric Functions Defined Using Sides of a Right Triangle

$\sin\theta = \dfrac{\text{opposite}}{\text{hypotenuse}}$	$\csc\theta = \dfrac{\text{hypotenuse}}{\text{opposite}}$
$\cos\theta = \dfrac{\text{adjacent}}{\text{hypotenuse}}$	$\sec\theta = \dfrac{\text{hypotenuse}}{\text{adjacent}}$
$\tan\theta = \dfrac{\text{opposite}}{\text{adjacent}}$	$\cot\theta = \dfrac{\text{adjacent}}{\text{opposite}}$

Consider what happens if we combine the labeling of the triangle in Figure 8.1 with the Pythagorean Theorem and then bring in some relationships from Table 8.2. By Pythagoras, we know that the following is true:

$$(\text{adjacent})^2 + (\text{opposite})^2 = (\text{hypotenuse})^2$$

Now, for us to have a triangle at all, the hypotenuse must be nonzero, and therefore it makes sense to divide both sides of this expression by $(\text{hypotenuse})^2$, which gives us

$$\frac{(\text{adjacent})^2}{(\text{hypotenuse})^2} + \frac{(\text{opposite})^2}{(\text{hypotenuse})^2} = 1$$

Note that, by properties of exponents, this is equivalent to

$$\left(\frac{\text{adjacent}}{\text{hypotenuse}}\right)^2 + \left(\frac{\text{opposite}}{\text{hypotenuse}}\right)^2 = 1$$

Based on our results of Table 8.2, we can now present one of the most useful of the trigonometric relationships:

$$\cos^2\theta + \sin^2\theta = 1$$

Because of its derivation from the Pythagorean Theorem, this is called a **Pythagorean trigonometric identity** or, more simply, a Pythagorean identity. It holds for all angles θ, whether or not the angle is acute.

The Pythagorean formula form where opposite, adjacent, and hypotenuse were substituted can also be rewritten in other ways, and these generate other

Note

Be aware of the convention on writing the powers of the trigonometric functions! The notation $\cos^2\theta$ is equivalent to and the accepted way of writing $(\cos\theta)^2$. Similar conventions apply to all the trigonometric function powers!

Pythagorean identities. In order for us to have a triangle at all, both the adjacent and the opposite sides must be nonzero, and consequently the equation

$$(\text{adjacent})^2 + (\text{opposite})^2 = (\text{hypotenuse})^2$$

can be divided by either $(\text{adjacent})^2$ or by $(\text{opposite})^2$. Let's see what those divisions would produce.

EXAMPLE 8.1

Divide the Pythagorean form $(\text{adjacent})^2 + (\text{opposite})^2 = (\text{hypotenuse})^2$ by $(\text{adjacent})^2$ and simplify the result. Deduce, based on the result, a new Pythagorean identity.

SOLUTION

Beginning with

$$(\text{adjacent})^2 + (\text{opposite})^2 = (\text{hypotenuse})^2$$

and dividing through by $(\text{adjacent})^2$, we obtain

$$\frac{(\text{adjacent})^2}{(\text{adjacent})^2} + \frac{(\text{opposite})^2}{(\text{adjacent})^2} = \frac{(\text{hypotenuse})^2}{(\text{adjacent})^2}$$

Simplifying as we did earlier, we find

$$1 + \left(\frac{\text{opposite}}{\text{adjacent}}\right)^2 = \left(\frac{\text{hypotenuse}}{\text{adjacent}}\right)^2$$

Consultation of Table 8.2 reveals this is equivalent to

$$1 + \tan^2\theta = \sec^2\theta$$

This is another Pythagorean identity.

EXAMPLE 8.2

Divide the Pythagorean form $(\text{adjacent})^2 + (\text{opposite})^2 = (\text{hypotenuse})^2$ by $(\text{opposite})^2$ and simplify the result. Deduce, based on the result, a new Pythagorean identity.

SOLUTION

Starting with

$$(\text{adjacent})^2 + (\text{opposite})^2 = (\text{hypotenuse})^2$$

and dividing through by $(\text{adjacent})^2$, we obtain

$$\frac{(\text{adjacent})^2}{(\text{opposite})^2} + \frac{(\text{opposite})^2}{(\text{opposite})^2} = \frac{(\text{hypotenuse})^2}{(\text{opposite})^2}$$

Simplifying as we did earlier, we find

$$\left(\frac{\text{adjacent}}{\text{opposite}}\right)^2 + 1 = \left(\frac{\text{hypotenuse}}{\text{opposite}}\right)^2$$

Consultation of Table 8.2 reveals that this is equivalent to

$$\cot^2\theta + 1 = \csc^2\theta$$

This is yet another Pythagorean identity!

Note

By commutativity of addition, we can also express those identities, respectively, as

$\tan^2\theta + 1 = \sec^2\theta$ and

$1 + \cot^2\theta = \csc^2\theta$

or as any other rearrangement of the terms. All are valid Pythagorean identities and may prove useful to us in the future!

We include the following remark, prior to moving on to the use of trigonometric identities: *each of those Pythagorean identities can be rewritten algebraically in another form, and those alternate forms are also considered Pythagorean identities (but are suppressed from the general list).* For instance, the first identity could be restated as $\sin^2\theta = 1 - \cos^2\theta$. At times, it is useful to think of the Pythagorean identities from this alternate perspective, as we'll see shortly.

Identities Used to Simplify Trigonometric Expressions

A trigonometric expression is any mathematical statement involving one or more of the trigonometric functions. For instance, $\cos\theta - 3\sin^2\theta\tan\theta$ is a trigonometric expression. If we examine such statements, it is possible that one of the identities we have presented can be used to rewrite the expression in a different form. The form may be aesthetically more pleasing, or it may be that later, when we attempt to prove further trigonometric identities, the alternative forms will be easier to work with. Why that would be the case is absolutely unclear at the moment, but we will develop an appreciation for that statement in the near future.

Consider, as an early illustration, the following expression:

$$\cos^4\theta - \sin^4\theta$$

Note that this is not an equation, since there is no equal sign present! It is a common mistake to use the terms "equation" and "expression" interchangeably, but we really should be precise here. It is possible that we may, on examination, discover a way to apply the identities we have established to this point as a means to rewrite this expression in a simpler, possibly more elegant form.

At first glance, you may not see how this is possible, but let's recall some elementary algebra results (when we first state this, you should not be discouraged if it is not something that would come to your mind on your own because this method of analysis is not obvious) and see how they might help us.

In a macro sense, the expression here is a difference of squares. That is, it can be viewed as being equivalent to

$$(\cos^2\theta)^2 - (\sin^2\theta)^2$$

At this point, we recall the formula for the factorization of the difference of squares:

$$A^2 - B^2 = (A + B)(A - B)$$

Applied to this situation, we obtain

$$(\cos^2\theta)^2 - (\sin^2\theta)^2$$
$$(\cos^2\theta + \sin^2\theta)(\cos^2\theta - \sin^2\theta)$$

The expression in the first set of parentheses, by our first Pythagorean identity in Table 8.3, is equal to 1, and thus the product expression is equivalent to

$$(1)\,(\cos^2\theta - \sin^2\theta)$$
$$\cos^2\theta - \sin^2\theta$$

Note

It is often useful, when considering trigonometric expressions that we wish to rewrite in an equivalent forms, to think in a "macro" sense. That is, what is the general algebraic structure presented in the expression, and what known algebraic methods can be applied to that structure?

TABLE 8.3 The Pythagorean Trigonometric Identities

The Pythagorean identities:
$\sin^2\theta + \cos^2\theta = 1$
$\tan^2\theta + 1 = \sec^2\theta$
$\cot^2\theta + 1 = \csc^2\theta$

That is, we have the rather surprising result that $\cos^4\theta - \sin^4\theta = \cos^2\theta - \sin^2\theta$.

Our experience with working with polynomial equations has been that equations become progressively more difficult to solve as the exponents on the variables increase, and hence it appears that the ability to lower the powers on the trigonometric functions might be a useful tool. Just how this might help us is, as we mentioned, not clear at the moment, but please be patient.

Notice that with every step, we expand our knowledge of trigonometric identities. That is, the expression $\cos^4\theta - \sin^4\theta = \cos^2\theta - \sin^2\theta$ is a new trigonometric identity! Since it was derived using known identities, it is added to our list and is known true for all angles θ.

EXAMPLE 8.3

Simplify the expression $\sin^4\theta + \cos^4\theta$.

Note

The list of trigonometric identities is *endless*! You should attempt to memorize those you use most frequently, which we identify specifically as essential to know. If new identities prove particularly useful, you will naturally memorize them as a consequence of their repeated use!

SOLUTION

While we saw in our introduction to this section an example involving the difference of squares and exploited our ability to factor such an expression, we must be cautious in attempting to repeat that method here. The expression involved is a sum of squares, but there is no factorization formula that holds for the sum of squares.

We do notice that, if we were to compute $(\sin^2\theta + \cos^2\theta)^2$, we would obtain $\sin^4\theta + 2\sin^2\theta\cos^2\theta + \cos^4\theta$, and consequently we can say that

$$\sin^4\theta + \cos^4\theta$$

is equivalent to

$$(\sin^2\theta + \cos^2\theta)^2 - 2\sin^2\theta\cos^2\theta$$

Compare this statement to the original expression very carefully. The two statements are equivalent, though this is possibly not obvious at first reading! The motivation for attempting to rewrite the expression in this way is simply that, in so doing, we can apply one of the Pythagorean identities because the expression in parentheses is known to be equal to 1. Thus, our expression is equivalent to

$$1 - 2\sin^2\theta\cos^2\theta$$

and we obtain yet another identity, $\sin^4\theta + \cos^4\theta = 1 - 2\sin^2\theta\cos^2\theta$.

EXAMPLE 8.4

Simplify the expression $\dfrac{\cos\theta}{\sin\theta} + \dfrac{\sin\theta}{1 + \cos\theta}$.

SOLUTION

Every example of trigonometric simplification should be viewed as a fresh problem, possibly related to previous examples, but possibly a novel problem that will require an innovative approach. Here, we have a pair of fractions having different denominators, and our experience from algebra suggests that we should obtain a common denominator and then add.

Note here that we don't *know* that this idea is the "right one" because there is no single "right method" to employ. In fact, this idea may be a terrible one that will lead us to a horrific expression best suited for immediate erasure. We will know only by trying, and we encourage you to experiment in your own work.

The common denominator appears to be the product of the existing denominators, so we obtain

$$\frac{\cos\theta(1 + \cos\theta)}{\sin\theta(1 + \cos\theta)} + \frac{\sin\theta(\sin\theta)}{\sin\theta(1 + \cos\theta)}$$

Simplifying and combining into a single fraction, we find

$$\frac{\cos\theta + \cos^2\theta + \sin^2\theta}{\sin\theta(1 + \cos\theta)}$$

Observe that, in the numerator, we can replace the expression $\cos^2\theta + \sin^2\theta$ by 1, using the Pythagorean identity. Thus, our expression is equivalent to

$$\frac{\cos\theta + 1}{\sin\theta(1 + \cos\theta)}$$

and this reduces to

$$\frac{1}{\sin\theta}$$

It so happens, referring to our reciprocal identities in Table 8.1, that this is, in turn, equivalent to

$$\csc\theta$$

and this is our final, simplified version. We have, ultimately, shown that the original expression

$$\frac{\cos\theta}{\sin\theta} + \frac{\sin\theta}{1 + \cos\theta}$$

is equivalent to $\csc\theta$.

A legitimate question to ask regarding the final conclusion of Example 8.4 is whether the final step of simplification is truly necessary and consequently whether $\frac{1}{\sin\theta}$ is an acceptable answer. This question is difficult to answer, since the expressions obtained at each stage of the simplification are actually all equivalent and hence have a reasonable claim as being "the simplified answer." With that thought in mind, we will accept the general practice that *if* a fundamental identity *can* be used, then it *must* be used. Simplification of a trigonometric expression should be viewed, in a sense, like a roller-coaster ride: no one gets off until the ride is over!

EXAMPLE 8.5 Simplify the expression $(\csc\theta - \sin\theta)(\sec\theta - \cos\theta)(\tan\theta + \cot\theta)$.

SOLUTION

This example will lead us to a rather surprising result. Prepare yourself for a rather shocking final form!

We notice that there are many trigonometric functions here. We can apply the reciprocal identities to the expressions in parentheses to halve the number of functions involved, and that (possibly) may represent some forward progress:

$$(\csc\theta - \sin\theta)(\sec\theta - \cos\theta)(\tan\theta + \cot\theta)$$
$$\left(\frac{1}{\sin\theta} - \sin\theta\right)\left(\frac{1}{\cos\theta} - \cos\theta\right)\left(\tan\theta + \frac{1}{\tan\theta}\right)$$

The expression appears to be moving toward greater complexity, but let's press onward. We use our approach of Example 8.4 and find common denominators:

$$\left(\frac{1 - \sin^2\theta}{\sin\theta}\right)\left(\frac{1 - \cos^2\theta}{\cos\theta}\right)\left(\frac{\tan^2\theta + 1}{\tan\theta}\right)$$

Then we apply the Pythagorean identities:

$$\left(\frac{\cos^2\theta}{\sin\theta}\right)\left(\frac{\sin^2\theta}{\cos\theta}\right)\left(\frac{\sec^2\theta}{\tan\theta}\right)$$
$$(\cos\theta)(\sin\theta)\left(\frac{\sec^2\theta}{\tan\theta}\right)$$

Here we reduced a factor of cosine and sine from the first two expressions. Now we can rewrite the final factor using the reciprocal identities once more:

$$(\cos\theta)(\sin\theta)\left(\frac{1}{\cos^2\theta}\cdot\frac{\cos\theta}{\sin\theta}\right)$$

Simplification of the result leads to the surprising result that all the trigonometric functions reduce away, and the expression simplifies to 1. That is, we have established that $(\csc\theta - \sin\theta)(\sec\theta - \cos\theta)(\tan\theta + \cot\theta)$ is equal to 1.

We will close out with a few more identities (remember, we told you that the list of identities is endless, so we could go on and on with examples!) that are sometimes referred to as the **symmetry identities** but are also called the **even-odd identities**. These identities will be proven in Section 8.2 (some of them in the exercises!), but it is convenient to close our list of elementary identities with this list.

Shown in Table 8.4, they are called "symmetry identities," because they can be established using the symmetry of the graphs of the trigonometric functions, and they are called "even-odd identities" because the symmetry properties hold because the individual trigonometric functions are even or odd, as the case may be [recall that a function f is called even if $f(-x) = f(x)$ for all x in the function's domain and odd if $f(-x) = -f(x)$ for all x in the function's domain].

TABLE 8.4 The Symmetry Identities, or the Even-Odd Identities

$\sin(-\theta) = -\sin\theta$
$\cos(-\theta) = \cos\theta$
$\tan(-\theta) = -\tan\theta$
$\cot(-\theta) = -\cot\theta$
$\sec(-\theta) = \sec\theta$
$\csc(-\theta) = -\csc\theta$

Exercises

Answer the following questions using proper grammar, complete sentences, and correct terminology.

1. What is meant by the term "trigonometric identity"?
2. Why do we refer to the identity $\cos^2 x + \sin^2 x = 1$ as a "Pythagorean identity"?
3. Explain how the other Pythagorean identities can be obtained from the one mentioned in Problem 2.
4. Is it enough to demonstrate that a given trigonometric formula holds for particular, known angles (such as $\frac{\pi}{6}, \frac{\pi}{4}, \frac{\pi}{3}$) to establish that the formula is a trigonometric identity? Explain.

Verify the following identities.

5. $\cos t \sec t = 1$
6. $\dfrac{\cos^2\theta}{1 + \sin\theta} = 1 - \sin\theta$
7. $\cos^2 x - \sin^2 x = 2\cos^2 x - 1$
8. $\csc x - \sin x = \cos x \cot x$
9. $\sin t \cot t = \cos t$
10. $\csc x \sec x = \dfrac{\csc^2 x}{\cot x}$
11. $\dfrac{\cot^3\theta}{\csc\theta} = \cos\theta \csc^2\theta - \cos\theta$
12. $\dfrac{1}{\tan x} + \tan x = \sec^2 x \cot x$
13. $\cos t \cot t = \dfrac{1}{\sin t} - \sin t$
14. $\dfrac{1}{\sin\theta + 1} + \dfrac{1}{\csc\theta + 1} = 1$
15. $\dfrac{\cos x}{\tan x - \tan x \sin x} = 1 + \csc x$

16. $\dfrac{1 + \sin v}{\cos v} + \dfrac{\cos v}{1 + \sin v} = \dfrac{2}{\cos v}$

17. $\dfrac{\sec(-x)}{\csc(-x)} = -\tan x$

18. $\sec t - \dfrac{\cos t}{1 - \sin t} = -\tan t$

19. $(1 + \sin y)[1 + \sin(-y)] = \dfrac{1}{\sec^2 y}$

20. $\tan x + \cot y = (\tan y + \cot x)(\tan x \cot y)$

21. $\cos x \tan x \csc x = 1$

22. $\cot y \tan y \sin y \csc y = 1$

23. $\sec^2 \theta \cot \theta = \tan \theta + \cot \theta$

24. $\dfrac{\csc x + \sec x}{\sin x + \cos x} = \cot x + \tan x$

8.2 More Trigonometric Identities: Sums and Differences of Angles, Double Angles, Half Angles, and the Quotient Identities

The Sum and Difference of Angle Identities and the Double-Angle Identities

There are other very useful trigonometric identities, and we will present a series of them in this section. However, the proofs of the accuracy of these formulas can be rather technical and will take us rather far afield from their use, which is our primary focus in this book. Nonetheless, we would be remiss in not proving at least one of the results to give you a hint of how the others are established.

Let's start with the identities that discuss the values of the trigonometric functions for sums of angles. When we refer to a sum of angles, we are referring to a pair of angles α and β and their sum $\alpha + \beta$. The difference of angles, in a similar way, is $\alpha - \beta$. A sum of angles identity is an expression that allows us to simplify one of the trigonometric functions of $\alpha + \beta$, while a difference of angles identity does likewise for $\alpha - \beta$.

We'll establish the identity for $\cos(\alpha - \beta)$ as a demonstration. Since the angles are arbitrary, for convenience we will suppose $0 < \beta < \alpha < 2\pi$; so that $\alpha - \beta$ is positive. Consider the unit circle construction of Figure 8.2 (recall the unit circle is a circle of radius 1 centered at the origin of the plane).

Let's identify the points on the unit circle at the terminal point of the shown rays as A, B, and C and let D be the point $(1, 0)$ as shown. Note that the x-coordinate of each point is precisely equal to the cosine of the associated angle and that the y-coordinate of each point is precisely equal to the sine of the associated angle. That is, if we identify $A = (x_A, y_A)$, then $\cos \alpha = x_A$, and $\sin \alpha = y_A$. Similar observations hold for the other points, and we encourage

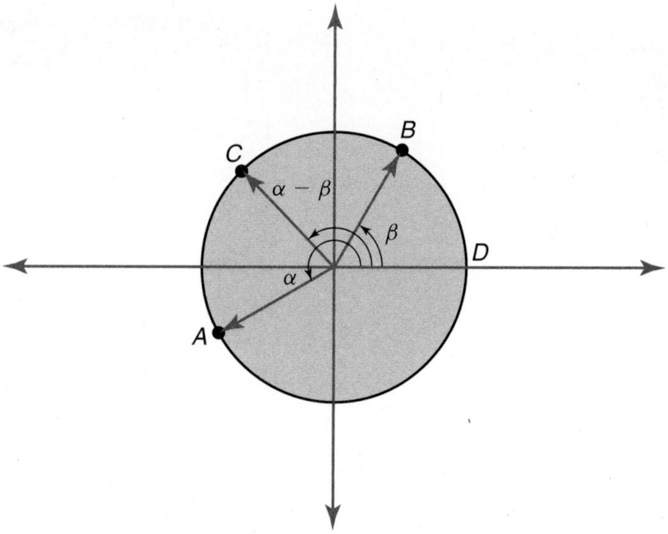

Figure 8.2 Unit circle construction for the difference of angles

you to draw right triangles to assure yourself of this. The reason this is true is seen if you keep in mind that the circle is a unit circle and hence has radius 1, which is then the hypotenuse measurement for the right triangles you would construct to determine the values of the trigonometric functions.

On consideration, it becomes clear that the line segments *AB* and *CD* are precisely equal to one another in length. If we employ the distance formula for the plane, we find the following rather interesting result to be true:

$$\sqrt{(x_A - x_B)^2 + (y_A - y_B)^2} = \sqrt{(x_C - 1)^2 + (y_C)^2}$$
$$x_A^2 - 2x_Ax_B + x_B^2 + y_A^2 - 2y_Ay_B + y_B^2 = x_C^2 - 2x_C + 1 + y_C^2$$
$$x_A^2 + y_A^2 + x_B^2 + y_B^2 - 2(x_Ax_B + y_Ay_B) = x_C^2 + y_C^2 - 2x_C + 1$$

Since the points are on the unit circle, $x_i^2 + y_i^2 = 1$ in all cases, so we get significant simplification:

$$x_A^2 + y_A^2 + x_B^2 + y_B^2 - 2(x_Ax_B + y_Ay_B) = x_C^2 + y_C^2 - 2x_C + 1$$
$$1 + 1 - 2(x_Ax_B + y_Ay_B) = 1 - 2x_C + 1$$
$$x_Ax_B + y_Ay_B = x_C$$

Reversing our substitutions, we find that

$$\cos\alpha\,\cos\beta + \sin\alpha\,\sin\beta = \cos(\alpha - \beta)$$

That result is one of the difference of angles identities. Note that it was heavily reliant on an understanding of our fundamental trigonometric concepts as well as the distance formula for the plane and some algebraic manipulations. There is nothing particularly difficult about it, but it is, admittedly, rather tricky and nonobvious.

We'll suppress the proofs of the other relationships for this section but will ask you to perform a similar analysis for the other identities using sine and cosine during the exercises. We encourage you to take a visual approach to establishing the identities because that can assist you with your understanding of the relationships and strengthen your ability to visually interpret angles.

Without proof, we will now present the other identities (Table 8.5), and then we'll turn our attention to using them to establish other relationships.

TABLE 8.5 Additional Trigonometric Identities

Identity Name	Identity Statement
Cosine of the sum of angles	$\cos(\alpha + \beta) = \cos\alpha\cos\beta - \sin\alpha\sin\beta$
Cosine of the difference of angles	$\cos(\alpha - \beta) = \cos\alpha\cos\beta + \sin\alpha\sin\beta$
Sine of the sum of angles	$\sin(\alpha + \beta) = \sin\alpha\cos\beta + \cos\alpha\sin\beta$
Sine of the difference of angles	$\sin(\alpha + \beta) = \sin\alpha\cos\beta - \cos\alpha\sin\beta$
Tangent of the sum of angles	$\tan(\alpha + \beta) = \dfrac{\tan\alpha + \tan\beta}{1 - \tan\alpha\tan\beta}$
Tangent of the difference of angles	$\tan(\alpha - \beta) = \dfrac{\tan\alpha - \tan\beta}{1 + \tan\alpha\tan\beta}$
Sine of a double angle	$\sin(2\alpha) = 2\sin\alpha\cos\alpha$
Cosine of a double angle	$\cos(2\alpha) = \cos^2\alpha - \sin^2\alpha$

The identities for tangent and for the double angles are interesting relationships to prove because they depend only on the identities for sine and cosine given earlier in the table rather than on geometry (although, since the identities for sine and cosine depend on geometry, the results for tangent and for the double angles also can be derived geometrically). It would be instructive to establish one of these identities here and leave the others for you to establish in the exercises.

EXAMPLE 8.6

Verify the symmetry identity $\sin(-t) = -\sin t$.

SOLUTION

We presented the symmetry identities at the end of Section 8.1 and stalled the presentation of the proofs of the relationships, but the time has now come to address this issue. The key idea is to recognize the nonobvious fact that $-t$ can be expressed as $0 - t$ and then apply the difference of angle identity for the sine function. That is,

$$\begin{aligned}\sin(-t) &= \sin(0 - t)\\ &= \sin 0\ \cos t - \cos 0\ \sin t\end{aligned}$$

Now, recalling $\sin(0) = 0$ and $\cos(0) = 1$, we have

$$\begin{aligned}\sin(-t) &= 0 - \sin t\\ \sin(-t) &= -\sin t\end{aligned}$$

which verifies the first of the symmetry identities.

EXAMPLE 8.7

Confirm the validity of the identity $\tan(\alpha + \beta) = \dfrac{\tan\alpha + \tan\beta}{1 - \tan\alpha\tan\beta}$.

SOLUTION

We recall that tangent is (from Table 8.1) the ratio of sine to cosine. Therefore, we can say that

$$\tan(\alpha + \beta) = \frac{\sin(\alpha + \beta)}{\cos(\alpha + \beta)}$$

From Table 8.5, we can now state that

$$\tan(\alpha + \beta) = \frac{\sin\alpha\,\cos\beta \times \cos\alpha\,\sin\beta}{\cos\alpha\,\cos\beta - \sin\alpha\,\sin\beta}$$

At this point, we can divide numerator and denominator by cos α cos β. If we do so, we find

$$\tan(\alpha + \beta) = \frac{\dfrac{\sin\alpha\,\cos\beta}{\cos\alpha\,\cos\beta} \times \dfrac{\cos\alpha\,\sin\beta}{\cos\alpha\,\sin\beta}}{\dfrac{\cos\alpha\,\cos\beta}{\cos\alpha\,\cos\beta} - \dfrac{\sin\alpha\,\sin\beta}{\cos\alpha\,\cos\beta}} = \frac{\tan\alpha \times \tan\beta}{1 - \tan\alpha\,\tan\beta}$$

which was the desired identity.

EXAMPLE 8.8

Prove the validity of the identity $\cos(2\alpha) = \cos^2\alpha - \sin^2\alpha$.

Note

Keep in mind that the notation $\cos^2 x$ is intended to convey $(\cos x)^2$.

SOLUTION

In this case, we can take the view $2\alpha = \alpha + \alpha$ and use the sum of angles formula for cosine. That is,

$$\cos(2\alpha) = \cos(\alpha + \alpha) = \cos\alpha\,\cos\alpha - \sin\alpha\,\sin\alpha$$
$$\cos(2\alpha) = \cos^2\alpha - \sin^2\alpha$$

EXAMPLE 8.9

Establish that $\cos(2x) = 1 - 2\sin^2 x$.

SOLUTION

The existence of this identity would demonstrate that there are additional expressions that are equivalent to cos(2x). Using the known identities from Table 8.3, we have

$$\begin{aligned}\cos(2x) &= \cos^2 x - \sin^2 x\\ &= (1 - \sin^2 x) - \sin^2 x \text{ (by the Pythagorean identities)}\\ &= 1 - 2\sin^2 x\end{aligned}$$

EXAMPLE 8.10

Using the sum of angles formula for sine, find an expression for sin(3x) that uses only sin$x$.

SOLUTION

If we take the view that $\sin(3x) = \sin(2x + x)$, we can establish a formula for sin(3x) using only the expression sin$x$:

$$\begin{aligned}\sin(3x) &= \sin(2x + x)\\ &= \sin(2x)\cos x + \cos(2x)\sin x\end{aligned}$$

$$= (2\sin x\cos x)\cos x + (1 - \sin^2 x)\sin x \text{ (by Example 8.9)}$$
$$= 2\sin x\cos^2 x + \sin x - \sin^3 x$$
$$= 2\sin x(1 - \sin^2 x) + \sin x - \sin^3 x$$
$$= 2\sin x - 2\sin^3 x + \sin x - \sin^3 x$$
$$= 3\sin x - 3\sin^3 x$$

Note

When you see an expression like sin(3*x*), it is a natural (and mistaken!) thought that the "3" can be moved in front of the expression and conclude that $\sin(3x) = 3\sin x - 3\sin^3 x$.

To this point, our work has been mainly theoretical, so it is natural to wonder if there is any use for these formulas. The primary use for the formulas lies in their applicability to the solution of trigonometric equations, which are (in turn) useful to solve applications in which harmonic motion or resonance occurs.

To this point, we haven't seen enough to solve trigonometric equations, but we can investigate an application where a double angle occurs.

EXAMPLE 8.11

The range R (in feet) of a projectile (an airborne object that is not self-propelled) fired at an angle A with the horizontal and having initial velocity v_0 feet per second is given by $R = \frac{1}{32}v_0^2\sin(2A)$. Suppose a jai alai player throws a ball with an initial velocity of 150 miles per hour at an angle of 2° with the horizontal. What is the range of the ball?

SOLUTION

The first thing we notice here is that the initial velocity is given in miles per hour, so we must convert that velocity to feet per second, which we can do through dimensional analysis:

$$\frac{150\text{ miles}}{\text{hour}} \times \frac{\text{hour}}{3600\text{ seconds}} \times \frac{5280\text{ feet}}{\text{mile}} = \frac{220\text{ feet}}{\text{second}}$$

Now, inserting that value into the formula, we find

$$R = \frac{1}{32}(220)^2\sin[2(2^\circ)]$$
$$R = \frac{1}{32}(48{,}400)\sin(4^\circ)$$
$$R = 1{,}512.5(0.0698)$$
$$R = 105.5725 = 105.6\text{ feet}$$

Thus, the range of the projectile is about 105.6 feet.

The Half-Angle Identities

The half-angle formulas allow us to find the values of the trigonometric functions of angles having the form $\frac{\alpha}{2}$ for some angle α. They can be derived from the double-angle formulas, provided that we look at things in a clever manner.

As was our practice earlier, we'll establish the result for one of the functions and then list the other results and ask you to prove their correctness in the exercises.

The claim is that $\sin\frac{\alpha}{2} = \pm\sqrt{\frac{1 - \cos\alpha}{2}}$. We can establish this by changing every occurrence of α in the formula for $\cos(2\alpha)$ by $\frac{\alpha}{2}$. Doing so, we find

$$\cos\left[2\left(\frac{\alpha}{2}\right)\right] = \cos^2\left(\frac{\alpha}{2}\right) - \sin^2\left(\frac{\alpha}{2}\right)$$

$$\cos\alpha = 1 - \sin^2\left(\frac{\alpha}{2}\right) - \sin^2\left(\frac{\alpha}{2}\right)$$

$$\cos\alpha = 1 - 2\sin^2\left(\frac{\alpha}{2}\right)$$

Now, if we solve for $\sin\left(\frac{\alpha}{2}\right)$, we find

$$2\sin^2\left(\frac{\alpha}{2}\right) = 1 - \cos\alpha$$

$$\sin^2\left(\frac{\alpha}{2}\right) = \frac{1 - \cos\alpha}{2}$$

$$\sin^2\left(\frac{\alpha}{2}\right) = \pm\sqrt{\frac{1 - \cos\alpha}{2}}$$

The value of + or − is determined based on the positive or negative nature of the sine function, which is determined by the quadrant in which the angle $\frac{\alpha}{2}$ lies. This will be discussed immediately after we present the remaining half-angle identities (Table 8.6), which we ask you to confirm in the exercises of this section.

TABLE 8.6 The Half-Angle Identities

Identity Name	Identity Statement
Half-angle formula for sine	$\sin\left(\frac{\alpha}{2}\right) = \pm\sqrt{\frac{1 - \cos\alpha}{2}}$
Half-angle formula for cosine	$\cos\left(\frac{\alpha}{2}\right) = \pm\sqrt{\frac{1 + \cos\alpha}{2}}$
Half-angle formula for tangent	$\tan\left(\frac{\alpha}{2}\right) = \pm\sqrt{\frac{1 - \cos\alpha}{1 + \cos\alpha}} = \frac{1 - \cos\alpha}{\sin\alpha} = \frac{\sin\alpha}{1 + \cos\alpha}$

The Signs of the Trigonometric Functions of Sine, Cosine, and Tangent Based on the Quadrant of the Angle's Terminal Side

To this point, we have taken what is called a right-triangle view of the trigonometric functions. This construction is extremely useful as a starting point for the exploration of trigonometry, but it requires us to limit our investigation to acute angles. We can consider angles that are not acute, and if we do so, we must expand our definition of the trigonometric functions in such a manner that we will not contradict our initial definitions.

Let's suppose that the angle is drawn in standard position, so its initial side lies along the positive x-axis of the plane and its vertex lies at the origin. Further, let (x, y) be any point on the terminal side of the angle α and $r = \sqrt{x^2 + y^2}$. Then the trigonometric functions can be defined as shown in Table 8.7.

TABLE 8.7 Alternate Definitions of the Trigonometric Functions

Trigonometric Functions
$\sin\alpha = \dfrac{y}{r}$
$\cos\alpha = \dfrac{x}{r}$
$\tan\alpha = \dfrac{y}{x}$
$\csc\alpha = \dfrac{r}{y}$
$\sec\alpha = \dfrac{r}{x}$
$\cot\alpha = \dfrac{x}{y}$

If we take this perspective, we can identify the signs of each of the trigonometric functions based on the quadrant in which the terminal side of the angle lies. The reason for this is that r, by construction, is positive, and the signs of x and y are completely determined by the coordinate values of points within the quadrants. The organization of the table may, at first glance, appear to be rather unorthodox, but it is motivated by the arrangement of the quadrants in the plane (Table 8.8).

TABLE 8.8 The Signs of the Trigonometric Functions, by Quadrant

Quadrant II	Quadrant I
Sine is positive, cosine and tangent are negative	Sine, cosine, and tangent are positive
Quadrant III	**Quadrant IV**
Tangent is positive, sine and cosine are negative	Cosine is positive, sine and tangent are negative

You may have noticed that we listed the signs results only for sine, cosine, and tangent. The remaining three functions are the reciprocals of these, and reciprocal expressions share sign value, so we can find the sign of cosecant, secant, and cotangent by referencing their reciprocal function sign values.

Let's tie this in with the half-angle identities through the following example.

EXAMPLE 8.12 Given that $\sin\alpha = \frac{7}{10}$ and that is in quadrant II, find the exact value of $\sin\left(\frac{\alpha}{2}\right)$.

SOLUTION

This problem is somewhat deceptive because it appears to ask a relatively innocent question, reliant only on the half-angle identity for sine. However, it will involve a fair amount of resourcefulness. Recall first that the identity is $\sin\left(\frac{\alpha}{2}\right) = \pm\sqrt{\frac{1-\cos\alpha}{2}}$ and contains the expression $\cos\alpha$, which is not provided in the problem. Moreover, the issue of sign (positive or negative) must be resolved.

There are several means that we could employ to uncover the value of cosine, but we will use the information given in Table 8.7. Based on the formula for the sine function, we can deduce that $y = 7$ and $r = 10$ and use the fact that $r = \sqrt{x^2 + y^2}$ to find the value of x.

By substitution, we find

$$10 = \sqrt{x^2 + 7^2}$$

$$100 = x^2 + 49$$

$$51 = x^2$$

$$\pm\sqrt{51} = x$$

The sign of x is unclear at the outset, but we recall that α is in quadrant II, in which the coordinate values of x are negative, and hence we determine $x = -\sqrt{51}$, from which it follows that $\cos\alpha = -\sqrt{\frac{51}{10}}$.

We are now ready to insert the value of cosine into the formula, which tells us

$$\sin\left(\frac{\alpha}{2}\right) = \pm\sqrt{\frac{1-\cos\alpha}{2}}$$

$$\sin\left(\frac{\alpha}{2}\right) = \pm\sqrt{\frac{1-\left(-\frac{\sqrt{51}}{10}\right)}{2}}$$

$$\sin\left(\frac{\alpha}{2}\right) = \pm\sqrt{\frac{10+\sqrt{51}}{20}}$$

$$\sin\left(\frac{\alpha}{2}\right) = \pm\frac{1}{2}\sqrt{\frac{10+\sqrt{51}}{5}}$$

We'll relax the condition that the radicand be rationalized here and content ourselves with accepting the radical expression (nasty as it is) as being simplified. It remains to determine the sign of the overall expression on the right side of the equation.

We were given that the angle α was in quadrant II, and that permits us to find the location of the angle $\frac{\alpha}{2}$:

$$90° < \alpha < 180°$$

$$45° < \frac{\alpha}{2} < 90°$$

Therefore, $\frac{\alpha}{2}$ is in quadrant I, within which sine is positive, and therefore our final result is that

$$\sin\left(\frac{\alpha}{2}\right) = \frac{1}{2}\sqrt{\frac{10 + \sqrt{51}}{5}}$$

The Quotient Identities

In Table 8.5, we presented the alternative definitions of the trigonometric functions using x, y, and r. Closer examination of the entries of that table reveals two rather interesting and highly useful relationships.

We saw that $\cos\alpha = \frac{x}{r}$ and $\sin\alpha = \frac{y}{r}$. Suppose we formed the possible quotients of these two relationships, $\frac{\cos\alpha}{\sin\alpha}$ and $\frac{\sin\alpha}{\cos\alpha}$. In the former case, replacing sine and cosine by their alternative definitions, we discover the following:

$$\frac{\cos\alpha}{\sin\alpha} = \frac{x/r}{y/r} = \frac{x}{r} \div \frac{y}{r} = \frac{r}{y} = \frac{x}{y} = \cot\alpha$$

In a similar manner, we can establish that $\frac{\sin\alpha}{\cos\alpha} = \tan\alpha$. These results are significant and are widely used in the work to come. Their presentation completes the list of what we'll call the fundamental list of identities (Table 8.9).

TABLE 8.9 The Quotient Identities

The Quotient Identities
$\frac{\sin\alpha}{\cos\alpha} = \tan\alpha$
$\frac{\cos\alpha}{\sin\alpha} = \cot\alpha$

Exercises

Answer the following questions using complete sentences, proper grammar, and correct terminology.

1. What is meant by the sum of angles? What is meant by the difference of angles?
2. What is meant by a double angle?
3. What is meant by a half angle?
4. What does it mean to say a function is an even function, and what does it mean to say that a function is an odd function?

Verify the following identities.

5. $\cos\left(\frac{\pi}{2} - x\right) = \sin x$
6. $\cos\left(\theta - \frac{\pi}{2}\right) - \sin\theta = 0$
7. $\tan(\theta + 5\pi) = \tan\theta$
8. $\sin\left(\frac{\pi}{6} + \theta\right) = \frac{\cos\theta + \sqrt{3}\sin\theta}{2}$
9. $\sqrt{2}\cos\left(\frac{5\pi}{4} - x\right) = -\cos x - \sin x$
10. $\cos(\pi - x) = \sin\left(x - \frac{\pi}{2}\right)$
11. $\sec(2x) = \frac{\sec^2 x}{2 - \sec^2 x}$
12. $2\cos\theta\csc(2\theta) = \csc\theta$
13. $(\sin x + \cos x)^2 = \sin(2x) + 1$
14. $\cos(2x) = \cos^4 x - \sin^4 x$
15. $\sec\frac{\theta}{2} = \pm\sqrt{\frac{2\tan\theta}{\tan\theta + \sin\theta}}$
16. $\tan\frac{\theta}{2} = \csc\theta - \cot\theta$
17. $\cos^2(2\theta) - \sin^2(2\theta) = \cos(4\theta)$
18. $\cos^4 t - \sin^4 t = \cos(2t)$

Verify the following symmetric identities.

19. $\cos(-t) = \cos t$
20. $\sec(-t) = \sec t$
21. $\tan(-t) = -\tan t$
22. $\cot(-t) = -\cot t$
23. $\csc(-t) = -\csc t$
24. Using a half-angle identity, find the exact value of $\cos(15°)$
25. Using a half-angle identity, find the exact value of $\sin(22.5°)$

26. Using a half-angle identity, find the exact value of $\tan(67.5°)$
27. Using a half-angle identity, find the exact value of $\sin(-67.5°)$

In the following exercises, use the half-angle identities to find the exact values of the following: $\sin\frac{x}{2}$, $\cos\frac{x}{2}$, and $\tan\frac{x}{2}$. Use the variable given in the problem.

28. $\cos x = \frac{12}{13}, \; \frac{3\pi}{2} < x < 2\pi$
29. $\tan u = \frac{1}{3}, \; 0 < u < \frac{\pi}{2}$
30. $\sec t = -5, \; \frac{\pi}{2} < t < \pi$

8.3 Verification of Further Identities

We have referred to the identities presented to this point as Pythagorean identities, reciprocal identities, sum and difference of angle identities, and so forth. These identities are given specific names because of their frequent use in applications as aids to solving trigonometric equations.

As we have said before, the list of trigonometric identities is endless. Any particular trigonometric expression can be rewritten in infinitely many different yet equivalent forms, and no single form is necessarily better than any other form. Which one you want to use is a matter of expedience, as a particular form might (in a given situation) be more useful to you than another form.

Were this to be a text on the broad field of trigonometry, we would spend significant time on the topic of the solution of trigonometric equations. However, that is not our focus here, so we will treat this topic as an exercise in strengthening mental acuity. That is, we will look on trigonometric identities and their use as a tool to practice rewriting expressions in a clever way, strengthening our ability to see patterns and relationships, and improving our factoring skills and knowledge of the basic rules of mathematics.

The construction of these verification problems is usually in the form of a proposed equality that you are then asked to "verify." You can use any known identity (by "known," we mean one of the fundamental identities given in the named sets or any identity you have already established yourself and happen to recall) to conduct the verification, and the objective is to take one side of the proposed equality and rewrite it so that it is identical to the other side. If this can be done, the identity is considered verified. If doing so is beyond your ability, then (on the face of it) it establishes nothing. The two sides of the proposed equality may, in fact, be identical, but you cannot see how to establish the relationship, or they may not be equal, in which case the alleged identity is not an identity at all.

We suggest the following algorithm for the **verification of identities**:

- Work to see if (through substitutions and simplifications) the left side of the equality can be rewritten so that it is identical to the right side of the equality. If it can, then the identity is verified. It is generally good practice to keep a list of the fundamental identities handy as an aid in your work.

- Useful techniques to consider are factoring expressions, adding fractions, or finding common denominators. Keep an eye on your target expression: the right side of the equality. See if the use of any fundamental identities will be helpful to create the trigonometric functions in the desired final expression.
- If you cannot come up with any ideas on what to do or if nothing seems to work, then convert all expressions to sines and cosines and work exclusively with those forms. Generally, this is a move bordering on desperation and is discouraged, but desperate times can call for desperate measures!

EXAMPLE 8.13

Verify the following identity: $\cos x + \sin x \tan x = \sec x$.

SOLUTION

Pressing on, we now start our work. Verification of identities is always an ad hoc process in that it is different in almost every example we encounter. On the left side of the equation, we see three trigonometric functions, and an immediate step toward simplification would be to reduce the number of functions to two by using the relationship $\tan x = \dfrac{\sin x}{\cos x}$. It may be that this is a terrible idea, but we will know only by trying it, so we take the plunge:

$$\cos x + \sin x \tan x = \sec x$$

$$\cos x + \sin x\left(\frac{\sin x}{\cos x}\right) = \sec x$$

It seems reasonable next to simplify the second term on the left side, which gives us

$$\cos x + \left(\frac{\sin^2 x}{\cos x}\right) = \sec x$$

Notice that there is nothing to suggest to us, at this point, that we are moving closer to a resolution of the problem. We are applying known identities and algebraic techniques, hoping for something favorable to turn up. On the left, we can rewrite the expressions with a common denominator, that will allow us to add, and that may be helpful:

$$\frac{\cos^2 x}{\cos x} + \frac{\sin^2 x}{\cos x} = \sec x$$

$$\frac{\cos^2 x + \sin^2 x}{\cos x} = \sec x$$

The numerator on the left side of the equation is equal to 1, by a Pythagorean identity, and this yields

$$\frac{1}{\cos x} = \sec x$$

On the left, we now can apply one of the reciprocal identities and conclude

$$\sec x = \sec x$$

which verifies the identity.

The preceding example is fairly tame but does require a certain degree of cleverness and persistence. There were a variety of trigonometric relationships and algebraic techniques in play here, and we simply had to be proficient at them in order to successfully establish the identity. We should reiterate, however, that this should not be viewed as *the* way to prove any other identity we might encounter! Each case should be approached much like a detective approaches a crime scene: as you investigate more and more cases, you gain experience with techniques that might work, but each situation is unique and may require new innovations!

EXAMPLE 8.14 Verify the identity $\dfrac{\csc^2\theta}{\cot\theta} = \csc\theta\sec\theta$.

SOLUTION

We note that cotangent is the reciprocal of tangent, and this allows us to rewrite the left side of the equality as

$$\csc^2\theta\tan\theta = \csc\theta\sec\theta$$

Comparing the two sides of the equality, we see that the left side has two factors of cosecant and that the right side has a single cosecant factor. Thus, we can consider the expression as

$$\csc\theta\csc\theta\tan\theta = \csc\theta\sec\theta$$

and now ponder if $\csc\theta\tan\theta$ can be rewritten as $\sec\theta$, which would complete the verification.

Using the quotient identity for tangent and the reciprocal identity for cosecant, we find

$$\begin{aligned}
\csc\theta\left(\frac{1}{\sin\theta}\right)\left(\frac{\sin\theta}{\cos\theta}\right) &= \csc\theta\sec\theta \\
\csc\theta\left(\frac{1}{\cos\theta}\right) &= \csc\theta\sec\theta \\
\csc\theta\sec\theta &= \csc\theta\sec\theta
\end{aligned}$$

This establishes the validity of the identity.

EXAMPLE 8.15 Verify the identity $\dfrac{\cos x - \cos y}{\sin x + \sin y} - \dfrac{\sin y - \sin x}{\cos x + \cos y} = 0$.

SOLUTION

The left side of the equality is a difference of fractions, so it seems reasonable to begin by finding a common denominator and then combining the fractions. The common denominator is the product of the existing denominators, and thus we find

$$\frac{(\cos x - \cos y)(\cos x + \cos y)}{(\sin x + \sin y)(\cos x + \cos y)} - \frac{(\sin y - \sin x)(\sin x + \sin y)}{(\cos x + \cos y)(\sin x + \sin y)} = 0$$

which can be rewritten as

$$\frac{(\cos x - \cos y)(\cos x + \cos y) - (\sin y - \sin x)(\sin x + \sin y)}{(\sin x + \sin y)(\cos x + \cos y)} = 0$$

If we now simplify the numerator, first multiplying out the products, we find

$$\frac{\cos^2 x - \cos^2 y + \sin^2 x - \sin^2 y}{(\sin x + \sin y)(\cos x + \cos y)} = 0$$

$$\frac{\cos^2 x - \sin^2 y - (\cos^2 y - \sin^2 y)}{(\sin x + \sin y)(\cos x + \cos y)} = 0$$

$$\frac{1 - 1}{(\sin x + \sin y)(\cos x + \cos y)} = 0$$

$$\frac{0}{(\sin x + \sin y)(\cos x + \cos y)} = 0$$

$$0 = 0$$

and this verifies the identity.

EXAMPLE 8.16 Verify the identity $\sec\theta + \tan\theta = \dfrac{\cos\theta}{1 - \sin\theta}$.

SOLUTION

The mantra we continue to preach is that every trigonometric identity verification stands alone from the rest, and you should resist the temptation to treat consecutive problems exactly alike. Sometimes this occurs, but many times a novel approach is called for.

Before jumping to the solution of this example, let's consider our options. We could convert the left side of the equality to fractions using the reciprocal and quotient identities, add the resulting fractions, and see where this takes us.

The right side of the equality has a two-term expression in the denominator, so it is possible that we could multiply numerator and denominator of that fraction by the **conjugate pairs** of the existing denominator. Recall that this would be the expression $1 + \sin\theta$, the same expression with the sign between the terms reversed.

It's unclear which one we should use, so we will choose the second option, since it offers a departure from what has been tried in the earlier illustrations and may prove instructional:

$$\sec\theta + \tan\theta = \frac{\cos\theta}{1 - \sin\theta}$$

$$\sec\theta + \tan\theta = \frac{\cos\theta(1 + \sin\theta)}{(1 - \sin\theta)(1 + \sin\theta)}$$

The denominator will simplify using the difference of squares formula and then one of the Pythagorean identities, yielding

$$\sec\theta + \tan\theta = \frac{\cos\theta(1 + \sin\theta)}{\cos^2\theta}$$

Note that we have not simplified the numerator on the right side of the equality. The reason for this is that, when this technique is being used, it is often possible to reduce the resulting expression, and hence it is useful to keep the expressions factored. Here, one of the cosines will reduce and give us

$$\sec\theta + \tan\theta = \frac{(1 + \sin\theta)}{\cos\theta}$$

Now, if we split the right side of the equality into the sum of fractions, the result is immediate:

$$\begin{aligned} \sec\theta + \tan\theta &= \frac{1}{\cos\theta} + \frac{\sin\theta}{\cos\theta} \\ \sec\theta + \tan\theta &= \sec\theta + \tan\theta \end{aligned}$$

You might be curious if the alternative approach would have worked out easier for us. The answer to that question is, at the moment, completely unclear. The unfortunate fact of the matter is that we simply do not know what method would take us to the result in the swiftest, easiest manner. We literally are blazing the trail as we go, hoping to find our way through this mathematical forest.

EXAMPLE 8.17

Verify the identity $\sec^6(\sec t \tan t) - \sec^4 t(\sec t \tan t) = \sec^5 t \tan^3 t$.

SOLUTION

In this case, we are faced with some rather large exponents and a somewhat horrific expression to the left of the equal sign. However, we note that there is a common factor staring us in the face, the $(\sec t \tan t)$, which seems to beg to be factored from the left side of the equation. This may be a lure to draw us into a rather messy trap, but it seems to be the most obvious step to take, so we'll follow that path. Remember, the very worst thing that could happen to us is that we reach a point at which we become stuck, and then we'll start over again and try something else:

$$\begin{aligned} \sec^6 t(\sec t \ \tan t) - \sec^4 t(\sec t \ \tan t) &= \sec^5 t \ \tan^3 t \\ (\sec t \ \tan t)(\sec^6 t - \sec^4 t) &= \sec^5 t \ \tan^3 t \end{aligned}$$

In this form, we can see that we can actually factor the left side a bit further, since we can remove a factor of $\sec^4 t$ from the second parentheses:

$$(\sec t \tan t)\sec^4 t\,(\sec^2 t - 1) = \sec^5 t \tan^3 t$$

Now observe the opportunity to combine the first two parentheses on the left and that the third parentheses contains an alternate form of one of the Pythagorean identities. This means that we can rewrite the expression as

$$\begin{aligned} (\sec^5 t \tan t)(\tan^2 t) &= \sec^5 t \tan^3 t \\ \sec^5 t \tan^3 t &= \sec^5 t \tan^3 t \end{aligned}$$

which completes the verification of the identity.

We have shown you a variety of examples to illustrate some of the ideas you can use to verify trigonometric identities, but the listing we've given you is by no means exhaustive. Literally, any mathematical technique you have ever seen is open to be used in the verification of trigonometric identities, and you should expect to be significantly challenged by some of the examples you'll encounter in the exercises.

Keep in mind that, if all else fails, your fallback position is to convert every single expression you encounter into one using only sines and cosines (this is always possible). By doing so, you may increase the complexity of the expression to an uncomfortable extent, but you will gain the benefit of thereby being limited to a single identity that can be used:

$$\cos^2\theta + \sin^2\theta = 1$$

Be confident as you begin the exercises! None of the problems is beyond your ability to solve, so they should all be within your grasp.

Exercises

Verify each of the following identities. Use any appropriate identity.

1. $\sin x \tan x + \cos x - \sec x = 0$
2. $\sin^2 x(\csc x + 1) = \dfrac{1 - \sin^2 x}{\csc x - 1}$
3. $\cot^4\theta = \cot^2\theta\csc^2\theta - \cot^2\theta$
4. $\sin x + \cos x = \dfrac{\cot x + 1}{\csc x}$
5. $(\sec x - \tan x)^2 = \dfrac{1 - \sin x}{1 + \sin x}$
6. $\dfrac{1}{\sec x - \tan x} = \sec x + \tan x$
7. $\dfrac{\cos(2x) - \cos(4x)}{\sin(2x) + \sin(4x)} = \tan x$
8. $\dfrac{\sin x}{1 - \cos x} - \dfrac{\sin x \cos x}{1 + \cos x} = \csc x\,(1 + \cos^2 x)$
9. $\left(\dfrac{\cos x}{\sin x}\right)(\sec x) = \csc x$
10. $\sin x + \cos x \cot x = \csc x$
11. $\left[\dfrac{\csc x}{1 + \csc x} - \dfrac{\csc x}{1 - \csc x}\right] = 2\sec^2 x$
12. $\sin x\,(\cot x + \cos x \tan x) = \cos x + \sin^2 x$
13. $\sin^2 x - \cos^2 x = \sin^4 x - \cos^4$
14. $\dfrac{1}{1 - \cos x} + \dfrac{1}{1 + \cos x} = 2\csc^2 x$

15. $\cos x \sin y = \frac{1}{2}\sin(x + y) - \frac{1}{2}\sin(x - y)$

16. $\tan\frac{\theta}{2} = \frac{1 - \cos\theta}{\sin\theta}$

17. $\cot x + \tan x = \frac{2}{\sin 2x}$

18. $\frac{\sin(x + y)}{\sin(x - y)} = \frac{\tan x + \tan y}{\tan x - \tan y}$

19. $\csc 2x = \frac{\sec x}{2\sin x}$

20. $\frac{1 + \sec x}{\csc x} = \sin x + \tan x$

21. $\tan\left(\frac{\pi}{4} + x\right) = \frac{\cos^2 x - \sin^2 x}{1 - \sin 2x}$

22. $\sin^2\left(\frac{x}{2}\right) = \frac{\sec x - 1}{2\sec x}$

Summary

In this chapter, you learned about:

- Simplifying expressions using trigonometric identities.
- The definition and derivation of the major trigonometric identities: reciprocal and Pythagorean identities, symmetry and even-odd identities, sum and difference of angle identities, double-angle, half-angle, and the quotient identities.
- How to verify other trigonometric identities.

Glossary

conjugate pairs Two binomial expressions differing only in the sign between the terms of the expressions.

even-odd identities A set of trigonometric identities expressing relationships between functions reliant on the properties of symmetry displayed by their graphs. Also called symmetry identities.

pythagorean trigonometric identity $\cos^2\theta + \sin^2\theta = 1$

reciprocal identities A set of trigonometric identities expressing relationships between functions reliant on the reciprocal relationships.

symmetry identities A set of trigonometric identities expressing relationships between functions reliant on the properties of symmetry displayed by their graphs. Also called even-odd identities.

trigonometric identity A statement expressing the equality of two trigonometric expressions.

verification of identities A process of algebraic manipulation and substitution through which one can establish the equality of two trigonometric expressions.

List of Equations

The reciprocal trigonometric identities for an angle θ:

$$\sin\theta = \frac{1}{\csc\theta} \qquad \csc\theta = \frac{1}{\sin\theta}$$

$$\cos\theta = \frac{1}{\sec\theta} \qquad \sec\theta = \frac{1}{\cos\theta}$$

$$\tan\theta = \frac{1}{\cot\theta} \qquad \cot\theta = \frac{1}{\tan\theta}$$

The alternate trigonometric relationships, based on the sides of a right triangle:

$$\sin\theta = \frac{\text{opposite}}{\text{hypotenuse}} \qquad \csc\theta = \frac{\text{hypotenuse}}{\text{opposite}}$$

$$\cos\theta = \frac{\text{adjacent}}{\text{hypotenuse}} \qquad \sec\theta = \frac{\text{hypotenuse}}{\text{adjacent}}$$

$$\tan\theta = \frac{\text{opposite}}{\text{adjacent}} \qquad \cot\theta = \frac{\text{adjacent}}{\text{opposite}}$$

The Pythagorean Identities:

$$\sin^2\theta + \cos^2\theta = 1$$

$$\tan^2\theta + 1 = \sec^2\theta$$

$$\cot^2\theta + 1 = \csc^2\theta$$

The Symmetry Identities
$\sin(-\theta) = -\sin\theta$
$\cos(-\theta) = \cos\theta$
$\tan(-\theta) = -\tan\theta$
$\cot(-\theta) = -\cot\theta$
$\sec(-\theta) = \sec\theta$
$\csc(-\theta) = -\csc\theta$

The Sum and Difference of Angles Identities:

Identity Name	Identity Statement
Cosine of the sum of angles	$\cos(\alpha + \beta) = \cos\alpha\,\cos\beta - \sin\alpha\,\sin\beta$
Cosine of the difference of angles	$\cos(\alpha - \beta) = \cos\alpha\,\cos\beta + \sin\alpha\,\sin\beta$
Sine of the sum of angles	$\sin(\alpha + \beta) = \sin\alpha\,\cos\beta + \cos\alpha\,\sin\beta$
Sine of the difference of angles	$\sin(\alpha - \beta) = \sin\alpha\,\cos\beta - \cos\alpha\,\sin\beta$
Tangent of the sum of angles	$\tan(\alpha + \beta) = \dfrac{\tan\alpha + \tan\beta}{1 - \tan\alpha\,\tan\beta}$

Identity Name	Identity Statement
Tangent of the difference of angles	$\tan(\alpha - \beta) = \dfrac{\tan\alpha - \tan\beta}{1 + \tan\alpha \tan\beta}$
Sine of a double angle	$\sin(2\alpha) = 2\sin\alpha \cos\alpha$
Cosine of a double angle	$\cos(2\alpha) = \cos^2\alpha - \sin^2\alpha$

Alternate Definitions of the Trigonometric Functions, based on a point (x, y) on terminal side of angle, in standard position, with $x^2 + y^2 = r^2$:

Trigonometric Functions
$\sin\alpha = \dfrac{y}{r}$
$\cos\alpha = \dfrac{x}{r}$
$\tan\alpha = \dfrac{y}{x}$
$\csc\alpha = \dfrac{r}{y}$
$\sec\alpha = \dfrac{r}{x}$
$\cot\alpha = \dfrac{x}{y}$

End-of-Chapter Problems

Verify the following identities.

1. $\sin t \csc t = 1$
2. $\dfrac{\sin^2\theta}{1 + \cos\theta} = 1 - \cos\theta$
3. $\tan\theta + \cot\theta = \csc\theta \sec\theta$
4. $\cos\theta \tan\theta = \sin\theta$
5. $\csc\theta \tan\theta = \sec\theta$
6. $\sec^2\theta + \csc^2\theta = \sec^2\theta \csc^2\theta$
7. $\dfrac{1}{\cos\theta + 1} + \dfrac{1}{\sec\theta + 1} = 1$
8. $\dfrac{\sin(-\theta)}{\cos(-\theta)} = \tan(-\theta)$
9. $\sec^2\theta - \tan^2\theta = 1$
10. $\sec^4\theta - \tan^4\theta = 1 + 2\tan^2\theta$

Verify the following identities.

11. $\sin(\alpha + \beta) = \sin\alpha \cos\beta + \cos\alpha \sin\beta$
12. $\tan(\alpha + \beta) = \dfrac{\tan\alpha + \tan\beta}{1 - \tan\alpha \tan\beta}$

13. $\sin(2\alpha) = 2\sin\alpha\cos\alpha$

14. $\cos\left(\frac{\alpha}{2}\right) = \pm\sqrt{\frac{1+\cos\alpha}{2}}$

15. $\tan x \csc x \cos x = 1$

Using a half-angle identity, find the exact value of the following expressions.

16. $\sin(15°)$

17. $\cos(45°)$

18. $\tan(22.5°)$

19. $\sin(75°)$

20. $\cos(60°)$

In the following exercises, use the half-angle identities to find the exact values of the following: $\sin\frac{x}{2}$, $\cos\frac{x}{2}$, and $\tan\frac{x}{2}$. Use the variable given in the problem.

21. $\sin\theta = \frac{5}{6}, \frac{\pi}{2} < \theta < \pi$

22. $\cos x = \frac{7}{9}, 0 < x < \frac{\pi}{2}$

23. $\sec x = 5, 0 < x < \pi$

24. $\csc(x) = 2, \pi < x < 2\pi$

25. $\sin(x) = \frac{1}{2}, 0 < x < \frac{\pi}{2}$

Verify each of the following identities. Use any appropriate identity.

26. $\sin(3\theta) = 3\cos^2\theta\sin\theta - \sin^3\theta$

27. $\cos 3\theta = \cos^3\theta - 3\sin^2\theta\cos\theta$

28. $\tan 3\theta = \frac{3\tan\theta - \tan^3\theta}{1 - 3\tan^2\theta}$

29. $\sin^2\theta = \frac{1-\cos 2\theta}{2}$

30. $\cos^2\theta = \frac{1+\cos 2\theta}{2}$

31. $\cos\theta\cos\varphi = \frac{\cos(\theta-\varphi) + \cos(\theta+\varphi)}{2}$

32. $\sin\theta\sin\varphi = \frac{\cos(\theta-\varphi) - \cos(\theta+\varphi)}{2}$

33. $\tan(x) + \sec(x) = \tan\left(\frac{x}{2} + \frac{\pi}{4}\right)$

34. $\sin x + \sin y = 2\sin\frac{x+y}{2}\cos\frac{x-y}{2}$

35. $\sin\theta = \frac{2\tan(\theta/2)}{1+\tan(\theta/2)}$

Objectives

When you have successfully completed the materials of this chapter, you will be able to:

- Understand the definition of complex numbers and simplify radical expressions that lead to complex numbers.
- Write complex numbers in rectangular, polar, and trigonometric form.
- Perform addition, subtraction, multiplication, division, and exponentiation of complex numbers.
- Solve application problems involving complex numbers.

9.1 Defining the Complex Numbers

A Reminder about Radicals

The applications of the complex numbers are varied and widespread, but in order for us to develop an understanding of those applications, we must first ensure that we have a foundational knowledge of these numbers. Therefore, before proceeding further, we retreat to reminding you of the definition of the principal square root:

DEFINITION: Principal of the square root:

$$\sqrt{A} = B \text{ if and only if } B \text{ is nonnegative and } A = B^2$$

The value of A is referred to as the **radicand** of the square root, and the initial limitations on the definition require that A be nonnegative.

From this definition, we can determine that $\sqrt{49}$ is equal to 7, since 7 is nonnegative and such that $49 = 7^2$. While it is true that $49 = (-7)^2$, the technical requirement that B should be nonnegative compels us to choose the value of 7 as the "principal" square root of 49.

Compare this problem to the search for $\sqrt{-49}$. The situation is not quite the same as in the first example because the radicand value is now negative, and the definition of the principal square root would tell us that we were looking for a nonnegative number B such that $-49 = B^2$.

Immediately, we see the dilemma on whose horns we have fallen and the reason the initial definition called for radicands to be nonnegative. Rules of signs require that the square of a number be positive, and therefore the search for the value of B must end in failure. We might suspect that this is not an untenable situation because we have seen examples of equations in our mathematical history that have "no solution" and could, therefore, conclude that $\sqrt{-49} = B$ would fall into that category.

Ultimately, and for reasons that are at present completely unclear and beyond the scope of our conversation, it turns out that this position proves unacceptable. For many applications, we find it necessary to produce an answer to equations such as this, so the problem must be examined in greater detail. We are led to the conclusion that the real numbers with which we have been working are inadequate for our purposes, so we expand our concept of numbers to incorporate a new entity: an "imaginary" number.

The development of this concept can be driven through recollection of one of the rules for working with radicals, particularly square roots. One of the rules, known as the "product rule," states,

$$\sqrt{ab} = \sqrt{a}\,\sqrt{b}$$

This rule holds that each of the individual square roots is real, and as such it does not help us in the current problem. We can, if we introduce one new concept, extend the product rule of square roots to cover the situation in which we now find ourselves. If the rule were to apply to square roots of negative numbers, then we could rewrite our example $\sqrt{-49}$ as $\sqrt{49}\,\sqrt{-1}$, and the problem would then be resolved if we were to determine how we should treat $\sqrt{-1}$. The idea could then be applied to the square root of any negative number, since a similar factorization could be used in any such case.

The Definition of the Imaginary Unit

We now present a definition, which we state in two equivalent forms:

DEFINITION: The imaginary unit:

$$\sqrt{-1} = i, \text{ or, equivalently, } -1 = i^2$$

The symbolic "number" i is referred to as the **imaginary unit** and appears "from nowhere" at this juncture.

You may find it troubling that we are using the letter i to represent a number, but keep in mind that this is not the first time that we have done this. The Greek letter π is already well known to us and is a canonical example of this concept. You may also feel obligated to ask where that "comes from," but we must allow our curiosity to be satisfied with the realization that definitions do not "come from" anywhere; rather, they are created. Consequently, this definition emerges entirely from abstraction and so cannot be derived in any manner.

The emergence of i permits us to simplify $\sqrt{-49}$, since we can then say

$$\sqrt{-49} = \sqrt{49}\,\sqrt{-1} = 7i$$

As was mentioned, the square root of any negative number can be handled in this manner, but for completeness, we'll do one more example (which may help you recall the process of simplification of square roots!):

$$\sqrt{-128} = \sqrt{128}\,\sqrt{-1} = \sqrt{128}\,i$$

Note that 128 is not a perfect square, so the simplification from this point is not as straightforward as our first illustration. Keeping in mind that simplification of a square root is *not* synonymous with decimal approximation, we search for the largest perfect square that happens to be a factor of 128. If no perfect squares are factors of 128, the expression cannot be simplified.

The perfect squares are the infinite list that begins with 4, 9, 16, 36, 49, 64, 81, 100, 121, 144, and continues forever. We've stopped at 144, since that exceeds the value of our radicand. From this list, we might observe that 4 is a factor of 128, but we're interested in the largest factor we can find, and that would be 64. Using that perfect square to decompose our radicand, we continue as

$$\sqrt{128}\,i = \sqrt{64}\,\sqrt{2}\,i = 8\sqrt{2}\,i$$

That is the "simplified" form of our result.

With the imaginary unit i now in our hands, we define the concept of a complex number:

DEFINITION: Complex Number

A complex number is any number of the form $a + bi$, where a and b are real numbers.

The value of a is called the **real part** of the complex number, and the value of b is called the **imaginary part** of the complex number. If $a = 0$ and thus our number has the form bi, we say the number is **pure imaginary**. If $b = 0$ and thus our number has the form a, we say the number is real. Another term (probably in more common use) for pure imaginary numbers is imaginary numbers. An **imaginary number** is any number whose square is a negative real number.

Note that this supports the original statement that began this chapter, that is, that the new system of numbers we were introducing would be an "extension" of the real numbers with which we were familiar. The real numbers are a subset of the complex numbers.

Examples of complex numbers would be any real number, such as 3, -7, 14, and so on; any pure imaginary number, such as $-2i$, $4i$, $8\sqrt{2}\,i$, and so on; or any combination, such as $5 + 3i$, $7 - 16i$, $1 + \sqrt{5}\,i$, and so on.

The Standard Form of a Complex Number

The form $a + bi$ is referred to as the **standard form**, or the **rectangular form**, of a complex number. There are other manners in which complex numbers can be represented, such as graphical, polar, and exponential form, but for computational simplicity (as we'll see), we will use this form for the time being. Note also that, while it is not prohibited that we might rearrange the expression in such a manner that the real part was given last, $bi + a$, this is generally not done, and therefore we will follow the common practice of positioning the real part first.

Two complex numbers are considered "equal" if they have precisely the same real and imaginary parts as one another. That is, $(a + bi) = (c + di)$ precisely if $a = c$ and $b = d$.

The Powers of the Imaginary Unit

Although it is not obvious, there are many implications spawned by the definition we have produced for the imaginary unit i. We'll explore some of them in the remainder of this chapter, but let's look at a curious property that emerges immediately.

Note

There are only four distinct powers of i, and all other powers of i can be simplified to one of those four forms.

This is a fairly elementary result in the grand scheme of things, but it may appear shocking at first glance. Why should there only be four powers of i? Let's investigate, and we'll see that this is a natural consequence of the definition we have created.

We can start with a few immediate statements. Note that since we have decided that any number raised to the power 0 should be equal to 1, $i^0 = 1$. Moreover, any number raised to the first power is the number itself, and thus $i^1 = i$, and $i^2 = -1$ (by definition!).

Keeping these facts in mind, we now turn our attention to i^3. The properties of exponents tell us $i^3 = i^2 i^1 = (-1)i = -i$. This is a surprising result, since the power of 3 has "simplified away."

Notice now that $i^4 = i^3 i^1 = (-i)(i) = -i^2 = -(-1) = 1$. We are now in a truly unusual position, indeed, because any higher power can be simplified down to one of these four we have already presented, using a similar technique. Consider $i^5 = i^4 i^1 = (1)\, i = i$. You could continue this process for any power of i by clever use of the properties of exponents.

Of course, it would be impractical to follow that process exactly for really large powers, like i^{73}, for instance, since it would involve a rather tedious procedure of reducing the powers by 1 over and over again until we reached one of our four "known" powers. Using the power rule of exponents, however, the expression can be simplified in the following manner: $i^{71} = i^{68} i^3 = (i^4)^{17}(-i) = -i$.

There is a way around that, which we will introduce as a rule, and then turn our attention to other matters:

Note

For any whole number power n, the expression $i^n = i^R$, where R is the remainder obtained when n is divided by 4.

Consequently, i^{117}, for instance, can be computed by finding the remainder of 117/4. Elementary division reveals the result to be 29, with a remainder of 1. Therefore, i^{117} simplifies to $i^1 = i$. Why does the rule work? If we use the result we "know" from the rule, we can, in a sense, reverse-engineer the problem to uncover the explanation.

Since 117/4 = 29 with a remainder of 1, we can say 117 = (29)(4) + 1. This tells us (although possibly not obviously—you may have to review your rules of exponents!)

$$i^{117} = i^{(4)(29)+1} = (i^4)^{29} i^1 = (1)^{29} i = i$$

While the rule specifies that n be nonnegative, we can extend the concept to encompass all integer powers of i, which we will leave to the exercises.

A Connection between the Complex Numbers and Matrices

We worked very hard in Chapter 7 to develop the concept of matrices, and to understand how to operate with them. There is, as we can readily point out, a connection between that topic and the complex numbers. This connection is nonobvious, but it does exist, and it has rather deep and surprising consequences.

Consider the following matrix: $i = \begin{bmatrix} 0 & 1 \\ -1 & 0 \end{bmatrix}$. You may note that we have named the matrix i, and that deliberate choice may rouse your suspicions immediately.

Recalling our method for multiplication of matrices, compute i^2:

$$\begin{bmatrix} 0 & 1 \\ -1 & 0 \end{bmatrix}\begin{bmatrix} 0 & 1 \\ -1 & 0 \end{bmatrix} = \begin{bmatrix} -1 & 0 \\ 0 & -1 \end{bmatrix} = -1\begin{bmatrix} 1 & 0 \\ 0 & 1 \end{bmatrix}$$

Since the final matrix shown is the identity matrix, which serves as the multiplicative identity for matrices, note that the square of the matrix we have called i behaves precisely in the same manner as the imaginary unit! That is, its square is equal to the matrix analog of -1!

This is a surprising result and suggests that a good deal of our work with matrices may apply to complex numbers. We will not explore this further at this time, but it is a surprising result.

The Principal Complex Square Root

We end this section with a brief consideration of a rather nonobvious complication arising from the introduction of the complex numbers.

For the real number square roots, such as $\sqrt{9}$, we recall that there was a debate over the result, since both 3^2 and $(-3)^2$ were equal to the radicand, and that this was resolved by assigning the positive value 3 to be the **principal square root** of 9 and that the square root symbol would be associated only with the principal square root.

Does the same situation arise with complex numbers? That is, if we consider $\sqrt{-9}$, will both $3i$ and $-3i$ satisfy the definition of square roots and consequently force the assignment of a principal square root?

$$(3i)(3i) = 9i^2 = 9(-1) = -9$$

while

$$(-3i)(-3i) = 9i^2 = 9(-1) = -9$$

The answer is yes, and therefore we shall again take the positive case to be the principal (complex) square root.

Exercises

For each of the following questions, answer in your own words, using complete sentences, proper grammar, spelling, and punctuation.

1. In your own words, explain what is meant by a "complex number."
2. Explain why it is not true that $\sqrt{36} = \sqrt{-36}\,\sqrt{-1} = 6\,i \cdot i = 6\,i^2 = (6)(-1) = -6$.
3. How does the imaginary part of a complex number differ from the real part?
4. In your own words, explain how one would find the simplified form of i^{43}.

Simplify the following radicals completely (remember, this does not mean decimal approximate!).

5. $\sqrt{-98}$
6. $\sqrt{-108}$
7. $\sqrt{-18}$
8. $\sqrt{-125}$
9. $\sqrt{-144}$
10. $\sqrt{-162}$

Identify the real and imaginary parts of the following complex numbers.

11. $8 + 5i$
12. $-12 + 7i$
13. $13 - 8i$
14. $-9 - 6i$
15. 5
16. $-8i$
17. $9i$
18. 21

Simplify the following powers of the imaginary unit i.

19. i^7
20. i^9
21. i^{14}
22. i^{25}
23. i^{48}
24. i^{70}
25. i^{27}
26. i^{594}
27. i^{199}

Using matrices, verify the following relationships.

28. $i^2 = -1$
29. $i^4 = 1$
30. $i^5 = i$

9.2 Algebraic Operations with Complex Numbers

When we use the term "algebra," we are referring to the four basic operations of mathematics: addition, subtraction, multiplication, and division. As we will observe, these operations applied to complex numbers mimic those used with some mathematical objects already familiar to us.

Recalling that complex numbers have the form $a + bi$, where a and b are real numbers, we can define the algebraic operations on the numbers in a reasonably natural manner. You will probably notice that the operations are closely related to operations on polynomials, at least in practice. Keep in mind that although the process mimics the work we've done with polynomials, the expressions involved are not polynomials but complex numbers. This will become important when we move on to examine the operations of multiplication and division.

Consider first the problem of adding first-degree polynomials and compare the outcome to the related matter of adding complex numbers.

EXAMPLE 9.1

Simplify this sum of polynomials $(5 + 7x) + (8 - 2x)$.

SOLUTION

We recall that the problem calls for the combination of "like terms," those terms that involve the same variable raised to the same power. In this case, the like terms are 5 and 8, and $7x$ and $-2x$. If we wish (though this is not necessary), we can reorganize the expression so the like terms are gathered together:

$$\begin{aligned}(5 + 7x) + (8 - 2x) &= (5 + 8) + (7x - 2x)\\ &= 13 + 5x\end{aligned}$$

The terms now present (not being like terms) cannot be combined, so our result is considered simplified.

Compare the previous example to the one that follows, which involves complex numbers as opposed to polynomials.

EXAMPLE 9.2

Simplify this sum of complex numbers: $(5 + 7i) + (8 - 2i)$.

SOLUTION

In a manner not unlike what we saw for addition of polynomials, addition of complex numbers is accomplished by adding the real parts and the imaginary parts. By reorganizing the terms, we obtain (much as we saw before)

$$\begin{aligned}(5 + 7i) + (8 - 2i) &= (5 + 8) + (7i - 2i)\\ &= 13 + 5i\end{aligned}$$

Let's make it concrete, with a formal definition for the process:

DEFINITION: Sum of two complex numbers in rectangular form:

$$(a + bi) + (c + di) = (a + c) + (b + d)i$$

In some examples, it is conceivable that either the real or the imaginary parts of the sum would vanish. In such cases, it is acceptable to omit the part that has vanished, unless the instructions specifically call for presentation of the solution in standard form.

As an illustration, consider the following example:

$$(5 - 7i) + (-5 + 3i) = 0 - 4i = -4i$$

If the instructions call for standard form, then the correct solution would be $0 - 4i$, but if no such instruction is given, then either form would be acceptable and considered correct.

Subtraction is handled in a similar manner, again mimicking the process we use in the subtraction of first-degree polynomials. As always, we must be cautious when encountering possible sign complications, but this is something with which we are already quite familiar.

EXAMPLE 9.3

Subtract the complex numbers $(11 - 9i) - (13 - 5i)$.

SOLUTION

As is the case with polynomials, the subtraction is distributed through the subtrahend expression, with the following result:

$$(11 - 9i) - (13 - 5i) = 11 - 9i - 13 + 5i$$

If we wish, we can gather the like terms together, but by now this should probably be apparent to us, and therefore this step ought to be easily suppressed from the written work:

$$= -2 - 4i$$

As we did with addition, we make the definition explicit:

DEFINITION: Difference of two complex numbers in rectangular form:

$$(a + bi) - (c + di) = (a - c) + (b - d)i$$

Note

In all computations involving complex numbers, if i^2 should appear, it should immediately be replaced by its equivalent form, (-1), and simplification should continue.

Multiplication and Division of Complex Numbers

Multiplication of complex numbers will also follow in the footsteps of polynomial operations, incorporating either the distributive law or the process of binomial multiplication (often referred to with the acronym FOIL). A slight complication arises at this point, which we should explicitly point out and accommodate.

Let's take a look at a few examples, and the situation should become clear.

EXAMPLE 9.4

Simplify $(9 - 3i)(5 + 7i)$.

SOLUTION

Pretending that these are binomials, we can multiply using FOIL, to obtain

$$(9 - 3i)(5 + 7i) = 45 + 63i - 15i - 21i^2$$

We note the appearance of i^2 and immediately convert it to (-1), which yields

$$45 + 63i - 15i + 21$$

Now, combining like terms, we obtain

$$66 + 48i$$

As we did with addition and subtraction, we present a formal definition of the process, which is sometimes useful in theoretical problems:

DEFINITION: Product of two complex numbers in rectangular form:

$$(a + bi)(c + di) = (ac - bd) + (ad + bc)i$$

In practice, however, the process is typically done using FOIL, as indicated in the example preceding the formal definition.

In the case where either a real number or a pure imaginary number is to be multiplied against a complex number, the distributive law is used. As in the previous cases, this is reminiscent of the work done with polynomials.

EXAMPLE 9.5

Simplify $8i(3 + 11i)$.

SOLUTION

Using the distributive law, we obtain

$$24i + 88i^2$$

which simplifies to

$$24i + 88(-1)$$
$$24i - 88$$

Technically, there is nothing wrong with the result, as stated. However, as mentioned earlier, the general agreement among mathematicians is that complex numbers will be given with the real part appearing first, so we would apply commutativity to switch the order and obtain

$$-88 + 24i$$

You may feel that this is not aesthetically pleasing, with the negative term coming first, but this is a displeasure with which we must live.

EXAMPLE 9.6

Simplify $(4 + 2i)^2$.

SOLUTION

This is a good point to remind ourselves of the common misconception known to math teachers around the world as the freshman's dream. Recall that when a binomial is raised to the second power, we do not simply square each of the terms! This expression is equivalent to $(4 + 2i)(4 + 2i)$ and so should be multiplied out using binomial multiplication:

$$(4 + 2i)(4 + 2i)$$
$$16 + 8i + 8i + 4i^2$$

Again, we recall that i^2 is equivalent to (-1) and make that substitution, obtaining

$$16 + 8i + 8i - 4$$

which we simplify to

$$12 + 16i$$

EXAMPLE 9.7

Simplify $-5i(2 - 7i)$.

SOLUTION

In this situation, we distribute the $-5i$ through the parentheses, being sure to watch our signs:

$$-10i + 35i^2$$
$$-10i + 35(-1)$$
$$-10i - 35$$

Although, as we've pointed out, there is nothing technically incorrect about this result, the general practice is to organize so that the real part is given first, so we obtain

$$-35 - 10i$$

You are, we hope, now feeling comfortable with the first three operations on complex numbers. If you have a bit of familiarity with polynomials and the algebraic operations on them, the new concepts may seem to be more than a little like a review. The final operation, division, is where we depart from the method of polynomials and introduce a new approach.

Before doing so, however, we present one more definition:

DEFINITION: Complex Conjugate Two numbers are complex conjugates if they differ only in the sign of their imaginary part. That is, $a + bi$ and $a - bi$ are complex conjugates.

For instance, $(3 + 2i)$ and $(3 - 2i)$ are conjugates, and $(4 - 7i)$ and $(4 + 7i)$ are conjugates. The expressions are duplications of one another, apart from the sign of the imaginary part of the complex number. When dividing complex numbers, the conjugate will play a critical role, as we will see in a moment.

Division problems involving the complex numbers are commonly presented in fraction form, such as this:

$$\frac{4 + 5i}{3 - 2i}$$

The process of dividing complex numbers follows this algorithm:

- Multiply the numerator and denominator by the complex conjugate of the denominator.
- Simplify numerator and denominator completely.
- Decompose the result into $(a + bi)$ form.

Let's see how this works, using the example shown above. Note that the first step requires us to identify the complex conjugate of the denominator, which is found by changing the sign of the imaginary part of the denominator, obtaining $3 + 2i$. Thus,

$$\frac{4 + 5i}{3 - 2i} \bullet \frac{3 + 2i}{3 + 2i}$$

We proceed to simplify, obtaining

$$\frac{12 + 8i + 15i + 10i^2}{9 + 6i - 6i - 4i^2}$$

Combining like terms and recalling that i^2 should be changed immediately to (-1), we simplify and find

$$\frac{2 + 23i}{13}$$

We are nearly done now. All that remains is to decompose the result into the standard form $a + bi$. We do this using the ordinary simplification rules for fractions,

$$\frac{2}{13} + \frac{23}{13}i$$

and we are finished.

As you can see, the situation is a bit more involved with division than it is with the other algebraic operations!

We'll do another example, but before doing so, we should make an important observation that is sometimes missed: complex numbers that are "pure imaginary" (have no real part) have a complex conjugate! The number $7i$, for instance, can be thought of as having the form $0 + 7i$, and therefore its conjugate is $0 - 7i$, or simply $-7i$.

EXAMPLE 9.8 Divide $\dfrac{11 - 6i}{5i}$.

SOLUTION

Following the process outlined earlier, we multiply numerator and denominator by $-5i$, the complex conjugate of the denominator:

$$\begin{aligned}\frac{11 - 6i}{5i} &= \frac{11 - 6i}{5i} \bullet \frac{-5i}{-5i}\\ &= \frac{-55i + 30i^2}{-25i^2}\\ &= \frac{-55i - 30}{25}\end{aligned}$$

Note that we have yet to decompose the expression into real and imaginary parts, but hopefully the steps to this point are apparent. We have suppressed the replacement of i^2 by -1 and have moved directly to the simplification. You may need a moment of examination to see just what was done, but be sure you follow the work!

$$\begin{aligned}\frac{-30 - 55i}{25} &= \frac{-30}{25} - \frac{55}{25}i\\ &= \frac{-6}{5} - \frac{11}{5}i\end{aligned}$$

Exercises

For each of the following questions, answer in your own words, using complete sentences, proper grammar, spelling, and punctuation.

1. Explain the process for adding two complex numbers.
2. Explain the process for multiplying two complex numbers.

3. Explain the process for dividing two complex numbers
4. What is meant by the term "complex conjugate"?

In each of following, simplify the expression completely and give the answer in standard form.

5. $(3 + 2i) + (9 + 7i)$
6. $(4 + 11i) + (7 + 5i)$
7. $(9 - 2i) + (5 + 3i)$
8. $(6 - 5i) + (19 + 8i)$
9. $(2 - 4i) - (8 + 3i)$
10. $(5 - 7i) - (10 + 5i)$
11. $(3 - 5i) - (7 - 5i)$
12. $(10 - i) - (4 - i)$
13. $(4 + 6i) - (7 + 3i)$
14. $(22 + 8i) - (10 + 5i)$
15. $(3 - 8i)(5 + 6i)$
16. $(9 - 3i)(1 + 4i)$
17. $(5 - 6i)(5 + 6i)$
18. $(3 + 2i)(3 - 2i)$
19. $(5i)(6 - 5i)$
20. $(-7i)(2 + 8i)$
21. $\dfrac{4 - 7i}{3i}$
22. $\dfrac{9 - 3i}{2i}$
23. $\dfrac{4i}{5 + 7i}$
24. $\dfrac{11i}{1 - 4i}$
25. $\dfrac{4 + 9i}{3 - 7i}$
26. $\dfrac{19 + 4i}{4 + 11i}$
27. $(3 + 2i)^2$
28. $(4i - 5)^2$

9.3 Graphical Representation of a Complex Number

We have seen that there are occasions when a graphical approach to problems will yield results that are obscured by computation. Consequently, it would be desirable if there were a visual method for representing complex numbers.

Before beginning the discussion, we should mention what we mean by "representation." We are referring to an alternative method of looking at the complex numbers through which we might gain insight into more of their properties. Just as we can look at rational numbers as decimal numbers, we can investigate the complex numbers in alternative forms, such as the matrix form mentioned earlier. In this case, the form will be graphical.

The Argand Diagram

To accomplish this, we will revise our view of the Cartesian plane, changing the x-axis into the **real axis** and the y-axis into the **imaginary axis**. When we do so, we find that we can associate the complex number $a + bi$ with the arrow (or ray) from the origin to the point in this **complex plane** having coordinates (a, b). This representation concept is due to Jean-Robert Argand, and consequently such a construction is often referred to as an **Argand diagram**.

As an example, consider the complex number $5 + 3i$. We associate this complex number with the arrow from the origin to the point (5, 3) in the complex plane and can visualize it by including an arrow from the origin to that point within the plane (Figure 9.1).

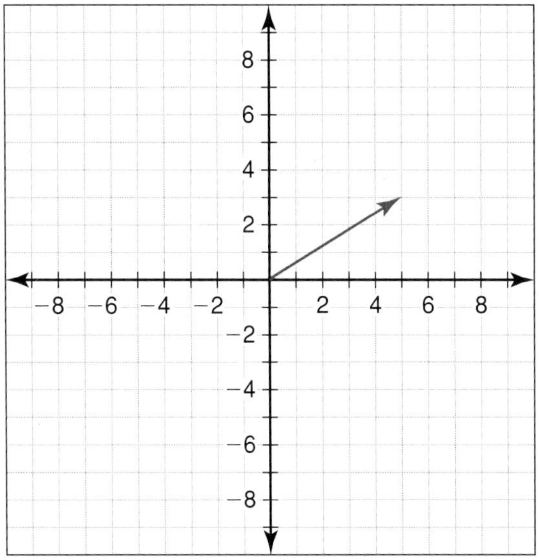

Figure 9.1 Graphical representation of $5 + 3i$

Now consider a second complex number, such as $(-2 + 4i)$, which we will add to the first. Our rules of complex number arithmetic tell us the result would be $3 + 7i$, but what does this mean from a graphical perspective?

Suppose we draw our second complex number in a "nontraditional" manner, pretending that our origin is now sitting at the terminal point of the arrow we just drew. If you think of rectangular coordinates, then to sketch $(-2 + 4i)$, we would move left two units and upward four units. Note where the "terminal point" of this motion would place us (Figure 9.2). It would place us precisely at the point (3, 7), which is the analog of the complex number $3 + 7i$. This method of addition for complex numbers presented graphically in the Argand diagram is referred to as the "tip-to-tail" addition method, or the "triangular" addition method.

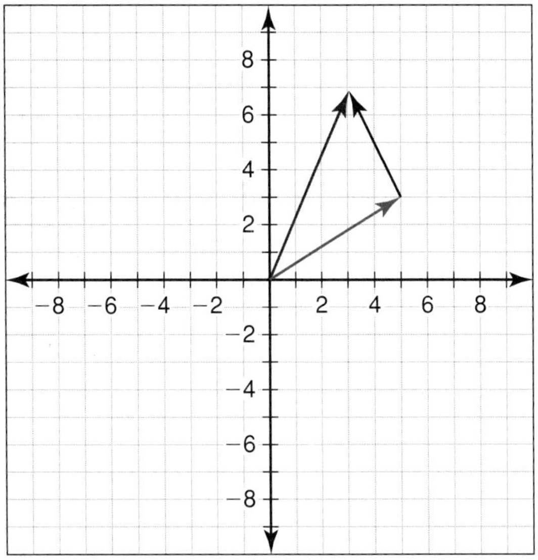

Figure 9.2 Adding vectors using the tip-to-tail method

EXAMPLE 9.9

Perform the complex addition $(-1 + 4i) + (5 + i)$ and then show the equivalent result graphically.

SOLUTION

Performing the addition as a purely arithmetical problem is a triviality, and by inspection we can assert the solution to be $4 + 5i$. Consider the situation from the graphical perspective, where we perform the process by visualizing arrows in the complex plane.

Here, we take the view that the first complex number is associated with the arrow connecting the origin to the point $(-1, 4)$ in the plane, to which we append a second arrow (which is the representation of the complex number $5 + i$ shifted to a position where its initial point coincides with the terminal point of the first arrow). We see the terminal point of the second arrow is located at the point $(4, 5)$ and now draw the arrow depicting the sum of the complex numbers connecting the origin to that point (Figure 9.3).

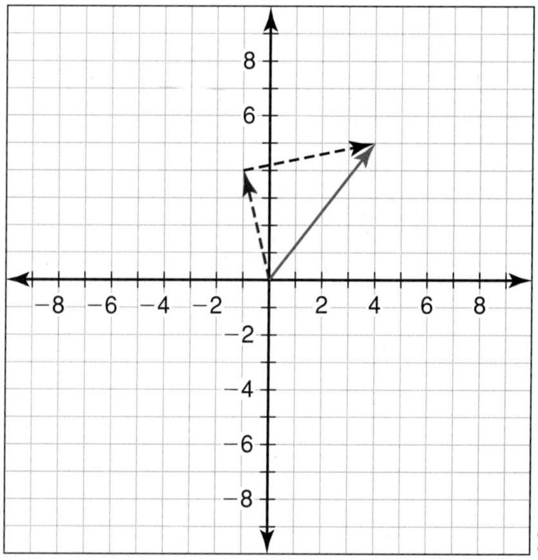

Figure 9.3 Adding vectors using the tip-to-tail method

When we add the complex numbers in this visual sense, it seems natural to wonder if there is any relationship between the angles of the arrows of the individual complex numbers and the angle of the arrow of the sum. Recalling some of our information from the chapter on trigonometry, let's investigate.

In the previous example, the first arrow had the point $(-1, 4)$ on its terminal side, and therefore (if we identify the angle the arrow makes with the positive x-axis as θ we can say that $\tan\theta = \dfrac{4}{-1}$ and use our results from trigonometry to conclude that $\theta \approx 1.326$ radians. The second arrow (if drawn in standard position) would have the point $(5, 1)$ on its terminal side, and (if we identify the angle the arrow makes with the positive x-axis as ϕ) we can conclude that $\tan\phi = \dfrac{1}{5}$, which implies $\varphi \approx 0$. The angle of the sum's arrow, by a similar computation, is approximately 0 radians. Thus, there seems to be no relationship between the angles of the individual arrows and the angle of their sum—at least, not any that we can discern at this time.

The preceding investigation illustrates a valuable mathematical and scientific technique we should always seek to foster: there are times when it appears that a particular operation may demonstrate some discernible properties. Experimentation may or may not reveal the existence of suspected relationships, but it is only through examination that we can gather evidence supporting or refuting those suspicions (and, we should point out, those experiments will not, themselves, prove the existence of a relationship; rather they will only give evidence to support the theory that the relationship may exist!).

EXAMPLE 9.10

Calculate the complex number product $(2 + i)(i)$ and then compare the individual complex numbers and their product from a graphical perspective.

Note

When a complex number is multiplied by *i*, the arrow representing the original complex number is rotated 90° degrees in a counterclockwise direction to obtain the arrow representing the product complex number, and the length of the product matches exactly the length of the original arrow. More will be said of this shortly.

SOLUTION

The product is easy enough to calculate, since it requires us merely to use the distributive property of multiplication. The result is $2i + i^2$, which simplifies to give $2i - 1$, or the more commonly given $-1 + 2i$.

Graphically, let's draw the individual arrows representing the complex numbers, in light brown and dark brown, respectively, and their product in black (Figure 9.4). The relationship between the arrows is not, perhaps, immediately clear unless we inspect the slope of the light brown arrow (representing $2 + i$), which is $\frac{1}{2}$, and the slope of the product arrow (representing $-1 + 2i$), which is -2. From our experience with the graphs of lines, we recall that lines whose slopes multiply together and produce a product of -1 are perpendicular to one another.

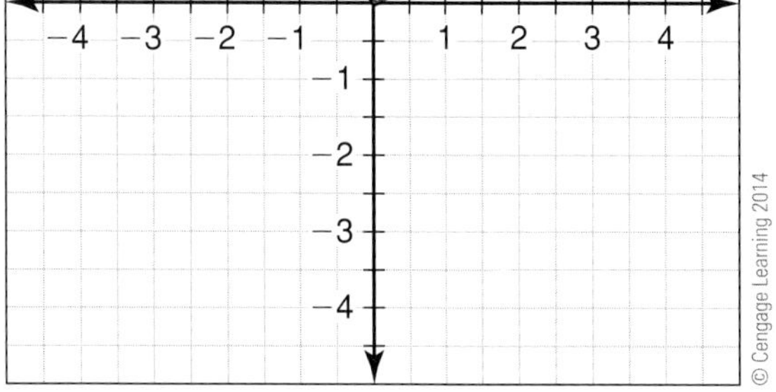

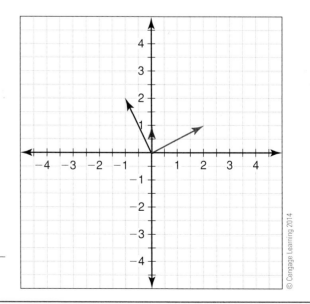

Figure 9.4 The vectors $-1 + 2i$ and $2 + i$

Exercises

Answer the following questions in your own words, using proper grammar, correct terminology, and complete sentences.

1. What is meant by an "Argand diagram"?
2. When using an Argand diagram, what is the meaning of the values on the horizontal axis?
3. When using an Argand diagram, what is the meaning of the values on the vertical axis?
4. Explain the tip-to-tail addition method of complex numbers and create an example of your own that illustrates the procedure.

Represent the following complex numbers on an Argand diagram.

5. $4i$
6. $-2i$
7. 3
8. 5
9. $3 + 7i$
10. $-2 + 5i$
11. $-7 - 3i$
12. $-4 - 5i$

In each of the following problems, calculate the indicated sum algebraically and then graphically in an Argand diagram. Show that the results agree.

13. $(3 + 2i) + (4 + i)$
14. $(5 + i) + (-3 + 2i)$
15. $(5) + (6 - i)$
16. $(-4) + (4 + 3i)$
17. $(-6 - 2i) + (8 - 3i)$
18. $(-1 - 5i) + (5 + 4i)$

For each of the following, calculate the complex product. Sketch the individual complex numbers and their product on an Argand diagram. In each case, find the arguments of the individual complex numbers being multiplied and also the argument of the product. Describe the impact of multiplying the first complex number by the second.

19. $(5 - 2i)(i)$
20. $(7 + 4i)(i)$
21. $(5 + 6i)(-1)$
22. $(3 - 2i)(-1)$
23. $(4 + 5i)(-i)$
24. $(6 - 3i)(-i)$

9.4 Other Forms of Complex Numbers: Polar, Trigonometric, and Exponential

The Polar Form of a Complex Number

There are many applications of mathematics in which it is awkward to work with our familiar x- and y-coordinates. These are primarily the situations in which the graphs fail to pass the vertical line test and hence are not functions. In those cases, a new method for mapping out the plane is introduced, called **polar coordinates**.

In polar coordinates, every point in the plane is identified using an ordered pair (r, θ), where r is the **directed distance** from the origin to the point and θ is

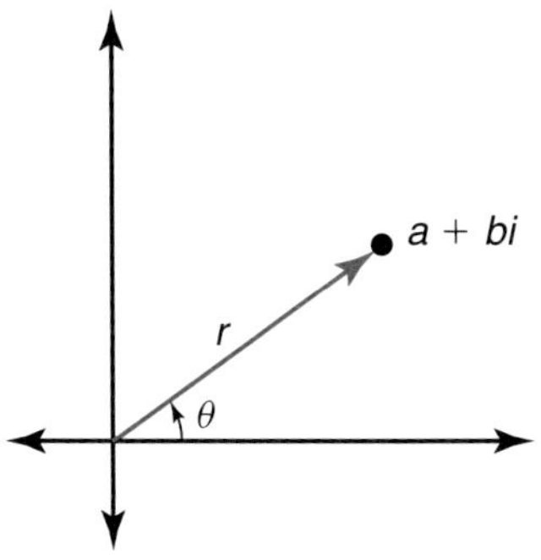

Figure 9.5 The argument and modulus of a complex number

the angle between the positive x-axis and the arrow connecting the origin to the point. In polar coordinates, the angle θ is typically given in radians, but when working with complex numbers, this is often relaxed, and we allow ourselves to use either degrees or radians as our form of angle measure.

Directed distance is a notion that arbitrarily assigns a particular direction to be considered to be "positive," implying that the precise opposite direction is to be considered "negative." In the present context, we imagine that we are sitting at the origin, facing out along the arrow pointing to the point in the plane, and the direction forward, toward the point, is considered positive. The direction precisely backward from that which we are facing is considered to be negative.

Consider now the diagram, which depicts the complex number $a + bi$ on an Argand diagram, with the angle θ displayed (Figure 9.5). If we consider the light brown arrow from the origin to (a, b) to have length r, then we can express the complex number in the form $r\angle\theta$, and this is called the **polar form of a complex number**. The representation is unique provided that we limit the angle θ to be in the interval $[0, 2\pi)$, or (in degrees) in the interval $[0, 360°)$. The value of r is also called the **modulus of a complex number**, and the angle θ is called the **argument of a complex number**.

EXAMPLE 9.11

Give the graphic representation of the complex number given in polar form by $10\angle 40°$.

SOLUTION

In this case, we merely need to draw an arrow, 10 units long, at an angle of 40° with the positive x-axis. Using a protractor, we can construct an angle of appropriate measure, and then with a ruler we can find the length of the arrow from the origin to the point representing the complex number in the Argand diagram (Figure 9.6). Note we have created a circular grid about the origin in this case in order to highlight the distance from the origin. The arrow shown has length 10 and is situated in such a manner that the angle it makes with the positive x-axis is 40°.

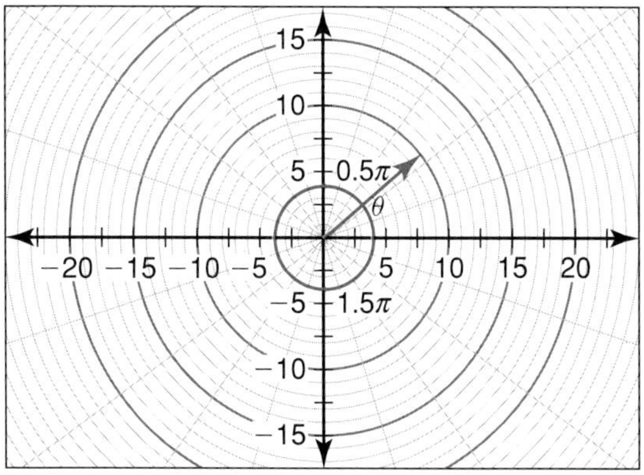

Figure 9.6 The polar representation of a complex number

The polar representation of the complex number is particularly convenient if we want to locate the point corresponding to the complex number in the complex plane. Naturally, if the angle is given in radians, we may have a slightly more difficult time, since radians are somewhat less familiar than degrees, but this can always be overcome by converting the angle to degrees, if desired, and then plotting.

We should address the problem of conversion from rectangular form to polar form and vice versa, and it is to this question we'll now turn. To accomplish these conversions, a certain amount of work must be done, and some care must be taken. The calculations will draw on our algebraic experience as well as our investigations of trigonometry, and we'll see that sometimes the situation is complicated by the realization that our calculators can lead us astray.

First, let's address the precise method for converting from rectangular form to polar form. This is generally the tougher of the two conversions, since this is where our calculators can (in a sense) betray us. The value of r is not where the complication arises, since r is the distance from the origin to the point (a, b), and the distance formula tells us that this is $r = \sqrt{(a - 0)^2 + (b - 0)^2} = \sqrt{a^2 + b^2}$, which we should give either exactly or as a decimal approximation (depending on what is called for, or the value implied by the context of the problem).

The source of potential trouble is the determination of the argument of the complex number. A natural method for calculating θ would be to use the trigonometric relationship that $\tan\theta = \frac{b}{a}$, and hence $\theta = \tan^{-1}\left(\frac{b}{a}\right)$. This is perfectly fine provided that the graphical representation of the complex number lies in quadrant I because for such complex numbers, the inverse tangent function will produce precisely the angular value desired. If the complex number's graphical representation lies in the second or third quadrant, then we must modify the result of inverse tangent, using $\theta = \left(\tan^{-1}\frac{b}{a}\right) + \pi$, while if the graphical representation lies in the fourth quadrant, we must use $\theta = \left(\tan^{-1}\frac{b}{a}\right) + 2\pi$.

The situation here is sufficiently complex that we present it in tabular form for reference.

TABLE 9.1 Method for Finding r and θ, Given the Rectangular Form of a Complex Number

Quadrant of Graphical Representation of Complex Number	Formula for r and θ of Polar Representation of the Complex Number
I	$r = \sqrt{a^2 + b^2},\ \theta = \tan^{-1}\left(\frac{b}{a}\right)$
II	$r = \sqrt{a^2 + b^2},\ \theta = \left(\tan^{-1}\frac{b}{a}\right) + \pi$
III	$r = \sqrt{a^2 + b^2},\ \theta = \left(\tan^{-1}\frac{b}{a}\right) + \pi$
IV	$r = \sqrt{a^2 + b^2},\ \theta = \left(\tan^{-1}\frac{b}{a}\right) + 2\pi$

EXAMPLE 9.12 For the complex number $-3 + 5i$, find the polar representation, measuring the argument in degrees.

SOLUTION

We can plot the point easily enough, since it corresponds to the point $(-3, 5)$ in an Argand diagram, and find that it lies in quadrant II of the plane.

Following the reasoning preceding the example, we can see that the value of r is found according to the computation $r = \sqrt{9 + 25} = \sqrt{34}$. The angle θ is found using $\theta = \left(\tan^{-1}\frac{b}{a}\right) + 180° \approx -59.03° + 180° \approx 120.97°$.

Therefore, the polar representation of $-3 + 5i$ is $\left(\sqrt{34},\ 120.97°\right)$.

EXAMPLE 9.13 Find the polar representation of the complex number $16 - 4i$, measuring the argument in radians.

SOLUTION

Again, we visualize the complex number in the Argand diagram and recognize that it lies in quadrant IV of the plane. Referencing Table 9.1, we see that

$$r = \sqrt{16^2 + (-4)^2} = \sqrt{256 + 16} = \sqrt{272} \approx 16.49 \text{ and that}$$

$$\theta = \left(\tan^{-1}\frac{-4}{16}\right) + 2\pi \approx 6.04$$

Therefore, the polar representation of $16 - 4i$ is approximately $16.49\angle 6.04$.

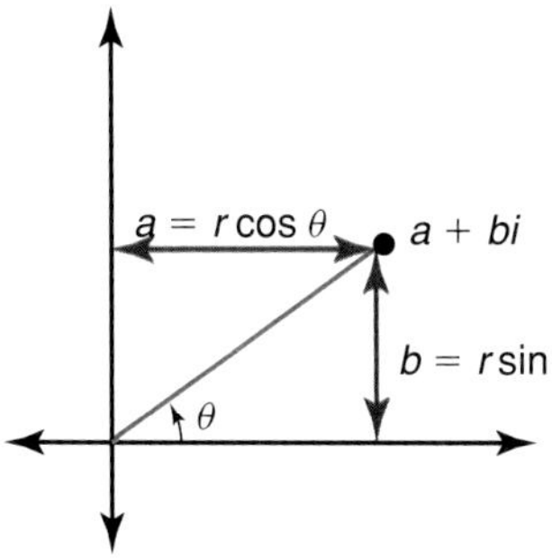

Figure 9.7 The relationship between the standard and the polar representations of a complex number

Conversion from polar representation to rectangular representation is less daunting, because we needn't worry about the eccentricities presented by the use of the inverse tangent function. If we extend a vertical line from the point (a, b) in the Argand diagram to the x-axis, right-triangle trigonometry tells us $a = r\cos\theta$, $b = r\sin\theta$, and therefore, the rectangular form of the complex number is found by substitution of those values of a and b into $a + bi$ (Figure 9.7).

EXAMPLE 9.14 For the complex number given in polar representation as $16\angle 110°$, find the standard form representation, rounding the values of a and b to two decimal places.

SOLUTION

Here only two computations are needed: $a = 16\cos 110° \approx -5.47$, and $b = 16\sin 110° \approx 15.04$. These values tell us that the standard representation of the complex number is approximately $-5.47 + 15.04i$.

EXAMPLE 9.15 For the complex number given in polar form by $13\angle\frac{2\pi}{3}$, find the exact rectangular form representation.

SOLUTION

The argument of this complex number is one of the known trigonometric angles, and this permits us to find the exact rectangular form of the complex number. You may have to review the section on trigonometry to see this, but $\cos\frac{2\pi}{3} = -\frac{1}{2}$ and $\sin\frac{2\pi}{3} = \frac{\sqrt{3}}{2}$, and thus the rectangular form of $13\angle\frac{2\pi}{3}$ is $-\frac{13}{2} + \frac{13\sqrt{3}}{2}i$.

The Trigonometric Form of a Complex Number

Once we discovered the formulas for calculating the values of a and b from the polar form of a complex number, we stated that we could substitute those values into $a + bi$ to obtain rectangular form, or standard form. However, if we consider the abstract version of that procedure, $(r\cos\theta) + (r\sin\theta)i$, we have a form that is referred to as the **trigonometric form of a complex number**. This is actually an adjunct to the polar form, but it is used so often that it is deserving of its own technical name. Occasionally, since there is a common factor of r that can be factored from the trigonometric form, we see this written as $r(\cos\theta + i\sin\theta)$, and this is in turn is sometimes shortened to read $r\cos\theta$.

The Exponential Form of a Complex Number

We already have four representations of complex numbers (rectangular, graphic, polar, and trigonometric), so it may seem to be overkill to present a fifth, but there are particular problems that are addressed more easily in yet another context from those already presented. This new form is called the exponential form and is a consequence of a result by the great Swiss mathematician Leonhard Euler, who made the following observation in the eighteenth century: $e^{i\theta} = \cos\theta + i\sin\theta$. The result is a consequence of the MacLaurin series for the sine and cosine functions, proof of which would take us far afield from our interests here.

Calling to mind the trigonometric form of the complex numbers, $r(\cos\theta + i\sin\theta$, application of Euler's formula produces the **exponential form of a complex number** to be $re^{i\theta}$. Note that, although polar and trigonometric forms of complex numbers had the freedom to use either radian or degree measure, the exponential form is valid only for radian measure of angles. The availability of this form will have significant consequences for applications of complex numbers.

EXAMPLE 9.16

Express $5(\cos 53° + i\sin 53°)$ in exponential form.

SOLUTION

Because the exponential form is valid only for radian measurement of angles, our first step is to convert the angle to radians. Recalling our method for conversion presented in Chapter 8, we see

$$\frac{53°}{1} \times \frac{\pi}{180°} \approx 0.29\pi$$

Therefore, direct substitution yields $5(\cos 53° + i\sin 53°) \approx e^{i(0.29\pi)}$.

EXAMPLE 9.17 Express the complex number $17e^{i(1.4\pi)}$ in rectangular, polar, and trigonometric forms.

SOLUTION

The trigonometric form is immediate, since we can directly substitute the values of r and θ to obtain $17(\cos 1.4\pi + i\sin 1.4\pi)$. The polar form is also easily obtained and is $17\angle 1.4\pi$. The rectangular form of the complex number is a bit harder to obtain, but we recall our conversions from polar to rectangular forms, $(r \cos \theta) + (r \sin \theta)i$, and merely need to insert the values of r and θ. This produces $17(\cos 1.4\,\pi) + (i \sin 1.4\,\pi)i \approx -5.25 - 16.17i$.

Multiplication and Division Using the Polar, Trigonometric, and Exponential Forms

It seems natural, once we've defined the polar, trigonometric, and exponential forms of the complex numbers, to ponder some of the operations we've performed on the numbers and to determine how those operations would be performed from these alternative points of view. It may be, after all, that the operations are simpler within another framework, and this is worthy of investigation.

Polar and trigonometric functions are so closely related that we may as well consider them together. The product of two complex numbers, $r_1\angle\theta_1$ and $r_2\angle\theta_2$, can be calculated readily—and with surprising results:

$$\begin{aligned}(r_1\angle\theta_1)(r_2\angle\theta_2) &= r_1(\cos\theta_1 + i\sin\theta_1)\,r_2(\cos\theta_2 + i\sin\theta_2)\\ &= r_1r_2(\cos\theta_1 + i\sin\theta_1)(\cos\theta_2 + i\sin\theta_2)\\ &= r_1r_2(\cos\theta_1\cos\theta_2 + i\cos\theta_1\sin\theta_2\\ &\quad + i\sin\theta_1\cos\theta_2 - \sin\theta_1\sin\theta_2)\\ &= r_1r_2[\cos\theta_1\cos\theta_2 - \sin\theta_1\sin\theta_2\\ &\quad + i(\cos\theta_1\sin\theta_2 + \sin\theta_1\cos\theta_2)]\end{aligned}$$

The sum of angles trigonometric identities for sine and cosine state the following:

$$\begin{aligned}\cos(\theta_1 + \theta_2) &= \cos\theta_1\cos\theta_2 - \sin\theta_1\sin\theta_2\\ \sin(\theta_1 + \theta_2) &= \sin\theta_1\cos\theta_2 + \cos\theta_1\sin\theta_2\end{aligned}$$

application of which yields

$$\begin{aligned}(r_1\angle\theta_1)(r_2\angle\theta_2) &= r_1r_2[\cos(\theta_1 + \theta_2) + i\sin(\theta_1 + \theta_2)]\\ &= r_1r_2\angle(\theta_1 + \theta_2)\end{aligned}$$

That is, to multiply two complex numbers, it is enough to multiply their moduli and add their arguments.

If we perform the multiplication of two complex numbers given in exponential form and apply the familiar rule for multiplication of exponential expressions ($a^m a^n = a^{m+n}$), we find

$$r_1e^{i\theta_1}r_2e^{i\theta_2} = r_1r_2\,e^{(i\theta_1 + i\theta_2)} = r_1r_2\,e^{i(\theta_1 + \theta_2)}$$

Note that this tells us, interestingly, that the polar, trigonometric, and exponential multiplication results agree!

EXAMPLE 9.18 Multiply the polar form complex numbers $17\angle 48°$ and $-3\angle 12°$.

SOLUTION

We simply multiply the moduli and add the angles to obtain $-51\angle 60°$.

EXAMPLE 9.19 Multiply the exponential form complex numbers $5e^{i(2.3)}$ and $2e^{i(0.9)}$.

SOLUTION

Applying the method for multiplication of complex numbers in exponential form, we find the product to be $10e^{i(3.2)}$.

If we carry multiplication to its extension of exponentiation with positive powers, we can generalize the results found to this point. This result is called DeMoivre's theorem:

$$\left[re^{i\theta}\right]^n = r^n e^{i(n\theta)}$$

or, equivalently,

$$[r(\cos\theta + i\sin n\theta)]^n = r^n[r(\cos(n\theta) + i\sin n\theta)]$$

EXAMPLE 9.20 Simplify $\left[2e^{i\left(\frac{\pi}{3}\right)}\right]^6$.

SOLUTION

Recalling the product and power rules for exponents, we find

$$\left[2e^{i\left(\frac{\pi}{3}\right)}\right]^6 = 2^6\left[e^{i\left(\frac{\pi}{3}\right)}\right]^6 = 64e^{i(2\pi)}$$

Since $e^{i(2\pi)} = \cos(2\pi) + i\sin(2\pi) = 1 + i(0) = 1$, the result simplifies further to 64. It may be somewhat surprising that a complex number raised to a particular power can be simplified to a real-numbered result! On reflection, this is not without precedent, since $i^2 = -1$ also has this property, but it is somewhat unexpected.

Division of complex numbers in these alternative forms can be examined in a similar manner, and we will leave the derivation for both forms to the exercises. The results for division are

$$\frac{r_1\angle\theta_1}{r_2\angle\theta_2} = \frac{r_1}{r_2}\angle(\theta_1 - \theta_2) \text{ or } \frac{r_1e^{i\theta_1}}{r_2e^{i\theta_2}} = \frac{r_1}{r_2}e^{i(\theta_1 - \theta_2)}$$

EXAMPLE 9.21 Divide the complex numbers $\dfrac{3\angle 48°}{6\angle 16°}$.

SOLUTION

Remember, the division is accomplished by dividing the moduli of the numbers and subtracting the angle measures. Therefore, the result is $\frac{1}{2}\angle 32°$.

EXAMPLE 9.22 Divide the complex numbers $\dfrac{14e^{i(3\pi)}}{2e^{i\left(\frac{\pi}{2}\right)}}$.

SOLUTION

In exponential form, we need to reduce the ratio of the moduli and subtract the angles according to the quotient rule of exponents. This results in $7e^{i\left(3\pi - \frac{\pi}{2}\right)} = 7e^{i\left(\frac{5\pi}{2}\right)}$.

Exercises

Answer the following questions in your own words, using proper grammar, correct terminology, and complete sentences.

1. Describe what is meant by "polar coordinates," and explain how they differ from rectangular $[(x, y)]$ coordinates.
2. What is the difference between the argument and the modulus of a complex number?
3. How does the trigonometric form of a complex number differ from the polar form of the complex number, and how do we obtain one form from the other?
4. Explain the exponential form of a complex number and how it is related to the trigonometric form of the complex number.

In each of the following, the polar form of a complex number is given. Give the graphic representation of the complex number.

5. $13\angle 50°$
6. $10\angle 120°$
7. $25\angle 210°$
8. $8\angle 270°$
9. $20\angle 300°$
10. $5\angle 45°$

Find the polar representation of the following complex numbers, measuring the angle in both degrees (rounded to the nearest tenth of a degree or radian, as needed).

11. $5 - 2i$
12. $4 + 3i$
13. $8 - 4i$
14. $-2 + 3i$
15. $-5 - 8i$
16. $-1 + i$
17. $3 - i$
18. $\frac{1}{2} - \frac{2}{3}i$
19. $\frac{3}{5} + \frac{4}{5}i$
20. i
21. $4i$
22. 5
23. -7
24. $-6 + 8i$

For the following complex numbers, given in polar representation, find the standard form representation, rounding the values of a and b to two decimal places, as needed.

25. $8\angle 50°$
26. $9\angle 135°$
27. $25\angle 180°$
28. $12\angle 90°$
29. $26\angle 220°$
30. $15\angle 245°$
31. $50\angle 300°$
32. $45\angle 315°$

For the following complex numbers, given in polar representation, find the exact rectangular form representation.

33. $5\angle\frac{\pi}{4}$
34. $4\angle\frac{3\pi}{4}$
35. $9\angle\frac{\pi}{6}$
36. $11\angle\frac{5\pi}{6}$
37. $17\angle\frac{5\pi}{3}$
38. $2\angle\frac{7\pi}{6}$

The following problems reference previous problems. Give the trigonometric form of the complex numbers in those problems.

39. Problem 25
40. Problem 26
41. Problem 27
42. Problem 28
43. Problem 29
44. Problem 30
45. Problem 31
46. Problem 32

The following problems reference previous problems. Give the exponential form of the complex numbers in those problems.

47. Problem 25
48. Problem 26
49. Problem 27
50. Problem 28
51. Problem 29
52. Problem 30
53. Problem 31
54. Problem 34

Express the following complex numbers, given in exponential form, in the corresponding rectangular, polar, and trigonometric forms (rounding components of the respective forms to two decimal places, as needed).

55. $4e^{i\pi}$
56. $3e^{i\left(\frac{3\pi}{2}\right)}$
57. $7e^{i\left(\frac{2\pi}{3}\right)}$
58. $11e^{i\left(\frac{7\pi}{6}\right)}$
59. $5e^{i(2.4\pi)}$
60. $6e^{i(3.1\pi)}$

In the following problems, find the product of the given complex numbers.

61. $5\angle 25°$ and $7\angle 45°$
62. $9\angle 105°$ and $4\angle 60°$

63. $9\angle 75^\circ$ and $4\angle 100^\circ$

64. $3\angle 0^\circ$ and $15\angle 180^\circ$

65. $4e^{i(4.5)}$ and $6e^{i(5.25)}$

66. $7e^{i(3.14)}$ and $15e^{i(1.57)}$

Simplify the following, using DeMoivre's theorem.

67. $\left[3e^{i\left(\frac{\pi}{3}\right)}\right]^5$

68. $\left[7e^{i\left(\frac{\pi}{3}\right)}\right]^2$

69. $\left[2e^{i\left(\frac{2\pi}{3}\right)}\right]^4$

70. $\left[3e^{i\left(\frac{5\pi}{4}\right)}\right]^3$

Divide the following complex numbers.

71. $\dfrac{5\angle 72^\circ}{10\angle 38^\circ}$

72. $\dfrac{20\angle 100^\circ}{5\angle 45^\circ}$

73. $\dfrac{35\angle 30^\circ}{7\angle 15^\circ}$

74. $\dfrac{45\angle 68^\circ}{6\angle 28^\circ}$

9.5 Applications of Complex Numbers

Imaginary Numbers and Their Place in the Real World

When we refer to the complex numbers with terms such as "imaginary" to distinguish them from the familiar "real" numbers we've always known, it is natural to wonder whether these new numbers have any application to our very real world. It is possible, of course, that they could be esoteric creations of mathematicians, interesting to study because of their patterns and behaviors but not terribly useful as a matter of practicality.

While that concern is justifiable, it turns out to be unwarranted because the complex numbers turn out to have a variety of applications in our real world. The concept appears within the study of electrical current, heat flow, structural design, and the flow of liquids around obstacles, just to name a few. In fact, many applications in which a particular quantity fluctuates in a regular manner can be modeled through the use of complex numbers (in a way that is easier to work with than if we were to restrict our attention to only real numbers). We'll take a look at a couple of examples and see how our work with imaginary numbers can tell us something about a real-life situation.

Electrical Impedance

When we use information that expresses a single dimension, such as distance, we say that information is a scalar quantity. Examples of scalar quantities would be the temperature in a room, the length of a beam, or the amount of current passing through a wire.

During investigation of alternating current (AC) circuits, engineers found that the quantities of voltage, current, and impedance (the AC analog of resistance) were not the one-dimensional quantities observed in the direct current circuits because they now possessed other dimensions (phase shift and frequency) that changed their nature. Consequently, to study AC circuits, higher-dimensional mathematics had to be used in order to express the two dimensions of frequency and phase shift simultaneously.

It turned out that the formula relating voltage, current, and impedance (whose derivation is beyond the scope of this text) was given by an alternative form of Ohm's law, $E = IZ$, where E, I, and Z represent the voltage, current, and impedance, respectively. Each quantity is complex because of its very nature; for instance, if we consider the impedance, it has complex representation $R\angle\theta$, where R is the magnitude of the impedance (the ratio of the voltage amplitude to the current amplitude) and θ is the phase shift by which the current is ahead of the voltage.

EXAMPLE 9.23 Suppose the current in a circuit is $3 + 2i$ amps and the impedance is $3 - i$ ohms. What is the voltage?

SOLUTION

Using the formula $E = IZ$, we find

$$E = (3 + 2i)(3 - i)$$
$$E = 9 - 3i + 6i - 2i^2$$
$$E = 9 + 2 + 3i$$
$$E = 11 + 3i$$

Thus, the voltage is $11 + 3i$ volts. In polar form, this equates to $\sqrt{130}\angle 15.26°$, which tells us (in a real sense) that the voltage is $\sqrt{130}$ volts, with a phase shift of 15.25°.

EXAMPLE 9.24 If the voltage in a circuit is $25 - 5i$ volts and the impedance is $1 + 5i$ ohms, what is the current?

SOLUTION

Using $E = IZ$, we can solve for the current as $E/Z = I$, and thus

$$\frac{25 - 5i}{1 + 5i} = I$$
$$\frac{25 - 5i}{1 + 5i}\,\frac{1 - 5i}{1 - 5i} = I$$
$$\frac{25 - 125i - 5i + 25i^2}{1 - 25i^2} = I$$
$$\frac{25 - 25 - 130i}{1 + 25} = I$$
$$\frac{-130}{26}i = I$$
$$\frac{-65}{13}i = I$$
$$I = -5i$$

so the current is $-5i$ amps.

Note that the last example tells us that the current is imaginary! That means not that there is no current but rather that the current has a phase shift of $-90°$ from the voltage and that the magnitude of the current is 5 amps.

Finding All *n*th Roots of a Real Number

The fundamental theorem of algebra tells us that a polynomial equation of degree n must have precisely n solutions if we include complex numbers and multiplicities. In particular, an equation of the form $x^n = c$ (where c is some real number) has precisely n solutions, and solving this equation is tantamount to finding the nth roots of the real number c.

The trouble with this rather simple-sounding problem is that solving this equation requires us to rewrite the equation in the form $x^n - c = 0$ and then produce a factorization of the nth-degree polynomial to the left of the equals sign. This, as you may recall from algebra, can be a nightmarish experience, and the mere suggestion of such an undertaking has driven even the best of students into a frenzy.

Fortunately, the problem (when viewed from the perspective of complex numbers) is a near triviality now that we have DeMoivre's theorem available. Recall that the result told us how to exponentiate complex numbers in exponential form:

DeMoivre's Theorem (Exponential Form)

$$\left[re^{i\theta}\right]^n = r^n e^{i(n\theta)}$$

Nothing was said in the theorem about the exponent n being integer valued, so we can apply it to the case in hand, creating a surprisingly simple solution to what had appeared to be a very complex problem once we make one more nonobvious observation.

We want to find the six roots of 64, so we will take six equivalent views of that number—$64 = 64\,e^{i(0)} = 64\,e^{i(2\pi)} = 64\,e^{i(4\pi)} = 64\,e^{i(6\pi)} = 64\,e^{i(8\pi)} = 64\,e^{i(10\pi)}$—and now raise the number (in all its forms) to the $\frac{1}{6}$ power using DeMoivre's theorem:

$$\left[64e^{i(0)}\right]^{\frac{1}{6}} = 64^{\frac{1}{6}}\,e^{i\left(\frac{0}{6}\right)} = 2e^{i(0)} = 2[\cos(0) + i\sin(0)]$$
$$= 2(1 + 0i) = 2$$

$$\left[64e^{i(2\pi)}\right]^{\frac{1}{6}} = 64^{\frac{1}{6}}\,e^{i\left(\frac{2\pi}{6}\right)} = 2e^{i\left(\frac{\pi}{3}\right)} = 2\left[\cos\left(\frac{\pi}{3}\right) + i\sin\left(\frac{\pi}{3}\right)\right]$$
$$= 2\left(\frac{1}{2} + \frac{\sqrt{3}}{2}i\right) = 1 + \sqrt{3}i$$

$$\left[64e^{i(4\pi)}\right]^{\frac{1}{6}} = 64^{\frac{1}{6}}\, e^{i\left(\frac{4\pi}{6}\right)} = 2e^{i\left(\frac{2\pi}{3}\right)} = 2\left[\cos\left(\frac{2\pi}{3}\right) + i\sin\left(\frac{2\pi}{3}\right)\right]$$

$$= 2\left(-\frac{1}{2} + \frac{\sqrt{3}}{2}\right) = -1 + \sqrt{3}i$$

$$\left[64e^{i(6\pi)}\right]^{\frac{1}{6}} = 64^{\frac{1}{6}}\, e^{i\left(\frac{6\pi}{6}\right)} = 2e^{i(\pi)} = 2[\cos(\pi) + i\sin(\pi)]$$

$$= 2(-1 + 0i) = -2$$

$$\left[64e^{i(8\pi)}\right]^{\frac{1}{6}} = 64^{\frac{1}{6}}\, e^{i\left(\frac{8\pi}{6}\right)} = 2e^{i\left(\frac{4\pi}{6}\right)} = 2\left[\cos\left(\frac{4\pi}{3}\right) + i\sin\left(\frac{\pi}{3}\right)\right]$$

$$= 2\left(-\frac{1}{2} - \frac{\sqrt{3}}{2}i\right) = -1 - \sqrt{3}i$$

$$\left[64e^{i(10\pi)}\right]^{\frac{1}{6}} = 64^{\frac{1}{6}}\, e^{i\left(\frac{10\pi}{6}\right)} = 2e^{i\left(\frac{5\pi}{3}\right)} = 2\left[\cos\left(\frac{5\pi}{3}\right) + i\sin\left(\frac{5\pi}{3}\right)\right]$$

$$= 2\left(\frac{1}{2} - \frac{\sqrt{3}}{2}i\right) = 1 - \sqrt{3}i$$

Note that this produces all six of the roots of 64, including the two real roots. We can further simplify the process by recognizing that the arguments of the nth roots can be found easily by dividing 2π by n, starting at 0 and adding $\frac{2\pi}{n}$ to it until one revolution around the complex plane is completed.

EXAMPLE 9.25 Find the fifth roots of 32.

SOLUTION

We know, from the fundamental theorem of algebra, that there are precisely five roots to find, and we do not know (at the outset) how many of them are real and how many of them are complex. Our roots will be (by DeMoivre's theorem) given by

$$2e^{i(0)},\ 2e^{i\left(\frac{2\pi}{5}\right)},\ 2e^{i\left(\frac{4\pi}{5}\right)},\ 2e^{i\left(\frac{6\pi}{5}\right)},\ 2e^{i\left(\frac{8\pi}{5}\right)}$$

If we allow ourselves to decimal approximate the values of sine and cosine, we can express these in standard form as

2, $(0.309 + 0.951i)$, $(-0.809 + 0.588i)$, $(-0.809 - 0.588i)$, and $(0.309 - 0.951i)$

Fractal Image Generation

A full discussion of fractals would take us extremely far from our intent here, but informally we can say that a fractal is a geometrical construction that can be subdivided into parts, each of which is a nearly identical copy of the original shape on a reduced scale. That is, if you focus your attention on a particular region of a fractal image and magnify that region, you will observe a replication of the overall image on a smaller scale.

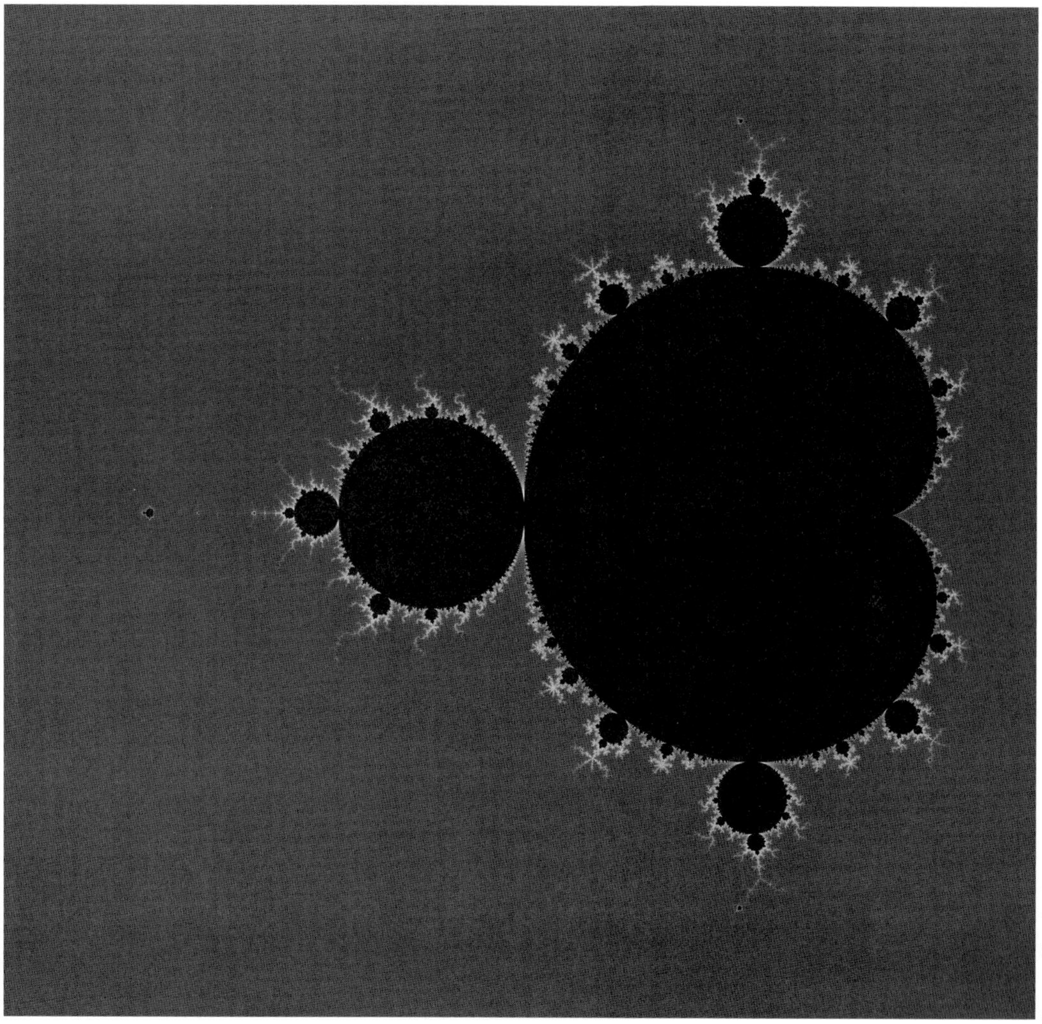

Figure 9.8 The Mandelbrot set

One of the more famous mathematical fractals is the Mandelbrot set, named in honor of Benoit Mandelbrot (1924–2010), a mathematician who conducted extensive studies of fractals. The Mandelbrot set (Figure 9.8) is constructed by applying an iterative, or repetitive, process to an arbitrarily chosen complex number and determining if the iterative process leads to a particular value (this result, in advanced mathematics courses, is referred to as convergence) or not (this condition is the antithesis of convergence, called divergence).

The iterative process involved in the construction of the Mandelbrot set is defined by the recursive sequence: $z_n = (z_{n-1})^2 + c$, where c is some specified complex number and z_0 is given. Suppose, for illustrative purposes, $c = 0.2 + 0.4i$ and $z_0 = 0$. Then we get the following progression:

$$z_0 = 0$$
$$z_1 = 0.2 + 4i$$
$$z_2 = (0.2 + 4i)^2 + (0.2 + 4i) = -15.76 + 5.6i$$

and so on. At each stage, determine the modulus of the complex number and maintain a record of these values. If the values diverge, which can be shown (in this case) to mean that they become larger than 1, then the process is stopped, and the point on the Argand diagram corresponding to c is colored white. If, on the other hand, all the modulus values remain less than 1, the point is colored black.

Repeat the process for another value c and determine if it should be colored black or white and continue for every point in the Argand diagram. The result is the Mandelbrot set.

Other fractals can be generated using other iterative rules, but you can see the idea. By defining specific rules for the value of the number to which the moduli converge, one can transform a black-and-white fractal into a colored image, and the results are uncanny images displaying tremendous elegance and beauty.

Exercises

In the following exercises, a particular current (in amps) and impedance (in ohms) are given. Find the exact voltage in each case and the phase shift (rounded to the nearest hundredth of a degree, as needed).

1. Current: $4 + 3i$ amps, Impedance: $2 - 4i$ ohms
2. Current: $7 + i$ amps, Impedance: $1 + 2i$ ohms
3. Current: $5 + 4i$ amps, Impedance: $4 + 5i$ ohms
4. Current: $120 + 0i$ amps, Impedance: $45 + i$ ohms
5. Current: $18 - 2i$ amps, Impedance: $15 + 0i$ ohms
6. Current: $2 + 2i$ amps, Impedance: $-3 + 3i$ ohms

In the following exercises, a particular voltage and current is given. Find the impedance in each case, rounding to two decimal places as needed.

7. Voltage: $30 - 10i$, Current: $4 + 2i$
8. Voltage: $1 - i$, Current: $3 + 4i$
9. Voltage: $125 + 10i$, Current: $5 + i$
10. Voltage: $35 + 20i$, Current: $8 + 5i$

In the following exercises, use DeMoivre's theorem to find the *n*th roots of the real number. Round your results to two decimal places, as needed.

11. The fourth roots of 2
12. The sixth roots of 10
13. The eighth roots of 1
14. The twelfth roots of 12
15. The third roots of 8
16. The fourth roots of 81

Find all solutions to the following equations, rounding the values in your answers to the nearest hundredth.

17. $x^5 - 3 = 0$
18. $x^7 + 10 = 0$
19. $2x^3 + 32 = 0$
20. $5x^4 - 42 = 0$

Summary

In this chapter, you learned about:

- Simplifying radical expressions.
- The definitions of a complex number and an imaginary unit.
- The algebraic operations of complex numbers.
- The forms and representation of complex numbers: rectangular, polar, trigonometric, and exponential and the Argand diagram.
- Some applications involving complex numbers.

Glossary

Argand diagram A Cartesian-type plane structure in which the x-axis is designated as the real axis, the y-axis is designated as the imaginary axis, and the complex number $a + bi$ is associated with the point (a, b).

argument of a complex number The value of theta in $r\angle\theta$.

complex plane An Argand diagram.

directed distance A distance measurement in which sign indicates direction.

exponential form of a complex number $re^{i\theta}$, where r is the modulus and θ is the argument of the complex number.

imaginary axis The vertical axis in an Argand diagram.

imaginary number Any number whose square is a negative real number. Equivalent to pure imaginary number.

imaginary part In an imaginary number $a + bi$, the number b.

imaginary unit The mathematical concept number i, defined by the rule $i = \sqrt{-1}$.

modulus of a complex number The value of r in the polar form $r\angle\theta$.

polar coordinates A system of representing location within the plane using a directed distance and an angle measured off a polar axis.

polar form of a complex number The notation $r\angle\theta$, where r is the directed distance from the origin to the point representing the complex number in an Argand diagram, and θ is the angle between the polar axis and the line segment connecting the origin to the point representing the complex number in that Argand diagram.

principal square root The nonnegative square root of a number.

pure imaginary A complex number whose real part is zero, or (equivalently) any number whose square is a negative real number. Another term for imaginary number.

radicand A number occurring within a radical symbol.

real axis The horizontal axis of an Argand diagram.

real number An imaginary number whose imaginary part is zero.

real part In an imaginary number $a + bi$, the number a.

rectangular form of an imaginary number The mathematical form $a + bi$. Also called the standard form.

standard form of an imaginary number The mathematical form $a + bi$. Also called the rectangular form.

trigonometric form of a complex number $(r\cos\theta) + (r\sin\theta)i$, where r and theta are the modulus and argument of the complex number, respectively.

End-of-Chapter Problems

Simplify the following radicals completely (remember, this does not mean decimal approximate!).

1. $\sqrt{-37}$
2. $\sqrt{-169}$

Identify the real and imaginary parts of the following complex numbers.

3. $17 + 5i$
4. $-8 + 7i$

Simplify the following powers of the imaginary unit i.

5. i^{13}
6. i^{-273}

Using matrices, verify the following relationships.

7. $i^6 = -1$
8. $i^7 = -i$

In each of the following, simplify the expression completely and give the answer in standard form.

9. $(4 - 7i) + (-7 + 6i)$
10. $(-3 + 6i) + (3 + 6i)$
11. $(18 - 7i) - (-13 + 15i)$
12. $(17 + 8i) - (12 + 13i)$
13. $(14 - 7i)(2 + i)$
14. $(17 + 3i)(3 + 17i)$
15. $\dfrac{-7+4i}{-8-3i}$
16. $\dfrac{8+4i}{2+2i}$
17. $(13 + 7i)^2$
18. $(5 + 7i)^2$

Represent the following complex numbers on an Argand diagram.

19. $3 + 2i$
20. $-5 + 3i$

In each of the following problems, calculate the indicated sum algebraically and then graphically in an Argand diagram. Show that the results agree.

21. $(-4 + i) + (3 - 2i)$
22. $(5 - 3i) + (-3 - 2i)$

For each of the following, calculate the complex product. Sketch the individual complex numbers and their product on an Argand diagram. In each case, find the arguments of the individual complex numbers being multiplied and also the argument of the product. Describe the impact of multiplying the first complex number by the second.

23. $(5 + 3i)(-1 - 4i)$
24. $(4 + 5i)(-1)$

In each of the following, the polar form of a complex number is given. Give the graphic representation of the complex number.

25. $5\angle 30°$
26. $7\angle 180°$

Find the polar representation of the following complex numbers, measuring the angle in both degrees and radians (rounded to the nearest tenth of a degree or radian, as needed).

27. $-5 - 3i$
28. $7 + i$

For the following complex numbers, given in polar representation, find the standard form representation, rounding the values of a and b to two decimal places, as needed.

29. $5\angle 30°$

30. $7\angle 22.5°$

For the following complex numbers, given in polar representation, find the exact rectangular form representation.

31. $13\angle\frac{\pi}{2}$

32. $3\angle\frac{\pi}{3}$

Give the trigonometric form of the following complex numbers.

33. $5\angle 30°$

34. $7\angle 22.5°$

Give the exponential form of the following complex numbers.

35. $13\angle\frac{\pi}{2}$

36. $3\angle\frac{\pi}{3}$

Express the following complex numbers, given in exponential form, in the corresponding rectangular, polar, and trigonometric forms (rounding the components of the respective forms to two decimal places, as needed).

37. $5e^{i2\pi}$

38. $13e^{i(3\pi)}$

In the following problems, find the product of the given complex numbers.

39. $4\angle 35°$ and $6.5\angle 40°$

40. $3\angle 75°$ and $4\angle 180°$

41. $6e^{i(3)}$ and $7e^{i(.14)}$

42. $13e^{i(3\pi)}$ and $3e^{i(5\pi)}$

Simplify the following, using DeMoivre's theorem.

43. $\left[4e^{i(\frac{\pi}{2})}\right]^5$

44. $\left[7e^{i(\frac{\pi}{4})}\right]^7$

Divide the following complex numbers.

45. $\dfrac{4\angle 72°}{2\angle 36°}$

46. $\dfrac{10\angle 182°}{4\angle 23°}$

Answer the following questions, rounding as indicated.

47. If the current in a circuit is $7 + 3i$ amps and the impedance is $3.5 + 4i$ ohms, what is the exact voltage, and what is the phase shift (rounded to the nearest hundredth of a degree if necessary)?

48. If the voltage across an element in a circuit is $100 + 100i$ volts and the current is $4 + 3i$ amps, what is the impedance of the element? Round to two decimal places as needed.

49. Use DeMoivre's theorem to find the seventh roots of 128. Round your results to two decimal places, as needed.

50. Find all solutions to $3x^3 + 81 = 0$. Round the values in your answer to two decimal digits if necessary.

Chapter 10 Vectors

A **vector** is a quantity that includes two components: magnitude and direction. Because of the nature of this definition, the only way we can fully specify a vector is by producing two numbers, one quantifying the vector's magnitude and the other its direction.

There are many instances in which a concept can be described only in this way. For instance, consider how we might attempt to describe the wind on any given day. We could state that the wind is blowing at 30 kilometers per hour, but this does not give a full description of the nature of how the wind is blowing. The speed of the wind is, in a sense, another way of expressing the *magnitude* of the wind's velocity. In order for us to get meaningful information, we also must know the *direction* of the wind. Consequently, we would consider the wind velocity to be a vector quantity.

Another illustration would be a representation of the velocity of information a squadron of jet airplanes. The motion of the jets can be represented by specifying the direction in which the jets are flying as well as the velocity with which they are traveling. This concept is what we can call the velocity vector of the squadron.

In this chapter, we will investigate these vector quantities that can be effectively described using only two measurements, a magnitude and a direction, and consider how mathematics can be performed on vectors.

Objectives

When you have successfully completed the materials of this chapter, you will be able to:

- Understand vectors and the various ways in which they can be represented.
- Resolve a vector into its components.
- Calculate the resultant of two vectors.
- Solve word problems involving vectors.

10.1 Vectors and Their Representation

Visual Representation of a Vector as a Directed Line Segment

A vector is a quantity whose magnitude and direction must be specified in order to fully capture its essence. The examples provided in the introduction to this chapter, wind velocity and the path of a squadron of jet airplanes, are illustrations of such quantities, but they are far from the only instances where this two-quantity description arises. Vectors enable us to reason about problems in space without use of coordinate axes, and this freedom is (as examination of the concept reveals) tremendously liberating.

Returning to the situation of the jets in formation and the wind speed, we'll attempt to gain an introductory understanding of the situation we're considering. Suppose the jets are flying on a heading of N20°W, with an air speed of 600 kilometers per hour. In the airspace through which they are flying, a headwind is blowing at 80 kilometers per hour, from due north (observe that wind direction is typically stated by reference to the direction from which the wind is coming, so the direction of the wind is actually due south!). We can imagine the motion of the jets to be depicted by an arrow in the Cartesian plane of length 600 at an angle of 110° with the positive x-axis (Figure 10.1).

The wind can also be depicted by such an arrow (or, as is common in airflow diagrams, by a field of arrows) in the same plane. As you can imagine, the wind is, in a sense, pushing against the jets, diminishing their speed slightly and working to push them slightly off course (Figure 10.2).

Through an investigation of the interaction of these two concepts, the motion of the jets and the impact on them by the wind, we can determine the true direction and speed of the jets relative to the ground. The use of arrows to depict vectors in a plane is fairly common, and the representation is referred to as the method of **directed line segments** (Figure 10.3). A directed line segment is typically depicted as a line segment having a fixed length and inclination to horizontal whose direction is displayed through the use of an arrowhead at one

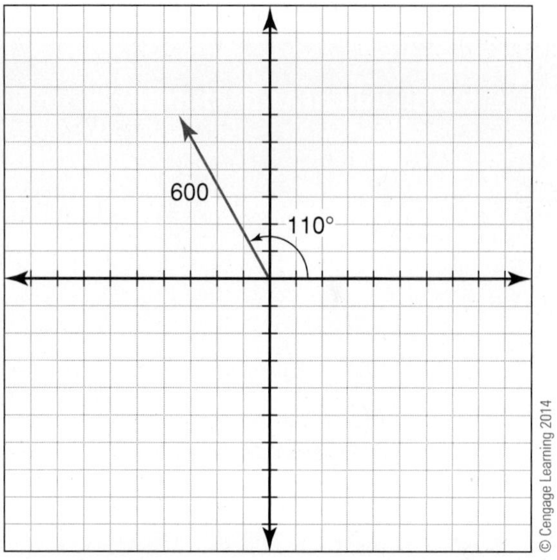

Figure 10.1 Vector depiction of speed and direction of jets

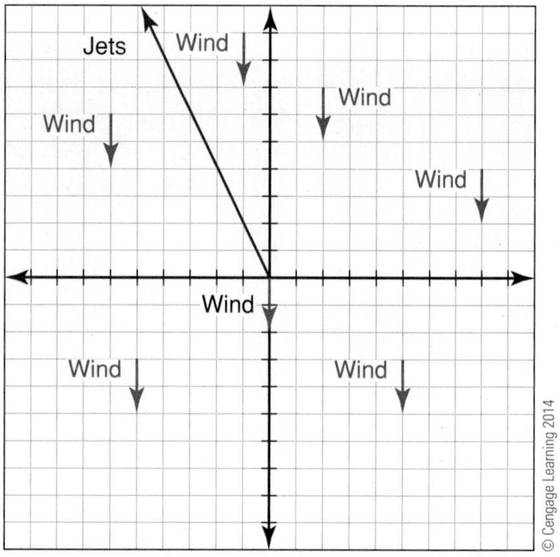

Figure 10.2 Vector depiction of speed and direction of jets and wind

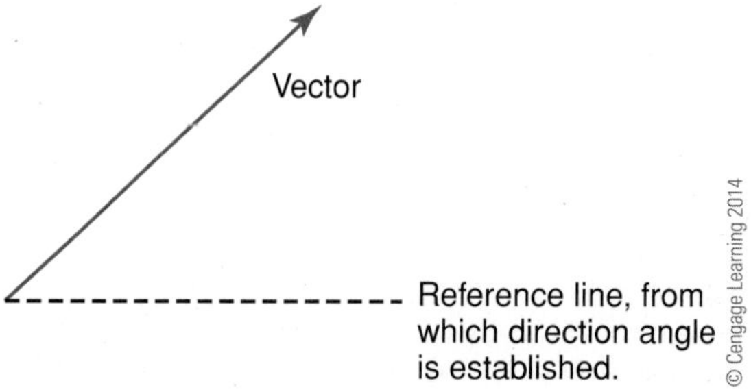

Figure 10.3 Vector shown in relation to reference line

end of the line segment. There is an uncanny resemblance between a directed line segment and a ray, but do not confuse the two: the directed line segment has finite length, while the ray extends forever in one direction. From this perspective, the length of the directed line segment conveys the magnitude of the vector, and the angle formed by the arrow indicating the vector and some arbitrary line of reference can be used to capture the vector's direction.

Two vectors are said to be equal if they have the same magnitude and direction, that is, if they have the same length and the same angle of inclination, no matter where situated in the plane. If the end points of the directed line segment are identified by letters P and Q, then we can use the notation $\overrightharpoon{PQ}$. Observe that this notation bears a superficial resemblance to the notation for a ray but that the arrowhead is more of a "half" arrowhead in the case of a vector. If that notation should become cumbersome, we could give the vector a single-letter name, such as $\bar{v}$, provided that we describe the vector with sufficient clarity. An alternative notation that is suitable for textbooks (but not so good for purposes of writing) is boldfaced lettering, such as **v**, to represent the vector.

A key notion to keep in mind is this: a vector has what we might describe as freedom of motion. As long as its length and inclination are not changed, we are free to move the directed line segment anywhere in the plane. This is sometimes useful, and we will exploit that property when we do some of our vector algebra.

When the vector is named in the manner already described, we can introduce a notation for the vector's magnitude. The notation appears identical to the familiar absolute value bars with which we are familiar, but in this context it is understood to represent vector magnitude. So, if our vector is **v**, then its magnitude is given by $|\mathbf{v}|$.

Polar Representation of a Vector

Since we have already used the concept of an angle of inclination to capture the concept of the direction of a vector, it seems natural to describe that angle through reference to the positive x-axis. Imagine that we positioned the vector in such a manner that its initial point was at the origin of the plane. The positive angle with the positive x-axis (in degrees) could be used as a descriptor for the angle of inclination of the vector, and in this way (if we call the angle θ) we could use the notation $\bar{v}\angle\theta$ to describe the vector.

EXAMPLE 10.1

Consider a vector drawn in a position where its initial point is at $(-2, 3)$ and its terminal point is at $(3, 7)$. Find the polar representation of the vector.

SOLUTION

First, let's give a graphic representation of the vector in the plane (Figure 10.4). Now the vector can be moved, if we wish, to the origin (recall that we are free to move the directed line segment about the plane as long as we don't change its angle of inclination or its length). When the vector is so positioned, we say the vector is in **standard position** (Figure 10.5).

Observe that, in this new location, the terminal point of the directed line segment now lies at the point $(5, 4)$ of the plane. The magnitude of the vector can be found rather easily using the Pythagorean Theorem and is equal to

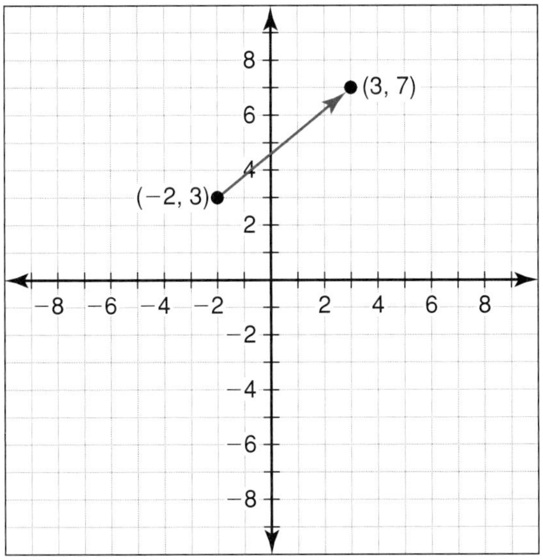

Figure 10.4 Graphic representation of a vector in the plane

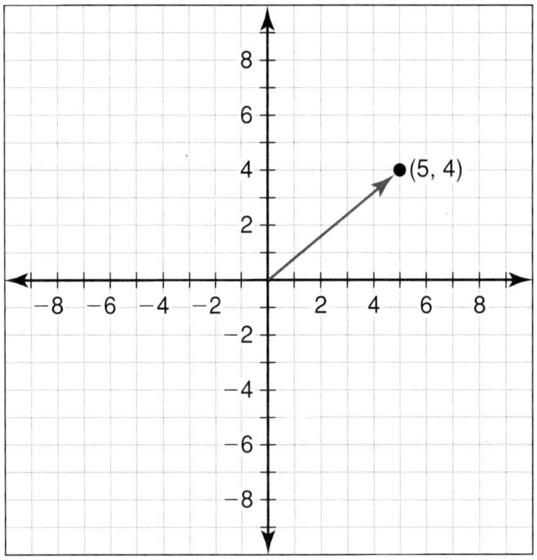

Figure 10.5 Vector shown in standard position

$\sqrt{5^2 + 4^2} = \sqrt{41}$. The angle of inclination can be found using the inverse tangent function and a decimal approximation: $\theta = \arctan\left(\frac{4}{5}\right) \approx 38.7°$. This tells us that the polar representation of the vector is $\sqrt{41}\angle 38.7°$.

Rectangular Representation of a Vector

Yet another representation of vectors is referred to as **rectangular form**. This form is essentially a description of the standard position of the vector and takes the form $\langle a, b\rangle$. The notation indicates that the vector, when placed in standard position, has its terminal point at (a, b). Referring to the illustration of the vector in Figure 10.5, the rectangular form of the vector displayed is seen, by inspection, to be $\langle 5, 4\rangle$.

EXAMPLE 10.2

Sketch the vector whose rectangular form is $\langle -3, 7\rangle$.

SOLUTION

Since the vector is to be drawn in standard position, all we need to produce is a directed line segment connecting the origin to the point $(-3, 7)$. Our familiarity with the plane makes this a relatively straightforward process (Figure 10.6).

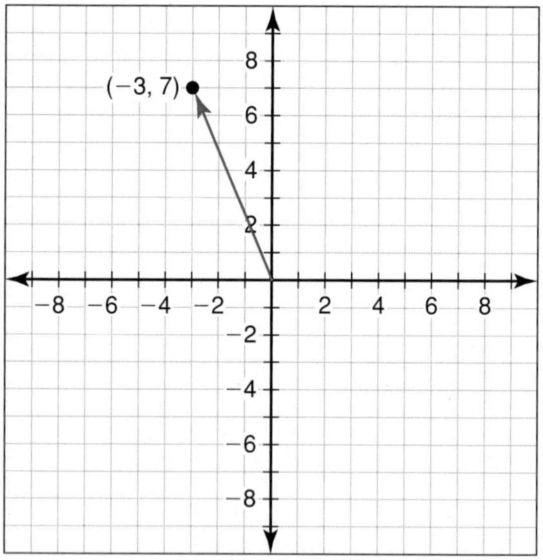

Figure 10.6 Vector <−3, 7> shown in standard position

Exercises

Answer the following questions in your own words, using proper grammar, correct terminology, and complete sentences.

1. In your own words, describe what is meant by a vector and give examples of vector quantities.
2. What is the difference between a vector and a directed line segment?
3. What facts about a vector must we know in order for us to create the polar representation?
4. How is the polar representation of a vector transformed into the rectangular representation?

The following exercises reference problems earlier in this problem set. For each referenced problem, give the rectangular form of the representation of the vector.

5. $(2, -5)$ and $(-3, 5)$
6. $(1, 1)$ and $(-2, 4)$
7. $(9, 0)$ and $(3, 7)$
8. $(-8, -3)$ and $(3, 4)$
9. $(-9, -7)$ and $(0, 0)$
10. $\left(\frac{1}{2}, 5\right)$ and $\left(\frac{-1}{3}, 2\right)$
11. $\left(\frac{2}{3}, 1\right)$ and $\left(\frac{1}{6}, 5\right)$
12. $(-3, 6)$ and $\left(5, \frac{1}{2}\right)$

The following exercises give the initial and terminal points of a vector. Give the polar representation for the vector.

13. Problem 5
14. Problem 6
15. Problem 7
16. Problem 8
17. Problem 9
18. Problem 10
19. Problem 11
20. Problem 12

In the following problems, the rectangular representation of a vector is given. Convert the representation to the polar representation. Do not approximate the magnitude but do round degrees to the nearest tenth of a degree, as needed. Sketch the given vector in the plane in standard position.

21. $\langle 2, 6 \rangle$
22. $\langle 4, 5 \rangle$
23. $\langle -1, 4 \rangle$
24. $\langle -5, -7 \rangle$
25. $\langle -3, -8 \rangle$
26. $\langle -1, 9 \rangle$
27. $\left\langle \frac{1}{3}, \frac{2}{3} \right\rangle$
28. $\left\langle \frac{-2}{5}, \frac{4}{5} \right\rangle$

10.2 Resolution of Vectors

We saw in Section 10.1 that a particular vector has three representations: the directed line segment, polar, and rectangular forms. The relationship between the latter can be seen as a consequence of our results from trigonometry applied to the directed line segment representation. Calling to mind the visual representation of the vector in standard position (Figure 10.6), we can superimpose a right triangle on the situation and then use trigonometry to uncover the relationship (Figure 10.7).

The rectangular representation of the vector is $\langle a, b\rangle$, and our results from right-triangle trigonometry tell us that and $a = |\vec{v}|\cos\theta$ and $b = |\vec{v}|\sin\theta$. Therefore, we have a direct relationship between the two representations of vector $\vec{v}$.

Note

If the vector $\vec{v}$, when drawn in standard position, has terminal point (a, b), then the rectangular form of $\vec{v}$ is $\langle a, b\rangle$, where $a = |\vec{v}|\cdot\cos\theta$ and $b = |\vec{v}|\cdot\sin\theta$, and θ is the direction angle of vector $\vec{v}$. Therefore, $\langle a, b\rangle = \left\langle |\vec{v}|\cos\theta, |\vec{v}|\sin\theta\right\rangle$.

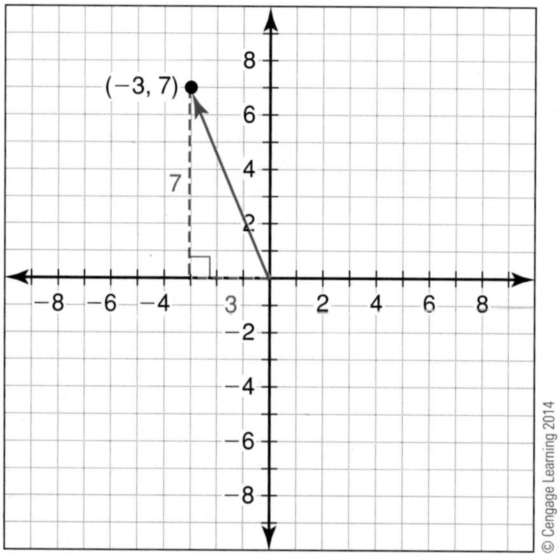

Figure 10.7 Right triangle superimposed upon a vector shown in standard position

This formulation of the representation of the vector arises frequently in applications and merits a special name, the **rectangular components** of vector $\vec{v}$, and a breakdown of a vector in the manner described is referred to as a resolution of the vector into its **x-components** and **y-components**. The x-component is typically expressed in the form $R_x = \left(|\vec{v}|\ \cos\theta\right)\vec{i}$ and the y-component as $R_y = \left(|\vec{v}|\ \sin\theta\right)\vec{j}$, where $\vec{i}$ and $\vec{j}$ are what we refer to as standard unit vectors; these will be discussed in greater detail shortly. Let's examine a few vectors and see if we can derive their resolution.

EXAMPLE 10.3 Consider the vector $\vec{v} = \langle 5, 3\rangle$ and resolve the vector into its rectangular components.

SOLUTION

It is sometimes helpful to visualize the vector (Figure 10.8), as this can help us to calculate the resolution. Since the rectangular components depend on the direction and the magnitude of the vector, we can begin by calculating the magnitude. Using the Pythagorean Theorem as before, we can find the magnitude of the vector: $|\vec{v}| = \sqrt{5^2 + 3^2} = \sqrt{34}$. The direction angle of the vector can be found using inverse tangent, since $\tan\theta = \frac{3}{5}$ implies $\theta \approx 30.96°$, and therefore the vector can be expressed as $\langle\sqrt{34}\cos 30.96°, \sqrt{34}\sin 30.96°\rangle$, and hence the horizontal and vertical components of the vector are $R_x = \left(\sqrt{34}\cos 30.96°\right)\vec{i}$ and $R_y = \left(\sqrt{34}\cos 30.96°\right)\vec{j}$ respectively.

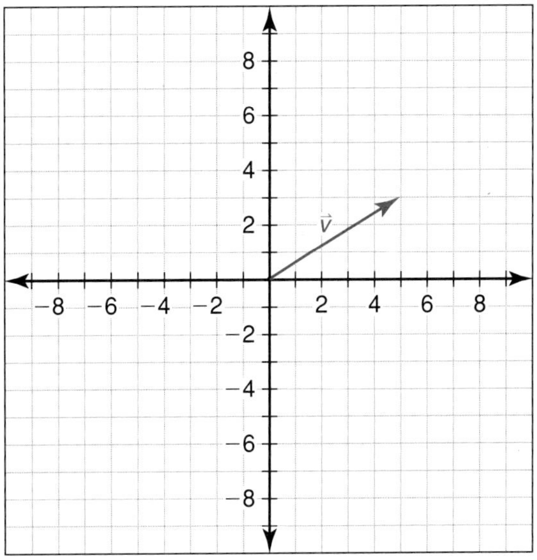

Figure 10.8 Vector <5, 3>, shown in standard position

EXAMPLE 10.4

Resolve the vector $60\angle 30°$ into its horizontal and vertical components.

SOLUTION

Recalling the formulas for the components of the vector, $R_x = |\vec{v}|\cos\theta$ and $R_y = |\vec{v}|\sin\theta$, substitute the magnitude and angular values to obtain

$$R_x = (60\cos 30°)\vec{i} \approx 60(0.866)\vec{i} \approx 51.96\vec{i}$$
$$R_y = (60\sin 30°)\vec{j} = 60(0.5)\vec{j} \approx 30\vec{j}$$

EXAMPLE 10.5

In vector graphics, the drawings of a graphic designer are saved as a series of points that are connected by directed line segments in a manner that results in files of significantly smaller size. Suppose that two successive points in such a design are located at (4, 1) and (−2, 7) in the plane. Resolve the vector connecting the first point to the second into its horizontal and vertical components.

SOLUTION

In this case, we are not given the rectangular form of the vector, so it would be to our advantage to obtain this first. Subtracting components, we find

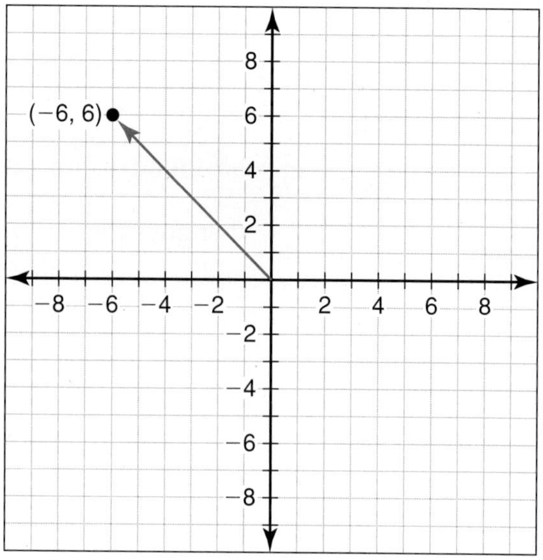

Figure 10.9 Vector <−6, 6>, shown in standard position

$\vec{v} = \langle -2 - 4, 7 - 1 \rangle = \langle -6, 6 \rangle$. As a reference, we sketch the vector in standard position (Figure 10.9).

The magnitude of the vector is $\sqrt{(-6)^2 + 6^2} = \sqrt{72} = 6\sqrt{2}$, which we find using the Pythagorean Theorem. Note that if we use inverse tangent to find the direction angle, we encounter a rather surprising result: $\tan\theta = \frac{6}{-6} \Rightarrow \theta = -45°$. This is a surprising result, perhaps, since that angle lies in quadrant IV, and we know that the vector lies in quadrant II. This is a consequence of the definition of inverse tangent, which yields angular values only in quadrants I and IV, and from experience we can recognize that this is actually 180° degrees from the correct angle, and thus we conclude $\theta = 135°$. Consequently, we find $R_x = 6\sqrt{2}\cos 135°$ and $R_y = 6\sqrt{2}\sin 135°$.

EXAMPLE 10.6

A block of weight 20 N (newtons) rests on an inclined plane of 45°. Resolve the weight W of the block into two components, one along the plane and the other perpendicular to it.

SOLUTION

Taking the x-axis along the direction of the inclined plane and the y-axis perpendicular to it, the components are as shown in the Figure 10.10. Observe that the placement of the axes is completely arbitrary and that we are free to position things in whatever manner we feel to be convenient. This practice is referred to as imposing a coordinate system on a physical situation and is quite common in practice.

The components are

$$W_x = (20 \sin 45°)\vec{i}$$

$$W_y = (20 \cos 45°)\vec{i}$$

Therefore, the weight of the block has components of 14.14 N down the plane and perpendicular to it (Figure 10.10).

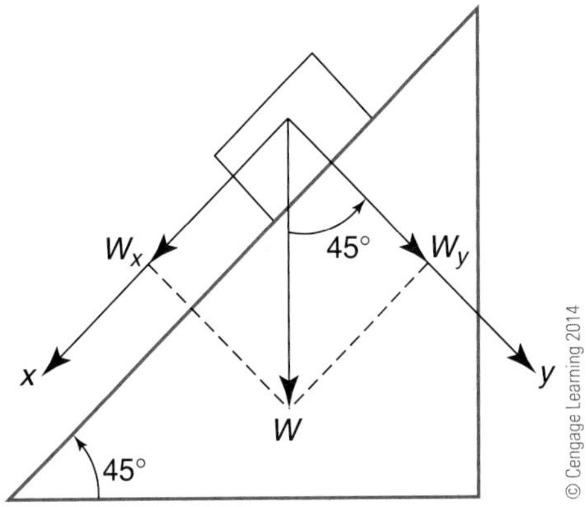

Figure 10.10 Block on an inclined plane

Exercises

In your own words, using correct grammar, spelling, and terminology, answer the following questions.

1. What is meant by the rectangular components of vector $\vec{v}$?
2. What is meant by a resolution of a vector into x- and y-components?
3. If a vector is given in a nonstandard position, with its initial and terminal points given, how do we find the resolution of the vector?

The following problems give a vector in polar form. Give the rectangular components of the vector, rounded to the nearest tenth of a unit.

4. $13\angle 120°$
5. $8\angle 73°$
6. $3\angle 240°$
7. $1.25\angle 50°$
8. $50\angle 10°$
9. $\sqrt{21}\angle 300°$
10. $\sqrt{7}\angle 270°$
11. $\sqrt{47}\angle 180°$

Resolve the following vectors into their rectangular components, rounding the angles to the nearest hundredth of a degree, as needed. Sketch the vector in standard position.

12. $\langle 4, 1\rangle$
13. $\langle 7, 3\rangle$
14. $\langle -3, 7\rangle$
15. $\langle -2.5, 11\rangle$
16. $\langle -4.8, -3\rangle$
17. $\langle -7, -9.9\rangle$
18. $\langle \pi, 1\rangle$
19. $\langle 2, \pi\rangle$
20. $\langle 4, -3\pi\rangle$
21. $\langle -5\pi, 2\pi\rangle$

The following problems describe a situation on which vectors can be imposed. Determine the rectangular form of the vector and then produce the resolution of the vector.

22. A computer mouse is moved across a mouse pad on which a coordinate system has been overlaid. The mouse is moved from an initial position on the coordinate system of $(5, -2)$ to a final position of $(-7, -4)$. The vector describes the motion of the mouse from its initial to terminal positions.

23. A block of weight 60 N rests on an inclined plane of 70°. Resolve the weight W of the block into two components, one along the plane and the other perpendicular to it.
24. A sailor is steering a boat in a heading of 210° at 8 miles per hour. Draw the vector representing the movement of the boat and resolve that vector into horizontal and vertical components.

10.3 The Resultant of Two Vectors

Two vectors in the plane can be combined through a process of addition, and the sum of those vectors is a new vector called the **resultant**. In a sense, calculation of the resultant is the reversal of the process of resolving a vector into its components because it involves the calculation of a new vector that has the same effect as the combined influence of the two original vectors.

The Algebraic Process of Vector Addition

If the vectors are given to you in component form, $\vec{u} = \left\langle u_x, u_y \right\rangle$ and $\vec{v} = \left\langle v_x, v_y \right\rangle$, then we can add the vectors by adding the individual components. That is, $\vec{u} + \vec{v} = \left\langle u_x + v_x, u_y + v_y \right\rangle$. Observe that the symbol for vector addition is precisely the same as is used in regular addition.

EXAMPLE 10.7 Find the resultant of the vectors $\langle 3, -2 \rangle$ and $\langle -4, 3 \rangle$.

SOLUTION

When given the vectors in component form, calculation of the resultant merely requires that we add the vectors component by component. Thus, our vector sum is $\langle 3 + (-4), -2 + 3 \rangle = \langle -1, 1 \rangle$.

EXAMPLE 10.8 Find the resultant of the vectors $\langle -4, 1 \rangle$ and $\langle 4, 5 \rangle$.

SOLUTION

Again, adding the components of the vectors, we obtain the resultant $\langle -4 + 4, 1 + 5 \rangle = \langle 0, 6 \rangle$. In a sense, vector addition is what we might call a "nice" operation because it is performed in precisely the manner you would expect it to be performed: addition of vectors is accomplished by adding the corresponding components of the vectors, and this is straightforward when we have the rectangular form of the vectors in our hands.

Suppose we wanted to find the resultant of two vectors given to us in polar form. Let's return to our first example of vector addition given above and investigate.

The polar form of $\langle 3, -2 \rangle$ is easily seen to be $\sqrt{13}\angle 326.3°$, and the polar form of $\langle -4, 3 \rangle$ is $5\angle 143.1°$. What about the sum of the vectors? We computed it to be $\langle -1, 1 \rangle$, the polar form of which is $\sqrt{2}\angle 135°$. On inspection, there is no discernible connection between the polar forms of the individual vectors and the polar form of the resultant, and in fact, there is no direct relationship. When adding vectors given in polar form, it is best to convert the vectors to rectangular form, perform the addition there, and then convert back to polar form, if desired.

EXAMPLE 10.9 Add the vectors $3\angle 50°$ and $6\angle 75°$.

SOLUTION

Converting each vector to rectangular form, we obtain the equivalent vectors $\langle 1.928, 2.298\rangle$, and $\langle 1.553, 5.796\rangle$. In this form, we add the components to obtain $\langle 3.481, 8.094\rangle$, which we then convert back to polar form. Choosing to call the resultant vector R, we find that $|\vec{R}| = \sqrt{3.481^2 + 8.094^2} = \sqrt{77.630} = 8.811$ and $\theta = 66.729°$, so the resultant vector has polar form $8.811\angle 66.729°$.

Addition of Vectors Visually: The Tip-to-Tail Method

We have seen that addition of vectors in rectangular form is nearly a triviality and that addition of vectors given in polar form requires the intermediate steps of translation to and from rectangular form. What about those vectors that are given to us as directed line segments? It seems natural to wonder if addition of directed line segments also requires translation into rectangular form, and it is this question we address now.

Happily, the answer is no, but at times we might wish to make the transition to rectangular form anyway, since that is not terribly difficult, given a directed line segment in standard position. There is an alternative method, however, that often proves useful when examining vectors in applications of physics.

The technique has various names, but the one we will employ here is the **tip-to-tail method of vector addition**. The addition is performed in the following manner: to find the resultant of $\vec{u}$ and $\vec{v}$, first draw vector $\vec{u}$ in standard position. Next, position vector $\vec{v}$ so that its initial point coincides with the terminal point of vector $\vec{u}$. The resultant vector is then the vector obtained by connecting the origin to the terminal point of vector $\vec{v}$ positioned as indicated.

EXAMPLE 10.10 Find the resultant of $\vec{u} = \langle 5, -2\rangle$ and $\vec{v} = \langle -1, 6\rangle$.

SOLUTION

Using our techniques of vector addition presented thus far, the computation of the resultant is a triviality, and we see immediately that the vector sum is $\vec{u} + \vec{v} = \langle 5 + (-1), -2 + 6\rangle = \langle 4, 4\rangle$. However, let's do the addition visually and confirm that the same result is obtained. We'll first draw vector $\vec{u}$ in standard position (Figure 10.11).

Suppose now that we attach the vector $\vec{v}$ to vector $\vec{u}$, positioned such that the initial point of $\vec{v}$ coincides with the terminal point of $\vec{u}$. Since vector $\vec{v}$ has component form $\langle -1, 6\rangle$, we can draw this vector in its new orientation by moving from the initial point 1 unit left and 6 units upward (Figure 10.12).

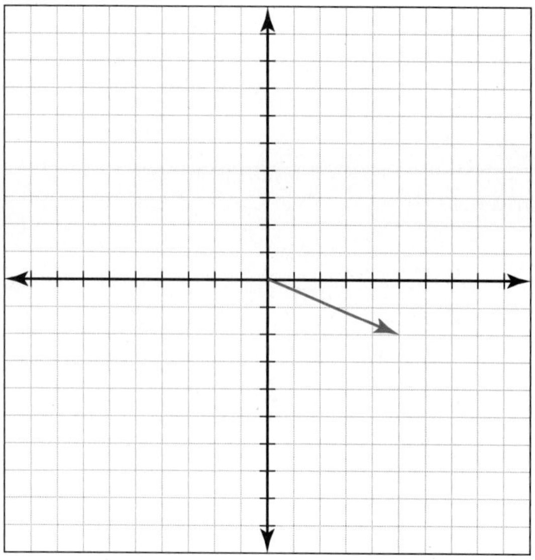

Figure 10.11 Vector <5, −2> shown in standard position

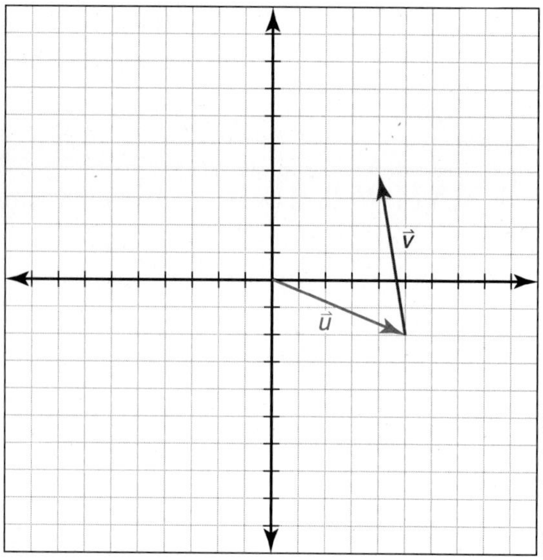

Figure 10.12 Vectors <5, −2> and <−1, 6>, aligned tip-to-tail

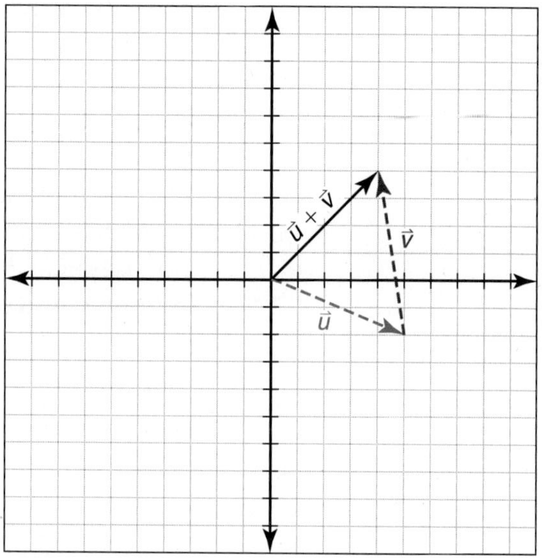

Figure 10.13 The resultant of vectors $\vec{u}$ and $\vec{v}$

We now construct the resultant vector by connecting the origin to the new terminal point of vector $\vec{v}$ (Figure 10.13).

Observe that the resultant vector is precisely the same as that which we received by algebraic computation, $\langle 4, 4 \rangle$.

Since the visual method of adding vectors yields the same resultant as the algebraic method, it is worth debating whether we even need the visual method. However, we should point out that quite a bit of work can be accomplished using the tip-to-tail addition method when tackling application problems. The interplay between forces, for instance, can be examined visually to gain insight into the physical dynamics of a situation, and this is sometimes not as apparent when calculating algebraically.

EXAMPLE 10.11 Add vectors $\vec{u}$ and $\vec{v}$ (Figure 10.14), first visually and then algebraically (as a confirmation).

SOLUTION

Leaving vector $\vec{u}$ as given, we will move vector $\vec{v}$ to align its initial point with the terminal point of $\vec{u}$ (Figure 10.15). The resultant vector is found by connecting the origin to the new terminal point of $\vec{v}$ (Figure 10.16).

By inspection, it appears that the rectangular form of the resultant is $\langle 1, 4\rangle$, but let's check it by algebraic computation. The original depiction of the vectors

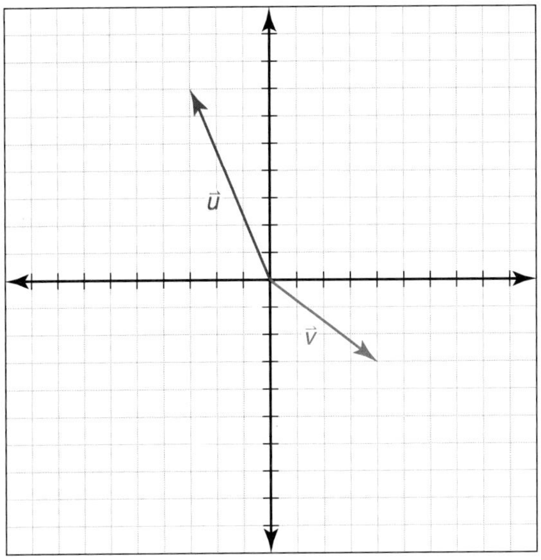

Figure 10.14 Vectors $\vec{u}$ and $\vec{v}$

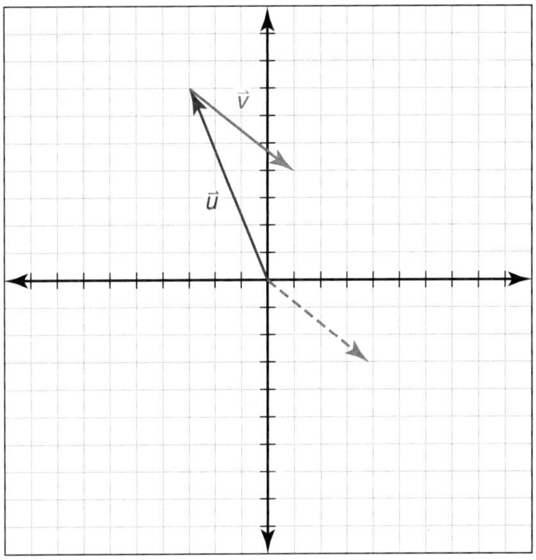

Figure 10.15 Vectors $\vec{u}$ and $\vec{v}$, positioned tip-to-tail

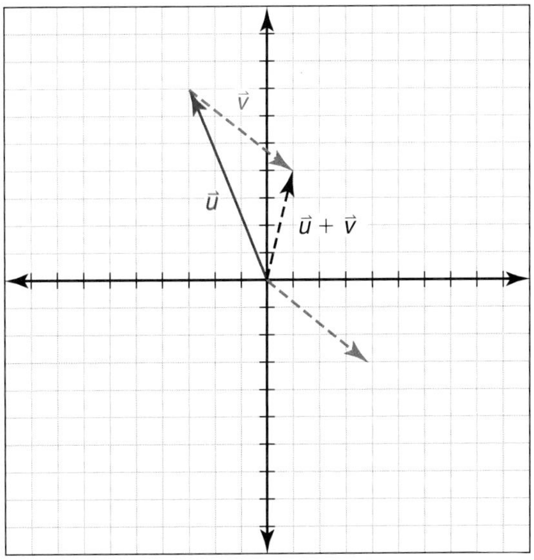

Figure 10.16 The resultant of vectors $\vec{u}$ and $\vec{v}$

in Figure 10.14 tells us the vectors are $\vec{u} = \langle -3, 7 \rangle$ and $\vec{v} = \langle 4, -3 \rangle$. Addition of the vectors component by component produces the resultant to be

$$\langle (-3 + 4), 7 + (-3) \rangle = \langle 1, 4 \rangle$$

which agrees with the visual result.

Exercises

Find the resultant of the following pairs of vectors.

1. $<3, -4>$ and $<5, 8>$
2. $<5, -13>$ and $<9, 2>$
3. $<-11, 5>$ and $<2, 3>$
4. $<-2, 7>$ and $<4, 2>$
5. $<-6, 5>$ and $<-3, 3>$
6. $<-14, 9>$ and $<-9, 4>$
7. $<-\frac{1}{2}, 3>$ and $<\frac{2}{3}, 5>$
8. $<-\frac{3}{4}, 9>$ and $<\frac{1}{5}, 7>$
9. $<\frac{2}{3}, \frac{5}{11}>$ and $<\frac{1}{4}, \frac{2}{3}>$
10. $<0.125, 0.9>$ and $<1.375, -2.4>$
11. $<0.919, 0.875>$ and $<1.2, -3.5>$

Convert the following vectors (given in polar form) to their corresponding rectangular representations and then find the sum of the vectors. Convert the sum of the vectors back to polar representation.

12. $5\angle 45°$ and $3\angle 90°$
13. $3\angle 110°$ and $8\angle -20°$
14. $15\angle 135°$ and $9\angle 45°$
15. $8\angle 60°$ and $14\angle 120°$
16. $7\angle 200°$ and $2\angle 70°$
17. $10\angle 210°$ and $11\angle 125°$

For the following vectors, perform the addition of vectors using the tip-to-tail method. First sketch each of the given vectors in standard position and draw the resultant vector also in standard position. Verify the answer by finding the rectangular representations of the individual vectors and showing the resultant matches that found with the tip-to-tail method.

18. $<5, 7>$ and $<3, -8>$
19. $<2, 5>$ and $<6, -4>$
20. $<-1, 6>$ and $<3, 5>$
21. $<-2, 7>$ and $<5, 3>$
22. $<-3, -1>$ and $<-4, -7>$
23. $<-1, -4>$ and $<-5, -3>$
24. $<6, -2>$ and $<5, -3>$
25. $<7, -6>$ and $<3, -5>$

10.4 Applications of Vectors

Many physical quantities are vector quantities, though you may not have thought of them as vectors in the past. For example, velocity, acceleration, alternating voltage, force, and the impedance of an electrical circuit are all vectors, since you must specify their magnitude and direction in order to give a full description of the concepts involved. An object can have multiple forces or other vector quantities acting on it, and in such a case, the techniques we developed for resultants could play a role (though there are other vector operations beyond the scope of our conversation, such as the cross-product of two vectors, that could also be involved). In such situations, you might be asked to calculate the resultant velocity, force, or voltage. Similarly, in a complex situation, you must know the component of velocity or force in a specific direction. In all such situations, you must have the skills to resolve a vector into its components or simultaneously find the resultant of a number of vectors. We'll now consider some examples of the use of vectors in application.

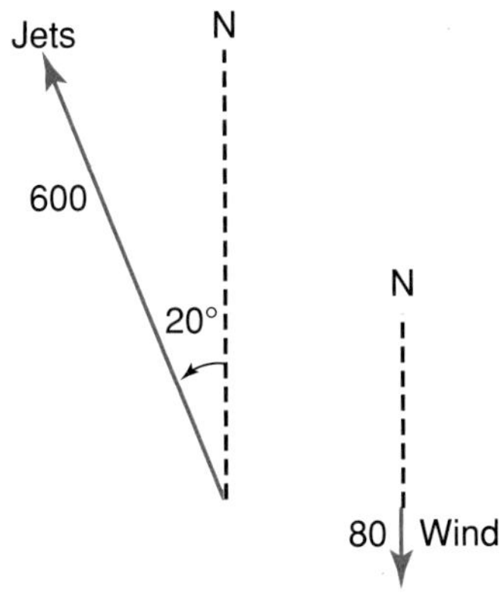

Figure 10.17 Vectors depicting speed and direction of jet squadron and wind

The Influence of the Current in a Medium

Suppose we now return to the question of the jets flying in formation presented at the beginning of the chapter. Recall, the situation was that a squadron was flying on a heading of N20°W with an air speed of 600 kilometers per hour. In the airspace through which they are flying, a wind is blowing at 80 kilometers per hour from due north (Figure 10.17). Let's now examine how the wind impacts the flight of the jets with regard to both the direction of flight and the speed of the squadron.

The question now arises as to how the wind impacts the motion of the jet squadron. We can imagine that, since the wind is (for the most part) working against the jets, the net result should be a slight "pushing" of the jets off course to the south and some negative impact on their speed. Since the two forces involved, the movement of the jets and the impact of the wind, are interacting directly, we can find the result of this interplay by calculating the resultant of the two vectors.

In this case, the motion of the jets is given using a bearing off true north, and we can restate the vector describing their motion by imposing a coordinate system on the situation and measuring the direction angle of the vector in that plane. As shown in Figure 10.18, the direction angle of the jets' vector is 110°, and the magnitude is 600. In polar form, therefore, the vector can be represented as $600\angle 110°$.

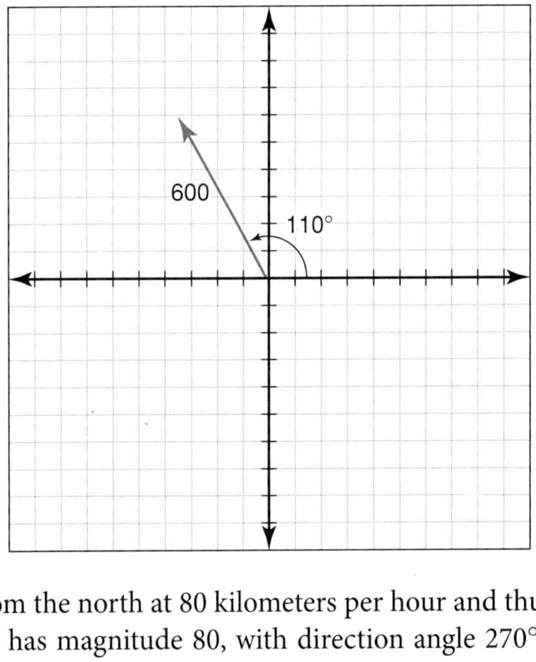

Figure 10.18 Vector showing speed and direction of jets, in standard position

The wind is blowing from the north at 80 kilometers per hour and thus the vector describing the wind has magnitude 80, with direction angle 270°. The polar form of the vector's representation is $80\angle 270°$, and we show the vector of the wind and that of the jet squadron on the same plane in Figure 10.19.

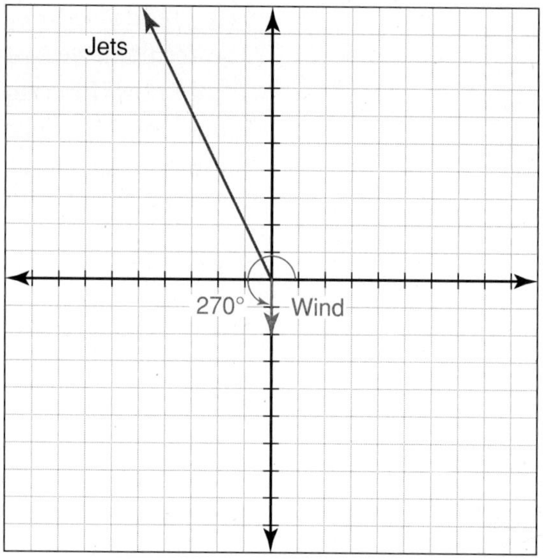

Figure 10.19 Vectors showing speed and direction of jets and wind, in standard position

In order for us to obtain the resultant, we convert these polar representations to rectangular form, obtaining (for the jets) $\vec{J} = \langle -205.21,\ 563.82\rangle$ and (for the wind) $\overrightarrow{W} = \langle 0,\ -80\rangle$. The resultant is now easily obtained through addition of the components, so $\vec{J} + \overrightarrow{W} = \langle -205.21,\ 483.82\rangle$. Converting back to polar representation (we'll suppress the work and assume that it's now familiar to you), $525.54\angle 112.98°$.

Analyzing this result, we see that the wind has definitely cut into the speed of the squadron of jets, reducing its speed to 525.54 kilometers per hour and pushing the jets to a heading of N22.98°W (Figure 10.20). Thus, to compensate for the impact of the wind, the jets will have to continuously make course adjustments in order to maintain the proper heading.

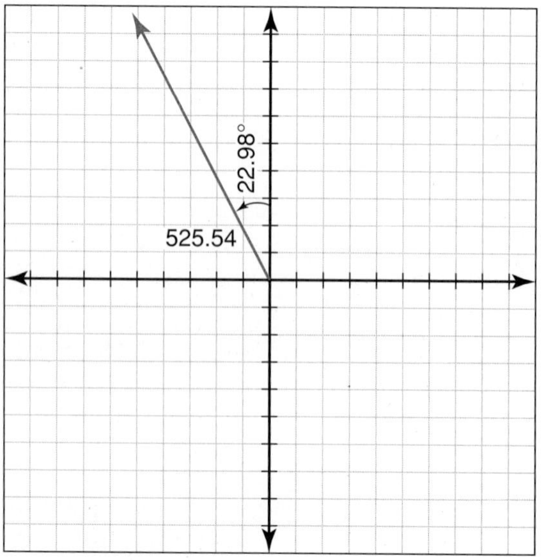

Figure 10.20 The resultant of the vectors representing the motion of the jets and the wind

EXAMPLE 10.12 A swimmer can swim in still water at a speed of 3 miles per hour. A river flows at the speed of 4 miles per hour. If the swimmer strikes out perpendicular to the banks of the river, find his resultant velocity in magnitude and direction.

SOLUTION

The diagram in Figure 10.21 shows the relationship between the velocity vectors involved in the problem. If $\vec{V}$ is the resultant velocity of the swimmer, then $|\vec{V}| = \sqrt{3^2 + 4^2} = 5$ miles per hour. The direction angle of the resultant vector is $\theta = \arctan \frac{3}{4} \approx 37°$, so the swimmer is actually moving at an angle of 37° relative to the banks of the river, with a resultant speed of 5 miles per hour.

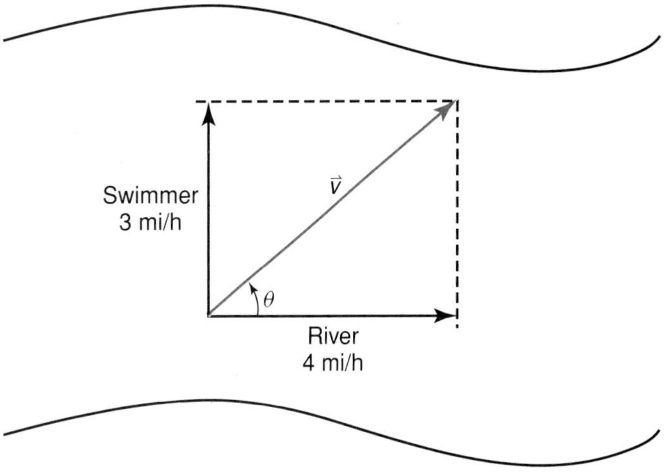

Figure 10.21 Motion of the swimmer and the river

Multiple Forces Acting to Move an Object

When forces are applied to objects, either to push them forward or to pull them along, those forces have a magnitude and a direction of application and thus are vector quantities. If we happen to know the magnitude and direction of the forces, then their resultant would give the magnitude and direction of the forces acting in combination, and if we know the net effect of the forces and the magnitude and direction of one of the forces and either the magnitude or the direction of the other force, we can determine the precise nature of that second force.

Let's suppose we have two individuals, a man and a boy, pulling a load using ropes attached to a ring set into the side of the object. Figure 10.22 depicts the situation. The man exerts a force of 10 N on the object, while the boy exerts an unknown amount of force. If the direction angles of the forces are as indicated in the figure, what is the amount of force the boy must be employing if the object is moving along the dashed line shown?

Using the methods we've already developed, we can show the rectangular form of vector $\vec{A}$, representing the force exerted by the man, to be $\langle 10\cos 30°, 10\sin 30°\rangle \approx \langle 8.66, 5\rangle$, and the rectangular form of the force vector applied by the boy to be $\langle F\cos 310°, F\sin 310°\rangle \approx \langle 0.64F, -0.77F\rangle$. Since the load moves directly along the x-axis, as indicated, it must be that the y-components of the

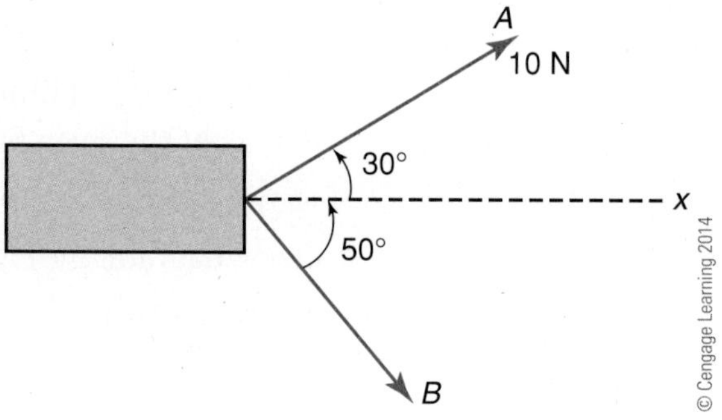

Figure 10.22 Forces pulling a block

vectors "cancel one another out." That is, we must have $5 + (-0.77F) = 0$, or that $F \approx 6.49$, so the boy must be exerting a force of 6.49 N in the direction indicated in order to balance the force applied by the man.

EXAMPLE 10.13 A technician places a 100-pound box of computer parts on a ramp having an angle of inclination 30° to the horizontal and pulls the box up the ramp using a wire harness attached to one end (Figure 10.23) by exerting a force of 75 pounds at an angle of 20° to the ramp. What is the force of friction restricting the movement of the box if the technician pulls it up the ramp at a constant speed?

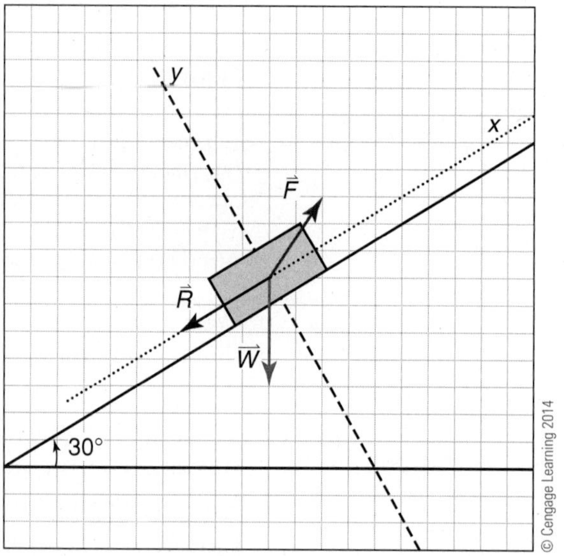

Figure 10.23 Pulling a box up an incline

SOLUTION

In the diagram, each of the forces involved is represented by a directed line segment. $\vec{F}$, $\vec{W}$, and $\vec{R}$ represent the force exerted by the technician, the downward force of the weight of the box, and the restricting force of the friction of the ramp's surface, respectively. Note that, for convenience, we've imposed a coordinate system that has the positive x-axis aligned with the direction of motion of the box's center of mass.

Newton's law of motion tells us the sum of the external forces on a body is equal to the product of the mass of the object and its acceleration ($\vec{F} = m\vec{a}$). Since the box is known to be moving at a constant speed, the acceleration is 0, and thus the resultant of the forces involved must be 0.

Because the frictional force on the object is along the negative x-axis (as assigned by us), we'll find the x-component of each of the three forces. $\vec{F}$ is a 75-pound force acting at 20° to the ramp, so the x-component of force is $\vec{F}$ is $75\cos 20° \approx 70.48$. The x-component of the weight of the box is $100\cos(-120°) = -50$, and the x-component of the resistance vector $\vec{R}$ is $-\left|\vec{R}\right|$. The sum of the x-components, therefore, is $70.48 + (-50) + \left(-\left|\vec{R}\right|\right) = 0$, which implies $\left|\vec{R}\right| \approx 20.48$ pounds.

Simulating Intelligence: Detecting Semantic Similarity

One of the challenges in constructing a machine that can simulate thought is the process of comparison. How can we have a machine determine if two items, such as written documents, are similar in form? This might be valuable for an instructor who was grading papers written by students and submitted electronically.

An idea that might prove useful would be semantic similarity, where a judgment was made regarding how closely related two documents would be. Unfortunately, this would require that we create a means of comparing two documents that a machine could employ to perform the task. A possible means of doing so can be created using vectors, although in a nonobvious manner.

Suppose we arbitrarily assign numerical values to words occurring within a sentence in a particular order. For instance, we can assign the values given in Table 10.1.

TABLE 10.1 Additional Trigonometric Identities

Occurrence Order	Word
1	the, a, an, I
2	pretty, attractive, glamorous
3	girl, woman, lady
4	walked, strolled, crept

For argument's sake, suppose we have an exemplar document, submitted by a student, that has the following sentence fragments:

The attractive woman walked . . . and A pretty girl ran . . .

We can assign a vector to each of the fragments, whose components express how many times a word from the list appears in the sentence in the specified order. From these two fragments, we can create the following vector: $\langle 2, 2, 2, 1\rangle$. The components of this vector indicate that the first word in the string matches a word from the table in both fragments, the second matches a word from the table in both fragments, the third matches a word from the table in both fragments, and the fourth matches a word from the table in only one fragment. Observe that this process creates a vector with four components, which is beyond the scope of our introduction to vectors, but this is shown for illustrative purposes only—if we restrict our attention to the first two words only (to limit the vectors to two components), we can get a primitive view of the problem of document comparison, so the exemplar document would correspond to the vector $\langle 2, 2\rangle$.

We suspect that the student has plagiarized his work from another document containing similar topics and that that document has within it the following fragments:

The glamorous woman walked ... and His pretty girl smiled ...

These two fragments can be expressed in vector form as $\langle 1, 2, 2, 1 \rangle$, or (if we again restrict our attention to the first two components only) the vector $\langle 1, 2 \rangle$. Of course, by restricting our attention to the first two components of the vector (and hence the first two words of the sentences), we are producing an extremely crude comparison between the fragments, but we are limited, for now, by our experience with two-dimensional vectors.

How can we use the vectors $\langle 2, 2 \rangle$ and $\langle 1, 2 \rangle$ to compare the documents? One possible method is to calculate the angle between the vectors. The closer the two fragments are to one another semantically, the more similar the representative vectors will be and hence the smaller the angle between the vectors. In this case, the direction angle of the first vector is easily seen to be 45°, while the direction angle of the second is 63.4°. The difference between the two direction angles, 18.4°, provides us with a means of defining the semantic similarity between the two documents, from which we can determine if that similarity rises to the level of plagiarism. All that remains is the imposition of an arbitrary scale with which we can determine if the level of similarity is too great to be permitted in an assignment.

Exercises

Answer the following problems involving applications of vectors.

1. An airplane flies due north at 600 kilometers per hour. It flies through a wind shear moving west at 160 kilometers per hour. Find the heading of the plane with respect to the ground, the speed of the plane with respect to the ground, and the distance traveled by the plane in 45 minutes.
2. Two paramedics lift a rugby player from the ground on a stretcher. One of the paramedics exerts a force of 400 N at an angle of 50° above the horizontal, while the other exerts a force of 300 N at 45° above the horizontal. What is the combined weight of the rugby player and the stretcher?
3. A river 0.6 kilometers wide flows at 10 kilometers per hour. Jose is traveling in a boat whose speed in still water is 30 kilometers per hour. In what direction should he travel (relative to the shore) if he wants to travel straight across the river (round your direction to the nearest tenth of a degree)? How many minutes will be required for Jose to cross the river, traveling in that direction?
4. A mass is suspended from a beam (Figure 10.24). The mass exerts a downward force of 1000 N. A rope is tied to the mass and pulls it horizontally with a force of 700 N. What force is exerted by the rope connected to the beam to hold the mass?
5. Billie can swim at a speed of 1.25 meters per second in still water. She plans to swim directly across a river whose downstream current has speed 0.8 meters per second. What heading must Billie take? What is

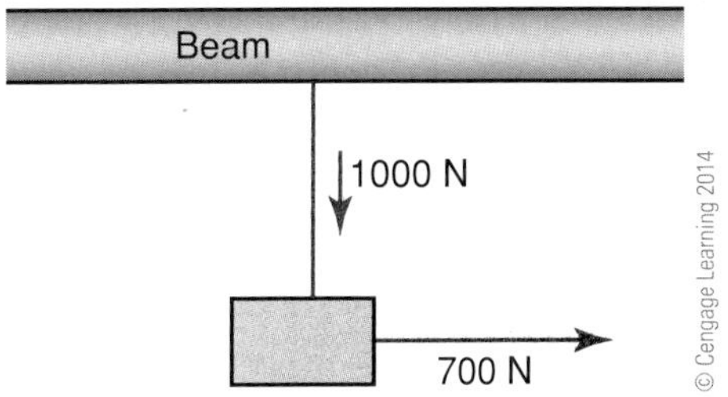

Figure 10.24 Mass suspended from a beam

her velocity (to the nearest tenth of a meter per second) relative to the bank of the river?

6. An airplane, heading due east with airspeed 300 miles per hour, is blown off course by a wind from the south blowing at 56 miles per hour. After two hours, how far was the airplane from its starting point? What was its actual speed relative to the ground? In what direction was it actually flying (to the nearest degree)?
7. Two farmers pull on ropes tied to a stubborn donkey (Figure 10.25). Find the single force equivalent to the sum of the two forces shown.

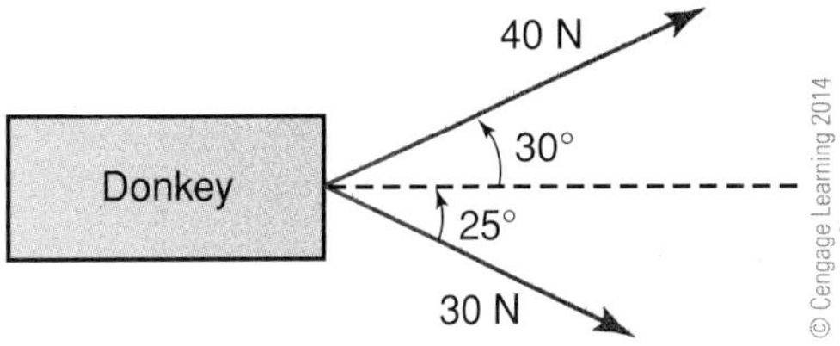

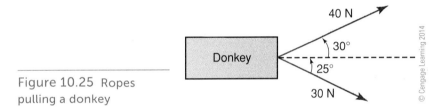

Figure 10.25 Ropes pulling a donkey

8. Three forces are applied to an object: force $\vec{A}$ is 74 N, 85° north of west; force $\vec{B}$ is 125 N, 64° south of west; and force $\vec{C}$ is 50 N, 80° south of east. What is the magnitude and direction of the sum of the forces?
9. Three forces act on an object at point O (Figure 10.26). What is the resultant force on the object?

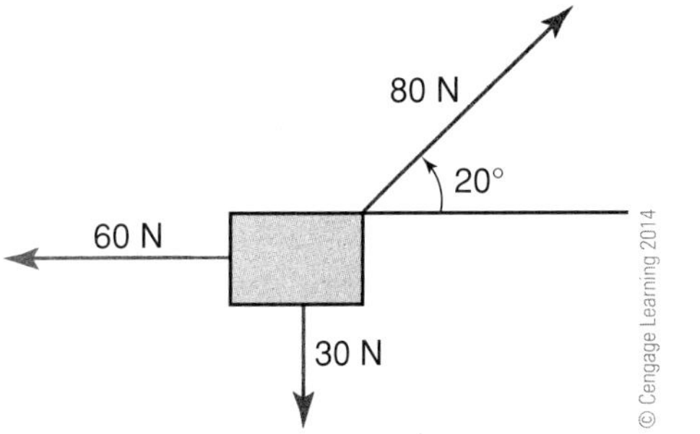

Figure 10.26 Three forces acting on an object

10. A tank travels due north at 15 meters per second and fires a shell whose velocity is 250 meters per second in a direction that appears to be due west to the gunnery officer aboard the tank. What is the true velocity and direction of the shell relative to the ground?
11. Two forces, 120 pounds and 200 pounds, act on a body and make a 62° angle with one another. What is the magnitude and direction of the resultant of the forces?

12. Two forces, 300 N and 58 N, act on a body and make an angle of 110° with one another. What is the magnitude and direction of the resultant of the forces?

Summary

In this chapter, you learned about:

- The definition of the term vector.
- Various ways of representing vectors: directed line segment, polar, and rectangular forms.
- Resolving a vector into its x- and y-components.
- Calculating the resultant of two vectors using algebraic addition and the tip-to-tail method.
- Some applications using vectors.

Glossary

directed line segment A line segment assigned a direction, indicated by an arrowhead at one end of the segment.

rectangular components of a vector The x- and y-components of a vector.

rectangular form of a vector $\langle a, b,\rangle$, where (a, b) is the terminal point of the vector when in standard position.

resultant The sum of two vectors.

standard position of a vector A sketch of the vector where its initial point is at the origin.

tip-to-tail method of vector addition A visual method of calculating the resultant of two vectors.

vector A quantity possessing both direction and magnitude.

x-component of a vector $R_x = (|\vec{v}| \cos\theta)\,\vec{i}$, where $\vec{i}$ is the standard unit vector $<1, 0>$ and θ is the angle between the positive x-axis and the vector when drawn in standard position.

y-component of a vector $R_y = (|\vec{v}| \sin\theta)\,\vec{j}$, where $\vec{j}$ is the standard unit vector $<0, 1>$ and θ is the angle between the positive x-axis and the vector when drawn in standard position.

End-of-Chapter Problems

The following exercises give the initial and terminal points of a vector. Give the polar representation for the vector.

1. $(3, 5)$ and $(4, 7)$
2. $(-8, 4)$ and $(4, 3)$
3. $(-7, -15)$ and $(12, -3)$
4. $(8, -15)$ and $(-12, 35)$
5. $(-3, 7)$ and $(6, 3)$

The following exercises reference problems earlier in this problem set. For each referenced problem, give the rectangular form of the representation of the vector.

6. Exercise 1
7. Exercise 2
8. Exercise 3
9. Exercise 4
10. Exercise 5

In the following problems, the rectangular representation of a vector is given. Convert the representation to the polar representation. Do not approximate the magnitude but do round degrees to the nearest tenth of a degree, as needed. Sketch the given vector in the plane, in standard position.

11. $\langle -3, 4 \rangle$
12. $\langle -3, -5 \rangle$
13. $\langle -2, 4 \rangle$
14. $\langle 4, -4 \rangle$
15. $\langle 5, 3 \rangle$

The following problems give a vector in polar form. Give the rectangular components of the vector, rounded to the nearest tenth of a unit.

16. $5\angle 75°$
17. $\sqrt{32}\angle 135°$
18. $43\angle 125°$
19. $7\angle 355°$
20. $15\angle 225°$

Resolve the following vectors into their rectangular components, rounding the angles to the nearest hundredth of a degree, as needed. Sketch the vector in standard position.

21. $\langle 4.75, -3.5 \rangle$
22. $\langle -3.5, 5 \rangle$
23. $\langle 2.5, 4 \rangle$
24. $\langle -4.5, -2 \rangle$
25. $\langle 2.5, 3.5 \rangle$

The following problems describe a situation upon which vectors can be imposed. Determine the rectangular form of the vector and then produce the resolution of the vector.

26. A shipping carton rests on a ramp that is inclined 25° to the horizontal. Gravity is exerting a force of 75 N on it straight down. Resolve the force on the block into two components, one along the ramp and the other perpendicular to it.
27. A graphics tablet has a resolution of 1200 dots per inch, and an *x*-*y* coordinate system. A line is drawn from $(-2412, 37)$ to $(317, 245)$. The vector describes the motion of the stylus from its initial to terminal positions. In addition to the resolution, determine how far the stylus moved.
28. A man is steering a car and is heading 20° west of north at 60 miles per hour. How fast is he moving to the west? How fast is he moving to the north?
29. A bicycle is rolling down a hill that is sloped at 15°. Gravity is exerting a force of 1000 N on it straight down. What are the resultant forces down the hill and into the hill?
30. A sailboat is constrained by its keel and rudder to go straight north. A wind is blowing from the southeast at 20 miles per hour. Resolve the wind into its northerly and westerly components. (Note: In navigation, true north is 0° and east is 90°.)

Find the resultant of the following pairs of vectors.

31. $\langle -3, 5 \rangle$ and $\langle 3, 5 \rangle$
32. $\langle 18, -4 \rangle$ and $\langle -2.75, 8 \rangle$
33. $\langle 15, -43 \rangle$ and $\langle 37, 9 \rangle$
34. $\langle -5, -3 \rangle$ and $\langle -3.5, 12 \rangle$
35. $\langle 13, 13 \rangle$ and $\langle -1, -3 \rangle$

Convert the following vectors (given in polar form) to their corresponding rectangular representations and then find the sum of the vectors. Convert the sum of the vectors back to polar representation.

36. $5\angle 60°$ and $7\angle 120°$
37. $-\frac{7}{8}\angle 35°$ and $1.2\angle 193°$
38. $12\angle 53°$ and $7\angle 225°$
39. $7\angle 115°$ and $3\angle 272°$
40. $5.5\angle 195°$ and $8.3\angle 98°$

For the following vectors, perform the addition of vectors using the tip-to-tail method. First sketch each of the given vectors in standard position, then draw the resultant vector, also in standard position. Verify the answer by finding the rectangular representations of the individual vectors and showing the resultant matches that found with the tip-to-tail method.

41. $\langle 2, -2\rangle$ and $\langle 3, 5\rangle$
42. $\langle -3.5, -2.5\rangle$ and $\langle 4.25, 4.5\rangle$
43. $\langle -4, -4\rangle$ and $\langle 1, 5\rangle$
44. $\langle 3.25, -4.5\rangle$ and $\langle -3, 5\rangle$
45. $\langle -2, -2\rangle$ and $\langle -2, 2\rangle$

Solve the following applications of vectors. Use sketches, as needed, to form the basis of your solution.

46. A small aircraft has a cruising speed of 100 miles per hour. A wind is blowing from N20°E at 20 miles per hour. What heading should the aircraft take in order to be actually moving due north, and how fast will it be moving?
47. Three forces are applied to an object. The first is 75 N exerted in direction 17° east of north, the second is 30 N exerted in direction 33° west of south, and the third is 45 N exerted in direction 25° east of south. What is the resultant force on the object?
48. A plant pot weighing 10 kilograms is hanging from a hook in the ceiling. Gravity is exerting a force on it of 981 N straight down. The hook is rated at 1200 N. Some enterprising soul has pulled the pot out of the way of tall people with a string that is extending horizontally from the pot to the right. The string is exerting a force of 400 N on the pot. The pot is not moving. Is the hook in danger of failing?
49. A car starts out from point A in the desert and drives 45 miles northwest. The driver then changes direction to 20° north of east and drives for an additional 30 miles. Once again she changes direction to head due north and drives an additional 60 miles. How far and in what direction from point A is the car?
50. The main nozzle on a spaceship is adjusted slightly incorrectly and is thrusting 5° to the right of the centerline of the ship. It generates a force of 10,000,000 N. What force does the thruster on the other side of the ship have to apply in order to keep the craft flying straight along its centerline? Assume that the side thruster thrusts at 90° to the centerline.

Chapter 11 Exponential and Logarithmic Equations

Early in the computer age, Gordon E. Moore (one of the founders of Intel) observed that the number of components in integrated circuits (IC) per minimum cost had doubled every year from the invention of the IC in 1958 through 1965. Moore predicted that this trend would continue, at least for the 10 years following his analysis, and this conjecture proved to have a high degree of accuracy. Many aspects of technological advancement appear to follow a similar pattern of growth, and this progression has (in its various incarnations) come to generally be referred to as Moore's law.

This type of change, where a value from year to year is a constant multiple from the predecessor, is referred to as exponential change. If the change represents an increase, we refer to the situation as exponential growth, while decreases are said to indicate exponential decay. Equations where such change occurs are said to be exponential equations, and it is to this new type of equation that we now turn our attention.

Closely related to the concept of an exponential is that of the logarithm, which will be seen to be the mathematical inverse of the exponential operation. As we examine equations involving these two topics, we will see that they appear in a wide variety of common "real-world" applications and that developing an understanding of them will provide us with an important analytical tool.

Objectives

When you have successfully completed the materials of this chapter, you will be able to:

- Graph an exponential function and identify it as being representative of exponential growth or decay.
- Solve problems involving exponential growth or decay.
- Convert between exponential and logarithmic forms.
- Evaluate logarithms and antilogarithms.
- Utilize the properties of logarithms.
- Solve logarithmic and exponential equations.
- Solve applied problems leading to exponential or logarithmic equations.

11.1 Exponential Functions

Let's examine Moore's observation about the increase in the number of components on an IC a bit more closely. Suppose, for the sake of simplicity, that the number of components on an IC that could be produced for minimum cost was 5 in 1958. The pattern of growth described by Moore would suggest that this number would grow to 10 in 1959, then to 20 in 1960, 40 in 1961, and so on.

Using time reference $t = 0$ to correspond to 1958, it follows that the growth in the number of components that can be used in an IC would obey the function $f(t) = 5(2^t)$. Note that evaluation of this function for $t = 0, 1, 2,$ and so on yields the correct value corresponding to the number of components on an IC for minimum cost in a given year.

A function of this type, having the form $f(x) = ca^x$, where c is a constant of any nonzero value and a is a nonnegative constant not equal to 0 or 1, is referred to as an **exponential function**. Such functions are characterized by the amount of change in the value of the function being proportional to the amount of the quantity present at a particular time.

The exponent in $f(x) = ca^x$ is shown as x, but it need not be that simple and could actually be some expression involving x. The constant a is referred to as the **base of an exponential function**, and the expression occurring in the exponent is referred to as the **variable exponent**. When the variable exponent is a linear expression in x, the graph of the function will either increase or decrease steadily, as we will see.

When graphing an exponential function, we can either make a table of values, from which we can sketch the curve, or use some sort of graphing utility.

In either case, there are two particularly common graph shapes that arise: that of the exponential growth curve and that of the exponential decay curve.

An exponential growth curve would (from left to right) increase at a progressively faster rate as x increases, creating a curve shaped something like the one shown in Figure 11.1. An exponential decay curve, on the other hand, would fall from left to right but at a progressively slower rate, as shown in Figure 11.2.

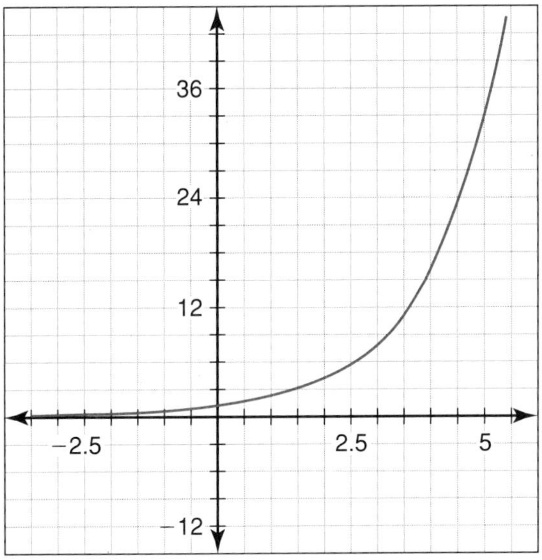

Figure 11.1 An exponential growth curve

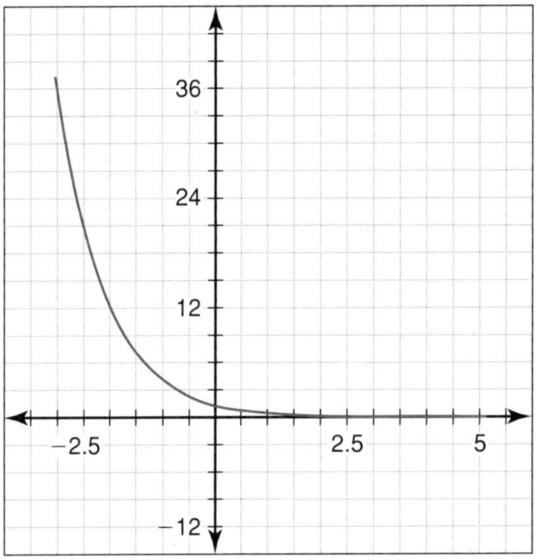

Figure 11.2 An exponential decay curve

EXAMPLE 11.1

Graph the simplified model of Moore's law, $f(t) = 5(2^t)$.

SOLUTION

In this context, the implied domain of the function is $t > 0$, so we can create a table of values beginning with zero and increasing (Table 11.1). We can plot the points by hand, or we can use a graphing utility to obtain a highly accurate graph of the function (Figure 11.3). Observe that the graph in Figure 11.3 rises rapidly on its far right end. This is indicative of what we refer to as an exponential graph and is a characteristic of all such graphs. The growth is so rapid that, when we examine a graph produced using a computer, we can be misled into believing that the graph possesses a vertical asymptote at some location on the horizontal axis, but this is not the case.

TABLE 11.1 Table of Values for $f(t) = 5(2^t)$

t	$f(t)$
0	5
1	10
2	20
3	40
4	80
5	160

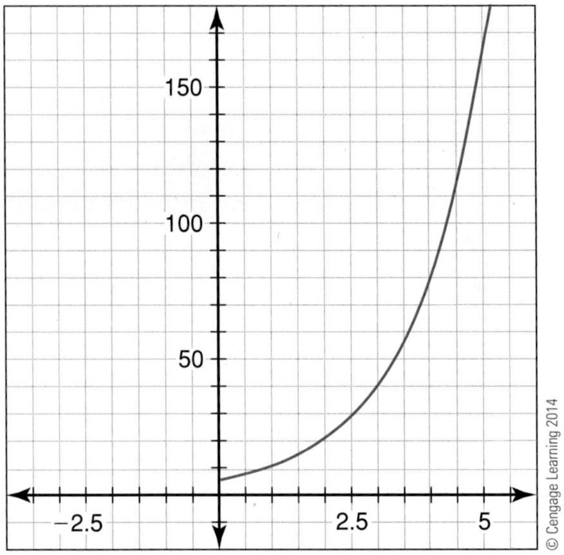

Figure 11.3 Graph of Moore's law

EXAMPLE 11.2 Graph the function $g(x) = e^{-0.5x}$, where $e \approx 2.718$.

SOLUTION

As before, we can construct a table of values and plot the points by hand, but we will rely on a graphing utility to produce the graph here (Figure 11.4). Here, we seem to have the reverse of the condition we saw in Example 11.1. The graph falls as we move from left to right, and levels off to the right. This behavior is what we referred to as exponential decay, and is typically seen in problems where a value is decreased by some fractional proportion over regular periods of time. This is commonly seen in applications such as the radioactive decay of unstable elements.

In Example 11.2, we encountered the base value e, which we approximated as 2.718 for the purposes of the example. The value e is an extremely important

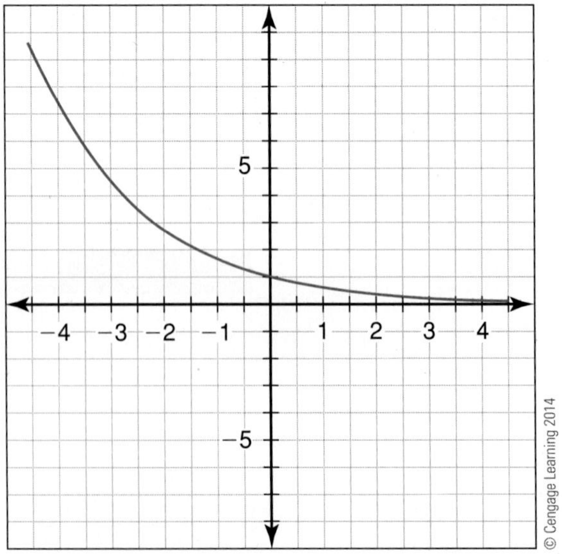

Figure 11.4 Graph of an exponential function

constant in applications and is referred to as the **natural exponential base.** The motivation for this name is that the value arises naturally in a wide variety of applications, from finance to optics to acoustics to population growth. The derivation of the value of e takes us away from our focus in this textbook but is an interesting excursion you may encounter in other courses. Like π, the value of e is an irrational number having an infinite, nonterminating, nonrepeating decimal representation. A partial representation of e is 2.71828182845904523560287, but a common approximation is 2.718. Your calculator has a stored value of e, and you should use that value unless asked specifically to use another approximation. Any exponential function whose base is the mathematical constant e is referred to as a **natural exponential function.**

EXAMPLE 11.3

A study is made to ensure that the temperature of a room does not become uncomfortably high during the course of a meeting. A technician finds that the temperature within the meeting room can be modeled by the equation $T(t) = 85 - 14e^{-t}$, where t is the time in minutes. Produce a graph of the exponential function and draw a conclusion about the temperature within the room during the meeting if it is known the meeting must not exceed two hours in length.

SOLUTION

Using a graphing utility, we can produce the graph of the exponential function, from which we can draw our conclusions. The graph is shown in Figure 11.5. Based on the graph, we can see that the temperature in the room will start slightly below 75°, will rise quickly, and will then level off. Noting that the scale on the vertical axis is 5 units per hash mark, we conclude that the temperature in the room will level off near 85°. From this, the technician can determine if such a temperature is uncomfortably high and thereby require an adjustment to the air-conditioning system.

Note that this graph does not display either exponential growth or decay. It evidently demonstrates some sort of asymptotic behavior, since the room

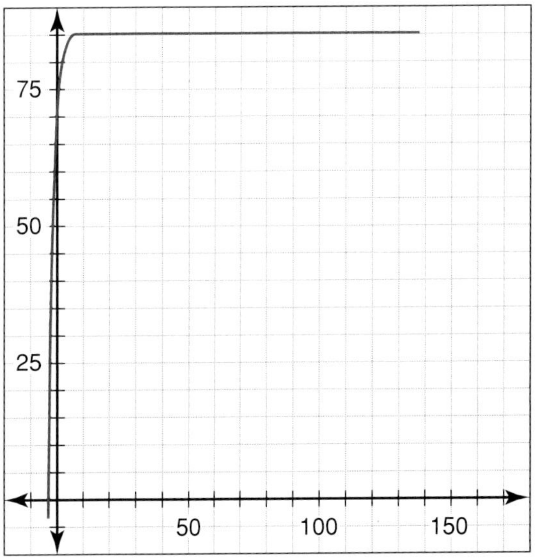

Figure 11.5 Graph of $T(t)$

temperature levels out, but since it is not decreasing, the change is not exponential decay.

It is difficult, in general, to determine by inspection if a particular exponential function's graph will demonstrate exponential growth or decay. At times, it's tempting to suspect that there is something intrinsic about the value of the base (whether it is either less than 1 or greater than 1) or the sign of the variable exponent (positive or negative), but this can be challenging to determine.

For instance, in Example 11.2, we saw an illustration of exponential decay, and the value of the base was e. This might lead us to conclude that that base value creates exponential decay, but this is refuted by Example 11.3. It is better (at least until more experience is gained) to produce the graph of the exponential function first and then base a conclusion of growth or decay on the curve generated.

EXAMPLE 11.4

Using a splitter in a cable line will cause the signal going back to the cable company to weaken, which can be particularly troublesome if using a cable modem. If the signal the modem returns to the company is too weak, your service may not work. Suppose the signal strength percentage for a splitter is modeled by the following function, where x is the number of ports in the splitter: $L(x) = 100(0.68)^x$. Graph the exponential function and determine if the situation represents a growth curve or a decay curve.

SOLUTION

The graph of $L(x)$ is given in Figure 11.6. Clearly, the graph demonstrates exponential decay, indicative of a rapid return signal strength diminution as the number of ports in the splitter increases. By the time we have used a three-port splitter, the graph suggests that a return signal strength of less than 35% would occur, which would possibly cause a failure in our cable service.

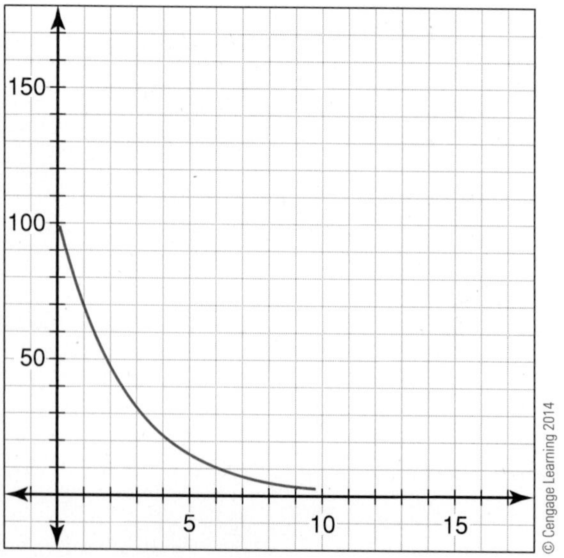

Figure 11.6 Graph of $L(x)$

Exercises

Find the function value.

1. Let $f(x) = 6^x$. Find $f(2)$.
2. Let $f(x) = \left(\frac{1}{6}\right)^x$ Find $f(2)$.
3. Let $f(x) = 4^x$. Find $f(-3)$.
4. Let $f(x) = \left(\frac{1}{3}\right)^x$ Find $f(-3)$.
5. Let $f(x) = 2^x$. Find $f(5)$.
6. Let $f(x) = 2^x$. Find $f(1.5)$.
7. Let $f(x) = 2^x$. Find $f(2.96)$.
8. Let $g(x) = \left(\frac{1}{3}\right)^x$ Find $g(1.5)$.
9. Let $f(x) = 4^x$. Find $f\left(\frac{5}{2}\right)$.
10. Let $f(x) = \left(\frac{1}{3}\right)^x$ Find $f\left(\frac{5}{2}\right)$.

Graph the function.

11. $f(x) = 4^x$
12. $f(x) = \left(\frac{1}{5}\right)^x$
13. $f(x) = 4^{-x}$
14. $f(x) = 2^{|x|}$
15. $f(x) = \left(\frac{9}{2}\right)^x$
16. $f(x) = \left(\frac{4}{5}\right)^x$

Solve the equation.

17. $2401^x = 7$
18. $\left(\frac{1}{7}\right)^x = 49$
19. $\left(\frac{5}{4}\right)^x = \frac{16}{25}$
20. $2^{-x} = \frac{1}{8}$
21. $4^{(9-3x)} = 64$
22. $2^{(5-3x)} = \frac{1}{16}$
23. $4^{(5+3x)} = \frac{1}{256}$
24. $36 = b^{\frac{2}{5}}$
25. $a^{\frac{3}{5}} = 64$
26. $\left(\frac{64}{27}\right)^{x+1} = \left(\frac{3}{4}\right)^{x-1}$

11.2 LOGARITHMS

Logarithms, the development of which is generally attributed to John Napier in 1614, were originally defined in Germany by Michael Stifel in 1544 in his textbook *Arithmetica Integra*. The creation of logarithms allowed mathematicians and scientists of the day to perform extremely difficult computations in applied fields and contributed greatly to the advance of scientific knowledge.

Definition of Logarithms

A logarithm is, essentially, an exponent, though in practice it is rare that one thinks of a logarithm in that way. In this section, we'll introduce the definition of logarithms, explore their relationship to exponentials, and introduce and use some of the logarithmic properties.

We'll start with the definition: If we know that N is a number such that $A^N = B$, where A is positive but not 1, then we say that N is the **logarithm**, base A, of the number B. That is, N is the power to which A must be raised in order to obtain the value B. Any expression involving a logarithm of any particular base is referred to as a **logarithmic expression.**

For instance, since $4^2 = 16$, we can say "2 is the logarithm base 4 of 16." The notation introduced is $2 = \log_4 16$, which we read as "2 is equal to the logarithm base 4 of 16," or (more briefly) "2 is the log base 4 of 16." Often, the value of the logarithm (that is, the value of the exponent N) is unknown and is in need of calculation. In some cases, this can be done by inspection.

EXAMPLE 11.5

What is the value of $\log_3 81$?

SOLUTION

When attempting to calculate a value of a logarithm (which, remember, is an exponent), it is generally best to set up a "dummy" variable, such as x, equal to the logarithm and then use the definition of logarithms to reorganize the equation into a more familiar form.

That is, we would say $x = \log_3 81$, and then (according to the definition) we can restate this to read $3^x = 81$. From our experience, we recall that 81 is the fourth power of 3, and therefore $x = 4$. This transition back and forth using the definition of logarithms is so common that the two equalities within that definition are given specific names. $A^N = B$ is referred to as the **exponential form of an equation**, while $N = \log_A B$ is called the **logarithmic form of an equation.** According to the definition, these two expressions are interchangeable, and one can be exchanged for the other whenever it proves convenient.

Note

If $A^N = B$, then $N = \log_A B$.

Before proceeding, let's go through a few additional conversions between the forms to ensure that we fully understand this equivalence.

EXAMPLE 11.6

Convert the following exponential expressions to the equivalent logarithmic form:

$$2^5 = 32$$

$$4^{-3} = \frac{1}{64}$$

$$10^3 = 1000$$

SOLUTION

In this case, the base of the exponential is 2, and this will also be the base of the logarithm. Using the equivalence relation given previously, the converted logarithmic form is $5 = \log_2 32$. In the second case, the base value is 4 for both the exponential and the associated logarithm, and our equivalent logarithmic form will be $-3 = \log_4\left(\frac{1}{64}\right)$. The base of the third exponential is 10, and this will also be the base for the associated logarithm. Thus, the equivalent form will be $3 = \log_{10} 1000$.

EXAMPLE 11.7

Convert the following logarithmic expressions to the equivalent exponential form:

$$\log_5 625 = 4$$
$$\log_7\left(\frac{1}{49}\right) = -2$$
$$\log_2 512 = 9$$

SOLUTION

As was the case in the conversions from exponential to logarithmic form, the key idea is identification of the logarithmic base and using that same value as the base of the exponential.

In the first case, the base is 5, and the equivalent exponential form will be $625 = 5^4$. In the second case, the base of the logarithm is 7, and this will also be the base for the exponential form, which is $\frac{1}{49} = 7^{-2}$. The base for both the logarithm and the associated exponential in the third case is 2, and the equivalent exponential expression is $512 = 2^9$.

With the ability to convert freely between the forms now in hand, we can calculate the approximate value of particular logarithms whose value cannot be determined through inspection. This can be done either by using a table of values or through the use of a calculator (and a particular mathematical rule, yet to be presented!).

EXAMPLE 11.8

What is the value of $\log_5 75$?

SOLUTION

Here, if we repeat the process of Example 11.5, we find that we immediately run into difficulties. We can set up the expression $x = \log_5 75$ and then convert to exponential form, where we obtain the equivalent expression $5^x = 75$. Here, however, we note that we are not as fortunate as we were before, since 75 is not a power of 5.

We can play with the expression, however, to attempt to find a "ballpark" value of x. We know that $5^2 = 25$ and $5^3 = 125$, so it must be that x, the power on 5 we seek, is between 2 and 3, somewhere.

At random, we can try $5^{2.5}$, which yields approximately 55.9, from which we conclude that the actual value of x must be greater than 2.5. Table 11.2 helps us organize our work a bit. Clearly, we can continue to make successive guesses, homing in on the desired value of x, attaining whatever degree of accuracy we wish. This is, obviously, not an efficient method of finding the value of x, since it involves considerable calculation, but it can be done, and in just the few steps we've shown that we've established that the true value must be between 2.68 and 2.685, which is already reasonably accurate. Shortly, we'll see a method for finding such a value of x quickly.

TABLE 11.2 Estimates for the Logarithm

x	5^x
2.6	65.7 (too low)
2.7	77.12 (too high)
2.65	71.2 (too low)
2.675	74.1 (too low)
2.685	75.3 (too high)
2.68	74.7 (too low)

EXAMPLE 11.9 What is the value of $\log_3 100$?

SOLUTION

Setting up the equality $x = \log_3 100$ and converting to exponential form, we obtain $3^x = 100$. As was the case in Example 11.8, 100 is not a power of 3, and therefore we are down to using a table of values and attempting to home in on the value of the logarithm by trial and error once again.

Realizing that $3^4 = 81$ and $3^5 = 243$, we see that the value of x must lie between 4 and 5, probably closer to 4. Thus, we can create a table of values and experiment (Table 11.3). Based on this work, we can conjecture that $\log_3 100$ is approximately 4.1918.

TABLE 11.3 Estimates for the Logarithm

x	3^x
4.2	100.904 (too high)
4.18	98.711 (too low)
4.19	99.802
4.195	100.351 (too high)
4.193	100.131 (too high)
4.192	100.021
4.1919	100.010
4.1918	99.999

Clearly, this method of finding approximate values of logarithms is not difficult, but it is rather clumsy and inefficient. Let's now give the rule for decimal approximating logarithms using a calculator, and then we'll examine an application or two and ultimately return to introduce other properties of logarithms.

Note

For common logarithms, where the base is 10, the notation is $\log B$.

Note

For natural logarithms, where the base is e, the notation is $\ln B$.

Common and Natural Logarithms

If the logarithmic base happens to be 10 or e, our calculators contain keys programmed to return the value of the logarithms. For base-10 logarithms, also known as common logarithms, the calculator key we need is labeled LOG. For base e logarithms, also known as **natural logarithms**, the calculator key to use is labeled LN or ln x. The identification of these keys is closely tied to the notation used for the common and natural logarithms, which is somewhat different from that of other logarithm bases.

When the base is 10, it is typically suppressed, so it is common to see $\log_{10} B$ written simply as $\log B$. When the base is e, we usually write not $\log_e B$ but rather $\ln B$.

EXAMPLE 11.10 Using the appropriate calculator keys, give the decimal approximation to the following logarithms and then check your answer by converting to exponential form:

a. $\log 15$
b. $\ln 15$
c. $\log 138$
d. $\ln 138$
e. $\log(-200)$
f. $\ln(-200)$

SOLUTION

a. Using the LOG key of the calculator, we find the approximate value of $\log 15$ to be 1.176. Conversion to exponential form yields $10^{1.176} \approx 15$, and if we approximate the left side, we find $14.997 \approx 15$, which verifies the result.

b. Using the LN key of the calculator, we find $\ln 15 \approx 2.708$. Observe the importance of using the proper calculator key, since the different logarithmic keys yield substantially different numerical results! Converting to exponential form, we find $e^{2.708} \approx 14.999 \approx 15$, and this confirms the result.

c. Again using the LOG key of the calculator, we find $\log 138 \approx 2.140$, which we confirm by converting to exponential form and seeing $10^{2.140} \approx 138.038$. Note that the exponential value is very close to the expected result; we should be within a very small fraction of the desired value when we perform our check.

d. With the LN key of the calculator, we find $\ln 138 \approx 4.927$. Conversion to exponential form verifies this result, as $e^{4.927} \approx 137.965$.

e & f. Use of either calculator key for the computation of these logarithms yields some form of error message on your calculator. Why would this be the case? Consider $\log(-200) = x$, converted to exponential form: $-200 = 10^x$. This is true only if the positive number 10 can be raised to some power that would yield a negative value. By the properties of real numbers, this is impossible, and therefore no such value of x can exist. A similar observation is made in the case of natural logarithms, which also yield an error message.

The Change of Base Theorem

If the logarithmic base is neither 10 nor e, we cannot calculate the value of the logarithm directly using a calculator key. This seems very unfortunate, since an infinite list of potential logarithmic bases (any positive value other than 1) is left to be addressed. However, the two calculator keys we have prove sufficient for the computation of any other logarithmic base because of a calculation rule called the change of base theorem.

Change of Base Theorem

To calculate $\log_A B$, for any positive value of A not equal to 1, we have $\log_A B = \dfrac{\log_t B}{\log_t A}$, for any positive value of t, not equal to 1. In particular, we have $\log_A B = \dfrac{\log B}{\log A}$, and also $\log_A B = \dfrac{\ln B}{\ln A}$.

EXAMPLE 11.11 Calculate the approximate values of the following logarithms and compare the results to Examples 11.8 and 11.9:

$$\log_5 75$$
$$\log_3 100$$

SOLUTION

Using the change of base theorem, we find $\log_5 75 = \dfrac{\log 75}{\log 5} \approx \dfrac{1.875}{0.699} \approx 2.683$. Recall that, in Example 11.8, we found (through trial and error) the approximate value of the logarithm to be between 2.68 and 2.685, though we worked no further to find an approximation at that time.

Again using the change of base theorem, we find $\log_3 100 = \dfrac{\log 100}{\log 3} \approx \dfrac{2}{0.477} \approx 4.192$. In Example 11.9, we used a table to approximate the value of the logarithm as 4.1918, which required extensive computation. Note that the actual calculator value from the change of base theorem was about 4.1918065, which is even more accurate than the table value we found.

When we say $\log_A B = N$, we refer to N as the logarithm, base A, of the number B. The number B, on the other hand, also is given a name, and we refer to it as the **antilogarithm of N, base A**. When seeking an antilogarithm, base A, we convert the equation to exponential form and make a direct computation.

EXAMPLE 11.12 What is the antilogarithm, base 8, of -1.35?

SOLUTION

Another way of stating this is (if we view the unknown value of the antilogarithm, base 8, of -1.35 as x) to say that $\log_8 x = -1.35$. Converting to exponential form, we have $x = 8^{-1.35}$, which we can decimal approximate as 0.0604, and this is the value of the antilogarithm.

Applications of Logarithms

Just as many real-life situations can be modeled by exponential expressions, so can others be described using logarithms. Logarithmic quantities are characterized by growth that continues more and more slowly as time progresses (though there are other types of growth, nonlogarithmic, that also have this quality).

For instance, consider a capacitive-resistive circuit, which consists of a capacitance, C, and a resistance, R, in series with a voltage supply V. When the circuit is switched off, the current flowing at time t seconds is given by $C = \dfrac{t}{R(\ln V - \ln R - \ln I)}$. If it is known that $V = 100$ and that $R = 500$ and that 0.01 seconds after the circuit is switched off, the current flowing is 0.198 amperes, we can find the value of the capacitance. A rather straightforward calculation reveals that, in this case, the capacitance is $C = 0.002$, which we encourage you to verify.

What other sorts of situations might lead to slow but continuing growth in a quantity? An example that is possibly closer to our everyday experience might be the growth in staffing of an organization's personnel department. When the company is fairly small and growing, the staffing of the personnel department is apt to grow quickly to accommodate the pressing needs of corporate expansion. As these individuals gain experience and are able to widen the scope of their responsibilities, it is conceivable that they would be able to automate a variety of processes in such a way that each individual in the department could effectively tend to the needs of larger sets of employees. Thus, we might expect the growth of the staffing of the personnel department to follow a curve similar to that shown in Figure 11.7. Here, x represents the employment of the company, in hundreds, and y represents the staffing of the personnel department. This situation is examined in more detail in Example 11.13.

EXAMPLE 11.13 The growth of a particular personnel department at a corporation is modeled by the equation $y = 3 + \ln(40x)$, where y is the number of persons in the department and x is the number of corporate employees, in hundreds. Use the formula to determine the personnel department staffing needs when the corporation has 100, 200, 300, and 1000 employees.

SOLUTION

When there are 100 employees, the value of x is 1, and therefore we compute $y = 6.68$, which we round down to 6 (there are no fractions of people, so we

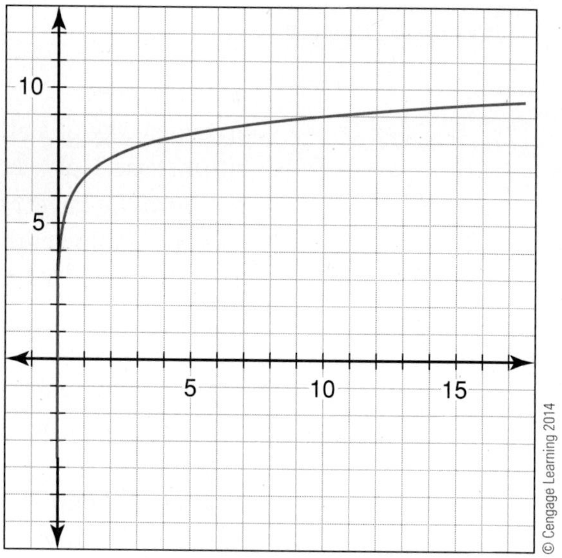

Figure 11.7 Graph depicting staffing in the personnel department

will choose to round down). When the number of employees is 200, we find $y = 7.38$, which we round down to 7, and therefore the increase of employee head count from 100 to 200 necessitates an increase of one person in the personnel department.

If the number of employees increases to 300, we find $y = 7.79$, which means that the 7 members of the department can handle the increased number of employees without additional help, and if the payroll increases to 1000, we find that $y = 8.99$, and hence 8 individuals are needed to staff the personnel department.

Notice we can use the formula describing the growth of the personnel department to anticipate its expansion on the basis of the number of employees. For instance, what if a particular vice president were interested in the number of employees that would necessitate a staffing level of 12 in the personnel department? We can substitute this value for y in the equation and solve by converting to exponential form:

$$\begin{aligned} 12 &= 3 + \ln(40x) \\ 9 &= \ln(40x) \\ e^9 &= 40x \\ 8103 &\approx 40x \\ 202.58 &\approx x \end{aligned}$$

Since x was in hundreds of employees, we conclude that the company would have to increase its payroll to a level of 20,258 in order to justify an expansion of the personnel department staffing to 12.

EXAMPLE 11.14 The impact of an advertising campaign for a new cell phone can be modeled by the formula $S(D) = 3500 + 80 \ln D$, where S is the number of units sold, in hundreds, after D thousand dollars have been spent on the campaign. Evaluate the function for the values of $D = 10$, 20, and 30 and interpret the results.

SOLUTION

When $D = 10$, we find $S = 3684.21$ and conclude that an expenditure of \$10,000 in advertising will yield sales of 368,421 cell phones. Similarly, the calculations for $D = 20$ and 30 yield results of 373,966 and 377,210 cell phones sold, respectively.

Note that increasing the expenditure from \$10,000 to \$20,000 produces a net change in sales of 5545 cell phones, while increasing another \$10,000 to reach an advertising cost of \$30,000 increases sales by only 3244 units. There are many possible reasons for this, but an examination of the graph of the sales function reveals that there is a diminishing return on sales increases as the expenditure on advertising continues to rise (Figure 11.8).

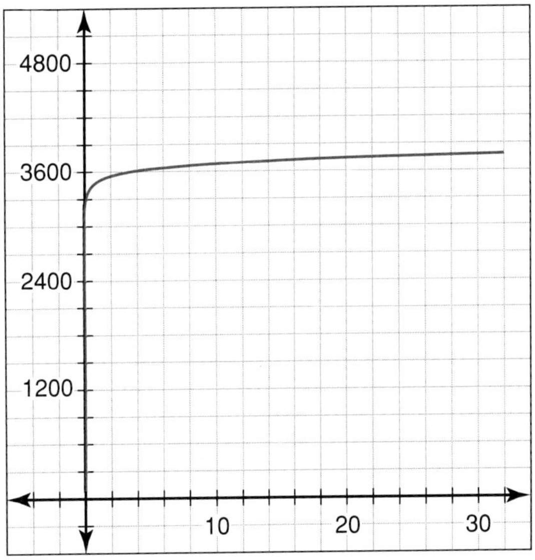

Figure 11.8 Diminishing return on sales

Exercises

Evaluate the following logarithms.

1. $\log_2\left(\frac{1}{2}\right)$
2. $\log_9\left(\frac{1}{81}\right)$
3. $\log_{10} 1$
4. $\log_{\frac{1}{5}} 5$
5. $\log_7 \sqrt{7}$
6. $\log_7 343$
7. $\log_\pi \pi$
8. $\log_{20} 1$
9. $\log_{10}(1{,}000{,}000)$
10. $\log_{20}(-1)$

Write the following logarithmic form.

11. $6^3 = 216$
12. $2^3 = 8$
13. $2^{-3} = \frac{1}{8}$
14. $\left(\frac{3}{11}\right)^3 = \frac{27}{1331}$

15. $\left(\frac{3}{11}\right)^{-3} = \frac{1331}{27}$

16. $4^{\frac{3}{2}} = 8$

17. $10^4 = 10{,}000$

18. $10^{-5} = 0.00001$

Write the following in exponential form.

19. $\log_5 125 = 3$

20. $\log_2\left(\frac{1}{4}\right) = -2$

21. $\log_{10}(0.0001) = -4$

22. $\log_{256} 4 = \frac{1}{4}$

23. $\log_{10}(0.0000001) = -7$

24. $\log_2 64 = 6$

Solve the following equations.

25. $\log_3 27 = x$

26. $\log_9 \frac{1}{729} = x$

27. $\log_4 \sqrt{4^{10}} = x$

28. $\log_x 8 = 3$

29. $x = 3^{\log_3 11}$

30. $x = \log_{10}(0.00001)$

31. $x = \log_3 \sqrt[4]{81}$

32. $\log_x 36 = -2$

33. $\log_4 x = -3$

34. $\log_5 x = -2$

Graph the following functions.

35. $f(x) = \log_4 x$

36. $f(x) = \log_{\frac{1}{5}} x$

37. $f(x) = \log_4 x + 4$

38. $f(x) = \log_5 (x + 3)$

39. $f(x) = \log_{\frac{1}{2}} (x + 2)$

40. $f(x) = \log_5 (x - 4)$

41. $f(x) = \log_{\frac{1}{2}} (2 - x)$

42. $f(x) = \log_{\frac{1}{3}} (1 - x)$

43. $f(x) = \log_6 (x^6)$

44. $f(x) = |\log (x + 1)|$

11.3 The Properties of Logarithms

Before we attempt to solve equations that involve logarithms, we must introduce a series of rules that will allow us to manipulate logarithmic expressions. The rules are analogs of the properties of exponents we have already seen and are collectively referred to as the **properties of logarithms**. The properties of logarithms are a set of equivalent forms of logarithmic expressions used to rewrite such expressions in alternative forms. The first properties are consequences of the definition of logarithms and are sometimes called the fundamental properties of logarithms, and the final three (which are the more commonly used) are sometimes called the computational properties of logarithms.

TABLE 11.4 Some Fundamental Properties of Logarithms

Rule	Proof
$\log_b b = 1$	Using the definition of logarithms, if we set $\log_b b = x$ and convert to exponential form, we see $b = b^x$. Consequently, it is immediate that $x = 1$.
$\log_b 1 = 0$	Again by the definition of logarithms, if $\log_b 1 = x$, conversion to exponential form yields $1 = b^x$, and therefore $x = 0$.
$\log_b b^n = n$	The definition of logarithms again implies if $\log_b b^n = x$, then $b^n = b^x$, and therefore $n = x$.
$b^{\log_b n} = n$	Here, if = we set the exponent on the left equal to x, we find $\log_b n = x$, which implies $n = b^x$. Thus, $b^{\log_b n} = b^x \Rightarrow \log_b n = x$, which tells us $b^{\log_b n} = b^x = n$

The fundamental properties are presented in the following list shown in Table 11.4. They are straightforward consequences of the properties of exponents, as we will demonstrate.

These properties are occasionally useful but are less so than the computational properties of logarithms. In practice, the only two of the fundamental properties that appear with relative frequency are the first two in the list, so we encourage you to keep those in mind especially.

The computational properties are so important that they are assigned particular names: the product rule of logarithms, the quotient rule of logarithms, and the power rule of logarithms. Note that we must specifically include the term "of logarithms" in the name of each rule, since there already exists a product rule, quotient rule, and power rule for exponentials.

The product rule of logarithms states that the logarithm (or log for short) of the product of two numbers is equal to the sum of their individual logarithms. That is, $\log_a(uv) = \log_a u + \log_a v$. In that equation, the two expressions are equivalent to one another, with the form to the left of the equal sign referred to as the condensed form of the expression and the form to the right of the equal sign as the expanded form of the logarithm.

THE PRODUCT RULE OF LOGARITHMS: For any base $a > 0, a \neq 1$, we have that $\log_a(uv) = \log_a u + \log_a v$.

EXAMPLE 11.15 Use the product rule of logarithms to expand the given logarithm:

$$\log_2 3x$$
$$\log 7y$$

SOLUTION

Here, we compare the given expression to the condensed form $\log_a(uv)$ and recognize that we have $a = 2$, $u = 3$, and $v = x$. Using the formula for the product rule, we expand the expression to obtain $\log_2 3 + \log_2 x$.

The properties of logarithms extend to both common and natural logarithms, so the product rule works just as well in the second case. Here, $u = 7$ and $v = y$, and thus the expression expands as $\log 7y = \log 7 + \log y$.

EXAMPLE 11.16 Use the product rule of logarithms to condense the sum of logarithms into a single logarithm:

$$\log_2 5 + \log_2 m$$
$$\ln 10 + \ln x$$

SOLUTION

If we compare the given sum of logarithms to the expanded form $\log_a u + \log_a v$, we recognize that here we have $a = 2$, $u = 5$, and $v = m$. Using the formula for the product rule , we condense the sum into the single expression $\log_2 5m$.

In the second case, we are using the natural logarithm, but the process is exactly the same as we saw in the previous example. Here, $u = 10$ and $v = x$, and our condensed form is $\ln 10x$.

The second computational property of logarithms is the quotient rule of logarithms. This rule states that the logarithm of a quotient is equal to the difference of the individual logarithms. That is, $\log_a\left(\frac{u}{v}\right) = \log_a u - \log_a v$. As was the case for the product rule, the left side of this equality is referred to as the condensed form of the expression and the right side as the expanded form.

THE QUOTIENT RULE OF LOGARITHMS: For any base $a > 0, a \neq 1$, we have that $\log_a\left(\frac{u}{v}\right) = \log_a u - \log_a v$.

EXAMPLE 11.17 Use the quotient rule of logarithms to expand the given logarithms:

$$\log_4\left(\frac{7}{11}\right)$$
$$\ln\left(\frac{c}{5}\right)$$

SOLUTION

Comparing the first given expression to the condensed form $\log_a\left(\frac{u}{v}\right)$, we see that $a = 4$, $u = 7$, and $v = 11$. Substitution into the expanded form yields $\log_4 7 - \log_4 11$.

The quotient rule, as we saw with the product rule, applies to natural logarithms, and therefore, for the second expression we have $a = e$, $u = c$, and $v = 5$. The expanded form of the given logarithm, therefore, is $\ln c - \ln 5$.

EXAMPLE 11.18 Use the quotient rule of logarithms to condense the given logarithmic expressions:

$$\log x - \log t$$
$$\log_7 3 - \log_7 2$$

SOLUTION

Comparing the given expression to the expanded form of the quotient rule, $\log_a u - \log_a v$, and realizing that we are working with common logarithms (since no base value a is shown), we determine the values of $a = 10$, $u = x$, and $v = t$. Substitution into the condensed form of the quotient rule yields $\log \dfrac{x}{t}$.

In the second case, $a = 7$, $u = 3$, and $v = 2$, and therefore the condensed form of the logarithmic expression is $\log_7 \dfrac{3}{2}$.

If the expression $\dfrac{u}{v}$ can be reduced to lowest terms, it should be reduced but never changed to a mixed number, by general agreement. On occasion, we will work with expressions that involve decimals, but we won't introduce decimals to problems unless they are given in the original problem.

The final computational rule of logarithms is called the power rule of logarithms. This states that $\log_a(u^n) = n \log_a u$. In general terms, we can say that the logarithm of a power is equivalent to a multiple of the logarithm of the exponential base.

THE POWER RULE OF LOGARITHMS: For any base $a > 0$, $a \neq 1$, and any exponent value n, we have that $\mathbf{log}_a\,(u^n) = n\,\mathbf{log}_a\,u$.

EXAMPLE 11.19 Apply the power rule of logarithms to the following logarithmic expressions:

$$\log_2\left(x^8\right)$$

$$\log_5\left(\frac{1}{t^2}\right)$$

SOLUTION

In the first case, the logarithmic base is $a = 2$, and the other values in the formula for the power rule of logarithms are $u = x$ and $n = 8$. Substitution into the power rule formula yields the following: $\log_2(x^8) = 8 \log_2 x$.

In the expression $\log_5\left(\dfrac{1}{t^2}\right)$, if we want to apply the power rule of logarithms, we need to view the expression $\dfrac{1}{t^2}$ as t^{-2}, and thus we have $\log_5\left(\dfrac{1}{t^2}\right) = \log_5(t^{-2})$. From this form, we see that $a = 5$, $u = t$, and $n = -2$. With those values in mind, the power rule of logarithms tells us $\log_5\left(\dfrac{1}{t^2}\right) = \log_5(t^{-2}) = -2\log_5 t$.

We summarize the three rules of logarithms here, for the sake of reference:

THE PRODUCT RULE OF LOGARITHMS: For any base $a > 0$, $a \neq 1$, we have that $\mathbf{log}_a\,(uv) = \mathbf{log}_a\,u + \mathbf{log}_a\,v$.

THE QUOTIENT RULE OF LOGARITHMS: For any base $a > 0$, $a \neq 1$, we have that $\log_a\left(\frac{u}{v}\right) = \log_a u - \log_a v$.

THE POWER RULE OF LOGARITHMS: For any base $a > 0$, $a \neq 1$, and any exponent value n, we have that $\log_a(u^n) = n\log_a u$.

Naturally, we can envision situations in which the properties of logarithms might be used in conjunction within a single problem. As long as we keep in mind the precise requirements of the individual properties, we will find that this is not beyond our ability to master.

EXAMPLE 11.20 Apply the properties of logarithms to fully expand the logarithmic expressions:

$$\log_5\left(\frac{x^4}{5y}\right)$$

$$\ln\left(\frac{7x^3}{w^2}\right)^4$$

SOLUTION

The key to successfully simplifying the first expression is to think "big picture" when looking at the argument of the logarithm. Observe that the expression is enclosed in parentheses and that it is a fraction of the form $\frac{u}{v}$, where $u = x^4$ and $v = 5y$. Thus, the quotient rule of logarithms is applicable to this expression. Use of the quotient rule yields $\log_5 x^4 - \log_5 5y$, which we then attempt to simplify further. The first logarithm involves the variable x raised to a power, and thus the power rule of logarithms applies to this expression. The second logarithm has argument $5y$, which is a product, and thus the product rule of logarithms applies to that expression:

$$\begin{aligned}\log_5\left(\frac{x^4}{5y}\right) &= \log_5 x^4 - \log_5 5y && \text{by the quotient rule}\\ &= 4\log_5 x - (\log_5 5 + \log_5 y) && \text{by the power rule and the product rule}\\ &= 4\log_5 x - \log_5 5 - \log_5 y && \text{by the distributive property}\end{aligned}$$

Recall that use of the properties of logarithms is dictated by the maxim that if a rule *can* be used, then it *must* be used. When we consider the individual logarithms in this expression, we notice that the second logarithm, $\log_5 5$, can be simplified using one of the fundamental properties of logarithms, $\log_a a = 1$, and thus the expression simplifies once more to

$$= 4\log_5 x - 1 - \log_5 y$$

In the second logarithmic expression, the argument is, initially, an exponential expression, and thus the power rule of logarithms applies in this case. Applying the individual properties of logarithms as needed, we find the expression simplifies in the following manner:

$$\ln\left(\frac{7x^3}{w^2}\right)^4 = 4\ln\left(\frac{7x^3}{w^2}\right) \text{ by the power rule of logarithms}$$

The argument of the logarithm is now a quotient, and therefore the quotient property of logarithms appllies.

$$= 4[\ln(7x^3) - \ln(w^2)]$$

In the first logarithm, the argument is a product, and so, by the product rule of logarithms.

$$= 4[\ln 7 + \ln(x^3) - \ln(w^2)]$$

The final two logarithms contain exponentials, and thus we can use the power rule of logarithms to finish the simplification:

$$= 4(\ln 7 + 3\ln x - 2\ln w)$$

As we will see in the exercises, mastering the properties of logarithms requires a good deal of practice. However, with diligence, we will find that we'll be able to handle expressions of surprising complexity.

Exercises

Express the following as a sum, difference, or product of logarithms using the properties of logarithms.

1. $\log_a(4^x 2y)$
2. $\log_6\left(\frac{\sqrt{13}}{9}\right)$
3. $\log_6\left(\frac{4\sqrt{9}}{8}\right)$
4. $\log(8x + 13y)$
5. $\log_{13}\left(\frac{19m}{n}\right)$
6. $\log_{19}\left(\frac{4\sqrt{x}}{y}\right)$
7. $\log_3\left(\frac{x^6\ y^3}{3}\right)$
8. $\log_6\left(\frac{m^5\ p^3}{36t^8}\right)$
9. $\log_6\sqrt{\frac{4x^9}{z^4}}$
10. $\log_6\sqrt[3]{\frac{x^4}{y^6z^9}}$

Write the sum, difference, or product of logarithms as a single logarithm (if possible) using the properties of logarithms.

11. $\log_4 13 - \log_4 a$
12. $\log_3 w - \log_3 s$
13. $6\log_5 q - \log_5 r$
14. $(\log_a m - \log_a n) + 4\log_a p$
15. $\frac{2}{3}\log_7 x + \frac{5}{6}\log_7 x - \frac{1}{9}\log_7 x$

16. $\frac{1}{3}\log_3(x^6) + \frac{1}{6}\log_3(x^9) - \frac{4}{9}\log_3 x$

17. $7 \log_5 p - 6 \log_5 y$

18. $2 \log_6 (5y) + 4 \log_6 (2y^2)$

19. $6 \log_6 (6x - 1) + 4 \log_6 (2x - 7)$

20. $\frac{3}{4}\log_2(p^2q^8) - \frac{1}{2}\log_2(p^5q^2)$

Evaluate the following logarithms using a calculator (round to three decimal places). Compare the results to those obtained by using the properties of logarithms and the approximations $\log_{10} 2 \approx 0.301$ and $\log_{10} 3 \approx 0.477$.

21. $\log_{10} 6$

22. $\log_{10} 18$

23. $\log_{10} \frac{27}{4}$

24. $\log_{10} (\sqrt[3]{18})$

25. $\log_{10} \sqrt[3]{24}$

26. $\log_{10} (54)$

In the following, let $u = \ln a$ and $v = \ln b$. Write the given expressions in terms of u and v (not using the natural logarithm function).

27. $\ln\left(\frac{a^4}{b^2}\right)$

28. $\ln(b^9a^4)$

29. $\ln \sqrt[4]{\frac{a^9}{b^8}}$

30. $\ln (\sqrt[5]{ab^7})$

Evaluate the following.

31. If $f(x) = \log_3 x$, find $f(3^2)$.

32. If $f(x) = \log_5 x$, find $f(125)$.

33. If $f(x) = 4^x$, find $f(\log_4 8)$.

34. If $f(x) = 6^x$, find $f(\log_6 9)$.

11.4 Exponential and Logarithmic Equations

An equation is called logarithmic if some or all of its terms involve logarithms, and an equation is called exponential if some or all of its terms involve exponentials. It is possible that an equation could involve a mixture of such expressions, of course, but such equations are typically impossible to solve computationally, and we will not be examining them here.

We'll start out by considering the logarithmic equations, finding exact solutions when we are able and approximate solutions otherwise. As we'll see, there is a general strategy to follow when attempting to solve a logarithmic equation exactly, and it is when this strategy fails that we must turn to approximation techniques.

A remark that needs to be made before setting out to solve logarithmic equations is that the solutions must, in all cases where exact solutions can be found, be checked for validity. That is, we must confirm that the solutions we find satisfy the original equation. As we'll see, the solution techniques can produce false, or extraneous, solutions, and it is against these that we must be on our guard.

In the logarithmic equations we will consider, every logarithm will contain the same base value. If the bases are not the same, we could force them into agreement by using the change of base rule, but this creates a level of complexity beyond the scope of our conversation here and is unworthy of examination at this time.

Our technique will be to attempt to rewrite the equation so that one of two situations arises: we have one logarithm equal to another logarithm, or we have one logarithm equal to a constant.

To solve a logarithmic equation, rewrite the equation to obtain one of the following forms:

$$\log_a u = \log_a v$$

or

$$\log_a u = C$$

In the first case, we can use the property of logarithms, which states

$$\log_a u = \log_a v \Rightarrow u = v$$

while in the second case, we can use the definition of logarithms to convert the equation to exponential form:

$$\log_a u = C \Rightarrow u = a^C$$

In either event, we should have an equation we are capable of solving by known methods. We should note out that u and v could be horrific expressions, but this is something beyond our control. In our illustrations and in the exercises, the expressions will never become so terrible that we will not be able to deal with them effectively.

Note that there is no suggestion as to how we would obtain one of the two forms mentioned. Doing so will require that we know and use the properties of logarithms to rewrite the particular equation we are solving in one of the desired forms, and this may require significant practice and effort.

EXAMPLE 11.21 Solve $\log_3 x - \log_3 (x + 1) = 2$ and check the solution for validity.

SOLUTION

We had indicated that the technique for solving was to rewrite the equation in one of two specific forms and then solve the resulting equation. Our first concern is to identify which of the two forms we should obtain, $\log_a u = \log_a v$ or $\log_a u = C$. Because our equation involves some terms with logarithms and some terms without, we choose to work toward the latter form. If every term involved a logarithm, we would opt to reach the form where we have one logarithm equal to another.

By the quotient rule of logarithms, we can rewrite the equation as

$$\log_3 \left(\frac{x}{x+1} \right) = 2$$

Converting from logarithmic to exponential form, we obtain

$$\frac{x}{x+1} = 9$$

Be certain that you understand the steps just performed. We have allowed a tiny portion of the work to lie "behind the curtains" to see if you can follow the reasoning.

With the logarithms removed from the problem, we are now free to solve the equation using any of our known techniques. Here, we can multiply both sides of the equation by $(x + 1)$ to clear the fractions, and this yields $x = 9x + 9$, which we proceed to solve:

$$-8x = 9$$
$$x = -\frac{9}{8}$$

The technique required for solving the nonlogarithmic equation we obtained will be our choice, and it is this uncertainty that can cause considerable angst. Literally, once the logarithms are removed from the problem, you are at the mercy of your own experience and must hope that you recall enough mathematics to solve the equation by whatever means are necessary.

It is tempting, at this point, to move on to the next problem, but keep in mind that all equations we are able to solve computationally must have their answers checked for validity. We must insert our solution into the original equation and confirm that it satisfies that equation. Note carefully that we must use the *original* equation to conduct our check!

Substituting, we find

$$\log_3\left(-\frac{9}{8}\right) - \log_3\left(-\frac{9}{8} + 1\right) = 2$$

Observe that, in this case, both of the logarithms are undefined, since we are not permitted to use negative numbers as the argument of a logarithm. Therefore, the solution we found does not satisfy the original equation and must be discarded. Since it was our only solution, we conclude that the equation has no solution.

It is a mistake to jump to the conclusion that the solution found was invalid because it was negative. This is a tempting, understandable, and completely reasonable assumption to make. Like many tempting, understandable, and completely reasonable assumptions, however, it is flawed.

It is not the negativity of the solution that causes it to be invalid, or extraneous, but rather that substitution of that value into the original logarithms produces undefined expressions that cause the answer to be discarded. It is conceivable that, through algebraic conditions existing within the problem, a negative solution could be valid.

EXAMPLE 11.22 Solve $\log_6(5 - x) + \log_6(-x) = 1$ and check the solution for validity.

SOLUTION

Recognizing that we do not have a logarithm in every term of the equation, we realize we must attempt to rewrite the equation in the form $\log_a u = C$, as was the case in Example 11.21. Using the product rule of logarithms, we obtain

$$\log_6 [(5 - x)(-x)] = 1$$

which we simplify to

$$-5x + x^2 = 6$$

Using our techniques for solving quadratic equations, we find

$$\begin{aligned} x^2 - 5x - 6 &= 0 \\ (x - 6)(x + 1) &= 0 \\ x - 6 = 0 \quad x + 1 &= 0 \\ x = 6 \quad x &= -1 \end{aligned}$$

Observe that, through a quirk in the problem, substitution of the positive result, $x = 6$, into the original equation yields an undefined logarithm, which tells us that the result $x = 6$ is extraneous and must be discarded.

On the other hand, substitution of $x = -1$ gives

$$\begin{aligned} \log_6 [5 - (-1)] + \log_6 [-(-1)] &= 1 \\ \log_6 6 + \log_6 1 &= 1 \\ 1 + 0 &= 1 \end{aligned}$$

which verifies that the negative solution we found, $x = -1$, is the valid solution to the equation.

The moral we should take from Example 11.22 is clear: we cannot jump to conclusions regarding the validity of solutions we find for logarithmic equations! Every solution we find must individually be checked for validity. In some cases, one or more of the solutions may turn out to be invalid, but there is no way to know this ahead of time, so the checks must be performed.

EXAMPLE 11.23 Solve $\log_4 (x - 3) + \log_4 (x + 5) = \log_4 20$ and check the solution for validity.

SOLUTION

In this situation, every term within the equation contains a base-4 logarithm. Therefore, we should attempt to rewrite the expression in the form $\log_a u = \log_a v$. Note that we could also choose to reach the form $\log_a u = C$ by moving all terms to the left side of the equation, but this is typically not desirable in the situation where all terms involve logarithms at the outset.

Applying the product rule of logarithms to the left side of the equation, we obtain

$$\log_4[(x - 3)(x + 5)] = \log_4 20$$

Now, we have two base-4 logarithms equal to one another, so the arguments of the logarithms must agree, and therefore

$$(x - 3)(x + 5) = 20$$

Using our known algebraic techniques, we solve this equation:

$$\begin{aligned} x^2 + 2x - 15 &= 20 \\ x^2 + 2x - 35 &= 0 \\ (x + 7)(x - 5) &= 0 \\ x = -7 \quad x &= 5 \end{aligned}$$

Note that we are omitting a step near the end of the problem, where the individual factors are set equal to 0, but in this case the expressions are fairly simple, so there is no harm in doing so.

Substitution of our solutions into the original equation reveals that, in this case, the first solution, $x = -7$, is extraneous, while the second solution, $x = 5$, is valid. We will leave this as an exercise for you to verify, but the checks tell us that the solution to the equation is the value $x = 5$.

EXAMPLE 11.24 In a network cable producing a signal loss of 12 decibels with an input of 6 volts, the voltage V at the end of the cable is given by the following formula:

$$-12 = 20(\log V - \log 6)$$

Solve the equation for V and interpret your result.

SOLUTION

This equation involves a logarithm with no base value showing, and thus we recognize it to be a common logarithm problem, with suppressed base value 10. Moreover, since some (but not all) terms in the equation involve logarithms, we judge that we should attempt to rewrite the equation in the form $\log_a u = C$.

We can combine the logarithms within the parentheses using the quotient rule of logarithms, producing

$$-12 = 20 \log\left(\frac{V}{6}\right)$$

Dividing both sides of the equation by 20, we obtain

$$-0.6 = \log\left(\frac{V}{6}\right)$$

At this point, we are ready to convert to exponential form and solve the equation:

$$\begin{aligned} 10^{-0.6} &= \frac{V}{6} \\ 6(10^{-0.6}) &= V \\ 1.5071 &\approx V \end{aligned}$$

In context, this tells us that the voltage at the end of the network cable is just over 1.5 V, which indicates that there was a drop of nearly one-quarter the original voltage through the cable.

In Example 11.24, note that we used a decimal approximation to the actual solution. In applications, it is seldom useful to present a solution in the exact form, since the numerical value of that solution is likely to be unclear and therefore not particularly useful. In the exercises, we will indicate when approximate values should be given and the degree to which those approximations should be rounded.

Exponential Equations

Much as was the case in the logarithmic equation examples, there are several possible approaches to solving exponential equations. Which one we employ depends on the structure of the problem itself, so we will have to analyze the equation before proceeding.

It is possible, though admittedly rather unlikely, that we will encounter an exponential equation that is in or that can be rewritten in the form $a^u = a^v$. Note that the two exponential bases are the same, and no other terms appear in the equation. In this scenario, we can apply the properties of exponents to conclude $u = v$ and solve the resulting equation using whatever methods are necessary.

In other exponential equations, we will need to reorganize the equation to obtain the form $a^u = C$, where C is a constant. With that form in hand, we can convert the equation to exponential form and solve, again using whatever methods are required.

EXAMPLE 11.25 Solve the exponential equation $3^{x^2 - 4x} = 3^5$.

SOLUTION

Clearly, this situation involves a pair of exponential expressions having the same base, so we can immediately set the variable exponent expressions equal to one another and solve.

Here, that would give

$$\begin{aligned} x^2 - 4x &= 5 \\ x^2 - 4x - 5 &= 0 \\ (x - 5)(x + 1) &= 0 \\ x = 5 \quad x &= -1 \end{aligned}$$

A check of both solutions reveals them to be valid results, and hence our equation has two solutions.

A slight variation on the theme exhibited in Example 11.25 would be the case where the two sides of the equation did not possess the same exponential base but where a clever observation would allow us to rewrite the expressions so that the same bases occurred.

EXAMPLE 11.26 Solve the exponential equation $2^{5x} = 8^{x - 4}$.

SOLUTION

Here, the bases are clearly unequal, being 2 and 8, and therefore the technique used before cannot be employed. Observe, however, that we have the good fortune that the base value 8 just happens to be a power of the other base, 2. That is, we note that $8 = 2^3$. While not appearing to be remarkably helpful at first glance, this coincidence turns out to be the key to solving the equation.

Using the properties of exponents, we see

$$\begin{aligned} 2^{5x} &= 8^{x - 4} \\ 2^{5x} &= (2^3)^{x - 4} \\ 2^{5x} &= 2^{3x - 12} \end{aligned}$$

Note that this transforms the equation with disparate bases into an equation having the same bases, whose method for solution we have already indicated:

$$\begin{aligned} 5x &= 3x - 12 \\ 2x &= -12 \\ x &= -6 \end{aligned}$$

Substitution of this value into the original equation establishes that this solution is correct.

Since exponential expressions do not have the same domain restrictions placed on logarithmic expressions, we do not technically have to check our results for validity, since evaluation of exponential expressions will not produce undefined results. Nonetheless, checking solutions is always good practice, and we encourage you to do so, if time permits.

EXAMPLE 11.27 Solve the exponential equation $5^{2x-7} = 30$.

SOLUTION

This problem stands apart from the previous two illustrations, since the right-hand side is not an exponential expression and cannot be transformed into an exponential expression because 30 is not a power of 5. Consequently, we will have to employ a different approach to find the solution.

Noting that our equation has the form $a^u = C$, we can convert to logarithmic form and attempt to solve there. This will yield an answer that is difficult to interpret on casual examination, so we will adopt the practice of giving both the exact answer and a decimal approximation to the solution, rounded to three decimal places.

The equivalent logarithmic form to $5^{2x-7} = 30$ is $2x - 7 = \log_5 30$, which we can solve for x to find the exact solution to the equation:

$$\begin{aligned} 2x - 7 &= \log_5 30 \\ 2x &= 7 + \log_5 30 \\ x &= \frac{1}{2}(7 + \log_5 30) \end{aligned}$$

The value of this expression is somewhat unclear, so we will use the change of base rule to decimal approximate the logarithm and then the value of x:

$$\begin{aligned} x &\approx 0.5(7 + 2.11328) \\ x &\approx 0.5(9.11328) \\ x &\approx 4.557 \end{aligned}$$

This solution, being somewhat more involved than that of the previous example, merits a demonstrated check for the purpose of illustration:

$$\begin{aligned} 5^{2(4.557)-7} &\approx 30 \\ 5^{2.114} &\approx 30 \\ 30.035 &\approx 30 \end{aligned}$$

Note that our check did not yield exact equality because of our use of a decimal approximation to the solution. However, we did get very nearly the exact value, and this suggests that our solution is correct.

EXAMPLE 11.28 Suppose the value of a particular brand of computer diminishes by 37% annually, according to a consumer services study. This suggests that the value V of the computer after t years is given by the formula $V = P_0(0.63)^t$, where P_0 is the initial price of the computer. Assuming the value of the computer has fallen from an initial cost of \$1386 to \$419, what is the age of the computer in months?

SOLUTION

The given values are for P_0 and V, respectively, and our task is to calculate the value of t. Since the formula uses time in years, our t value will then have to be converted to months through multiplication by 12:

$$419 = 1386(0.63)^t$$

Our equation has a single term involving an exponential, and that suggests we try to rearrange the expression into the form $a^u = C$. Note that, to accomplish this, we must divide each side of the equation by 1386:

$$\begin{aligned} 419 &= 1386(0.63)^t \\ \frac{419}{1386} &= 0.63^t \\ 0.3023 &\approx 0.63^t \end{aligned}$$

We can now convert the equation to logarithmic form and approximate the value of t using the change of base rule, as was done in Example 11.27, or we can take an alternative approach that will accomplish the same result.

Taking the natural logarithm of each side of the equation, we have

$$\ln(0.3023) \approx \ln(0.63^t)$$

and the power rule of logarithms yields

$$\ln(0.3023) \approx t\ln(0.63)$$

Now is as good a time as any to obtain decimal approximations for the logarithms, so we do that next:

$$-1.196 \approx t(-0.462)$$

which we solve as

$$2.589 \approx t$$

Recall that this value is the age of the computer in years, so we multiply the result by 12 to obtain the final result that the age of the computer, in months, is approximately 31.

We should now have a reasonable amount of preparation in solving both logarithmic and exponential equations, so this seems an appropriate place to bring the discussion to a close. The essential facts, however, should be restated: for each type of equation, there exist different classifications that will dictate the method of solution. You should take care to analyze the equations carefully and to choose your technique wisely. Should you choose an incorrect approach, do not despair because your attempt at solution will lead you quickly to a dead end, and this will suggest that you try another approach.

Finally, keep in mind that *all* solutions to logarithmic equations must be checked for validity because of the limitations on the domains of logarithmic expressions. While it is not necessary to check the solutions to exponential equations, it is strongly encouraged that you do so.

Exercises

Solve the following equations. Be sure to check your answers for validity.

1. $\ln(18x + 3) = \ln 11$
2. $\log(x - 3) = \log 2$
3. $\log(x - 3) = 1 - \log x$
4. $\log_9(x - 7) + \log_9(x - 7) = 1$
5. $\log(4x) - \log 5 = \log(x + 1)$
6. $\log(2 + x) = \log 5 + \log(x - 2)$
7. $\ln(2x) + \ln(8x) = \ln 17$
8. $\log(3x) + \log(2x) = \log 35$
9. $\ln(-x) + \ln 4 = \ln(3x - 9)$
10. $\log(x + 10) - \log(4x - 3) = 1$
11. $\ln(e^x) - \ln(e^7) = \ln(e^8)$
12. $\ln(e^{2x}) + \ln(e^5) = \ln(e^{20})$
13. $\log_2 \sqrt{2x^2} = \dfrac{7}{2}$
14. $\log_3 \sqrt{27x^6} = 48$

For the following applications, use logarithms or exponentials and give your answers in complete sentences. If necessary, round your answers to the nearest whole number.

15. The growth of the population of a particular city is modeled by the formula $P(t) = 9759e^{0.002t}$, where t is the number of years since 1984. Use the formula to find the population of the city in 1994.
16. The growth of the population of a particular city is modeled by the formula $P(t) = 12{,}965e^{0.006t}$, where t is the number of years since 1970. Use the formula to determine the time needed for the population to double.
17. The growth of the population of a particular city is modeled by the formula $P(t) = 10{,}226e^{0.005t}$, where t is the number of years after 1970. According to the formula, in which year would the population of the city reach 15,339?
18. The growth of the population of a particular city is modeled by the formula $P(t) = 18{,}920e^{0.008t}$, where t is the number of years after 1970. According to the model, in what year did the population of the city reach 100,000?

In the following, answer the application problems rounding as instructed.

19. Suppose $f(x) = 34.2 + 1.2\log(x + 1)$ models salinity of ocean water at a particular depth at a certain location on the globe. If x is the depth of water in meters and $f(x)$ the grams of salt per kilogram of seawater, what is the salinity at a depth of 72 meters, rounded to the nearest hundredth of a gram per kilogram.

20. Suppose $f(x) = 25.2 + 1.4\log(x + 1)$ models salinity of ocean water at a particular depth at a certain location on the globe. If x is the depth of the water in meters and $f(x)$ the grams of salt per kilogram of seawater, at what depth, rounded to the nearest meter, will the salinity equal 30 grams per kilogram?
21. The height in meters of the females of a particular tribe is approximated by the formula $H(t) = 0.52 + \log\left(\frac{t^2}{9}\right)$, where t is the female's age in years. Approximate the height of a 4-year-old female of the tribe, rounded to the nearest hundredth of a meter.
22. The population growth of a particular animal species in an area under study is described by $F(t) = 400\ln(2t + 3)$, where t is the time measured in months. Find the population of the species in an area six months after the species has been introduced.

Summary

In this chapter, you learned about:

- Exponential functions and how to graph exponential curves and distinguish between exponential growth and decay.
- Logarithmic functions, their properties, and how to convert between exponential and logarithmic forms and recognize and evaluate logarithms and antilogarithms.
- Logarithmic and exponential equations and their applications.

Glossary

antilogarithm of N, base A The number B in the equation $\log_A B = N$.

base of an exponential function In a function $f(x) = ca^x$, the value of a.

exponential form of an equation $A^N = B$, where A and B are positive constants, with A not 1.

exponential function A function of the form $f(x) = ca^x$.

logarithm If N is a number such that $A^N = B$, where A is positive but not 1, then we say that N is the logarithm, base A, of the number B.

logarithmic expressions Expressions involving logarithms with any particular base.

logarithmic form of an equation $N = \log_A B$, where A and B are positive constants, with A not 1.

natural exponential base The mathematical constant e, whose approximate value is 2.71828182845904523560287.

natural exponential function An exponential function whose base is the mathematical constant e.

natural logarithm A logarithm whose base value is the mathematical constant e.

properties of logarithms A set of equivalent forms of specific logarithmic expressions used to rewrite logarithmic expressions in alternative forms.

variable exponent The expression in the exponent of an exponential function.

List of Equations

The product rule of logarithms: For any base $a > 0$, $a \neq 1$, we have that

$$\log_a(uv) = \log_a u + \log_a v$$

The quotient rule of logarithms: For any base $a > 0$, $a \neq 1$, we have that

$$\log_a\left(\frac{u}{v}\right) = \log_a u - \log_a v$$

The power rule of logarithms: For any base $a > 0$, $a \neq 1$, and any exponent value n, we have that $\log_a(u^n) = n\log_a u$.

End-of-Chapter Problems

Find the function value.

1. Let $f(x) = \left(\frac{1}{4}\right)^x$. Find $f(3)$.
2. Let $g(x) = 3^x$. Find $g(3.5)$.
3. Let $f(x) = 2^{(2-x)}$. Find $f(4.5)$

Graph the function.

4. $f(x) = 2^{\frac{2}{3}x}$
5. $g(x) = \left(\frac{3}{4}\right)^{-x}$
6. $f(x) = 3^{-3x}$

Solve the equation.

7. $5^{(4-2x)} = \frac{1}{25}$
8. $c^{\frac{2}{7}} = 4$
9. $7^{\left(4-\frac{1}{2}x\right)} = 49$
10. $9^{(x+2)} = \frac{1}{81}$

Evaluate.

11. $\log_6 216$
12. $\log_{\frac{1}{2}} 4$
13. $\log_8 512$

Write in logarithmic form.

14. $\left(\frac{4}{9}\right)^{\frac{3}{2}} = \frac{8}{27}$
15. $\left(\frac{16}{81}\right)^{\frac{5}{4}} = \frac{32}{243}$
16. $13^3 = 2197$
17. $15^4 = 50{,}625$

Write in exponential form.

18. $\log_{32} 64 = \dfrac{6}{5}$

19. $\log_{\frac{3}{2}} \dfrac{16}{81} = -4$

20. $\log_{15} 225 = 2$

Solve the equation.

21. $\log_x 0.8 = 1.2$

22. $x = \log_5 \sqrt[5]{15625}$

23. $1 = \log_5 \sqrt[3]{x - 6}$

24. $\log_3(27) = \sqrt[x]{\dfrac{1}{243}}$

Graph the function.

25. $f(x) = \log_{\frac{5}{2}}(x + 4)$

26. $f(x) = \log_{\frac{1}{2}}(x - 3)$

27. $f(x) = \log_5\left(\dfrac{3}{2}x\right)$

Simplify the following expressions.

28. $\log_6(6^x 3^y)$

29. $\log_a\left(\dfrac{b^{2.5}}{c^3 d^4}\right)$

30. $\log_a\left(\dfrac{a^3 b^4}{c^a d^5}\right)$

31. $\log_3\left[9^4 \bullet \left(\dfrac{1}{81}\right)^5\right]$

Calculate the following logarithms.

32. $\log_{10}(\sqrt[5]{24})$

33. $\ln\left(17^{\frac{1}{7}}\right)$

34. $\log_8(\sqrt[3]{17})$

35. $\log_7(\sqrt[3]{343^4 \bullet 2401^3})$

In the following exercises, let $u = \ln(a)$ and $v = \ln(b)$. Write the expression in terms of u and v without using the function ln.

36. $\ln\left(a\sqrt[5]{\dfrac{b^5}{a^4}}\right)$

37. $\ln\left(b^7 a^{\frac{1}{4}}\right)$

38. $\ln\left(\dfrac{b^{3a}}{\frac{b}{a^2}}\right)$

39. $\ln\left(\dfrac{\sqrt[3]{b}}{a^5}\right)$

Simplify the expressions as much as possible.

40. If $f(x) = \log_5(x)$, find $f(5\sqrt{5})$.

41. If $f(x) = 7^x$, find $f[\log_7(33)]$.

42. If $f(x) = 9^{3x}$, find $f[\log_9(3)]$.

43. If $f(x) = \log_{13}(3x)$, find $f\left(\dfrac{13}{3}\right)$.

In the following exercises, solve for *x*.

44. $\ln(6x + 5) + \ln(3x - 4) = \ln(12)$
45. $\log(3 + x) - \log(x - 3) = \log(6)$
46. $\ln(6x - 4) + \ln(3 - x) = \ln(8)$
47. $\log\left(\dfrac{3x-2}{4-2x}\right) = 3$

Solve the following applications of logarithms.

48. If the Consumer Price Index starts at a base of 220.223 at the beginning of 2011 and increases by 3% per year, in what year will the index have doubled?
49. A particular cask of radioactive waste emits radiation at a rate of 10,000 mSv/h (millisieverts per hour). The waste has a half-life of 10,000 years, which means that the rate at which it will emit radiation decreases by $\frac{1}{2}$ in 10,000 years. If the safe level of radiation is assumed to be 3.5 mSv per year, how many years have to pass before the waste is safe?

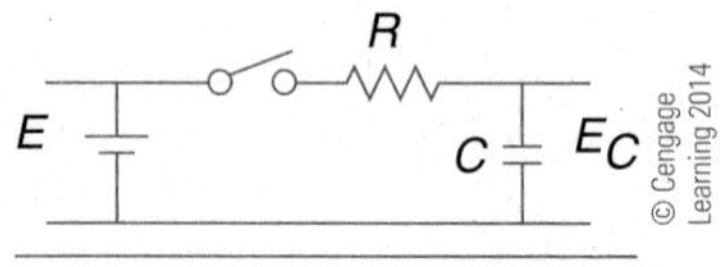

Figure 11.9 Electric circuit

50. Given an electric circuit as shown in Figure 11.9, with $R = 1$ kilohm and $C = 1\ \mu$F, and assuming that the capacitor starts fully discharged and the switch is closed at time $t = 0$, then the formula for E_C, the voltage across the capacitor, is given by $E_C = E - Ee^{(-t/\tau)}$, where $\tau = RC = 0.0001$. At what time does the voltage across the capacitor, E_C, equal 90% of the voltage of the battery, E?
51. A population of lions would naturally grow at 3% per year. Unfortunately, hunters are killing 5% of the population each year. How long will it be until the population of lions is halved?

Chapter 12 Probability

When we speak of probability, we are talking about techniques with which we can analyze particular situations involving outcomes that are uncertain or within which there lies risk. Probabilities tell us the extent of uncertainty or risk and permit us to make strategic choices in our actions.

It isn't difficult to envision situations where uncertainties might arise in the everyday world. Politicians might want to know the likelihood of an event in order to gauge its impact on the outcome of an election, a consumer might want to know the reliability of a particular appliance in order to make an informed choice regarding a purchase, and a network designer might be interested in the potential for failure of electronic components when designing a computer system.

Most of us have developed at least a basic understanding of probability in the course of our lives and possess at least a foundational knowledge of the principles involved. In this chapter, we will take a closer look at the concepts of probability and formalize this intuitive understanding we have developed through our life experience. In so doing, we will be able to analyze problems where uncertainty arises and make informed decisions regarding the situations within which they appear.

Objectives

When you have successfully completed the materials of this chapter, you will be able to:

- Understand and use the concepts of probability and odds.
- Calculate and apply the notion of expected value.
- Recognize and solve "or" and "and" problems.
- Utilize conditional probability and Bayes's rule.
- Understand and work with permutations and combinations.

12.1 Basics of Probability

The terms "probability" and "odds" are sometimes taken to be synonymous, but we must note that these two ideas are most definitely not the same. Probability tells us the likely long-term relative frequency of an event occurring, while odds offer a comparative statement of the number of times an event will occur contrasted to the number of times it will not or vice versa. As you can see, these concepts having significantly different meanings.

Before commencing with a study of probability and odds, let's lay out some of the fundamental terminology we'll be using. The landscape of probability and statistics is populated by the following notions: experiments, outcomes, sample spaces, and events.

Experiments, Outcomes, Sample Spaces, and Events

An **experiment** is some planned operation that is conducted under controlled conditions. This might be as basic as tossing a coin or rolling a die or as specific as a count of failures in a particular type of circuit used within a computer configuration. The key notion is that an experiment is some activity we are observing.

Outcomes are the results of experiments, and the **sample space** is the set of all possible outcomes of the experiment. The size of a sample space might be immense, and one particular challenge we will face is the computation of that size.

An **event** is some particular subset of the sample space. An event can be a particular outcome in the sample space (in which case we call it a **simple event**), or it could be a set of outcomes taken as a group. Usually, it is some specific occurrence in which we are interested and about which we wish to learn. Events are typically given uppercase letter names, such as event A or event B, for ease of reference.

EXAMPLE 12.1 If we roll a pair of tetrahedral (four-sided) dice labeled 1, 2, 3 and 4 on their faces, what is the sample space, and what are two examples of events?

SOLUTION

In an experiment of rolling a pair of four-sided dice (labeled 1, 2, 3, and 4 on their faces), we can list the sample space as a set of ordered pairs, where the first number indicates the value shown on the first die and the second number the value of the second die. Thus, the sample space is

$$\{(1, 1), (1, 2), (1, 3), (1, 4), (2, 1), (2, 2), (2, 3), (2, 4), (3, 1), (3, 2), (3, 3), (3, 4), (4, 1), (4, 2), (4, 3), (4, 4)\}$$

Observe the sample space has 16 possible outcomes within it. Since an event is any particular subset of the sample space, there is a vast array of possible events. Two such might be the specific event $\{(1, 1), (2, 2), (3, 3), (4, 4)\}$, which we might describe as "the two dice showing the exact same number," or $\{(1, 2), (2, 3), (3, 4)\}$, which could be described as "the first die showing a number exactly one less than that shown on the second die."

Theoretical Probability

At the outset, we need to acknowledge that there are two distinct types of probability. The first type is referred to as theoretical probability and the second as empirical probability. In this book, when we use the term "probability," we will be referring to theoretical probability. If we intend to discuss empirical probability, we will explicitly use that terminology, but this will be rare, since we are interested primarily in theoretical probability.

The **theoretical probability of event *A***, written as $P(A)$, is the defined to be the number of outcomes for event A divided by the total number of outcomes within the sample space. Since the number of possible outcomes must always be greater than or equal to the number of outcomes within any particular event, the value of $P(A)$ must lie between 0 and 1, inclusive. An event with probability 0 is said to be an **impossible event**, while an event with probability 1 is referred to as a certain event, or a **certainty**. If two distinct events have same probability value, we say they are **equally likely events**.

EXAMPLE 12.2 If two four-sided dice are rolled, what is the probability of having the value shown on the first die being precisely 1 less than the value shown on the second die?

SOLUTION

In Example 12.1, we pointed out that the size of the sample space in this experiment was 16 and also found that the number of outcomes within this particular event was 3. Therefore, if we designate this event with the name A, we can say $P(A) = 3/16$, or $P(A) = 0.1875$.

If we are considering two events, A and B, then an outcome is considered to be in the event A or B if the outcome is in either event A, event B, or both. This usage of the word "or" is what we referred to as the inclusive form during our discussion of logic. An outcome is considered to be in the event A and B if the outcome is in both A and B simultaneously.

EXAMPLE 12.3

If we roll two four-sided dice and consider the following pair of events,

$$A = \{(1, 3), (2, 1), (3, 3)\} \text{ and } B = \{(1, 1), (2, 2), (3, 3), (4, 4)\}$$

what are the events A or B and A and B?

SOLUTION

A or $B = \{(1, 3), (2, 1), (3, 3), (1, 1), (2, 2), (4, 4)\}$ This event contains all outcomes in event A, event B, or both.

A and $B = \{(3, 3)\}$ This event contains only those outcomes lying within events A and B simultaneously.

In Example 12.2, we determined the probability of the event A to be 0.1875. What does this value mean? The value of a probability allows us to predict the long-term relative frequency of an event, should an experiment be repeated many times. Of course, the word "many" is somewhat vague, but that can't be helped.

Let's suppose we revisit the example and apply this interpretation of probability to the experiment. We are rolling a pair of four-sided dice, and we are wondering about the event where the value shown on the first die is precisely 1 less than the value showing on the second die. Imagine that the pair of dice were thrown 100,000 times. The value of the probability of the event A being 0.1875 tells us that we expect that the value shown on the first die will be precisely 1 less than the value shown on the second die (100,000)(0.1875) = 18,750 times. Of course, in all likelihood, the number of times that event A occurs will probably not be exactly 18,750, but it should be fairly close to that value. If it is not, then we might suspect that there was something inherently wrong with the two dice, since the probability tells us what the relative frequency of the event should be if the experiment is conducted repeatedly.

Every event has its **complement**, which consists of all outcomes not in the original event. The complement of event A is denoted A' (read as "A prime"), and we observe that, since A and A', by definition, collectively contain all possible outcomes of the experiment, $P(A) + P(A') = 1$.

EXAMPLE 12.4

What is the probability of rolling a pair of six-sided dice and obtaining a result where the two dice do not show the same number?

SOLUTION

When two dice are thrown, there are 36 possible outcomes, as shown in Figure 12.1. To find the probability in which we are interested, we have two options. First, we could count the number of outcomes in the event A, where the two dice show different numbers. Second, we could count the number of outcomes in the complementary event A', which is the event where the two dice show precisely the same number. Since A' is an event having fewer outcomes, we choose to do that: $A' = \{(1, 1), (2, 2), (3, 3), (4, 4), (5, 5), (6, 6)\}$, and thus $P(A') = 6/36 \approx 0.167$.

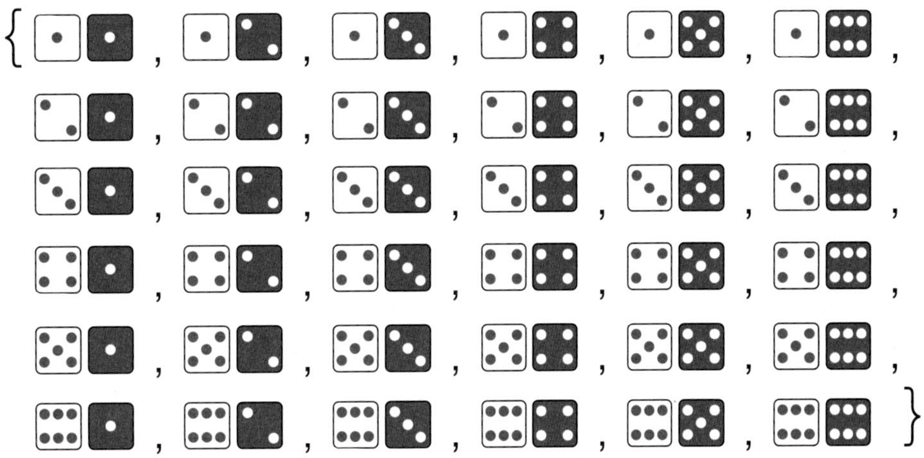

Figure 12.1 Set of outcomes for rolls of two six-sided dice

Since $P(A) + P(A') = 1$, we have $P(A) + 0.167 \approx 1$, so $P(A) \approx 0.833$, and thus we conclude that the probability of rolling two six-sided dice and not having the two dice show the same number is approximately 0.833.

Empirical Probability

We intend to focus our attention on theoretical probability, but empirical probability deserves at least some investigation. Empirical probability is determined entirely on the basis of observations rather than on investigation of the size of a sample space and particular events.

The **empirical probability of event A** is calculated in the following manner: an experiment is performed n times, and the number of times event A occurs, call it k, is recorded. The empirical probability of event A occurring is determined to be k/n.

EXAMPLE 12.5

A coin is tossed 10 times, and a result of heads is obtained 7 times. What is the empirical probability of tossing this coin and obtaining heads?

SOLUTION

Since the coin was tossed 10 times, we have $n = 10$, and since a result of heads occurred 7 times, we have $k = 7$. If the event is called A, we conclude that the empirical probability is $P(A) = 7/10 = 0.7$.

The flaws in empirical probability are self-evident, but we'll mention just one here. When we perform an experiment such as the one described in Example 12.5, we know intuitively that the coin has "no memory." That is, from flip to flip, if the coin is fair (not a "trick coin"), then the probability of obtaining a result of heads has no dependency on the number of times the coin has been flipped historically with a result of heads.

A significant observation, however, is expressed by what we call the **law of large numbers**, which states that if an experiment is repeated many, many times (allowing ourselves the use of the ambiguous word "many"), then the empirical probability will approach the value of the theoretical probability. In other words, if the coin is fair, then it is entirely possible that we would flip the coin 10 times and obtain a result of heads 7 of those times. However, if we were to repeat the experiment many times, possibly 1000 times, then we should find that the empirical probability should draw close to 0.5, the theoretical probability of flipping a coin and obtaining a result of heads.

Exercises

Answer the following questions using proper grammar, complete sentences, and correct terminology.

1. How does an outcome differ from a sample space? Give an example of each and indicate how the example illustrates the difference between the two.
2. How does empirical probability differ from theoretical probability?
3. Explain why probability values must be between 0 and 1 and why the values of 0 and 1 are possible probability values.

Probability Values

4. The value -0.85 cannot be the probability of some event. Why not?
5. The value 120% cannot be the probability of some event. Why not?

Sample Spaces

6. An experiment is to select a single date during a given year. This can be done, for example, by picking a random person and inquiring for his or her birthday. Disregarding leap years for the sake of simplicity, what is the sample space?
7. We have a bag with five balls in it. Two are numbered 1, two are numbered 2, and one is numbered 3. Our game is to randomly draw a ball from the bag one time. What is the sample space?
8. Eloise enjoys sitting in front of the TV and randomly grabbing two chocolates at a time from her snack jar. The snack jar contains a large number of nut log swirls, Turkish delights, and mocha surprises. Describe the sample space and also the event that Eloise selects at least one mocha surprise in her first grab.
9. You flip a fair coin and then choose a number from the bag containing only $\{1, 2, 3, 4, 5\}$. What is the sample space?
10. How many outcomes are in the sample space created by listing all possible strings of two letters drawn from a list of 10 distinct letters of the alphabet without repeating any letter?

    (Note: The letters are drawn one at a time)

Theoretical Probabilities

11. If you roll a pair of eight-sided fair dice and count the total number showing on the top faces, what is the sample space of all possible

outcomes? Are the outcomes equally likely? Assign probabilities to the sample space of possible outcomes and determine the probability of rolling a 6 or 7 on a single toss of the dice.

12. When a pair of 10-sided fair dice are thrown and the total of the numbers on the top faces calculated, what is the sample space of possible outcomes? Are the outcomes equally likely? What is the probability of rolling a total less than 4 on a single toss of the dice?
13. In a particular study of productivity for workers in a particular company, it is found that the employees vary in their time of highest productivity during the day. It is found that 310 are most productive between 9:00 a.m. and noon, 280 are most productive between noon and 3:00 p.m., and 175 are most productive between 3:00 p.m. and 6:00 p.m. Assuming that the time intervals include the first minute but not the last of the time window, calculate the probability that a randomly selected employee will be most productive in each of the particular time windows.
14. Suppose it is standard policy at a particular company for employees to run the automatic update program on their desktop PCs once per month. A study by the information technology department reveals that of 1375 employees, 1180 run the program in a given month. What is the empirical probability that a particular employee, chosen at random, will have run the update program?
15. A particular bag contains a red chip, a blue chip, and a yellow chip. A second bag contains a red chip and a yellow chip. If an experiment consists of drawing a chip from each bag at random, what is the sample space? What is the probability of drawing two chips that are both the same color?
16. Four cards are dealt from a shuffled deck of 52 cards. How many hands are possible that satisfy each condition? (a) The hand contains all black cards. (b) The hand contains three jacks. (c) The hand contains all diamonds. What is the probability of the event described in (b) occurring?
17. Suppose that two six-sided dice are thrown: one red and one blue. What is the probability you will roll a 2 on the red die and a 4 on the blue die? What is the probability you will roll an even number on the red die and an odd number on the blue die? What is the probability you will roll a 1 on both dice or a 2 on both dice?

Complementary Events

For the following events, what is the complementary event? Also, what is the probability of the event given and its complement?

18. Rolling a 4 or 5 on a six-sided die.
19. Selecting a letter of the alphabet at random and choosing a vowel.
20. Selecting a day of the week whose name ends in "y."
21. Selecting a month of the year whose name begins with "J."

Empirical Probabilities

22. In a standard text written in English, the most commonly occurring letters are e, t, and a. It is known that the probability of these letters occurring are, respectively, 0.131, 0.104, and 0.081. If a particular page of a computer science text written in English is examined and it is known that the text of the page contains 480 letters, how many e's, t's, and a's would you expect to find? Examine the words in this problem statement (all the way to the end), counting all the letters, and determine how many e's occur. What is the empirical probability of a randomly selected letter in this problem being an e, and how does this compare to the theoretical probability?
23. At a particular company, data are compiled to determine the number of employees who arrive at work near the start of the workday (9:00 a.m.), and the data shown in Table 12.1 are found. What is the probability that a randomly selected worker will arrive between 9:00 a.m. and 9:05 a.m.? What is the probability that a randomly selected worker will arrive at or before 9:05 a.m.?

TABLE 12.1 Table for Exercise 23

Arrival Time	Number of Workers
Before 9:00 a.m.	65
From 9:00 a.m. to 9:05 a.m.	42
After 9:05 a.m.	18

24. Suppose a computer store is visited by 128 customers in a particular day, of whom 37 were Mac owners. Based on this observation, what is the likely number of 900 customers in a week who would be Mac owners?
25. On a given day, you find that of the 284 emails in your in-box, 31 are junk mail. What is the probability, based on this observation, that a randomly selected email in your in-box the next day will be junk mail?

12.2 Odds

If you've ever been to a gambling establishment such as a casino or participated in a lottery, you've seen references to "odds," usually pertaining to the "odds of winning." Winning, of course, is a relative event, since what constitutes "winning" depends on which side of the gaming table you stand, but let's assume that the situation is being observed from the player's point of view. The outcome of a game of chance, "win" or "lose," is something we refer to as an "event." That is, an event is an outcome of some particular activity under consideration.

Odds in Favor of and Odds against an Event

When we talk about odds in a mathematical sense, there are two major categorizations we use: **odds in favor of an event** and **odds against an event**.

The notation for each is identical, so we'll need to be a bit careful to pay attention to the terminology in use so that the two are not confused.

Let's consider odds in favor of an event first. Suppose we have a particular situation we are observing, and there are n possible outcomes to the situation. Note that, in order for us to examine and understand the activity, it is an essential starting point that we know the number of possible outcomes for the activity, so we ultimately will have to address the problem of counting the number of possible outcomes.

For instance, imagine we are flipping a fair coin twice. Once again, the word "fair" refers to the neutrality of the coin and that the coin is not a "trick coin" and that every possible outcome is as likely to occur as any other. There are four possible outcomes—HH, HT, TH, and TT—where the letters H and T refer to flip results of "heads" or "tails," and the first letter shown is the result of the first flip the second letter that of the second flip. Thus, in this situation, n would be 4. We are interested, for whatever reason, in one particular event: the result where two heads are flipped. Note that there is one such outcome: the HH outcome. This "desired" outcome we can refer to as a favorable outcome, meaning that it satisfies the condition we are looking for in the experiment. The other, "undesirable" outcomes we can refer to as unfavorable outcomes.

The odds in favor of an event are given as $a{:}b$, which we read as "a to b." The value of a is the number of favorable outcomes, the value of b is the number of unfavorable outcomes, and since these encompass all possible results of the experiment, it must be the case that $a + b = n$. In our example, the odds in favor of flipping a coin twice and obtaining heads on both tosses is 1:3, which means that of the possible outcomes, there is 1 favorable outcome and there are 3 unfavorable outcomes.

The odds against an event happening are the exact reversal of the odds in favor, where we list the number of undesirable outcomes first and the number of desirable outcomes second. Thus, in the same situation, the odds against flipping a coin twice and obtaining heads on both tosses is 3:1.

Some points must be made before moving forward as matters of clarification and standardizing our notation. In the notation $a{:}b$, the values of a and b should be integer valued and such that they have no common factors (also described as the condition of a and b being relatively prime to one another). If a and b should have common factors, we can divide both by that common factor (much like we were reducing fractions) to simplify the odds statement. That is, if we were told that in a particular experiment there were 14 favorable outcomes and 10 unfavorable outcomes for our event, then we might (at first) say that the odds in favor of the event were 14:10, but since those have a common factor of 2, we can divide each number by the common factor and conclude that the odds in favor of the event are 7:5. This "simplified" form can be construed as meaning that for every 7 favorable outcomes, there are 5 unfavorable outcomes possible.

EXAMPLE 12.6

Suppose it is known that two bins of computer hard drives (one containing defective drives and the other containing correctly functioning drives) are inadvertently mixed together. Visual inspection does not permit an observer to determine which drives are defective, and in the process of combining the drives, they were well mixed. If there were 74 defective drives and 26 correctly functioning drives, what are the odds in favor of randomly selecting a defective hard drive?

SOLUTION

At the outset, we must determine what constitutes a "favorable" result. In this case, though it seems counterintuitive, a favorable result is the selection of a defective hard drive. There are, in this situation, 74 favorable outcomes and 26 unfavorable outcomes, and thus the odds in favor of selecting a defective hard drive can initially be stated as 74:26. Since those numbers have a common factor of 2, we can reduce the expression to the final form 37:13, which are the odds being sought.

EXAMPLE 12.7

It is known that in a company employing 86 individuals, 24 of the employees are female. If an employee is to be selected at random to be the subject of a productivity review, what are the odds against the selected individual being female?

SOLUTION

In this situation, because we are interested in the odds against randomly selecting a female employee, we can begin by finding the odds in favor of that choice and then reverse the expression to obtain the odds against. Since there are a total of 86 employees, we see that there are 24 female and 62 males, so the odds in favor of selecting a female employee at random would be 24:62, or 12:31. Reversing this result, we conclude that the odds against selecting a female employee would be 31:12.

The Relationship between Probability and Odds

Although the concepts are not synonymous, there is a relationship between probability and odds, and we can use one to find the other. Let's suppose we know the probability of a given computer at a company being a Linux machine is 37%. If we decide to choose a machine at random, what are the odds in favor of the chosen machine being a Linux machine?

The key to solving this problem lies in the recognition that the percentage can be expressed as an equivalent fraction or decimal and that fraction or decimal can be used to determine a number of favorable and unfavorable selections of a particular type of computer. The probability being 37% indicates that if 100 computers are present, 37 of them would be Linux machines (and, therefore, 63 would not). Thus, the odds in favor of randomly selecting a Linux machine would be 37:63.

RULE FOR CONVERSION OF PROBABILITY OF AN EVENT *E* TO ODDS IN FAVOR OF AN EVENT *E*

If the probability of event E is P%, then the odds in favor of event E are P:$(100 - P)$, where P and $(100 - P)$ are reduced in such a way that the values have no common factors. If the probability is given as a decimal or as a fraction P, then the odds in favor of event E are calculated as P:$(1 - P)$.

To convert in the opposite direction, given the odds and desiring the probability, we can take the odds in favor of event E, having the form a:b and calculate the probability of event E using the formula $P(E) = a/(a + b)$.

RULE FOR CONVERSION OF ODDS IN FAVOR OF AN EVENT E TO THE PROBABILITY OF EVENT E

If the odds in favor of event E occurring are a:b, then the probability of event E occurring is calculated as $a/(a + b)$.

EXAMPLE 12.8 If it is known that the probability of a rolling a total of 7 on two six-sided dice is $\frac{1}{6}$, what are the odds against rolling a total of 7 on two six-sided dice?

SOLUTION

Since the probability of rolling a 7 is given as a fraction, we can use the rule for conversion to odds in favor of rolling a 7 as $\left(\frac{1}{6}\right):\left(1 - \frac{1}{6}\right)$, or $\frac{1}{6}:\frac{5}{6}$, which simplifies to 1:5. Thus, the odds against rolling a total of 7 on the dice would be 5:1.

EXAMPLE 12.9 For a particular video game player console, we are told that the probability of purchasing a game containing a defective sample game disk is 0.035. What are the odds of purchasing a game containing a defective disk?

SOLUTION

Because of the manner in which the question is framed, the "favorable" outcome corresponds to the purchase of a game player console containing a defective disk, the probability of which is given to be 0.035. The fractional equivalent of this probability is 35/1000, which indicates that the odds are (on first glance) $0.035:(1 - 0.035) = 0.035:0.965$.

This answer is unacceptable, since the values involved are not integers. We can multiply both values by 1000 to overcome this and obtain the expression 35:965. Since these values have (at least) common factor 5, we can divide both by 5 to obtain 7:193. The numbers involved here do not have common factors, and thus we conclude that the odds in favor of purchasing a game player console having a defective game disk are 7:193.

Keep in mind the significance of the values presented in a statement of odds in favor of an event. In Example 12.9, we found the odds in favor of purchasing a game player with a defective game disk to be 7:193. This indicates that if we set about purchasing such game players, then for every 7 players containing a defective sample game disk, there will be 193 players not containing such a disk.

Exercises

Answer the following questions using proper grammar, complete sentences, and correct terminology.

1. In your own words, explain the difference between odds in favor of an event and odds against an event.
2. How do odds differ from probabilities?

3. What determines whether or not an outcome is a "success" when computing odds?
4. In your own words, explain how the odds in favor of an event are used to calculate the probability of that event occurring.

For each of the following problems, calculate the odds described.

5. What are the odds in favor of selecting a heart from a deck of cards containing only the red cards from a standard 52-card deck?
6. For any particular natural number represented in hexadecimal notation, what are the odds against the numbers farthest-right digit being a 2?
7. An event has even odds when the odds in favor of (or against) the event are 1:1. What is the probability of an event with even odds?
8. Thirteen out of 20 emails in your in-box are junk emails. If you open one of the emails at random, what are the odds in favor of you opening a junk email? What are the odds against you opening a junk email?
9. An inspector at a particular manufacturer estimates that 0.5% of MP3 players are defective. What are the odds in favor of randomly selecting an MP3 player made by this manufacturer and finding a properly functioning MP3 player?
10. Lionel collects coins from different countries. He has five from Canada, two from France, one from Russia, four from Great Britain, and one from Germany. If he accidentally loses one coin, what are the odds in favor of the lost coin being from Russia?
11. Suppose a weather forecaster states that the probability of rain today is 25%. What are the odds it will rain today?
12. The door prize at a company's Christmas party with 25 people present is given by writing numbers 1 through 25 on the bottom of the paper plates used. What is the probability that an individual had the winning plate? What are the odds against an individual having the winning plate?
13. If it is known that 13% of the population of the United States prefers to manipulate their computer mouse with their left hand, what are the odds of choosing a random individual in the United States and finding they do not prefer to manipulate their mouse with their left hand?
14. A miniature golf course offers a free game to golfers who make a hole in one on the last hole. Last week, 44 out of 256 golfers made a hole in one on the last hole. What are the odds against randomly choosing one of those golfers and finding they had made a hole in one on the last hole?
15. An individual is asked to randomly select a number from 1 to 30, inclusive. What are the odds in favor of the individual choosing a prime number?
16. You lost a ring in a rectangular field that is 110 yards by 65 yards. You search a rectangular section of the field that is 25 yards by 32 yards. What is the odds against the ring being in the section you search?

17. A train runs every 15 minutes. You arrive at the train station without consulting the train's time schedule. What are the odds in favor of you waiting 10 or more minutes for your train?
18. A traffic light is green for 17 seconds, yellow for 3 seconds, and red for 20 seconds. Suppose your approach to the light is not affected by traffic or other factors. What are the odds in favor of the light being green when you first see the light?
19. Many Web sites have ads whose appearance is based on probability. Advertisers pay the Web site based on the number of times the ad appears. If the probability that an ad appears when a particular Web site is loaded is 0.2, what are the odds against an advertiser visiting the Web site and seeing his ad?

Odds and Probabilities

20. If it is known the odds in favor of a particular event are 3:8, what is the probability of the event occurring?
21. Referencing the situation in Exercise 20, what is the probability of the event not occurring?
22. If it is known the odds against an event occurring are 11:5, what is the probability of the event occurring?
23. Referencing the situation in Exercise 22, what is the probability of the event not occurring?
24. Suppose that, in a batch of 4000 microprocessors, it is known that the odds in favor of randomly selecting a defective processor is 1:100. What is the probability of randomly selecting a defective processor?
25. A popular software package is known to have been installed incorrectly on some machines at a particular corporation. If it is known the odds in favor of an incorrect installation are 1:5, what is the probability a randomly chosen machine will have had the software incorrectly installed?
26. It is known that 2.4% of all men in the United States suffer from protanopia, a form of color blindness. What are the odds in favor of choosing a man in the United States and finding that he suffers from this form of color blindness?

12.3 Expected Value

Consider the following situation, in which we undoubtedly have all found ourselves in the past: you are taking a multiple choice examination, each question of which has four possible answers from which we can select. There is one, and only one, correct solution, which is worth 4 points, and three incorrect solutions. To discourage guessing, the instructor has told us he will assign a 1-point penalty to any incorrect or unanswered question.

On a particular question for which we cannot rule out any of the solution options and on which we have no idea regarding the correct answer, should we guess?

The answer depends on a concept known as the **expected value** of a guess, which will tell us the expected return in a game of chance, where we take the view that a "game of chance" corresponds to a situation where we must make a selection from a set of alternatives each having a predetermined value. Clearly, the situation of test taking satisfies this criterion and hence can be considered to be a game of chance.

In order to determine the expected value of a guess, we need to know two things: the probability involved in the various choices open to us and the value of each of those choices. Since some outcomes represent a favorable return and some an unfavorable, we arbitrarily choose to assign favorable results a positive value and unfavorable results a negative value.

Expected value need not apply to a game of chance. It can pertain to any situation in which a variety of possible outcomes exist and for which every outcome has some particular known chance of occurring and a value assigned to its occurrence. We'll see, in the examples, that there exist a wide variety of such circumstances, and thus the importance of the notion of expected value cannot be overstated.

Expected Value

If a situation or a game of chance has a set of n outcomes having values $x_1, x_2, \ldots, x_n$ and the probabilities of those outcomes are $P(x_1), P(x_2), \ldots, P(x_n)$, then the expected value of the game is defined to be

$$E = x_1P(x_1) + x_2P(x_2) + \cdots + x_nP(x_n)$$

Let's examine the test-taking illustration mentioned earlier. There are four possible answer choices, only one of which is the correct solution. From this, we conclude that the probability of randomly guessing the correct solution is $\frac{1}{4}$. There are three answers that are incorrect solutions, and thus we have a $\frac{3}{4}$ probability of randomly guessing an incorrect solution. We've been told that a correct solution will be awarded 4 points, while an incorrect solution will be assigned a 1-point penalty, and thus the value of the correct solution is 4 and that of the incorrect solution is -1.

Thus, our expected value is

$$E = 4\left(\frac{1}{4}\right) + (-1)\left(\frac{3}{4}\right) = \frac{4}{4} - \frac{3}{4} = \frac{1}{4}$$

What does this tell us? Since the expected value of a guess is $\frac{1}{4}$, this means that we expect a positive impact on our overall test score of $\frac{1}{4}$ points for every such guess we are forced to make, in the long run. Thus, if we were to guess on 20 of the questions of the exam, we could expect that guessing on those questions will result in an impact of $(20)\left(\frac{1}{4}\right) = 5$ points toward our score.

In other words, on this particular exam, with this grading scale, we should go ahead and guess on all the problems for which we have no idea regarding the solution, since the penalty for an incorrect guess is not heavy enough to produce an overall negative impact on our score. If the instructor really wanted to discourage guessing on the exam, a high enough penalty should be placed on the incorrect solutions to give us a negative expected value for questions on which we would be tempted to guess.

EXAMPLE 12.10

Suppose a textbook company determines, based on historical data, that a typical history textbook rough draft contains 5 errors per chapter and that a typical philosophy textbook rough draft contains 3 errors per chapter. If a collection of 11 such books is laid out on a table, 7 of which are philosophy books, and one is chosen at random, how many errors would the company expect to find in any given chapter?

SOLUTION

In this situation, the value of each outcome is the number of errors expected in the chapter, and here (since we are counting the number of errors) those values are positive in both cases. The probability of choosing a history textbook is $\frac{4}{11}$, and the probability of choosing a philosophy textbook is $\frac{7}{11}$, and therefore the expected value for the number of errors in the chosen chapter is

$$E = 5\left(\frac{4}{11}\right) + 3\left(\frac{7}{11}\right) = \frac{20}{11} + \frac{21}{11} = \frac{41}{11} \approx 3.7$$

EXAMPLE 12.11

Suppose an airline determines that 7% of all people making reservations on a particular flight will not show up. Consequently, it is their policy to sell 53 tickets for a flight that can hold only 50 passengers. Supposing that a flight where each ticket is sold for \$275 is completely sold out (through booked sales) and that persons failing to board their booked flight are charged a \$75 change-of-booking fee, what is the expected value of the flight (assuming that all no-shows transfer to another flight)?

SOLUTION

Note the airline has strategically chosen to ensure the number of expected "shows" will be below the capacity of the plane. In reality, airlines build more complicated profit models that account for "bump fees" they must pay to booked passengers who arrive to find there has been a higher-than-expected turnout, and hence no seat is available for them on the plane. That being said, we will take the view that (in this case) the historical trend holds true for the flight in question and that 93% of the booked passengers appear to board the plane. The expected value of this flight corresponds to the total of the ticket sales, plus the rebooking fees the airline will charge the late passengers. Here, we compute that the airline expected revenue will be

$$E = (275)(53)(0.93) + (275 + 75)(53)(0.07)$$
$$E = 13{,}554.75 + 1298.50$$
$$E = 14{,}853.25$$

EXAMPLE 12.12 A hospital wishes to determine the expected length of time a patient will be cared for in the intensive care unit (ICU). A study is made, and the data shown in Figure 12.2 are compiled. What is the expected stay in ICU, according to the data?

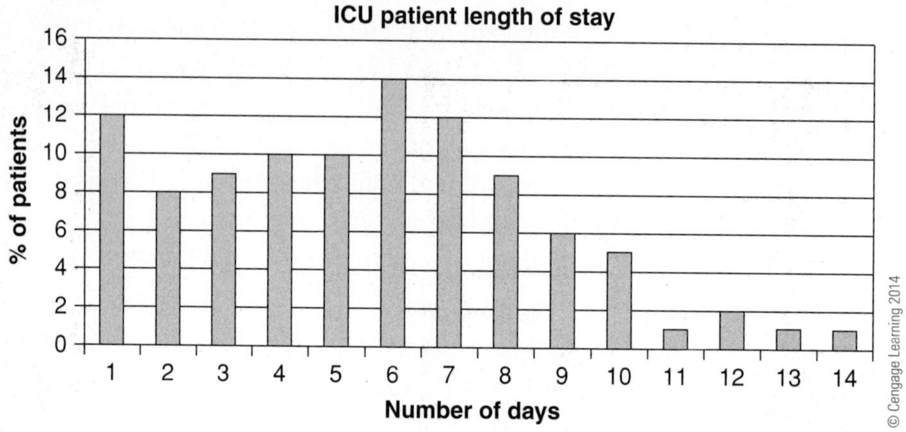

Figure 12.2 Length of ICU stay

SOLUTION

In this case, the value of each outcome is the number of days per stay, and therefore all the values for the outcomes will be positive.

Here, based on the information in the chart, we find

$$E = (1)(0.12) + (2)(0.08) + (3)(0.09) + (4)(0.10) + (5)(0.10) + (6)(0.14) + (7)(0.12) + (8)(0.09) + (9)(0.06) + (10)(0.05) + (11)(0.01) + (12)(0.02) + (13)(0.01) + (14)(0.01)$$

$$E = 0.12 + 0.16 + 0.27 + 0.40 + 0.50 + 0.84 + 0.84 + 0.72 + 0.54 + 0.50 + 0.11 + 0.24 + 0.13 + 0.14$$

$$E = 5.51$$

From this, we can conclude that the hospital expects the average stay in ICU to be 5.51 days.

EXAMPLE 12.13 A study is made on the number of times a typical baby can be expected to awaken during the night. It is found that, in a particular trial, the number of times a baby wakens and the probabilities corresponding to each possibility are

Table 12.2 Probability of Baby Wakenings

Awakenings	Probability
0	0.02
1	0.13
2	0.28
3	0.52
4	0.02
5	0.03

as given in Table 12.2. What is the expected number of times a typical baby will awaken during the night, according to this information?

SOLUTION

The number of awakenings can be considered the value of each occurrence, in this example, and thus, with the formula for calculating expected value, we find

$$E = (0)(0.02) + (1)(0.13) + (2)(0.28) + (3)(0.52) + (4)(0.02) + (5)(0.03)$$
$$E = 0 + 0.13 + 0.56 + 1.56 + 0.08 + 0.15$$
$$E = 2.48$$

From this, we can conclude that a typical baby will awaken between 2 and 3 times per night.

We should mention, in closing, that the expected value of an experiment really gives us the expectation should a procedure or situation be observed many, many times. For instance, in the awakening-baby example, we would predict the value we found to hold if a large number of children were observed. It would not be surprising to find a child chosen at random who would awaken five times in a given night, for example. The significance of expected value is that it allows us to make predictions in the presence of a large number of repetitions of an experiment.

Exercises

Answer the following questions using proper grammar, complete sentences, and correct terminology.

1. In your own words, explain the meaning of expected value and how it is calculated.
2. When calculating expected value, what distinction must be observed between favorable and unfavorable events?

In the following problems, calculate the expected value as indicated.

3. In a particular game, you roll a six-sided die. If the outcome is a 4, you win \$13. If the outcome is anything else, you lose \$2. What is the expected value of a single play of the game?
4. In a particular game, you roll a six-side die. If the outcome is a 4, you win \$13. If the outcome is a 1, you win \$6. On all other outcomes, you lose \$5. What is the expected value of a single play of the game?
5. In a particular game, you roll a six-sided die. If the outcome is a 1 or a 2, you win \$6. If the outcome is a 3, you win \$18. If it comes up a 5, you lose \$20, and on the other outcomes you win or lose nothing. What is the expected value of a single play of the game?
6. A card is drawn at random from a deck consisting of cards numbered 2 through 10. A player wins \$1 if the number on the card is odd and loses \$1 if the number if even. What is the expected value of his winnings?
7. A card is drawn at random from a deck of playing cards. If it is red, the player wins \$1; if it is black, the player loses \$2. Find the expected value of the game.

8. You have \$100 and play the following game. An urn contains two white balls and two black balls. You draw the balls out one at a time without replacement until all the balls are gone. On each draw, you bet half of your present fortune that you will draw a white ball. Assuming that on a draw of a white ball you receive an amount equal to your bet (plus the amount of your wager), what is your expected final fortune?
9. Exactly one of eight similar passwords opens a certain program. If you try the passwords, one after another, what is the expected number of passwords that you will have to try before having success?
10. A mechanic inspects your car and finds that the engine has either bad piston rings or bad valves. Without tearing the engine apart, she tells you she believes there is a 60% probability that the problem is the piston rings, a 40% probability that the problem is the valves, and an 80% chance that both the piston rings and the valves are bad. She tells you the piston rings will cost \$2300 to repair, the valves will cost \$1800 to repair, and the rings and valves combined will cost \$3300. If you know you can get a reliable used car for \$3000, should you repair the car or buy the used car?
11. Mario and John are playing a card game. Every time Mario draws a jack or a queen, John gives him \$4. When Mario draws a king or an ace, John gives him \$5. In turn, Mario gives John \$$X$ for any other card drawn. In order for John to expect a profit of \$1 per draw, what should be the value of X?
12. Sue and Jai are playing a coin-tossing game. Each flips a fair coin, and if two heads appear, Sue pays Jai \$5. If two tails appear, Sue pays \$1 to Jai. In other cases, Jai pays Sue \$2. What is the expected value of the game from Jai's perspective?
13. In training a new worker to perform a delicate mechanical operation, the probabilities that the worker will be successful on her first, second, and third attempt are 0.1, 0.2, and 0.7, respectively. Find the expected number of successes the worker will have in these three attempts.
14. A gambler who has \$10 plays the following system. At the first toss of a coin, he bets \$2 on heads and quits if he wins. If he wins, his payment is a return of his bet plus \$5. If he loses, he bets \$4 on heads on the second toss and quits if he wins. If he loses again, he bets his last \$4 on heads on the final toss. What is his expected final gain in this game with this strategy?
15. In the famous "St. Petersburg Paradox" game, a gambler wins 2 rubles with probability 0.5, 4 rubles with probability 0.25, 8 rubles with probability 0.125, and so on. Before making any computations, speculate how much you would be willing to pay to play this game so your expected value would be 0 (meaning the game would be "fair"). What is the expected value of the game?
16. In a survey of a three-block span of a particular street, one building is chosen from each block. Some of the buildings are apartments, as indicated in Table 12.3. What is the expected number of apartment buildings in the sample?

Table 12.3 Distribution of Homes in Exercise 16

Block	1	2	3
Number of apartment buildings	2	5	1
Number of private homes	11	4	9

17. A box contains five tickets numbered 1 through 5. They are drawn from the box one by one. What is the expected value of the number of times the number on the ticket matches the ordinal number of the draw (for instance, that on draw 3, we would select the ticket bearing number 3)?
18. In Las Vegas, a roulette wheel has 38 slots numbered 0, 00, 1, 2, . . . , 36. The 0 and 00 slots are green, and half of the remaining 36 slots are red and half are black. A croupier spins the wheel and throws an ivory ball. If you bet \$1 on red, you win \$1 if the ball stops in a red slot; otherwise, you lose \$1. What are the expected value of your winnings if you bet \$1 on red?
19. In Las Vegas, the roulette wheel has a 0 and a 00 and then the numbers 1 to 36 marked on equal slots; the wheel is spun, and a ball stops randomly in one slot. When a player bets \$1 on a number, he receives \$36 if the ball stops on this number for a net gain of \$35; otherwise, he loses his dollar bet. Find the expected value for his winnings.
20. The number of calls per hour for service by a representative of the information technology department at a particular company has been shown to follow the distribution displayed in Table 12.4. What is the expected number of calls for service per hour?

Table 12.4 Distribution of Calls in Exercise 20

Number of calls X	0	1	2	3	4	5
$P(X)$	0.08	0.10	0.15	0.40	0.25	0.02

21. The probability that a computer network will be inoperative the indicated number of hours per week during the initial installation phase of the network is shown in Table 12.5. What is the expected number of hours the network will be inoperative?

Table 12.5 Time Computer Network is Inoperative in Exercise 21

Number of hours X	4	5	6	7	8	9
$P(X)$	0.01	0.10	0.29	0.33	0.25	0.02

12.4 "AND" AND "OR" PROBLEMS AND CONDITIONAL PROBABILITY

Consider a situation where two events were being considered in unison. For instance, we might be thinking of event A as the failure of a desktop computer's hard drive and event B as being the failure of that same computer's sound card. Assuming we know the likelihood of each of these events, we can consider the likelihood of both event A and event B occurring together, or the likelihood of either event A or event B occurring. A combination of a pair of events A and B using the words "and" or "or" creates what is called a **compound event**.

It is to these situations we are referring when we discuss the "and" problem and the "or" problem. Computation of these probabilities depends on the relationship, if any, between the events A and B.

The "And" Problem

Let's start with consideration of the compound event involving the word "and." Before beginning, we will define what it means for a pair of events to be independent of one another. Events A and B are called **independent** if the occurrence of event A does not influence the occurrence of event B. Two events that are not independent are referred to as **dependent** events. Sometimes we can tell, based on our experience, if two events are independent of one another, but in other cases it is less obvious.

EXAMPLE 12.14 Suppose a six-sided die is tossed twice. Event A is the situation where the first toss produces a result of 4, while event B is where the second toss produces a result of 3. Are events A and B independent?

SOLUTION

Yes, they are. The result on any particular toss of a die has no influence (despite what "hunch" bettors might think!) on the result of any other die toss. Thus, the occurrence of event A will not impact the occurrence of event B.

EXAMPLE 12.15 Two cards are drawn in succession from a deck of 52 standard playing cards. Let event A be the first card drawn being an ace and event B be the second card drawn being an ace. Are the events A and B independent?

SOLUTION

No, they are not. The fact that the first card is not replaced before drawing the second card creates a dependence on the two events. Suppose, for instance, the first card drawn is not an ace, so event A has not occurred. Now four aces remain in the deck, but we have only 51 cards from which to choose, and therefore $P(B) = \frac{4}{51}$. Had event A occurred, only 3 aces would remain in the deck, and hence $P(B) = \frac{3}{51}$. Thus, the occurrence or nonoccurrence of event A impacts the occurrence of event B, and therefore the events are dependent.

If events A and B are independent, then the calculation of $P(A \text{ and } B)$ is relatively straightforward: the probability of A and B occurring is the product of $P(A)$ and $P(B)$. That is, for independent events, $P(A \text{ and } B) = P(A)P(B)$. This rule is sometimes called the multiplication rule.

MULTIPLICATION RULE FOR INDEPENDENT EVENTS If A and B are independent events, then $P(A \text{ and } B) = P(A) \cdot P(B)$

EXAMPLE 12.16 Suppose we modify the last example and consider the following situation. A card is drawn from a standard deck, its value is determined, and then it is replaced into the deck and a second card drawn. Let A be first card drawn being an ace and B be second card drawn being an ace. What is $P(A \text{ and } B)$?

SOLUTION

Since the first card drawn is replaced before the second card is drawn, the value of the first card has no impact on the value of the second card, and hence the events are independent of one another. Moreover, $P(A) = \frac{4}{52} = \frac{1}{13} = P(B)$, and therefore $P(A \text{ and } B) = P(A)P(B) = \left(\frac{1}{13}\right)\left(\frac{1}{13}\right) = \frac{1}{169}$. Note that it is surprisingly unlikely for this situation to actually occur!

EXAMPLE 12.17 Suppose it is known that the failure of a computer's hard drive is completely independent of the failure of its sound card and that on any given day, the probability of a hard drive failure is 0.001, while the failure of the sound card is 0.00035. What is the probability of (on a particular day) both the hard drive and the sound card failing?

SOLUTION

Let A be the computer's hard drive failing and B be the computer's sound card failing. Since we are assuming the events are completely unrelated, we can apply the multiplication rule to find $P(A \text{ and } B)$. In this case,

$$\begin{aligned} P(A \text{ and } B) &= P(A)\,P(B) \\ &= (0.001)(0.00035) \\ &= 0.00000035 \end{aligned}$$

We can see that it is extremely unlikely that both of these components would fail on the same day, since the occurrence of one has no bearing on the occurrence of the other.

Note

If $P(A \text{ and } B) = 0$, we say the events A and B are mutually exclusive.

If it should turn out that $P(A \text{ and } B) = 0$, then the events cannot occur simultaneously. Such events are said to be **mutually exclusive**. Examples of such events might be where event A is the event where your computer is turned off and event B is the event where your computer is turned on. Since these situations cannot exist together, the events A and B would be considered mutually exclusive.

To this point, we have limited our attention to the situation where events A and B are independent. What if the events A and B are dependent? In this case, the probability of the second event is affected by the occurrence or nonoccurrence of the first event. In order for us to devise a formula for this situation, it is necessary for us to first define and understand conditional probability, a topic we will introduce once we've discussed the "or" problem type. At that point, we'll revisit the "and" problem in the case where events are known to be dependent.

The "Or" Problem

Now let's turn our attention to the compound event involving the word "or." If we consider the compound event "*A* or *B*," then the event is judged to have occurred in the case where *A*, *B*, or both have occurred. This is, you may recall, what we described as the "inclusive or" in our examination of logic earlier in the text.

As an example, we can designate "my laptop is stolen" to be event *A* and "my identity is stolen" to be event *B*. Then the compound event "*A* or *B*" would be taken to mean "my laptop is stolen, or my identity is stolen." In this case, we consider the event "*A* or *B*" to have taken place if either my laptop is stolen, my identity is stolen, or both. That interpretation may seem to be a bit broad, but it is the view we will take here.

The probability for the occurrence of the compound event "*A* or *B*" is defined by the following rule:

$$P(A \text{ or } B) = P(A) + P(B) - P(A \text{ and } B)$$

In the particular case where *A* and *B* are mutually exclusive events, this formula simplifies to

$$P(A \text{ or } B) = P(A) + P(B)$$

EXAMPLE 12.18 A technician in charge of troubleshooting both the word processing and the spreadsheet software has checked the word processing software on half the machines in a computer lab and the spreadsheet software on two-thirds of the machines in the lab. Being particularly careless, he has not paid attention to which machines he has tested for which software. What is the probability that any given computer in the lab has had at least one software type tested?

SOLUTION

If we designate as event *W* the situation where a computer has had its word processing software tested and event *S* represents a computer having had its spreadsheet software tested, then we are intending to calculate $P(W \text{ or } S)$. The events are clearly not mutually exclusive, since it must be the case that at least some of the machines have had both software systems checked. Therefore, the formula we will need to use is $P(W \text{ or } S) = P(W) + P(S) - P(W \text{ and } S)$

Moreover, since one software system being checked has nothing to do with another being checked, the events are independent, and thus $P(W \text{ and } S) = P(W)P(S) = \left(\frac{1}{2}\right)\left(\frac{2}{3}\right) = \frac{1}{3}$.

By using the given probabilities, we can conclude

$$P(W \text{ or } S) = \frac{1}{2} + \frac{2}{3} - \frac{1}{3} = \frac{5}{6}$$

EXAMPLE 12.19 A student at a university is asked what type of computer she has in her home. The probability that the student has a Macintosh computer is 0.37, the probability that the student has a PC is 0.81, and that a student has both is 0.27. What is the probability that a student owns a Macintosh, a PC, or both?

SOLUTION

If M is the event where the student owns a Macintosh and P is the event where the student owns a PC, then the statement of the problem leads us to a direct substitution into the formula

$$P(M \text{ or } P) = P(M) + P(P) - P(M \text{ and } P) = 0.37 + 0.81 - 0.27 = 0.91$$

From this, we conclude that there is a 91% probability that the student owns either a Mac, a PC, or both.

Conditional Probability

A conditional probability is the probability of an event happening given that another event has already occurred. For instance, revisiting the situation described in Example 12.15, suppose we want to know the probability of drawing two cards in succession from a standard deck of 52 cards and having both be aces. If the events are named A and B, the notation employed is $P(B|A)$, which is read as "the probability of B occurring given that A has occurred," or, more briefly, "the probability of B given A."

In the situation of Example 12.15, the event $B|A$ is precisely the probability of drawing an ace on the second selection given that the first selection already was an ace. Note that, for this situation, $P(B|A) = \frac{3}{51} = \frac{1}{17}$, since (given the first card drawn was an ace) there are 3 aces remaining in a deck containing only 51 cards.

For dependent events, we can use the concept of conditional probability to define $P(A \text{ and } B)$ through a second multiplication rule:

MULTIPLICATION RULE FOR DEPENDENT EVENTS

If A and B are dependent events, then $P(A \text{ and } B) = P(A)\,P(B|A)$.

Note

Observe that this formula simplifies to the first multiplication rule in the situation where A and B are independent: if A and B are independent, then the occurrence or nonoccurrence of A has no impact on the occurrence of B, and thus $P(B|A)$ must equal $P(B)$.

Let's consider how a conditional probability might be calculated. As we'll see, it involves a somewhat closer examination of the situation in which we are interested. Suppose, as an illustration, we consider a population of 10,000 men and find that of these men, there is a 0.79 probability that any given man will live to age 60 and that there is a 0.58 probability that any given man will live to age 70. If we consider a man who is currently age 60, what is the probability that he will live to age 70?

This is an example of conditional probability. Here, the original population has 10,000 members, and let's assign the event names E to a man living to age 60 and F to a man living to age 70. Since the man is known to already be age 60, we are making the assumption that event E has already occurred, and (given that assumption) we now wish to know the probability of event F occurring.

In order to calculate this probability, we need to closely consider the portion of the population involved in the actual problem in which we are interested. Of the original 10,000 men, the number who will actually live to age 60 is $(0.79)(10{,}000) = 7900$, and the number who will live to age 70 is $(0.58)(10{,}000) = 5800$. Therefore, we are wondering what the probability would be that the man being considered (who, by assumption, is one of the 7900 who lived to age 60) will be one of the 5800 who live to age 70. Thus, $P(F|E) = 5800/7900 \approx 0.734$, and we conclude that there is a 73.4% probability (approximately) that the 60-year-old man will live to age 70.

EXAMPLE 12.20 Suppose we roll a six-sided die once. If we let A be the event where we roll a 1 on the die and B be the event that the total on the die is less than 3, then what is $P(A|B)$?

SOLUTION

If we are aware that the roll on the die is less than 3, then the two possible outcomes were a roll of 1 or 2, and thus the sample space has only two outcomes. Knowing that exactly one of these two outcomes occurred and the desired outcome is the event where a 1 is rolled, then the probability of rolling a 1, given that it is known that the roll was either a 1 or a 2, would be exactly $\frac{1}{2}$.

EXAMPLE 12.21 Suppose that an insurance company headquarters employs 900 people and that 47% of them are female. If it is known that 28% of the female employees are under the age of 30, what is the probability of an employee being under the age of 30 given that the employee is female?

SOLUTION

Of the 900 employees, since 47% of the employees are female, this tells us the company employs 423 females. Of those, 28% are under the age of 30, and therefore 118 of the female employees are under the age of 30.

Since the chosen employee is known to be female, the question amounts to finding the percentage that 118 is from 423. Thus, if F is the event where the chosen employee is female and T is the event where the chosen employee is under the age of 30, we can conclude that $P(T|F) = \frac{118}{423} = 0.279$.

We should comment that, *if* we happen to know $P(A \text{ and } B)$ and also $P(B)$, our second multiplication rule provides a method for finding $P(B|A)$:

$$\begin{aligned} P(A \text{ and } B) &= P(A)P(B\,|\,A) \\ \frac{P(A \text{ and } B)}{P(A)} &= P(B\,|\,A) \end{aligned}$$

This formula is one form of a rule known as Bayes's theorem, which generalizes to a much more complicated-looking result, applicable to independent events. Let's take a look at an example where this formula could be applied.

EXAMPLE 12.22 Suppose there are two small bags, labeled A and B. Bag A contains two clear marbles and three steel marbles, while bag B contains one clear marble and one steel marble. A bag is chosen at random, and a marble is pulled from it. What is the probability that we selected bag A given that a steel marble was drawn?

SOLUTION

If we identify event A as bag A being selected and S as a steel marble being selected, then what we are after is $P(A|S)$. From our formula above, we have

$$P(A \mid S) = \frac{P(A \text{ and } S)}{P(S)}$$

Here, consideration of the situation reveals that there are seven possible combinations of events and that the probability of drawing bag A and a steel marble is $\frac{3}{7}$, while the probability of drawing a steel marble is $\frac{4}{7}$. Therefore,

$$P(A \mid S) = \frac{3/7}{4/7} = \frac{3}{4}$$

EXAMPLE 12.23 In a sample of 1500 computers, 400 are owned by educational institutions, while 1100 are owned by individuals. Of those owned by educational institutions, 24 are faulty, while of those owned by individuals, 38 are faulty. If one computer is chosen at random, what is the probability that it is owned by an educational institution given that the computer is faulty?

SOLUTION

Let A be the event where the computer is owned by an educational institution and B the event that the computer is faulty. Then, from the given information, we see that $P(B) \approx 0.0413$ and also that $P(A \text{ and } B) = 0.016$. Using Bayes's theorem, we find that $P(A \mid B) = P(A \text{ and } B)/P(B) \approx 0.016/0.0413 \approx 0.387$. In other words, if a computer is selected at random and is found to be faulty, then there is a 38.7% probability that the computer belongs to an educational institution.

Permutations and Combinations

There are times when we wish to know the number of possible arrangements we can form from a particular set of objects. For instance, if I consider the set of letters $\{M, A, C\}$, we can rearrange these letters into a variety of three-letter orderings, and the question is, How many arrangements of the three letters can we form?

Since the set is small, it takes no great amount of effort for us to produce the list of arrangements by brute force: $\{MAC, MCA, AMC, ACM, CMA, CAM\}$. From this, we deduce that there are six such arrangements of the three letters. Each of these arrangements is called a **permutation** of the letters in the original set.

> **Note**
> A permutation of a set of objects is an arrangement of those objects into a particular order.

We can calculate the number of permutations that exist for a specific set of objects by considering the possible choices for each location within the arrangement. In the illustration using the set $\{M, A, C\}$, for the first letter in the arrangement we can choose any of the three letters. That choice being made, we have two remaining letters from which we can choose as our second letter, and then there remains a single letter for our third letter in the arrangement. That is, we have $(3)(2)(1) = 6$ possible arrangements of the letters in the set.

> **Note**
> If n is a natural number, then factorial notation $n!$ represents the product of the number n and all smaller natural numbers. The value of n! corresponds to the number of permutations of a set of n elements.

The product of a set of natural numbers, beginning at some specified value n and including all smaller natural numbers, can be expressed using **factorial notation**, $n!$:

$$n! = n * (n-1) * (n-2) \ldots 3 * 2 * 1$$

EXAMPLE 12.24 Calculate the following factorial values:

a. 5! b. 9! c. (3.5)!

SOLUTION

a. The notation 5! represents the product of 5 with all smaller natural numbers, and hence $5! = (5)(4)(3)(2)(1) = 120$.

b. The notation 9! represents the product of 9 with all smaller natural numbers, and hence $9! = (9)(8)(7)(6)(5)(4)(3)(2)(1) = 362{,}880$.

c. Factorials, by definition, can be computed only for natural numbers. Therefore, (3.5)! is an undefined expression.

If we wish to find the number of arrangements we can construct using some but not all of the elements of a set, then the formula is modified slightly. Suppose, for instance, we have 17 objects, and we wish to find out how many arrangements we can form, taking any 4 members of the set at a time. In such a case, we would have 17 choices for the first object, and having made that choice, we would have 16 choices for the second, 15 for the third, and 14 for the fourth, or $(17)(16)(15)(14) = 57{,}120$. possible arrangements.

In general, if we start out with n objects and we wish to find the number of possible permutations of any r of those objects, then that total is $n(n-1)(n-2)\ldots(n-r+1)$. Or, if we prefer, we can express the same product as $\frac{n!}{(n-r)!}$. Mathematically, we often describe this as the number of permutations of n objects, taking r at a time, denoted ${}_nP_r$.

EXAMPLE 12.25 A trifecta is a particular type of wager in a horse race, which is won by specifying the horses that finish first, second, and third in a race in the exact order they finish. If a particular derby has 19 horses competing, how many different trifecta bets are possible?

SOLUTION

The problem is really asking us the number of 3-horse permutations we can form from a set of 19 original horses, ${}_{19}P_3$. This is $= \frac{n!}{(n-r)!} = \frac{19!}{(19-3)!} = \frac{19!}{16!} = (19)(18)(17) = 5814$.

EXAMPLE 12.26 For a particular conference, some of the speakers will be arranged at a long table at the front of the presentation hall, while (because of size limitations) others will have to sit in the auditorium and be called up later. A total of eight speakers have been invited, with room for only five at the table. Because of personal prestige and ego issues, it is considered significant to the speakers where at the table they are positioned. How many permutations of the five speakers at the front table are possible?

SOLUTION

We have eight different speakers from whom to choose, and only five are able to be seated at the table. Therefore, since the ordering of the individuals at the table is significant (i.e., if the five individuals at the table are Adams, Baker, Charles, Diego, and Evans, then arranging them in that order is considered distinct from arranging them as Adams, Charles, Baker, Diego, Evans), then we must calculate the number of possible seating arrangements at the table as

$$ {}_8P_5 = \frac{8!}{(8 - 5)} = \frac{8!}{3!} = 6720 $$

One of the significant facts about a permutation is that every reordering is considered to be distinct from the others. When you take a group of objects from a collection of other items and the order is *not* considered important, then we say we are considering a combination of objects.

This matter is so subtle that we should illustrate the difference between the two concepts, using a single set. Suppose our set is $\{M, A, C\}$, and we want to form all possible two-letter permutations of the letters. We know that the number of such permutations will be ${}_3P_2. = \frac{3}{1!} = 6$, and we can list the permutations as $\{MA, MC, AM, AC, CA, CM\}$.

When we say that, in a permutation, the order is significant, we mean that the pairings MA and AM, for instance, are not considered to be the same despite the fact that the pairings involve the same two letters. For a combination, on the other hand, the order is not considered significant, only the grouping. So any grouping that contains the same members as another grouping is considered to be the same as that other grouping! Thus, the set of all two-letter combinations of the letters in the set $\{M, A, C\}$ would be $\{MA, MC, AC\}$.

Note

A combination of a set of objects is a collection of some or all of the objects into a grouping, in any order whatsoever. Two combinations are considered to be the same if they contain the same members, regardless of order.

Calculating the number of possible combinations is similar to though slightly different from the method used to calculate the number of permutations. The key idea is to use division to eliminate the repetition of groupings that have the same members but in a different order. The number of such similar groups is precisely the number of ways the members in the collection can be arranged, that is, the number of permutations of the members.

Therefore, the number of combinations of a set of r items taken from a collection of n objects, which we denote as ${}_nC_r$, is given by the formula $\frac{n!}{r!(n - r)!}$.

EXAMPLE 12.27 Suppose a company wishes to select 3 of its 14 employees to form an action group to develop a new logo for the organization. How many action groups is it possible to create?

SOLUTION

Because the ordering of the employees is not relevant (the group consisting of Ann, Bob, and Carol is the same as the group consisting of Bob, Carol, and Ann), this is a combination problem. The number of elements in the overall set (the company) is 14, and the number being used to form the group is 3, and therefore we must calculate

$$_{14}C_3 = \frac{14!}{3!11!} = \frac{14 \cdot 13 \cdot 12 \cdot 11!}{3 \cdot 2 \cdot 1 \cdot 11!} = 364$$

Thus, there are 364 possible action groups the company can form.

EXAMPLE 12.28 A particular lottery game asks a player to select 6 numbers from a set of 43. The winning number is selected during a televised procedure in which a set of 6 ping-pong balls is selected from a container holding 43 numbered ping-pong balls, well mixed. How many possible winning drawings exist (order does not matter)?

SOLUTION

Since the order of the numbers being drawn is not important, this is a combination problem, amounting to the computation of the number of sets of 6 numbers we can select from a collection of 43. This is $_{43}C_6 = \frac{43!}{6!37!} = \frac{(43)(42)(41)(40)(39)(38)37!}{(6)(5)(4)(3)(2)(1)37!} = 6{,}096{,}454$.

EXAMPLE 12.29 Recall that, earlier in the text, we introduced binary numbers (numbers represented only using the digits 0 and 1). How many nine-digit binary numbers have exactly two 1s in them?

SOLUTION

This problem is, surprisingly, rather complicated. In order to find the solution, we will need to devise a rather clever way to consider the representations of the numbers, since they will all consist of seven 0s and two 1s. The trick we'll use is to associate with each binary number a set of two numbers, $\{a, b\}$, where a and b indicate the position within the binary numbers where the 1s are located. Note that this allows us to think of the problem as a combination, since the ordering of the values a and b is irrelevant; that is, the number 1001 0000 would be equivalently represented by the pair {9, 5} and {5, 9}. Therefore, the problem really amounts to determining the number of combinations of pairs of numbers we can select from the digits {1, 2, 3, 4, 5, 6, 7, 8, 9}. This is $_9C_2 = \frac{9!}{2!7!} = \frac{(9)(8)7!}{(2)(1)7!} = 36$. Thus, we can conclude that there are exactly 36 nine-digit binary numbers having exactly two 1s within their representation.

Exercises

For the following questions, answer in your own words, using complete sentences and proper grammar, spelling, and punctuation.

1. What is the distinction between independent and dependent events? Give an example of each type and explain why your examples are of independent/dependent events.

2. What is meant by the concept of conditional probability?
3. How does a permutation differ from a combination?
4. What is meant by referring to two events as being "mutually exclusive"?

Are the following pairs of events (A and B) independent or dependent? Explain.

5. A: Your hard drive crashes. B: Your computer monitor fails.
6. A: Lightning strikes your home. B: Your computer is damaged.
7. A: Your computer memory has reached capacity. B: Your file save operation failed.
8. A: You buy a new computer. B: Your hard drive crashes.
9. A: You roll a 1 on the first toss of a six-sided die. B: You roll a 1 on the second toss of a six-sided die.
10. Two cards are drawn from a deck of 52 cards, with the first draw made and then replaced in the deck prior to making the second draw. A: You draw an ace on the first draw. B: You draw an ace on the second draw.

Suppose events A and B are known to be independent events. Find $P(A \text{ and } B)$ in the following problems.

11. $P(A) = 0.8, P(B) = 0.2$
12. $P(A) = 0.25, P(B) = 0.25$
13. $P(A) = \frac{1}{3}, P(B) = \frac{2}{5}$
14. $P(A) = \frac{6}{7}, P(B) = \frac{1}{3}$

Calculate the probabilities described in the following problems.

15. Suppose it is known that (on a given day) the failure of an automobile alarm system is independent of the failure of its satellite radio system. If the probability that the alarm system will fail is 0.025 and the probability that the satellite radio system will fail is 0.125, what is the probability that both systems will fail on any given day?
16. Suppose it is known that (on a given day at a particular production facility) the probability of an industrial accident occurring is independent of a visit from a federal safety official. If the probability of an industrial accident occurring on a given day is 0.05 and the probability of a visit by a federal safety official is 0.00125, what is the probability that an industrial accident occurs on the same day a federal safety official is on-site?
17. If the failure of a particular business in a given year is independent of the imposition of a new federal tax structure within that year and the probability of the business failing is 0.33 while the probability of an imposition of a new federal tax structure is 0.05, what is the likelihood that the business will fail in the same year a new federal tax structure is implemented?
18. If the purchase of a new car by a particular individual in a given month is independent of a change in the price of gasoline in that month and the probability a new car will be purchased by that individual during the month is 0.15 while the probability of a change in the price of gasoline is 0.25, what is the probability the individual will buy a car and the price of gasoline will change?

19. It is known that the probability of a virus attack on a computer network is independent of the installation of a new operating system within the network. The probability of a new operating system being installed is 0.33, while the probability of a virus attack occurring and the new system being installed is 0.012. What is the probability of a virus attack on the computer network?
20. The probability that a car will be painted yellow by a worker at a manufacturing plant on a given day is independent of the probability that black leather seats will be installed in that car on that day. If it is known that the probability of a yellow car having black leather seats being produced on a given day is 0.04 and the probability of a car being painted yellow on that day is 0.2, what is the probability that a car will have black leather seats installed on that day?
21. The probabilities that two persons are alive in 30 years' time are independent of one another. If it is known that person A has a 0.4 probability of being alive in 30 years while person B has a 0.25 probability, what is the probability that both will be alive in 30 years time?
22. Referring to the situation described in Problem 21, what is the probability that person A or person B is alive in 30 years' time?
23. The names of four directors of a company will be placed in a hat, and a two-member delegation will be selected at random to represent the company at an international meeting. Let *A*, *B*, *C*, and *D* denote the directors of the company. What is the probability that *A* or *B* is selected?
24. The letters of the words "GOOD" and "BYE" are written on individual cards and the cards placed into a bag. A card is picked at random. What is the probability of picking the letter O or the letter E?
25. A pair of dice is rolled. What is the probability that the sum of the numbers rolled is either 7 or 11?
26. Suppose that a high school consists of 25% juniors and 15% seniors and that the remaining 60% is students of other grades. If a student is selected at random from the high school, what is the probability the chosen student is either a junior or a senior?
27. If a coin is tossed twice, what is the probability that it will land either heads both times or tails both times?
28. If a coin is tossed twice, what is the probability that on the first toss the coin lands heads and that on the second toss the coin lands tails?
29. An inspector at a plant manufacturing computer keyboards has checked the wiring on two-thirds and the functionality on one-third of the machines produced by a particular production line in a given day. What is the probability that any given keyboard chosen from that production line has had either the wiring or the functionality of the keyboard tested by the inspector on that day?
30. A community college student is asked about how they are financing their education. The probability the student is receiving financial aid is 0.7, the probability the student is receiving money from their parents is

0.45, and the probability the student is receiving both types of assistance is 0.37. What is the probability the student is receiving one or both of these forms of financial assistance?

31. In a given region, a weather forecaster predicts that if it rains on a given day, then 10% of the time it will rain on the next day. If it is not raining on a given day, there is a 22% chance it will rain on the next day. If there is a 75% chance of rain on Monday, what is the probability of rain on Tuesday?

32. In Seattle during December, the probability of rain on a given day is 0.45. On a day on which it rains, the probability Kurt takes his umbrella is 0.75, and on a day on which it does not rain, the probability Kurt takes his umbrella is 0.2. On a randomly selected day in December, given that Kurt has taken his umbrella, what is the probability it rains? Round your answer to three decimal places.

33. A medical research lab proposes a screening test for a disease. To try out this test, it is given to 100 people, 30 of whom are known to have the disease and 70 of whom are known not to have the disease. A positive test indicates the disease, and a negative test indicates no disease. Unfortunately, such medical tests can produce two kinds of errors: a "false-negative test," where a person actually has the disease but the test indicates he does not, or a "false-positive test," where a person does not have the disease but the test indicates he does. Assuming the test is judged to be 99% effective in determining the presence of the disease (that is, the probability the test will give a positive result when the disease is present is 0.99), and that the test generates a false-positive result 7% of the time, what is the probability that a test will result in a false-negative result? Round your answer to three decimal places.

34. Assuming that giving birth to a boy is as likely as giving birth to a girl, what is the probability that a family of two children has two boys given that the first child born was a boy?

35. In a poker hand, John has a very strong hand and bets $5. The probability that Mischa has a better hand is 0.04. If Mischa had a better hand, she would raise with probability 0.8, but with a poorer hand she would raise only with probability 0.2. If Mischa raises, what is the probability that she has a better hand than John does?

36. Fifteen percent of a company's employees are smokers, and the rest are nonsmokers. For each nonsmoker, the probability of dying during the year is 0.02, while for a smoker the probability of dying during the year is 0.07. Given that an employee has died during the year, what is the probability the employee was a smoker?

37. Workplace accidents at Company X are classified as either minor or serious. Last year, 84% of Company X's accidents were minor, the rest serious. 12% of workers involved in minor accidents required a doctor's attention, and 76% of workers in serious accidents required a doctor's attention. For any Company X worker involved in an accident *not* requiring doctor's services, what is the probability the accident was minor?

38. An automobile insurance company divides its policyholders into two groups: good drivers and bad drivers. In a study of 3000 drivers holding a policy with the company, it is found that 2750 of those in the study were good drivers. A good driver has a 0.02 probability of making a claim in a given year, while a bad driver has a 0.12 probability of making a claim in the year. Given that a particular policyholder has made a claim, what is the probability the policyholder was a good driver?

39. Suppose you have a box containing 22 jelly beans: 10 red, 5 green, and 7 orange. You pick 2 at random, without replacement. What is the probability that the first is red and the second is orange?

40. A computer manufacturer buys cables from three firms. Firm A supplies 50% of all cables and has a 1% defective rate. Firm B supplies 30% of all cables and has a 2% defective rate. Firm C supplies the remaining 20% of cables and has a 5% defective rate. (a) What is the probability that a randomly selected cable that the computer manufacturer has purchased is defective (i.e., what is the overall defective rate of all cables they purchase)? (b) Given that a cable is defective, what is the probability it came from firm A? From firm B? From firm C?

41. Approximately 1% of women aged 40 to 50 have breast cancer. A woman with breast cancer has a 90% chance of a positive test from a mammogram, while a woman without breast cancer has a 10% chance of a false-positive result. What is the probability a woman has breast cancer given that she just had a positive test?

42. The probability that a randomly chosen male has a circulation problem is 0.25. When randomly selecting a male, there is a probability of 0.072 the male will be a smoker who has a circulation problem. What is the probability a male is a smoker, given they have circulation problems?

43. Suppose a voter poll is taken in three states. In state A, 50% of voters support the liberal candidate; in state B, 60% of the voters support the liberal candidate; and in state C, 35% of the voters support the liberal candidate. Of the total population of the three states, 40% live in state A, 25% live in state B, and 35% live in state C. Given that a voter supports the liberal candidate, what is the probability that he or she lives in state B?

44. A building has 60 occupants consisting of 15 women and 45 men. The men have probability $\frac{1}{2}$ of being color blind and the women have probability $\frac{1}{3}$ of being color blind. Suppose you choose uniformly at random a person from the 60 in the building. What is the probability that the person will be color blind?

45. There are 6! permutations of the 6 letters of the word "Galois." In how many of them is "G" the second letter?

46. Seven different books are on a shelf. In how many different ways could you arrange them?

47. How many license plates are there that start with 3 letters followed by 5 digits (no repetitions)?
48. In how many ways can you seat 7 persons at a circular dinner table?
49. From a group of 3 women and 7 men, how many different committees consisting of 2 women and 3 men can be formed? What if 2 of the men are feuding and refuse to serve on the committee together?
50. A computer password has 4 letters in it. How many different 4-letter combinations can be made? How many different combinations exist if the 4 letters must be different?
51. A consumer group plans to select 2 computer monitors from a shipment of 8 to check the picture quality. In how many ways can they choose 2 monitors?
52. How many different 12-person juries can be chosen from a pool of 20 jurors?
53. How many ways can we place 6 identical balls into 7 separate (but distinguishable) boxes?
54. A book publisher has 3000 copies of a discrete mathematics book. How many ways are there to store these books in their three warehouses if the copies of the book are indistinguishable?

Summary

In this chapter, you learned about:

- Identifying the difference between the terms "probability" and "odds."
- Solving problems using experiments, outcomes, sample spaces, and events.
- Calculating and applying the notion of expected value.
- Recognizing and solving "and" and "or" problems.
- Utilizing conditional probability and Bayes's theorem.
- Understanding and working with permutations and combinations.

Glossary

certainty An event whose probability of occurrence is 1.

complement The set of all outcomes in the sample space not in the event.

compound event A combination of a pair of events *A* and *B* using the words "and" or "or."

dependent events Two events, the occurrence or non-occurrence of one having no effect of the occurrence or nonoccurrence of the other.

empirical probability of event *A* Probability of occurrence based on observation of a succession of trials. If n trials are performed and event A occurs k times, the empirical probability of occurrence is k/n.

equally likely events Two events whose probabilities of occurrence are equal.

event A subset of the sample space of an experiment.

expected value The expected return in a game of chance or in an experiment.

experiment A planned operation conducted under controlled conditions.

factorial notation For a natural number n, the notation $n!$, defined to be the product of n with all smaller natural numbers.

impossible event An event whose probability of occurrence is 0.

independent events Two events, one of whose occurrence does not influence the occurrence or non-occurrence of the other.

law of large numbers That law stating that if an experiment is repeated a great number of times, the empirical probability will approach the theoretical probability for the event.

multiplication rule for dependent events If A and B are dependent events, then $P(A \text{ and } B) = P(A) \times P(B|A)$.

multiplication rule for independent events If A and B are independent events, then $P(A \text{ and } B) = P(A) \times P(B)$.

mutually exclusive events Two events A and B such that $P(A \text{ and } B) = 0$.

odds against an event The ratio b:a, where a is the number of favorable outcomes for the event and b is the number of unfavorable outcomes of the event.

odds in favor of an event The ratio a:b, where a is the number of favorable outcomes for the event and b is the number of unfavorable outcomes of the event.

outcome The result of an experiment.

permutation A rearrangement of the members of a set.

sample space The set of all possible outcomes of an experiment.

simple event A particular outcome in a sample space.

theoretical probability of event A The number of outcomes for event A divided by the total number of outcomes within the sample space.

End-of-Chapter Problems

Answer each of the following. Use complete sentences, proper grammar, and appropriate terminology.

1. An experiment involves throwing three fair six-sided dice. What is the sample space? Describe the event in which the sum of the dice is exactly four.
2. A bag has three balls in it: a red one, a green one, and a blue one. The experiment is to draw two balls from the bag, keeping track of the order in which you draw them. What is the sample space?
3. You flip three fair coins, keeping track of the order of the results. What is the sample space? Describe the event that you will get exactly two tails.
4. You throw two 20-sided dice keeping track of the order of the throws. What is the sample space? Describe the event that the sum of the two dice is at least 39.
5. You are given a spinner, such as you find in many games, where the arrow will come to rest on one of the numbers 1 to 10. If you spin the spinner twice and keep track of the order in which you get numbers, what is the sample space? Ignore the possibility that the spinner will end up pointing between numbers. Describe the event that the sum of the two numbers is exactly four. What is the probability of that event?
6. You are given two bags, each of which contains a red ball, a green ball, and a blue ball. The experiment is to draw one ball from each bag. What is the sample space? Describe the event that the colors of the two balls you draw are the same. What is the probability of that event?

7. A nursing home is surveyed and the ages and sexes of the residents are summarized in Table 12.6.

Table 12.6 Age and Gender in a Nursing Home

Age Range	Females	Males
61–70	155	143
71–80	85	56
81–90	42	18
Over 90	12	1

What is the probability that a resident chosen at random is female? What is the probability that a resident chosen at random is a man over 80?

8. A doughnut shop makes eight different kinds of doughnuts. Each night, they make them in the quantities shown in Table 12.7.

Table 12.7 Doughnut Shop Sales

Type of Doughnut	Quantity
Glazed jelly	144
Powdered jelly	180
Sugared jelly	72
Plain	96
Glazed chocolate	156
Blueberry	36
Lemon filled	72
Glazed	240

What is the probability that a doughnut drawn from the night's baking is some kind of jelly doughnut? What is the probability that it is a lemon-filled doughnut?

For the following events, what is the complementary event? Also, what is the probability of the event given and its complement?

9. Throwing three fair coins and getting all heads.
10. Selecting a month of the year and choosing one with 31 days.
11. You roll a six-sided die twice, and the sum of the rolls is 8.
12. A roulette wheel has 38 slots, 00, 0, 1, . . . , 36, of which 18 are red and 18 are black (the 00 and 0 slots are neither red nor black). The event is that the ball lands in a red slot.
13. You go to a produce wholesale market and count the number of cases of broccoli. There are 133 of them, of which 23 you find to be organic. What is the (empirical) probability that a randomly selected case of broccoli is organic? If you examined 972 cases, how many of them would you expect to be organic?
14. You poll the students at your school about their computers, and you find that 573 of them use Windows, 862 use Mac OS, and 37 of them run some form of Linux. Based on this, what is the probability that a

randomly selected student will use Mac OS? If you polled 10,000 students, how many of them would you expect to use Mac OS?

15. You have played an adventure game 432 times, and you have entered a particular room in the game 1286 times. You encountered an ogre in the room 255 times. What is the (empirical) probability of encountering an ogre on entering that room? If you play the game 726 more times and enter that room 1326 more times, how many times do you expect to encounter an ogre?

16. A batch of cars is examined after they leave the factory. Table 12.8 shows the number of defects found in the cars. What is the (empirical) probability that a car chosen at random from the batch has at least two defects? If you examined another 1000 cars, how many of them would you expect to have at least two defects?

Table 12.8 Car Defects at a Factory

Number of Defects	Number of Cars
0	327
1	86
2	45
3 or more	22

For each of the following problems, calculate the odds described.

17. The Boston Celtics have won 10 out of their last 16 games. Based on past performance, what are the odds that they will win the next game?

18. A car manufacturer estimates that 0.04% of the cars that they ship will be returned as "lemons." What are the odds against purchasing a lemon?

19. A horse has won 18 out of the last 20 races that it was entered in. Based on past performance, what are the odds against it winning the next race?

20. A factory makes wiring harnesses for automobiles. It is estimated that 0.01% of them are defective in some way. What are the odds against randomly choosing a defective wiring harness? What are the odds for it?

21. If it is known that the odds against a particular event are 7:13, what is the probability of the event occurring?

22. If it is known that the odds of a computer having a virus are 13:9, what is the probability that a randomly selected computer will have a virus?

23. If the odds are 10:11 that a marriage will end in divorce, what is the probability that a randomly chosen marriage will not end in divorce?

24. A greyhound has won 5 races and lost 17. Based on past performance, what are the odds against his winning the next race?

In the following problems, calculate the expected value as indicated.

25. In a particular game, a card is drawn from a standard deck. If the card is the jack of diamonds, you win $10; if it is the queen of spades, you lose $13. If the card is a heart, you lose $1. What is the expected value of a single play of the game?

26. A class of students is observed to have the grades shown in Table 12.9.

Table 12.9 Grade Distributions in a Class

Grade	Number of Students
A	12
B	27
C	24
D	5
F	3

Given the normal scale (i.e., A = 4.0, B = 3.0, C = 2.0, D = 1.0, and F = 0.0), what is the expected grade of a student chosen at random?

27. A factory makes widgets. Table 12.10 shows the number of widgets, along with the number of defects, found for a random sample of widgets. What is the expected value of the number of defects found when choosing a widget at random?

Table 12.10 Number of Defects in a Group of Widgets

Number of Widgets	Number of Defects
0	371
1	25
2	3
3	1

28. You are faced with a choice of whether to invest in a start-up. Your investment would be \$10,000, and there are three possible outcomes: (1) the firm will be incredibly successful, and you will get back \$1,000,000; (2) the firm will be moderately successful, and you will get back \$50,000; or (3) the firm will go belly up, and you will get nothing back. The probability of the first outcome is believed to be 0.1%, and the probability of the second is believed to be 2%. How much should you expect to get back, and is the investment worthwhile?

Are the following pairs of events (*A* and *B*) independent or dependent? Explain.

29. *A*: Your car breaks down. *B*: Your father receives a Nobel Prize.

30. You have a bag that contains 50 white balls and 50 black balls. You make two draws without putting the ball back in the bag between draws. *A*: You draw a white ball on the first draw. *B*: You draw a white ball on the second draw.

31. You are dealt a poker hand of five cards. *A*: You get dealt a flush (all five cards of the same suit). *B*: You get dealt a royal flush (A, K, Q, J, and 10 of one suit).

32. You have four coins in your pocket: a quarter, a dime, a nickel, and a penny. You take out one coin and then another, without putting the first one back. *A*: You take out a quarter as your first coin. *B*: You take out a quarter as your second coin.

Suppose events A and B are known to be independent events. Find $P(A \text{ and } B)$ in the following problems.

33. $P(A) = 0.10$, $P(B) = 0.30$
34. $P(A) = \dfrac{11}{13}$, $P(B) = \dfrac{13}{110}$

Calculate the probabilities described in the following problems.

35. You draw two cards from a standard deck without replacement. What is the probability that the second card will be a jack given that the first card was a jack?
36. Given a deck that contains only the 12 face cards from a standard deck, from which you draw nine cards, what is the probability that you will draw at least one jack?
37. At a dance, there are 91 men and 97 women. The probability that a man at the dance is over 60 is 25%, and the probability that a woman at the dance is over 60 is 35%. If a person picked at random from all of the people at the dance, what is the probability that they are over 60?
38. Suppose that you have a 47% chance of winning a particular game in a casino. If you win, you receive \$1; if you lose, you lose \$1. What are your expected winnings or losses after playing the game 10 times?
39. You are dealt a poker hand of five cards. What are the odds that you will get a flush (five cards of the same suit)?
40. You are dealt a poker hand of five cards, and it is almost a flush (five cards of the same suit). You have four cards of the same suit and one card of a different suit. What is the probability that you can "fill" the flush by discarding one card and drawing another and end up with a flush?
41. How many different five-card poker hands can be dealt from a standard deck? Remember that order does not matter. What is the probability of getting a royal flush (A, K, Q, J, and 10 of the same suit)?
42. You are given a bag of 12 tiles with the numbers 1 to 12 on them. How many different ways are there of choosing 3 of them given that the order matters? What is the probability that any given choice includes the number 3?
43. Pretend that there are 40,132 four-letter words in the English language. Ignore capitalization. What is the probability that a four-letter sequence chosen at random is an English word? What is the probability that a four-letter sequence chosen without repeating any letters is an English word? (Assume that there are 30,176 four-letter words with no repeated letters.)
44. Given a set of 10 tiles, numbered from 1 to 10, there are 10! different ways of arranging them (order matters). How many of these ways have the value 3 second? What is the probability that the 3 will be second?
45. Say that there are 100 million cars on the road in the United States and that 15 million of them are Japanese. Furthermore, say that 5 million of them are Hondas (a Japanese auto manufacturer). What is the probability that a car chosen at random is a Honda? What is the probability that a car chosen at random is a Honda given that it is a Japanese car?

46. We note, as a fact, that, for any two events A and B, $P(B) = P(A)P(B|A) + P(A')P(B|A')$. If we know that $P(A) = \frac{1}{3}$, $P(A \text{ and } B) = \frac{1}{5}$, and that $P(B) = \frac{1}{4}$, then what are $P(B|A)$, $P(B|A')$, and $P(A' \text{ and } B)$?
47. A grocery store stocks 15 kinds of breakfast cereal. They are on the shelves in some order. How many different orders are there? What is the probability that cereal A is the first? Given that cereal A is first, what is the probability that cereal B is second?
48. A company with 15 employees is forming a committee to examine benefits. The committee is to have 3 members. What is the probability that both Bob and Allen, who hate each other, will be on the committee?
49. A group of 1000 people is surveyed, and it is found that 10 of them are addicted to heroin. Of the 10 who are addicted to heroin, 2 also smoke marijuana. Of the people who are not addicted to heroin, 198 smoke marijuana. What is the conditional probability that a person is addicted to heroin given that they smoke marijuana? What is the probability that a person smokes marijuana? Are the two events (addicted to heroin and smoke marijuana) independent?
50. Of a group of 10,147 people, 3,032 are college graduates, 2,732 had parents who were college graduates, and 1,563 are college graduates and had parents who were college graduates. What is the conditional probability that a person had parents who were college graduates given that the person is a college graduate?

Chapter 13 Statistics

Statistics is a branch of mathematics that deals with the collection, analysis, assessment, use, and presentation of numerical data. Data consist of information about some particular thing or activity that is being observed and analyzed.

Statistics is broken into two broad types: descriptive statistics and inferential statistics. The former is where information is gathered and organized about some observed phenomenon, and the latter is where conclusions are drawn about the data that has been gathered. In this book, our main focus will be on descriptive statistics, and we will leave the exploration of inferential statistics to a proper course in the subject, since the complexity and nuance involved would exceed the scope of our work here.

Objectives

When you have successfully completed the materials of this chapter, you will be able to:

- Recognize and select from the various types of sampling techniques.
- Produce and interpret statistical graphs.
- Recognize and calculate the measures of central tendency.
- Recognize and calculate the measures of dispersion.
- Understand and perform calculations with the normal distribution.
- Understand and perform calculations with the binomial distribution.
- Establish a linear correlation and perform linear regression.

13.1 The Different Techniques of Sampling

The collection of objects or people we study in statistics are referred to as a **population**. In an ideal world, we would be able to gather the entire population together and take our measurements from every individual member, but this is typically not possible because of various constraints on us (time, convenience, expense, and practicality). For this reason, we make our examination of the population through the use of a **sample**, which is a subset of a population. When we construct a sample, the procedure is referred to as sampling the population or more simply **sampling**, and it is from the sample that we amass our information about the population, or our **data**.

When we consider the data, we obtain measurements, or numerical information, about the sample, and this information is called a **statistic**. Were we able to find the same information (somehow!) for the entire population, then the numerical information would be called a **parameter**. If we are able to construct a **representative sample**, a subset of the population that very closely mirrors the qualities and characteristics of the overall population, then we can deduce that the value of the statistic will serve as a predictor of the value of the parameter.

EXAMPLE 13.1

The government is interested in the reliability of a new version of a mobile phone and is (in particular) examining the reliability of its built-in Web browser. Since over 2 million of the phones have been sold, it is impractical to gather all the phones together and inspect them, so a collection of 5000 phones will be gathered and the number of defective Web browsers determined. In this situation, what is the population, what is the sample, what is the statistic, and what is the parameter?

SOLUTION

The overall collection of objects under study is the complete set of the particular version of the new mobile phone, and the sample is the 5000 phones that will be examined. Since the focus of the study is the reliability of the Web browser, the statistic will be the number of phones having a defective browser, while the parameter will be the number of phones (wherever located) that have a defective browser.

One of the key issues when attempting to gather information about a population is the method for constructing the sample we will use to calculate the statistic, from which we hope to make a prediction regarding the parameter. We need to use a method that is efficient but that also affords the opportunity to construct a representative sample. Of course, even though some methods may be preferable to others as a source of a representative sample, limitations dictated by circumstance may force us to choose a particular method because of expedience.

The various methods we can use to construct samples are referred to as sampling techniques. The sampling techniques we will investigate here are the convenience sample, the systematic sample, the random sample, the cluster sample, and the stratified sample. We'll define each of these types and then follow with some examples illustrating their use.

As a final remark before introducing the various types, we point out that the first type, the convenience sample, is the only example we'll examine that is referred to as a **nonprobability sampling method**. Such a technique is characterized by the trait that not all the individuals in the population have an equal chance to be chosen as part of the sample. The remaining types of sampling are probability sampling methods, which do allow every population member an equal chance of selection.

Convenience Sampling

Let's imagine we wanted to gather information that will allow us to determine the mean (the numerical average) income level of the residents of a particular community, such as New York City. Of course, we could track down every resident of the city, ask them, record the results, and then compute the mean (the average) of those salaries. This would be time consuming, difficult, and expensive. We decide to speculate on the mean salary of residents of New York City by finding the mean salary of a sample of the city's residents and extrapolating from that to estimate the value for the city as a whole.

A **convenience sample** is constructed using readily obtained information. No effort is made to find a representative sample, but instead the choice is made to use information close at hand.

We happen to be passing by a corner market in the city and see three young men in their early twenties standing on the front steps. These three men, we decide, will form our sample, since they are very convenient to question regarding their mean salary. We find that two of them work odd jobs and have an annual income of \$18,600 and \$12,000, respectively, while the third works as a waiter in a restaurant and has an income of \$32,000. Since the mean of these three salaries is approximately \$20,866.67, this is our sample statistic, and we conclude the value of the population parameter (the mean salary of all New York City residents) to have that same value.

You can see the weakness in the convenience sample quite readily. The members of our sample, except in cases of great good fortune, are unlikely to form a representative sample, and there exists the potential for the existence of bias within the group from which we are gathering data. The strength of the convenience sample lies in the ease with which the information is gathered and the speed with which we were able to draw our conclusions.

The only time a convenience sample statistic would be particularly useful as an accurate predictor of a population parameter would be in the case where a population under study was known to be **homogeneous**. That is, we had a population within which there was very little variation from member to member.

This sort of sampling is often used in pilot studies where a researcher is interested in gathering preliminary data. Since the data are known to be preliminary, it may be worth the time gained to use convenience sample as a starting point of research.

EXAMPLE 13.2

Marcus wants to determine the average age of a student at his university. Because it is impractical to survey all the students in the institution, he decides to use his economics class (which typically has full attendance) as his sample. This is a convenience sample, since it is a subgroup of the population readily at hand for Marcus and has no guarantee of being representative of the population as a whole.

Systematic Sampling

A **systematic sample** is created when members of a population are selected using some arbitrarily designed algorithm where the sample members are chosen from among the population in a periodic manner. Like the convenience sample, it is relatively easy to construct, and the periodic quality with which the information is gathered serves to overcome the evident weaknesses in the former method.

The members of the population are (in some manner) arranged and assigned identifying numbers, and then a starting number for the selection process is chosen. The researcher then selects an interval number, such as 6, which would determine that (beginning with the starting number) every sixth member of the population will be surveyed in order to form the sample.

Imagine the situation already described where we wish to determine the mean salary of residents of New York City. We could compile a list of all the residents' names and number them beginning with 1. Then an arbitrary starting value could be selected, such as 312 (which we could select randomly or through the use of a random number generator), and an interval number, such as 500. Then, our sample would consist of those residents having identification numbers 312, 812, 1312, 1812, ...

Like all sampling techniques, the method of systematic sampling has its advantages and disadvantages. The advantages are that the method is relatively simple, the population will have the sampling taken from a relatively broad spectrum of the membership, and the systematic nature of the selection tends to eliminate the possibility of randomly choosing a clustered subset of the population sharing common traits. The main disadvantage of the technique lies in the potential that the systematic nature of the sample selection could coincidentally interact with a concealed periodic trait existing within the population. If the sampling were to match the periodicity of the trait, the sample would be compromised by that common quality among the members.

EXAMPLE 13.3 In order to find how many children live in each house, on average, on Baker Street, a researcher decides to use systematic sampling. Baker Street is 40 blocks long and contains 400 homes. It is decided to number the homes from 1 to 400 and then survey a sample of 25 homes. If we randomly choose to start with house number 7 on the list, then, taking $\frac{400 - 7}{25} = 15.72$, we can obtain our sample by choosing every 15th house on the list, beginning with number 7. Thus, we would inquire at houses 7, 22, 37, 52, and so on.

Random Sampling

In a **random sample**, the researcher needs only to ensure that all members of a population are included in a membership list and then devise a wholly random method to select the desired number of members for the sample. The selection method can be any process that ensures that the selection procedure is random and can be as simple as drawing names from a well-mixed collection of strips of paper on which all names are listed to using a computer software system to select population members at random.

The advantages of random sampling are the ease with which we can form the sample and the fact that it is a "fair" method for choosing sample members. The technique is also likely to generate a representative sample of the population because (in theory) the only obstruction to the creation of such a sample using this method is luck. The disadvantages of random sampling are the need for an up-to-date listing of the population and the requirement that some truly randomized method be used to select the members of the sample.

EXAMPLE 13.4 The U.S. Army wishes to determine the average shoe size of a male U.S. Army enlisted soldier. Since the army is very well organized, it is evident that a complete list of the population of enlisted men is available and can be assembled in a computer database. It is decided to construct a sample of 5000 soldiers, and thus each soldier's name is associated with a number, and a random number generator is used to select 5000 numbers at random. The soldiers whose names correspond to the 5000 selected numbers will make up the random sample to determine the average shoe size.

Cluster Sampling

When the population size makes random sampling impossible, a common technique to use in its stead is called **cluster sampling**. In this sort of sampling, the overall population is subdivided in some evidently natural manner, such as by geographic region, creating clusters of the population. If our population were, for instance, the whole of the United Kingdom, then the clusters might be the individual counties within the United Kingdom. The researcher then randomly selects one or more of these clusters (being sure that each of the clusters in the population has an equal chance of being selected) and uses the entire set of individuals in those clusters (or a random sample of the individuals in those clusters) as the sample.

The hope, of course, is that the individual clusters possess a breakdown of qualities in proportions that closely resemble the population as a whole. In such a case, the clusters provide a useful microcosm of the entire population and serve as more manageable groups on which to conduct research.

For instance, let's suppose we wish to know what the population of the United States thinks of the economic policy of the current president. Obviously, the population of the United States is too large to inquire this information of everyone, so we choose to use cluster sampling to take the pulse of the nation. The country is naturally subdivided into states, which we can use as our clusters, and then we can use a random selection process to choose, say, Michigan as our sample. Since the population of Michigan is also rather large, we can choose to use a subset of that population as our sample, selecting residents of Michigan by forming a random sample and then inquiring of the individuals in that sample as to their opinion of the current economic policy.

Cluster sampling is often relatively cheap, quick, and easy to perform. Researchers can allocate limited resources to a few randomly selected clusters in order to gather their information rather than attempt to examine the population as a whole and may even be able to use a larger sample size than would be possible in random sampling, since surveyors would have a smaller area to cover and thus could contact more individuals. The disadvantage of cluster sampling is that it is apt to be nonrepresentative of the diversity of the population as a whole, since like-minded individuals are prone to live together, and thus a particular cluster could have a vastly different perspective and set of opinions than another cluster or the population at large. Additionally, cluster sampling, by design, leaves vast regions of the population unsampled and therefore overlooks huge (and possibly statistically significant) components of the population.

EXAMPLE 13.5

A particular nationwide corporation wishes to investigate the usage rate of mobile phones by its employees. The corporation first chooses to subdivide its employees into clusters according to state of residence. This breakdown will serve as the clusters to construct the sample. A random selection method is used to choose three states from which to select the sample members, and within those states a random sample is created for the employees from that state, and the cell phone usage average is found. The average cell phone use for the sample created from the three clusters is then used as the hypothesized value of the average cell phone use for the employees of the corporation as a whole.

Stratified Sampling

There are times when a researcher hopes to ensure that all particular subgroups, or **strata**, are represented in the sample. In such a case, we use **stratified sampling**. The population is subdivided into the particular subgroups of interest, and then individuals are randomly selected from each of the strata. In this text, we will discuss only proportionate stratified sampling, in which individuals are selected in a level proportional to the representation of their subdivision in the overall population. (A second type of stratified sampling, called disproportionate stratified sampling, ignores this criterion.)

It is essential, in this type of sampling, that the strata be nonoverlapping. That is, we must ensure that our subdivision of the population as a whole does not allow for a particular member of the population to satisfy the criteria for membership in multiple strata. The most common types of strata used would be age, gender, religion, nationality, education level, and socioeconomic status.

Because of the extra effort required to ensure that the strata are nonoverlapping and that the random sample obtained from the strata is proportional to the subdivisions by classification of the population as a whole, this type of sampling has a high degree of statistical precision and thus requires a relatively small sample size. This can save a researcher considerable time, money, and effort.

EXAMPLE 13.6

An effort is made by a computer gaming company to determine how many games are owned by any particular individual in the city of Los Angeles. Assuming that a complete registry of the citizens of Los Angeles exists, the population can be subdivided into strata according to age in increments of 10 years (ages 1 to 10, 11 to 20, 21 to 30, and so on). From each of these age brackets, a random sample is chosen, and each member of the sample is questioned as to whether they own the game in question. The number of individuals chosen in each bracket is such that the proportion of people within each age bracket in the sample matches the proportion in the age bracket in the city of Los Angeles.

Exercises

In each of the following, identify the type of sampling used: random, stratified, systematic, cluster, or convenience. Explain your answer.

1. The names of 54, 81, and 48 students are drawn from hats containing the names of the freshman, sophomore, and senior classes, respectively, at a particular high school.
2. A sample of students at a large community college is created by generating a list of the student body (in random order) and selecting every 71st student on the list.
3. A market researcher chooses 600 people to interview from each of the 10 largest cities in the United States.
4. An actuary studies the driving record of 500 drivers between the ages of 40 and 70 from within his company's client list.

5. A tax auditor selects every 1000th tax return received by the Colorado Department of Revenue.
6. A polling company assigns the voters in a small community numbers, then uses a computer to generate a set of 100 random numbers and interviews the viewers in a particular community corresponding to those numbers.
7. To assess the quality of automobiles manufactured at a particular plant, an inspector checks every 25th vehicle on the assembly line.
8. An education researcher randomly selects 100 elementary schools and interviews the teachers at each school.
9. A political scientist investigating the popularity of the president of the United States surveys 100 workers in his local Democratic office.
10. The names of contestants in a game show are written on separate cards, the cards are placed within a bag, and a name is drawn from the bag.

13.2 Statistical Graphs

Once we have collected our data, the natural question arises regarding what we should do with it. In other words, how can we represent the data in a manner that allows for its analysis and interpretation? For instance, suppose we were to examine the age of computers within a large company. Since the company is large, we may decide to examine a sample of the machines, but even so, looking at the ages of all of the computers could be a daunting task. A better idea might be to examine the median and the standard deviation of the ages. These are two measures we can use to describe data, and we will introduce and examine them shortly. Alternatively, we might wish to produce and examine a **statistical graph** of the data.

There exist many different methods for the presentation of data graphically, each of which holds its own merits. Whatever the type, the statistical graphs allow us to observe trends, detect patterns, and compare data values quickly.

Some of the graphical methods for depicting data include the dot plot, bar chart, histogram, stem-and-leaf plot, frequency polygon, box-and-whisker plot, and pie chart. Full examination of each of these types would better be reserved for a full course in statistics, so we will limit our attention here to three types in common use: the stem-and-leaf plot, the histogram, and the box-and-whisker plot.

The Stem-and-Leaf Plot

Let's imagine we are considering the number of workdays for each of a group of workers in a particular six-month period. It is found that there were 118 workdays during the period and that the number of workdays for each worker was as follows (listed lowest to highest):

92, 98, 107, 109, 112, 112, 112, 114, 114, 115, 116, 116,
116, 117, 117, 118, 118, 118, 118

The data, as presented, are organized, but it is likely that in a real-world situation, we wouldn't even have that going for us. We can further organize the data by subdividing each of the data values into two parts: a stem and a leaf. The leaf will be the final digit in the workday total, and the stem will consist of the preceding digit(s). Thus, in the data value 98, the stem is 9, and the leaf is 8. In the data value 115, the stem is 11, and the leaf is 5.

When constructing the stem-and-leaf plot, a table is made with the stems being listed vertically in the left column and the leaves (in increasing order and showing all repetitions) in the right column alongside their corresponding stems. In this case, we would have the plot shown in Table 13.1. The column headings of "Stem" and "Leaves" are not generally shown but are included here merely as an introductory reference. The plot reveals information about the data that was not evident in the original list (and that would have been even less apparent had the original data not been listed in increasing order), principally that most of the workers worked over 110 days.

TABLE 13.1 A Stem-and-Leaf Plot

Stem	Leaves
9	2, 8
10	7, 9
11	2, 2, 2, 4, 4, 5, 6, 6, 6, 7, 7, 8, 8, 8, 8

The stem-and-leaf plot offers a quick method for organizing the data and for obtaining a picture of the situation being examined. When examining data presented in this manner, our attention is drawn (or should be drawn) to **outliers**, also referred to as extreme values, which are data values lying far from the other data. In this situation, we notice the outlier value 92 and perhaps also 98, which lie well apart from the other data values. The situation of these workers merits investigation, since they appear to stand far from the norm of our other employees.

EXAMPLE 13.7

Suppose we wish to examine the ages of individuals within a particular university. We form a random sample of students and find that they have the following ages:

18, 18, 18, 19, 19, 19, 19, 19, 20, 20, 20, 21, 21, 21, 24, 24, 25, 25, 28, 31, 32, 46

Construct a stem-and-leaf plot of the data and remark on any outliers.

SOLUTION

The stem-and-leaf plot of the data, using stem values of 1, 2, 3, and 4 and one-digit leaves, is as shown in Table 13.2. Note that, much as we might expect at a university, the main group of ages appears to be clustered in the teens and twenties. There is an outlier value of 46, which indicates that a relatively small number of nontraditional students attend this particular university. It is somewhat surprising to note that the majority of the students were found to be in their twenties, which seems to run contrary to the concept of the university catering to the traditional college student. This may suggest to us that our data set is somewhat flawed, or it could suggest that our concept of the "typical college student" is incorrect.

TABLE 13.2 A Stem and Leaf Plot of Student Ages

1	8, 8, 8, 9, 9, 9, 9, 9
2	0, 0, 0, 1, 1, 1, 4, 4, 5, 5, 8
3	1, 2
4	6

The Histogram

A **histogram** is a very useful and common tool for representing data. Typically, it is used when you have a significant number of data values, possibly 100 or more. The data set need not have that many data values to construct a histogram, but generally this is the threshold.

A histogram is a chart consisting of contiguous boxes, arranged within a construction made up of a vertical axis and a horizontal axis. The vertical axis is labeled with the frequencies of the individual subsets of the data set and the horizontal with the quantity that the data represents (such as the age-groups of students in Example 13.7). When the data set is particularly large, the vertical axis may be relabeled with **relative frequency**, as we will demonstrate shortly.

Much like the stem-and-leaf plot, a histogram gives a good impression of the spread of the data, the value around which it is "centered," and the spread of the data. To construct the graph, we first determine how many bars (or intervals, or classes) we wish to use to represent the data. Typically, the number of bars spans from 5 to 15, but this is not written in stone.

A starting point is then selected for the first interval, chosen such that it is less than the smallest data value. For convenience, we usually choose a starting value that is relatively close to the smallest data value. The ending point of the last interval should be as far above the largest data value, as the starting point is below the smallest data value, at least approximately.

The width of each bar is found by calculating the difference between the ending point and the starting point and dividing by the number of bars desired. The bar width can be rounded, if desired, as long as the width chosen does not allow any data value to fall on the boundary delineating two successive bars.

Let's revisit Example 13.7 and construct a histogram depicting the data.

EXAMPLE 13.8

Suppose we wish to examine the ages of individuals within a particular university. We form a random sample of students and find that they have the following ages:

18, 18, 18, 19, 19, 19, 19, 19, 20, 20, 20, 21, 21, 21, 24, 24, 25, 25, 28, 31, 32, 46

Construct a histogram depicting the data, using 10 intervals.

SOLUTION

We note that the smallest data value is 18, and thus a convenient starting point would be any value less than 18. We *could* choose 17, but since all the data values are integer valued, making such a choice increases the possibility of a data value falling on the border between two bars. Thus, we will choose to use 17.5 as our starting point. Since the highest data value is 46, we will use as our ending value 46.5, since it is as far above the highest value, as 17.5 is below the lowest value.

We were asked to use 10 intervals, and thus our interval length should be

$$\frac{46.5 - 17.5}{10} = \frac{29}{10} \approx 3$$

The boundaries of our bars will be, therefore, 17.5, 20.5, 23.5, 26.5, 29.5, 32.5, 35.5, 38.5, 41.5, 44.5, 47.5. Note that our highest boundary level did not come out precisely to what we had decided would be our end value, but this is okay and the best we can manage (Figure 13.1).

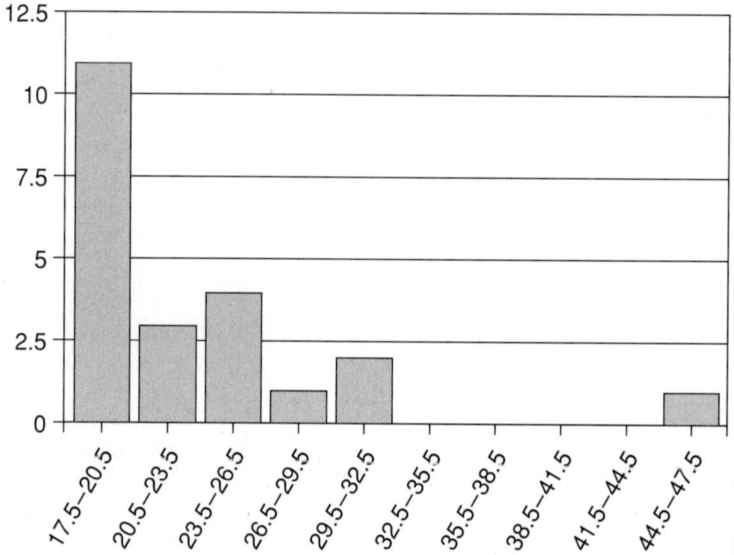

Figure 13.1 Histogram of student age data

EXAMPLE 13.9

It is suspected that the number of hours of sleep per night has an impact on productivity, and therefore a company decides to determine the number of hours of sleep for its employees. A sample of the employees is taken, and the information shown in Table 13.3 is obtained. Construct a histogram displaying the information on employee sleep patterns.

TABLE 13.3 Hours of Sleep per Employee per Night

Hours of Sleep	Number of Employees
4	2
5	5
6	8
7	11
8	13
9	6
10	2

SOLUTION

Here, it seems that a good idea (in order to obtain the best information) would be to use bars of width 1 unit. We can choose starting point 3.5, and then our bars would have boundaries 3.5, 4.5, 5.5, 6.5, 7.5, 8.5, 9.5, and 10.5, with which we could have a bar for each possible number of hours of sleep. The number of employees corresponds to the frequency for each data value, and hence we can construct the histogram shown in Figure 13.2.

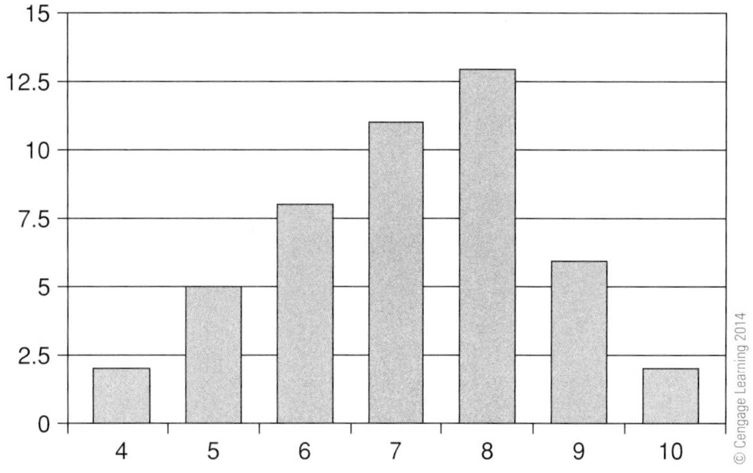

Figure 13.2 Histogram of hours of sleep data

The Box-and-Whisker Plot

The **box-and-whisker plot** is an excellent graphical tool to present the concentration of the data within a sample. The graphs also allow us to see the extreme values of the data and to gain a sense of the data's range, or spread. The plot is constructed with five pieces of information: the smallest and largest values, the first quartile, the median, and the third quartile.

Before we can advance, we need to discuss the final three information bits just mentioned. The **median** is a measure we will encounter again in the section on the measures of central tendency, but it is a fairly straightforward concept, and there is little harm in introducing it here. The first and third quartiles are other types of medians, so once we have a handle on the meaning of the median, the quartiles should be understandable to us as well.

The median is the central, or middle, value of the data set. It is that data value that has half the data values less than it and half the data values greater than it. It goes without saying that the data must be arranged in increasing order for us to find the median, but this poses little difficulty. It is somewhat surprising that, in some cases where there are an even number of data values, the median may not be one of the data values at all.

For instance, let's consider an example of a data set:

4, 5, 7, 11, 12, 13, 14, 15, 18, 19, 21

There are 11 data values in the set, and the median is that entry lying precisely in the middle of the set. If we choose the sixth value in order, the data point 13, then that value has precisely 5 entries below it and 5 entries above it. It splits the data evenly and therefore is the median.

Some people like to have a formula for locating the median, so we can say that if a data set has n data values, with n odd, and with the data values arranged in increasing order, then the median is the $\frac{n+1}{2}$ entry in the list.

For the example above, there were 11 entries, so the median was entry number $\frac{11 + 1}{2} = \frac{12}{2} = 6$ in the ordering, which corresponds to 13.

If the data set should have an even number of data values, then the median is the average of the middle two data values. That is, if our data set was

1, 1, 2, 2, 4, 6, 7, 9, 10, 12, 15, 15, 17, 18

which has 14 values, the two middle values (the seventh and eighth data values) are 7 and 9, and to find the median, we compute $\frac{7 + 9}{2} = \frac{16}{2} = 8$. Notice that the median is not one of the data values; this may seem surprising (or possibly inappropriate) to you, but there was never a requirement that the median would be one of the members of the data set. On occasion, we will find a value for the median that is not among the values of data in our set, but this is permissible.

EXAMPLE 13.10 Find the median of the following data set: 3, 5, 7, 12, 13, 14, 21, 23, 23, 23, 23, 29, 39, 40, 56.

SOLUTION

The data values are, for convenience, provided to us in increasing order. Were they not so arranged, our first step would be to reorder them so that the values were increasing. We next count the number of data values, which is 15.

Since this is an odd total, we know that the median will be in the data set and will be the eighth value in order, since $\frac{15 + 1}{2} = 8$. Here, the eighth value in the progression is 23, which we conclude to be the median of the set.

With the median in our hands, we next must find the first and third quartiles, and then we will be ready to construct our box-and-whisker plot. The first quartile (which is defined to be the value for which 25% of the data is less than or equal) is the median of the first half of the data (determined by the median), and the third quartile (defined to be the value for which 75% of the data is less than or equal) is the median of the second half of the data.

You may be wondering why there isn't a "second quartile," and the answer to this is that there is a second quartile—it's the same as the median of the data set.

Let's find the first and third quartiles of the data set in Example 13.10.

EXAMPLE 13.11 Find the first and third quartiles of the following data set:

3, 5, 7, 12, 13, 14, 21, 23, 23, 23, 23, 29, 39, 40, 56

SOLUTION

We had found that the median was the eighth data value in the progression, which was the first occurrence of 23 in the list. Since there are seven data values below the median, which is an odd number, we find that the first quartile is the fourth data value, since $\frac{7 + 1}{2} = \frac{8}{2} = 4$. Thus, the first quartile is 12.

The third quartile is the median of the second half of the data, and since there are seven data values in the second half of data, we find again that the median of that set is the fourth entry in the progression, which is 29, and thus the third quartile is 29.

To construct the box-and-whisker plot, we use a horizontal number line and a rectangular box. The smallest and largest data values should be visible on the number line, but that number line can and often does extend farther out to both sides. The first quartile marks the left end of the box and the third quartile the right end of the box. Note that this implies that 50% of all the data in the set must be inside the box. A vertical line is drawn within the box at the location of the median, and then "whisker" lines are extended from the box left and right to the extreme data values.

EXAMPLE 13.12 For the data discussed in Examples 13.10 and 13.11, the box-and-whisker plot would be as shown in Figure 13.3. A box-and-whisker plot offers an excellent visual representation of the data, permitting us to see the spread of the data and where the bulk of the data values are concentrated.

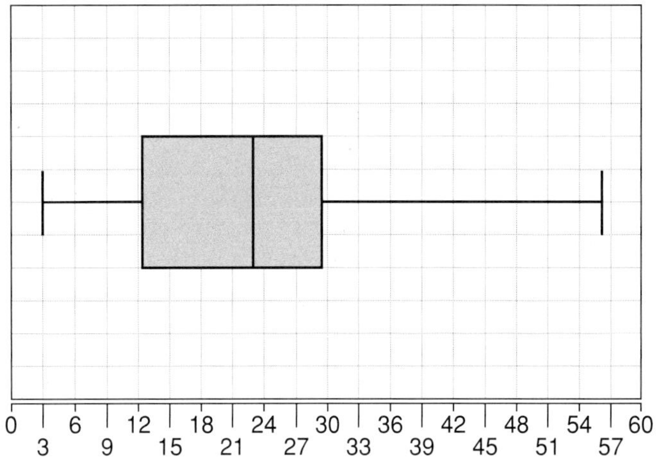

Figure 13.3 Box-and-whisker plot

In closing, we remark that the notion of the outlier has been left rather vague. What qualifies a data value to be an outlier? There is no general agreement on what constitutes the definition of an outlier data point, but there is a good "rule of thumb" we might keep in mind.

If we designate Q_1 as the first quartile and Q_3 as the third quartile, then the difference $Q_3 - Q_1$ is defined to be the **interquartile range** (IQR). A value is suspected to be an outlier if it is more than (1.5)(IQR) below the first quartile or more than (1.5)(IQR) above the third quartile.

In Example 13.11, the IQR $= 29 - 12 = 17$, and $(1.5)(\text{IQR}) = 25.5$. Any data values less than $(12 - 25.5) = -13.5$ or greater than $(29 + 25.5) = 54.5$ would be candidates as outliers. Thus, the only data value of the set that might be an outlier is 56.

Exercises

In the following exercises, construct a histogram.

1. The frequency table shown displays the number of vacation days in a given year for 30 computer programmers. Create a histogram of the data.

Days Off	Frequency
0–1	10
2–3	1
4–5	7
6–7	7
8–9	1
10–11	4

2. In a survey, 20 voters were asked to report their age. The frequency table below shows the results. Create a histogram of the data.

Age	Frequency
20–29	5
30–39	5
40–49	6
50–59	0
60–69	4

3. The ages of the voters were taken during a brief interview. Using five classes with a uniform class width of 10 years, create a histogram of the data.

35	29	48	63	64	38	21	23	41	68
61	42	43	47	33	37	46	27	23	30

Construct a histogram for the following problems:

4.

Life of a Lightbulb (in hours)	Number of Bulbs
400–499	50
500–599	75
600–699	125
700–799	100
800–899	50

5.

Number of Exam Scores	Number of Students
50–59	2
60–69	8
70–79	45
80–89	35
90–100	15

6.

Height in Centimeters	Number of Individuals
150–154	40
155–159	75
160–164	110
165–169	75
170–174	30

7.

Weight of Package in Grams	Number of Packages
1200–1249	50
1250–1299	75
1300–1349	125
1350–1399	100
1400–1449	50

8.

Weight in Pounds	Number of Students
101–130	25
131–160	75
161–190	55
191–220	86
221–250	14

9.

Number of Weeks of Vacation	Number of Employees
1–2	24
3–4	30
5–6	18
7–8	15
9–10	9
11–12	6

10.

Hours of Sleep Per Night	Number of Students
4–5	14
6–7	40
8–9	32
10–11	14

11.

Hours Worked Per Day	Number of Workers
2–3	5
4–5	20
6–7	45
8–9	50
10–11	15

For the following, construct both a stem-and-leaf and a box-and-whisker plot for the data set.

12. In a given class, the final exam scores were

88	93	62	67	90	45	83	83	92	94
68	72	81	93	53	74	80	95	93	53
78	99	74	51	96	97	94	51	88	48
84	69	66	60	57	78	74	68	62	75

13. Achmed wants to study the ages of his employees. He compiles a list of their ages, and finds

49	20	26	55	29	45	55	23	41	25
53	54	24	18	18	39	30	29	47	35
30	35	51	29	45	33	48	31	52	40

14. The average commuting time for students at a nonresidential college are given by

23	33	36	37	11	5	6	44	5	25
3	39	21	12	32	43	10	27	37	4
7	12	26	42	8					

15. At a major university, attendance at a particular course for a period of 15 weeks was given by

20	24	14	40	32	42	29	22	41	34
36	42	18	17	19	34	28	17	15	16
26	12	21	33	19	20	30	34	36	35

16. The ages of 35 instructors at a particular private technical college are

44	45	39	50	52	39	48	48	37	51
50	35	45	36	42	42	36	45	52	50
36	48	52	33	47	52	52	44	36	41
40	38	41	31	51					

17. A professional basketball game plays an 82-game schedule, and scores the following totals in their games:

113	98	84	85	95	83	116	108	92	86
110	106	107	117	100	106	107	108	104	87
97	85	102	117	98	102	102	112	109	83
110	98	87	90	102	117	98	97	89	111
90	110	89	116	110	86	104	86	85	115
108	110	85	104	88	90	83	100	106	98
102	93	90	102	85	102	105	114	113	113
91	102	117	112	87	85	115	111	102	88
97	106								

18. The number of attendees at a particular conference for the past 30 years is

236	204	214	217	212	214	226	243	205	215
209	247	218	201	219	228	212	215	231	201
206	219	250	234	209	216	205	229	200	218

In the following, create stem-and-leaf plots.

19. The number of laps run by 20 participants in an endurance study are

24	46	21	32	29	21	24	30	23	44
39	22	36	31	53	28	51	26	55	42

20. The attendance at a series of 16 soccer matches, in thousands, is as follows:

84	96	67	85	67	74	87	84	92	76
95	80	88	62	90	93				

21. The midterm test scores for 30 students in a college statistics course are as follows:

95	74	69	86	74	65	90	96	90	63
85	67	94	69	87	66	71	88	78	92
67	70	70	95	88	93	74	73	91	65

22. The weights of 22 members of the varsity football team are as follows:

144	152	142	151	160	152	131	164	141	153	140
144	175	156	147	133	172	159	135	159	148	171

23. Twenty-five workers were surveyed on how long it took them to commute each day to work by train. The results, in minutes, are as follows:

79	29	83	32	89	33	18	39	18	64
70	32	81	25	95	69	20	59	18	37
94	24	33	36	52					

24. The ages of 45 members of a local diving club are as follows:

25	37	40	27	18	50	29	45	49	43
27	49	33	27	49	32	48	42	42	44
19	24	38	21	29	48	21	18	35	45
46	26	36	41	47	23	24	38	34	44
24	22	42	20	41					

25. The normal monthly precipitation (in inches) for 39 U.S. cities (use 9 rows for this example) is as follows:

3.5	1.6	2.4	3.7	4.1	3.9	1.0	3.6	1.7	0.4	3.2	4.2	4.1
4.2	3.4	3.7	2.2	1.5	4.2	3.4	2.7	4.0	2.0	0.8	3.6	3.7
0.4	3.7	2.0	3.6	3.8	1.2	4.0	3.1	0.5	3.9	0.1	3.5	3.4

26. The weights (in pounds) of 25 randomly selected young men are as follows:

154	154	160	150	145	161	154	164	150	162
162	162	161	157	157	165	164	145	158	163
156	154	162	159	156					

27. A nurse measures the blood pressure of individuals visiting her clinic. Below is a relative frequency histogram of the blood pressure readings she obtained. The blood pressure is rounded to the nearest whole number. Estimate the percentage of people whose blood pressure was between 100 and 119, inclusive?

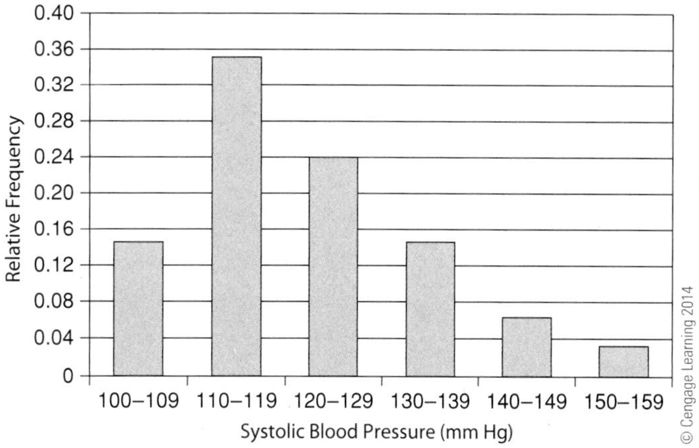

28. A nurse measures the blood pressure of individuals visiting her clinic. Below is a relative frequency histogram of the blood pressure readings she obtained. The blood pressure is rounded to the nearest whole number. Estimate the percentage of people whose blood pressure was between 120 and 139, inclusive?

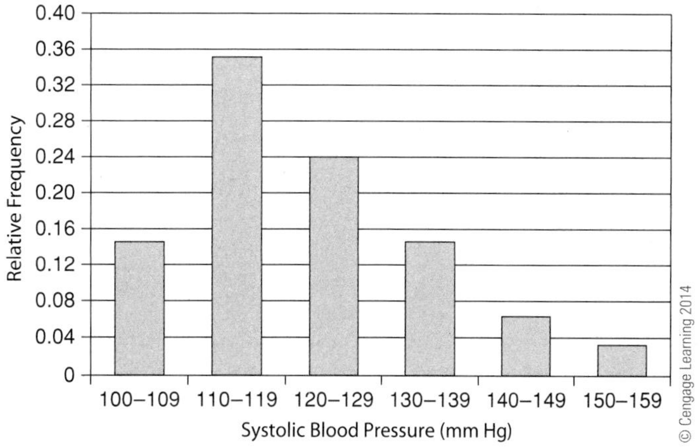

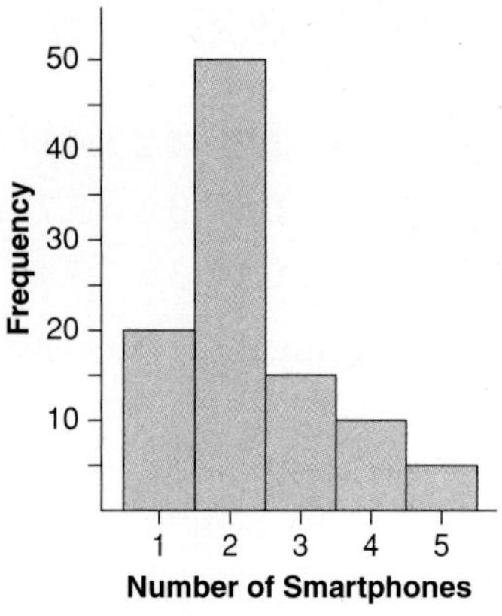

29. For problem 27, identify the class width of the histogram.
30. For problem 28, identify the class width of the histogram.
31. The histogram shown at left represents the number of households in which a resident owns a smart phone. How many households are represented by the graph?
32. For the histogram shown in 31, construct a relative frequency histogram.
33. In a survey, 26 voters were asked their ages. Construct a histogram to represent the data, using 5 classes).

33	43	42	50	30	41	48	53	54	42
39	29	28	47	50	51	30	26	29	50
35	29	41	31	44	25				

34. In a survey of 50 individuals, each was asked how many vehicles they have owned in their lives. The results were as shown; make a histogram for the data, using a minimum of 6 classes.

0	2	2	5	0	2	5	1	6	4
2	3	0	6	2	4	1	4	1	1
2	2	1	3	6	6	3	0	5	3
1	1	5	0	0	0	3	4	1	0
1	0	1	6	3	2	0	1	4	3

13.3 The Measures of Central Tendency

When we talk about the notion of "central tendency," we are really discussing a method for determining the average value of a set of data. You may suspect that the word "average" is well understood, but it is, as we'll learn, rather vague. When we look at a set of data, what do we mean by the concept of average? Intuitively, the word indicates "what most of the data values are like" or "what is typical." But that concept is, as we'll see, extremely flexible. There are at least three different concepts that can lay claim to establishing what is average in a data set, and each is used in its own way and has a legitimate claim to the designation as what is average. The trio of concepts we are discussing are the mean, the median, and the mode, and in this section we will learn how to calculate each of these measures of central tendency.

The median was already discussed in Section 13.2; it is the middle of the data when arranged in increasing order. You can visualize it as being like the median of a roadway, that strip of grass that divides the roadway in half (about which we are always being warned to keep off!).

EXAMPLE 13.13 For the following data set, find the median:

50, 53, 59, 59, 63, 63, 72, 73, 74, 75, 77, 81, 83, 88, 92, 93, 94

SOLUTION

There are 17 data points in the set, so (referencing the information in Section 13.2), we know that the median is entry $\frac{17 + 1}{2} = \frac{18}{2} = 9$ in order from the start of the set. Counting entries, we find that the median is 74. Observe that, if we put the data in order and indicate the median with an arrow, we see that the median splits the data in half (Figure 13.4).

Figure 13.4 Location of the median

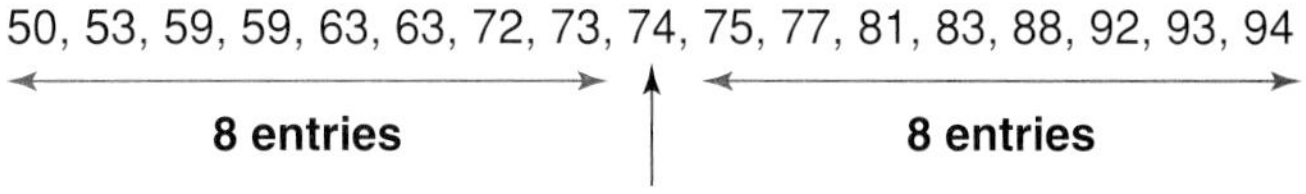

The median is a good indicator of the central, or average, value of a data set if there exist outliers within the data set. As we'll see when we examine the mean, that measure can be sensitive to outliers (and hence yield a somewhat unrealistic view of what is average in the data set).

EXAMPLE 13.14 A consumer magazine wishes to test the life of a wireless optical mouse and conducts an experiment where 25 mice are tested. The life span of the individual mice, in months, was found to be

3, 11, 11, 14, 15, 16, 16, 17, 21, 21, 21, 22, 23,
25, 28, 28, 29, 31, 33, 35, 35, 37, 41, 58, 73

What is the median life span of a wireless optical mouse, according to the data?

SOLUTION

Since there are 25 data points and the data are ordered, the median will be entry 13 in order from the beginning of the list. The 13th entry in the list is 23, which we conclude to be the median life span of the wireless optical mouse (Figure 13.5).

Figure 13.5 Location of the median

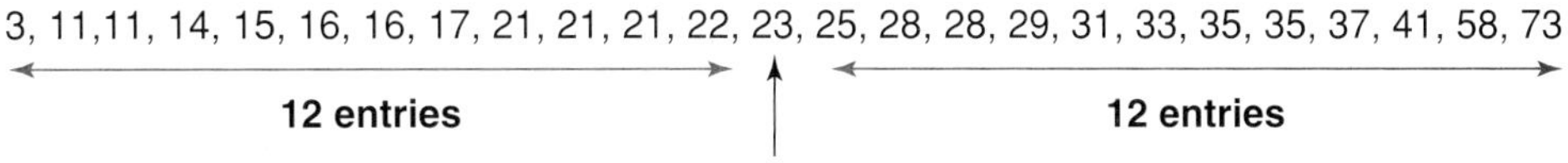

The Mean

The **mean**, also called the arithmetic mean, is what most people tend to think of when they contemplate calculating an average. It is found by adding the data values in the set and dividing by the number of data values in the set. The notation for the mean of a data sample is $\overline{x}$, which we read as "x bar," while the mean of a population is symbolized as μ, the Greek letter mu. When we use a random sample, the sample mean should provide a fairly good approximation of the population mean.

EXAMPLE 13.15 For the situation in Example 13.14, find the mean life span of a wireless optical mouse.

SOLUTION

We add the 25 data values and divide by 25 in order to find the mean. This is a tedious but straightforward calculation:

$$\bar{x} = \frac{3 + 11 + 11 + 14 + 15 + 16 + 16 + 17 + 21 + 21 + 21 + 22 + 23 + 25 + 28 + 28 + 29 + 31 + 33 + 35 + 35 + 37 + 41 + 58 + 73}{25}$$

$$\bar{x} = \frac{664}{25}$$

$$\bar{x} = 26.56$$

Thus, the mean life span of a wireless optical mouse, according to this survey, is 25.68 months.

Note that the median and the mean don't happen to agree in this situation. There are cases where they do happen to coincide, but in general this is somewhat unlikely for any particular sample.

Because the mean is calculated by adding the individual data points and dividing by the number of data points, it is a measure that is highly sensitive to the presence of outliers. When we locate the median, we are interested only in the middle value in a list of the data set, so it is of little importance if the extreme values in the set are outliers. However, consider what happens when we have an outlier in a data set and attempt to calculate the mean.

EXAMPLE 13.16 Consider a neighborhood consisting of seven homes. The homes are priced at \$190,000, \$200,000, \$210,000, \$210,000, \$215,000, \$220,000, and \$225,000. A new house is built in the same neighborhood and is priced at \$625,000. What was the mean price of the homes in the neighborhood before and after the new house was built?

SOLUTION

Before the new home was built, the mean is calculated as

$$\mu = \frac{190{,}000 + 200{,}000 + 210{,}000 + 210{,}000 + 215{,}000 + 220{,}000 + 225{,}000}{7}$$

$$\mu = \frac{1{,}470{,}000}{7}$$

$$\mu = 210{,}000$$

and we conclude that the mean price of a home in the neighborhood was \$210,000. Note that we chose to use μ to designate the mean, since we were taking not a sample but rather the entire population. This seems to be a reasonable measure of the average price of a home in the neighborhood, since all the prices of the homes were clustered around that value.

With the new home included, we find

$$\mu = \frac{190{,}000 + 200{,}000 + 210{,}000 + 210{,}000 + 215{,}000 + 220{,}000 + 225{,}000 + 625{,}000}{8}$$

$$\mu = \frac{2{,}095{,}000}{8}$$

$$\mu = 261{,}875$$

and we conclude the mean price of a home in the neighborhood was now $261,875.

Note that the new value of the mean is somewhat unreasonable, since almost all the homes have values significantly below this level, and hence this is a rather distorted view of what the average price of a home in the neighborhood would be. The reason for the radical departure from the majority of home prices is the existence of the new home, whose price of $625,000 is an outlier for the data set.

We should remark that, in the original data set of homes in Example 13.16, the median home price was $210,000, which actually agreed with the mean for this situation. The median home price in the set, including the new home, is found to be $212,500, which reflects a minor increase over that of the original set but not a significant one. This is, as we mentioned, because the median is relatively impervious to the existence of outliers. In fact, this is the reason that announcements regarding home prices in a particular region almost always refer to the median home price rather than the mean home price.

The Mode

Note

It is possible for a set of data to have one or two modes. If there are two modes, the set is called a bimodal set of data.

The third measure of central tendency is the **mode**. The mode is the most frequently occurring value in the data set. If a data set has two values that occur most often and the same number of times, then both values are considered the mode, and the data set is referred to as **bimodal**. Some books refer to sets having more than two modes as "multimodal," but others describe this situation as having no mode. We will adopt the latter convention here and describe a set having more than two values occurring most frequently as having no mode.

EXAMPLE 13.17 What is the mode of the data set of home prices in the original neighborhood described in Example 13.16: $190,000, $200,000, $210,000, $210,000, $215,000, $220,000, and $225,000?

SOLUTION

The value occurring most frequently is $210,000, and we conclude that this is the mode of the data set. Observe that, in this case, we have the unusual situation of the mean, median, and mode all being the same value.

EXAMPLE 13.18 Six students taking a real estate examination establish scores of 430, 430, 480, 495, 495, and 510. What is the mode of this set of data?

SOLUTION

The values 430 and 495 each appear twice on the list, which is the maximum occurrence of any data value. We conclude that this set is bimodal and that 430 and 495 are the modes.

We close this section with a question to ponder: which of the three measures of central tendency is best? The answer is, it depends on what you are trying to show. For instance, consider the situation of the houses in the neighborhood after the new house was built. For snob appeal, the old owners might wish to use the mean as their "average" home value in the neighborhood because it brings a bit more prestige to the locality. From a practical point of view, the new owners might wish to use the median, since that would establish the level of taxation they might have to pay for certain services to the city or for their insurance. The mode would also be a good choice for them, since it also yields a lower average value.

Finally, we should point out that the law of large numbers tells us that if we were to take samples of larger and larger size from any population, the mean of the sample, $\overline{x}$, will approach the mean of the sample space, μ. This is actually an important result, called the central limit theorem, which is explored in greater detail in classes dedicated to statistics.

Exercises

For the following data sets, find the mean, median, and mode (if they exist). If any of these measures does not exist, explain.

1. 10 10 1 0 7 10 0 6 6 8 6 7
   10 1 1 0 2 2 3 8 6 6 3 0
   5

2. 7 9 11 9 7 7 11 7 10 13 8 13
   13 9 13 13 10 10 13 10 8 10 8 9

3. 15 19 4 9 19 16 15 10 6 8
   11 11 6 5 12 17 11 20 17 6
   7 15 9 14 20 5 5 8 7 20

4. 11 14 11 13 12 16 11 16 11 13
   16 12 16 13 12 11 13 11 16 16

5. 12 8 5 7 13

6. −5 13 −1 3 5

7. 11 10 1 18 5 4 4 10

8. 230 220 212 212 212 212 220 215 216 216 220 220

9. 2.4 1.7 2.2 3.8 1.6 2.2 3.9 2.2 1.2 2.8 3.5 3.2 2.8
   4.0 3.7 3.5 4.0 3.0 2.8 2.6 3.7 2.6 1.6 1.2 3.3

10. 7.764 6.609 2.15 4.851 5.645

11. 1.1 1.3 0.5 2.0 1.0 1.5 1.7 1.2 5.0 1.5 1.4 0.6 0.8
    0.7 2.0 1.5 1.6 0.7

12. 3 5 19 24 38 38 45

13.	16	30	37	44	47	45	33						
14.	27	28	31	27	31	27	29						
15.	9	15	28	24	32	41							
16.	6	6	628	12	29	43	39	32					
17.	35	32	49	38	46	49	32	44	34	41	48	49	45
	43	49	50	50	46	34	50	36					
18.	43	45	44	44	41	47	50	49	45	49	42	44	49
	48	49	48	50	47	47	42						
19.	2.13	2.07	1.95	2.19	2.14	1.92	1.82	2.12	2.19	1.84	2.07	1.82	1.90
	2.09	1.88											
20.	2.9	2.2	2.2	2.2	3.0	2.5	2.1	2.9	2.1	2.1	2.7	2.9	2.8
	2.1	3.0											

13.4 The Measures of Dispersion

At times, it is of interest to find how far a set of data is "spread apart." Knowledge of this feature of data can tell us if there is a pattern within a set of data, since if the "spread" of the data was relatively small, we could assume that there was a certain amount of consistency within the set, while a larger spread would indicate that the data followed no discernible pattern.

The Range of a Set of Data

The most obvious (and, sadly, least informative) measure of the spread of a set of data is called the **range**. The range is found by subtracting the smallest data value from the largest. This gives a very rudimentary illustration of how far the values within the set are distributed, but it is also quite vulnerable to the presence of outliers in the data.

EXAMPLE 13.19 If we reexamine the set of data values giving the prices of homes in a neighborhood, which were

$190,000, $200,000, $210,000, $210,000,
$215,000, $220,000, $225,000, and $625,000

the range would be found by calculating the difference $625,000 − $190,000 = $435,000. This tells us that the prices of homes in the neighborhood (which was what the data represented) were spread across an interval of $435,000.

Unfortunately, the value of the range of the data set is not very informative. For instance, is the value $435,000 large or small? We might say, "Well, obviously, it's large!," but what if the same range were found for a set of luxury mansions each priced in the neighborhood of $20 million? The decision about whether this is a large value for the range is completely relative, as it depends on the quantity being examined.

The Standard Deviation

The most common measure of the dispersion of a data set is the **standard deviation.** This will play a crucial role when we examine the concept of the normal distribution, which we are approaching quickly. The normal distribution is one of the most important of the data distributions in mathematics, and so it is crucial that we have an understanding of the meaning of the standard deviation.

The concept of standard deviation is tied closely to the meanings of the individual words of which the term is comprised. "Standard" refers to that which is typical or usual or normal, and "deviation" is the departure from that standard or norm. Thus, the value of the standard deviation will indicate how far from the norm a typical data value in the set will lie.

The process of calculating the standard deviation by hand is a useful exercise, but in practice most people use computer software to calculate the standard deviation. Here, we'll indicate the method and work out some examples, keeping in mind that, for us, this will be a computational exercise that we know in advance we won't apply in the field and that the true import of the standard deviation lies in its application.

Standard deviation of a sample is designated by the lowercase letter s, while the standard deviation of a population is indicated by the Greek lowercase letter sigma, σ. Computation of the standard deviation by hand is not difficult (though it can be time consuming) for small data sets and is no harder (but much longer!) for large data sets. We'll restrict our computations by hand to data sets that are relatively small.

If x is a value within the data set and $\overline{x}$ is the sample mean, then $x - \overline{x}$ is called a **deviation.** Thus, for any data set, there are as many deviations as there are individual data values. We will need the complete set of deviations for the data set in order to calculate the standard deviation, and thus we should be certain before moving ahead that we are familiar with the computation of the mean. You may want to review that now before moving ahead.

To compute the standard deviation, our first step will be to compute the **variance** of the data set. The variance is the average of the sums of the squares of all those individual deviations we mentioned earlier, and it is designated using s^2 for the sample variance and "sigma squared," or σ^2, for the population variance. The choice of using the same lowercase English and Greek letters as was done for the standard deviation is not accidental, since the standard deviation is the square root of the variance. If the characteristics of the sample are the same as those of the overall population, then s will be a good estimate for the value of σ. When we calculate the sample variance, we will (for reasons beyond the scope of this book) divide the sum of the squares of the deviations by 1 less than the number of data values. This turns out to give a better estimation of the population variance, for reasons we will leave to a full class in statistics.

EXAMPLE 13.20 A study is made of the average cost of a private room per day in a nursing home in a variety of cities across the United States in an attempt to find the average value of the cost of a stay at all such facilities. It is found that in a nine-city survey, the average cost of a private room per day is \$200, \$123, \$110, \$195, \$163, \$179, \$195, \$106, and \$105. What is the standard deviation within this data set?

SOLUTION

In order to calculate the standard deviation, it is first necessary to find the sample mean, from which we can find the individual deviations. In this case, using the methods of Section 13.3, we find the mean to be approximately \$152.89. Using this value, we can now find the individual deviations, displayed in Table13.4, for convenience, suppressing the dollar signs. The variance was defined to be the sum of the squares of the individual variations, and therefore the variance will be

$$s^2 = \frac{(47.11)^2 + (-29.89)^2 + (-42.89)^2 + (42.11)^2 + (10.11)^2 + (26.11)^2 + (42.11)^2 + (-46.89)^2 + (-47.89)^2}{9 - 1}$$

$$s^2 \approx \frac{13{,}774.88888}{8}$$

$$s^2 \approx 1721.86$$

Now, the square root of the variance is the standard deviation, and therefore $s \approx 41.50$. Thus, the sample standard deviation is approximately \$41.50.

TABLE 13.4 Deviations from the Mean

Data Value	Deviation
200	47.11
123	−29.89
110	−42.89
195	42.11
163	10.11
179	26.11
195	42.11
106	−46.89
105	−47.89

Note

If we have in our possession all the data for the entire population, then we calculate the variance by dividing the sum of the squares of the deviations by the exact number of data values.

You can probably tell that the computation of the standard deviation by hand is a rather painstaking process. It is not difficult, merely time consuming, and must be done with a good deal of care. You will find that many computer software programs and graphing calculators can calculate the standard deviation, but it is always important to understand what the computer or calculator is doing "behind the scenes."

We now need to say a few words about what the standard deviation tells us. First of all, the standard deviation must, by definition, be greater than or equal to 0. If $s = 0$, there is no spread or dispersion of the data, and all the data values are the same. If s is much larger than 0, then the data are significantly spread apart from the mean.

When we first do a computation of the standard deviation, its meaning may not be clear. If we graph the data (such as in a histograph), we will find that most of the data for symmetric distributions (where the data is relatively evenly distributed on both sides of the mean) lies within 1 standard deviation of the mean, in either direction (Figure 13.6).

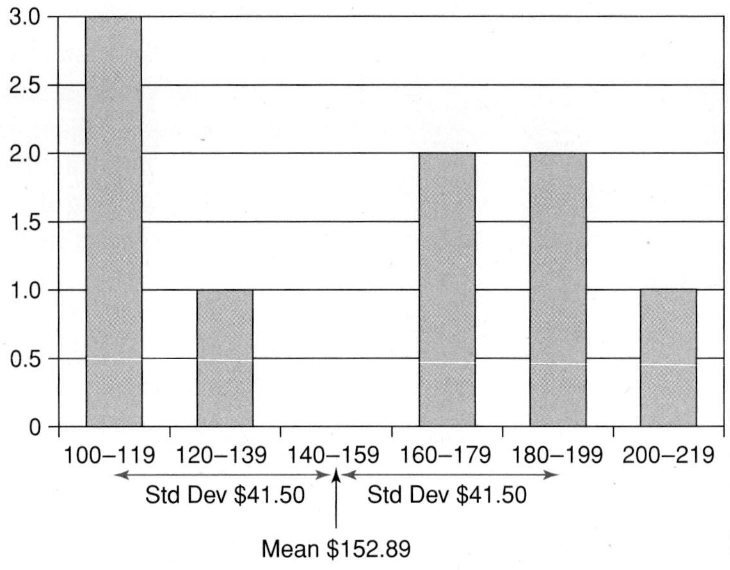

Figure 13.6 Histogram of nursing home cost data

The standard deviation will allow us, in addition, to compare two particular data values within distinct data sets. For instance, let's suppose two students are being compared on the basis of academic achievement. One student, Michel, has a grade-point average (GPA) of 2.77, while another student, Ramon, has a GPA of 3.15. Michel and Ramon attended different schools, and therefore it is difficult to determine on the basis of GPA alone which student ranks higher in comparison to his peers.

A study is made, and it is found that the mean GPA in Michel's school was 3.0, with a standard deviation of 0.7, while the mean GPA in Ramon's school was 3.46, with a standard deviation of 0.4. In order to assess how well Michel and Ramon performed relative to their peers, we can calculate the number of standard deviations from the mean that their GPAs stood.

For Michel, this would be $\frac{2.77 - 3.0}{0.7} \approx -0.33$, while for Ramon, it would be $\frac{3.15 - 3.46}{0.4} - 0.78$. The fact that both of these values are negative stems from the fact that each student's GPA was below the mean in his school, and the computation reveals that Michel's GPA was 0.33 standard deviations below the mean, while Ramon's was 0.78 standard deviations below the mean. Therefore, relative to their classes, Michel was actually ranked better than Ramon despite Ramon's having a higher GPA.

Exercises

In each of the following, find the range of the set of data.

1. Miguel teaches a class in chemistry, and he gave his students a quiz. Their scores were

46	31	35	45	39	45	29	47	36	27	33	25	25
42	28	49	30	39	49	35	32	41	30	29	30	

It turns out that, for these distributions (and this is a special feature of the normal distribution), the median and the mode always match the mean precisely. Parenthetically, we would be remiss if we did not mention that the equation that generates this curve is extremely complicated; you see it in a course on statistics that employs methods learned in calculus, which is a math class that incorporates some of the most remarkable intellectual achievements in the history of human thought. We won't need the technical formula, so we'll suppress it here, but you should be aware that there is a formula describing this curve!

Now, here's the interesting part of normal distributions, which some folks call the **empirical rule**.

RULE: Approximately 68% of all the data lie within 1 standard deviation (on either side) of the mean. Approximately 95% of all these data lie within 2 standard deviations (on either side) of the mean. Approximately 99.7% of all these data lie within 3 standard deviations (on either side) of the mean.

What does that tell us? Let's look at an issue that affects each of us as we shop at a department store: an individual's height, in particular, as a first example, the heights of men in the United States. Suppose a man is rather on the tall side, 6 feet 4 inches, or 76 inches, tall, and wears pants with "inseam" length 36 inches. Such a man, when searching for slacks in a typical store, has a difficult time finding the proper-size slacks. The question is, why? One would think that the store would stock the widest possible variety of sizes so as to accommodate its customers.

Height, like many physical traits of humans, is normally distributed. That is, if we constructed a histogram depicting the heights of human beings (and, in particular, men in the United States), that histogram would generate a bell curve, or normal distribution. Studies have shown that the mean height of all men in the United States is around 70 inches, with a standard deviation of 2.8 inches.

Returning to the man in question, his height of 76 inches is 6 inches above the mean. This may not sound like much, but since the standard deviation of the population is 2.8 inches, this says that the man's height is $\frac{76 - 70}{2.8} \approx 2.14$ standard deviations above the mean. Since 95% of all the men in the United States are, according to the empirical rule, within 2 standard deviations of the mean, this means that 95% of all men in the United States will have heights between 64.4 and 75.6 inches, with 2.5% of the men being shorter than 64.4 inches and 2.5% of the men being taller than 75.6 inches (remember, the population is symmetric about the mean!). Thus, the 6-foot-4-inch-tall man is among a minority of less than 2.5% of the men in the country, and therefore it is probably not financially worthwhile for the retailer to occupy shelf space for such a small percentage of the population.

The central idea in working with normal distributions (the process that, in fact, allows us to work with them easily) is the concept of the z-score. The ***z*-score** is a standardized value that allows us to transform any normal distribution into what we refer to as the standard normal distribution. The value we found in the illustration above, the number of standard deviations from the mean that our particular data value was, though we did not identify it as such, is a z-score.

The Standard Normal Distribution and z-Scores

The characteristic of the population we are studying is often referred to as a **random variable**, or more simply as a variable. The common notation for variables is an uppercase Latin letter, such as X, Y, or Z, while the specific values of that variable are denoted by lowercase Latin letters, such as x, y, or z. For instance, we could have identified the heights of men in the United States as X and the particular height of the individual under consideration (76 inches) as a value of x.

If the variable X is known to be normally distributed (this is something that a mass of research has indicated to the researcher) and the population mean and standard deviations are μ and σ, respectively, then for any specific value of the variable x, we define the z-score of the value x to be $z = \dfrac{x - \mu}{\sigma}$. The z-score tells us how many standard deviations the value x is above (or to the right of) or below (to the left of) the mean, μ. A positive z-score indicates that the value of the variable is above the mean and a negative z-score that the value of the variable is below the mean. A value of the variable equal to the mean has a z-score of 0. For reasons we will make clear shortly, a z-score is, by convention, rounded to two decimal places, as necessary.

EXAMPLE 13.21 A popular belief is that a person can safely lose 2 pounds per week, or 8 pounds per month, on average through improved diet and consistent exercise. If research establishes that the amount of weight lost by individuals attempting to lose weight is normally distributed and that the standard deviation of weight loss is 1.6 pounds, then what is the z-score for a weight loss of 11 pounds, and what does this tell us about this weight loss, relative to the average?

SOLUTION

The value $x = 11$ is the particular value of the variable in which we are interested, and the mean and standard deviation are given to us as $\mu = 8, \sigma = 1.6$. Since weight loss is known to be normally distributed, we find $z = \dfrac{11 - 8}{1.6} = \dfrac{3}{1.6} \approx 1.88$.

The positive sign of this result tells us that the value 11 is above the mean, and the value 1.88 tells us that it is 1.88 standard deviations above the mean.

EXAMPLE 13.22 The Serendipity Jump Drive manufacturer claims that the life of its jump drives is normally distributed with a mean of 3.4 years and a standard deviation of 0.25 years. A particular jump drive fails after 3 years. What is the z-score for this value of the variable, and what does this tell us about this particular jump drive, relative to the average?

SOLUTION

The value $x = 3$ is the particular value of the variable in which we are interested, and we are given the mean and standard deviation. Since the life span of the jump drives is claimed to be normally distributed, we calculate the z-score to be $z = \dfrac{3 - 3.4}{0.25} = -1.60$. Because this value is negative, we know that the life of this particular jump drive is below the level of the mean, and the value of the z-score tells us that the life span was 1.60 standard deviations below the mean.

The particular value of the z-score allows us to glean much more information than a simple number of standard deviations above or below the mean that our particular data value happens to be. We can use the value of the z-score and a table called a standard normal cumulative probability table (or standard normal table) to calculate the "standing" of our value of the random variable in relation to the rest of the population. Such tables are often computerized these days, but there are also relatively brief charts, such as the one shown in Table 13.5, that we can use to find the information we need.

Note

The entry in the standard normal distribution table tells us the percentage of the overall population that our data value is above!

The table, you will notice, has a reference column down the left edge and also a row along the top edge. The column along the left edge provides us with the first two digits of our z-score, while the row across the top yields the last digit. For instance, if our z-score were 1.37, we would find (in the vertical column) the line beginning with 1.3 and (in the top row across) the column headed with 0.07 and locate the intersection of this row and column. The table entry corresponding to this z-score would be 0.9147 (be sure that you understand how to read the table before proceeding because that is crucial!). Software or Web sites programmed with a standard normal table permit you to simply enter the z-score and return the table entry for you immediately.

The significance of this table entry cannot be overstated in problems involving the normal distribution. A data value with z-score 1.37, which (from the table) yields a standard normal table entry of 0.9147, is not only 1.37 standard deviations above the mean but also greater than 91.47% of the rest of the data in the population!

EXAMPLE 13.23 Let's revisit the problem that began the discussion of normal distribution, the 6-foot-4-inch-tall man searching for slacks in the department store. We had found (though we did not call the value by name at the time) that the man's height had a z-score of 2.14. The table entry corresponding to this z-score is (and we urge you to verify that you can find this value!) 0.9838, which indicates that the man in question is taller than 98.38% of the men in the United States.

We should observe that the table value also tells us something else: if we know the man in question is taller than 98.38% of the men in the United States, then it stands to reason that he must also be shorter than 1.62% of the men in the United States. That is, to find the percentage of our population that is greater than our data value, we should subtract the table entry from one.

EXAMPLE 13.24 The Graduate Management Aptitude Test (GMAT) is widely used to assess students who are applying to graduate schools of business. If it is known that, for a particular year, the mean GMAT score was 503, with a standard deviation of 89,

TABLE 13.5 The Standard Normal Distribution Table

z	0.00	0.01	0.02	0.03	0.04	0.05	0.06	0.07	0.08	0.09
−3.4	0.0003	0.0003	0.0003	0.0003	0.0003	0.0003	0.0003	0.0003	0.0003	0.0002
−3.3	0.0005	0.0005	0.0005	0.0004	0.0004	0.0004	0.0004	0.0004	0.0004	0.0003
−3.2	0.0007	0.0007	0.0006	0.0006	0.0006	0.0006	0.0006	0.0005	0.0005	0.0005
−3.1	0.0010	0.0009	0.0009	0.0009	0.0008	0.0008	0.0008	0.0008	0.0007	0.0007
−3.0	0.0013	0.0013	0.0013	0.0012	0.0012	0.0011	0.0011	0.0011	0.0010	0.0010
−2.9	0.0019	0.0018	0.0018	0.0017	0.0016	0.0016	0.0015	0.0015	0.0014	0.0014
−2.8	0.0026	0.0025	0.0024	0.0023	0.0023	0.0022	0.0021	0.0021	0.0020	0.0019
−2.7	0.0035	0.0034	0.0033	0.0032	0.0031	0.0030	0.0029	0.0028	0.0027	0.0026
−2.6	0.0047	0.0045	0.0044	0.0043	0.0041	0.0040	0.0039	0.0038	0.0037	0.0036
−2.5	0.0062	0.0060	0.0059	0.0057	0.0055	0.0054	0.0052	0.0051	0.0049	0.0048
−2.4	0.0082	0.0080	0.0078	0.0075	0.0073	0.0071	0.0069	0.0068	0.0066	0.0064
−2.3	0.0107	0.0104	0.0102	0.0099	0.0096	0.0094	0.0091	0.0089	0.0087	0.0084
−2.2	0.0139	0.0136	0.0132	0.0129	0.0125	0.0122	0.0119	0.0116	0.0113	0.0110
−2.1	0.0179	0.0174	0.0170	0.0166	0.0162	0.0158	0.0154	0.0150	0.0146	0.0143
−2.0	0.0228	0.0222	0.0217	0.0212	0.0207	0.0202	0.0197	0.0192	0.0168	0.0183
−1.9	0.0287	0.0281	0.0274	0.0268	0.0262	0.0256	0.0250	0.0244	0.0239	0.0233
−1.8	0.0359	0.0351	0.0344	0.0336	0.0329	0.0322	0.0314	0.0307	0.0301	0.0294
−1.7	0.0446	0.0436	0.0427	0.0418	0.0409	0.0401	0.0392	0.0384	0.0375	0.0367
−1.6	0.0548	0.0537	0.0526	0.0516	0.0505	0.0495	0.0485	0.0475	0.0465	0.0455
−1.5	0.0668	0.0655	0.0643	0.0630	0.0618	0.0606	0.0594	0.0582	0.0571	0.0559
−1.4	0.0808	0.0793	0.0778	0.0764	0.0749	0.0735	0.0721	0.0708	0.0694	0.0681
−1.3	0.0968	0.0951	0.0934	0.0918	0.0901	0.0885	0.0869	0.0853	0.0838	0.0823
−1.2	0.1151	0.1131	0.1112	0.1093	0.1075	0.1056	0.1038	0.1020	0.1003	0.0985
−1.1	0.1357	0.1335	0.1314	0.1292	0.1271	0.1251	0.1230	0.1210	0.1190	0.1170
−1.0	0.1587	0.1562	0.1539	0.1515	0.1492	0.1469	0.1446	0.1423	0.1401	0.1379
−0.9	0.1841	0.1814	0.1788	0.1762	0.1736	0.1711	0.1685	0.1660	0.1635	0.1611
−0.8	0.2119	0.2090	0.2061	0.2033	0.2005	0.1977	0.1949	0.1922	0.1894	0.1867
−0.7	0.2420	0.2389	0.2358	0.2327	0.2296	0.2266	0.2236	0.2206	0.2177	0.2148
−0.6	0.2743	0.2709	0.2676	0.2643	0.2611	0.2578	0.2546	0.2514	0.2483	0.2451
−0.5	0.3085	0.3050	0.3015	0.2981	0.2946	0.2912	0.2877	0.2843	0.2810	0.2776
−0.4	0.3446	0.3409	0.3372	0.3336	0.3300	0.3264	0.3228	0.3192	0.3156	0.3121
−0.3	0.3821	0.3783	0.3745	0.3707	0.3669	0.3632	0.3594	0.3557	0.3520	0.3483
−0.2	0.4207	0.4168	0.4129	0.4090	0.4052	0.4013	0.3974	0.3936	0.3897	0.3859
−0.1	0.4602	0.4562	0.4522	0.4483	0.4443	0.4404	0.4364	0.4325	0.4286	0.4247
−0.0	0.5000	0.4960	0.4920	0.4880	0.4840	0.4801	0.4761	0.4721	0.4681	0.4641

TABLE 13.5 The Standard Normal Distribution Table (*Continued*)

z	0.00	0.01	0.02	0.03	0.04	0.05	0.06	0.07	0.08	0.09
0.0	0.5000	0.5040	0.5080	0.5120	0.5160	0.5199	0.5239	0.5279	0.5319	0.5359
0.1	0.5398	0.5438	0.5478	0.5517	0.5557	0.5596	0.5636	0.5675	0.5714	0.5753
0.2	0.5793	0.5832	0.5871	0.5910	0.5948	0.5987	0.6026	0.6064	0.6103	0.6141
0.3	0.6179	0.6217	0.6255	0.6293	0.6331	0.6368	0.6406	0.6443	0.6480	0.6517
0.4	0.6554	0.6591	0.6628	0.6664	0.6700	0.6736	0.6772	0.6808	0.6844	0.6879
0.5	0.6915	0.6950	0.6985	0.7019	0.7054	0.7088	0.7123	0.7157	0.7190	0.7224
0.6	0.7257	0.7291	0.7324	0.7357	0.7389	0.7422	0.7454	0.7486	0.7517	0.7549
0.7	0.7580	0.7611	0.7642	0.7673	0.7704	0.7734	0.7764	0.7794	0.7823	0.7852
0.8	0.7881	0.7910	0.7939	0.7967	0.7995	0.8023	0.8051	0.8078	0.8106	0.8133
0.9	0.8159	0.8186	0.8212	0.8238	0.8264	0.8289	0.8315	0.8340	0.8365	0.8389
1.0	0.8413	0.8438	0.8461	0.8485	0.8508	0.8531	0.8554	0.8577	0.8599	0.8621
1.1	0.8643	0.8665	0.8686	0.8708	0.8729	0.8749	0.8770	0.8790	0.8810	0.8830
1.2	0.8849	0.8869	0.8888	0.8907	0.8925	0.8944	0.8962	0.8980	0.8997	0.9015
1.3	0.9032	0.9049	0.9066	0.9082	0.9099	0.9115	0.9131	0.9147	0.9162	0.9177
1.4	0.9192	0.9207	0.9222	0.9236	0.9251	0.9265	0.9279	0.9292	0.9306	0.9319
1.5	0.9332	0.9345	0.9357	0.9370	0.9382	0.9394	0.9406	0.9418	0.9429	0.9441
1.6	0.9452	0.9463	0.9474	0.9484	0.9495	0.9505	0.9515	0.9525	0.9535	0.9545
1.7	0.9554	0.9564	0.9573	0.9582	0.9591	0.9599	0.9608	0.9616	0.9625	0.9633
1.8	0.9641	0.9649	0.9656	0.9664	0.9671	09678	0.9686	0.9693	0.9699	0.9706
1.9	0.9713	0.9719	0.9726	0.9732	0.9738	0.9744	0.9750	0.9756	0.9761	0.9767
2.0	0.9772	0.9778	0.9783	0.9788	0.9793	0.9798	0.9803	0.9808	0.9812	0.9817
2.1	0.9821	0.9826	0.9830	0.9834	0.9838	0.9842	0.9846	0.9850	0.9854	0.9857
2.2	0.9861	0.9864	0.9868	0.9871	0.9875	0.9878	0.9881	0.9884	0.9887	0.9890
2.3	0.9893	0.9896	0.9898	0.9901	0.9904	0.9906	0.9909	0.9911	0.9913	0.9916
2.4	0.9918	0.9920	0.9922	0.9925	0.9927	0.9929	0.9931	0.9932	0.9934	0.9936
2.5	0.9938	0.9940	0.9941	0.9943	0.9945	0.9946	0.9948	0.9949	0.9951	0.9952
2.6	0.9953	0.9955	0.9956	0.9957	0.9959	0.9960	0.9961	0.9962	0.9963	0.9964
2.7	0.9965	0.9966	0.9967	0.9968	0.9969	0.9970	0.9971	0.9972	0.9973	0.9974
2.8	0.9974	0.9975	0.9976	0.9977	0.9977	0.9978	0.9979	0.9979	0.9980	0.9981
2.9	0.9981	0.9982	0.9982	0.9983	0.9984	0.9984	0.9985	0.9985	0.9986	0.9986
3.0	0.9987	0.9987	0.9987	0.9988	0.9988	0.9989	0.9989	0.9989	0.9990	0.9990
3.1	0.9990	0.9991	0.9991	0.9991	0.9992	0.9992	0.9992	0.9992	0.9993	0.9993
3.2	0.9993	0.9993	0.9994	0.9994	0.9994	0.9994	0.9994	0.9995	0.9995	0.9995
3.3	0.9995	0.9995	0.9995	0.9996	0.9996	0.9996	0.9996	0.9996	0.9996	0.9997
3.4	0.9997	0.9997	0.9997	0.9997	0.9997	0.9997	0.9997	0.9997	0.9997	0.9998

what is the probability that a randomly selected test taker's score will be greater than 480, assuming that the test scores were normally distributed?

SOLUTION

Using the variable value $x = 480$, we calculate the z-score to be $z = \dfrac{480 - 503}{89} \approx -0.26$. The entry from the standard normal distribution table corresponding to this z-score is 0.3974, which indicates that 39.74% of test takers scored below 480. Since the problem asked us for the probability of a score being higher than 480, it follows that 60.36% of the test takers scored higher than 480, and thus there is a 0.6036, or 60.36%, probability that a randomly selected test taker would score higher than 480 on this administration of the GMAT.

EXAMPLE 13.25 A tire manufacturer wants to devise a warranty length for its new midgrade tire and conducts a study of the life of the tires under typical use. It is found that the life of the tires is normally distributed, with a mean mileage value of 53,800 miles and a standard deviation of 1950 miles. If the manufacturer wants to construct its warranty in such a manner that no more than 3% of all tires will fail while under warranty, what guaranteed mileage should it advertise?

SOLUTION

Since the company wants to ensure that only 3% of the tires will fail within the warranty period, the only tires that should be covered would be those whose life span was less than 97% of all the tires sold. In other words, the life span of the tires replaceable under warranty should be greater than only 3% of the population, and we need to find the value of the random variable that would produce such a chart entry.

We examine the standard normal distribution table and look for the entry in the body of the table closest to 0.03. The closest entry to this is -1.80, which corresponds to chart entry 0.0359. If we insert that value into our z-score formula, we have an equation in the unknown x we can readily solve:

$$-1.80 = \frac{x - 53{,}800}{1950}$$

$$-3510 = x - 53{,}800$$

$$50{,}290 = x$$

The manufacturer should set the warrantee mileage limit of at least 50,290 miles if it wants to replace no more than 3% of the tires under warranty.

If we use a bit of ingenuity, we can actually say one thing more about the standard normal distribution table and its use. Since the table tells us the percentage of the population having a value of our random variable less than a particular value, we can use the table to find the percentage of the population having a value of the random variable *between* two particular values as well.

EXAMPLE 13.26 Let's use the same example of heights of men in the United States to answer another question. We know that the average height of a man in the United States is 70 inches, that the standard deviation is 2.8 inches, and that the heights are normally distributed. What percentage of the men in the United States have heights between 68 and 71 inches? That is, between 5 feet 8 inches and 5 feet 11 inches?

SOLUTION

In order to find this result, we will need the z-scores of each of the two heights involved. For the value $x = 68$, we find that $z = -0.71$, while for $x = 71$, we see that $z = 0.36$. Referencing the standard normal distribution table for these two z-scores, we find that the entries are 0.2389 and 0.6406, respectively.

This tells us that the percentage of men in the United States whose height is less than 68 inches is 23.89%, while the percentage whose heights are less than 71 inches is 64.06%. It follows that the difference between these two percentages must be the percentage of men in the United States whose heights are between 68 and 71 inches, or 40.17%.

EXAMPLE 13.27 A vending machine is designed to dispense an average of 7.8 ounces of hot chocolate into an 8-ounce cup. If the standard deviation of the amount of coffee dispensed is 0.2 ounces and the amount of hot chocolate dispensed is normally distributed, find the percentage of times the machine will dispense between 7.5 and 7.9 ounces.

SOLUTION

We are looking for the percentage that fall between two particular values of the random variable, and thus we need to find two z-scores and their corresponding chart entries and then calculate the difference between those chart values. For the value $x = 7.5$, we find $z = \dfrac{7.5 - 7.8}{0.2} = -1.50$, and for $x = 7.9$, we find that $z = \dfrac{7.9 - 7.8}{0.2} = 0.50$. The corresponding standard normal distribution table entries are 0.0688 and 0.6915, and the difference of these values is $0.6915 - 0.0688 = 0.6227$. This suggests that the vending machine will dispense between 7.5 and 7.9 ounces of hot chocolate 62.27% of the time.

EXAMPLE 13.28 Many households in the United States have at least one computer. Suppose the average number of hours spent playing games on a computer on any given day is normally distributed, with a mean of 1.75 hours and a standard deviation of 0.33 hours. What is the probability that the amount of time spent playing games on a randomly selected computer is between 1.75 and 2.25 hours?

SOLUTION

Again, we are interested in a particular interval of values of our random variable, and thus we will have to find a pair of z-scores, the corresponding standard normal distribution table entries, and then the difference between those entries. Observe a rather strange result for the value $x = 1.75$: the z-score is $z = \dfrac{1.75 - 1.75}{0.33} = 0$. At first, this might be surprising, but we should recall that we said that the z-score corresponding to a variable value equal to the mean would be 0. For $x = 2.25$, we find $z = \dfrac{2.25 - 1.75}{0.33} \approx 1.52$. Examination of the standard normal distribution chart reveals the corresponding chart entries to be 0.5000 and 0.9357, respectively. The difference of these values is 0.4357, from which we conclude that the probability is 43.57% that a randomly selected computer would be used to play games between 1.75 and 2.25 hours per night.

Exercises

For the following, find the area beneath the standard normal curve, between the given *z*-scores.

1. $z = 0$ and $z = 1.12$
2. $z = 0$ and $z = 1.89$
3. $z = -1.06$ and $z = 1.87$
4. $z = -1.5$ and $z = 2.62$

Use a table to find the *z*-score corresponding to the given conditions.

5. 33% of the area under the standard normal curve is to the right of the score.
6. 48% of the area under the standard normal curve is to the left of the score.
7. 15% of the area under the standard normal curve is to the left of the score.
8. 22% of the area under the standard normal curve is to the right of the score.

Find the area under the normal curve in the following situations.

9. Find the percentage of the area under the curve between the mean and 3.01 standard deviations from the mean.
10. Find the percentage of the area under the curve between the mean and 2.41 standard deviations from the mean.
11. Find the percentage of the area under the curve between the mean and 1.35 standard deviations from the mean.
12. Find the percentage of the area under the curve between the mean and 0.64 standard deviations from the mean.
13. Find the percentage of the total area under the curve between $z = 1.41$ and $z = 2.83$.
14. Find the percentage of the total area under the curve between $z = -0.05$ and $z = 1.92$.
15. Find the percentage of the total area under the curve between $z = -2.96$ and $z = 0.87$.
16. Find the percentage of the total area under the curve between $z = 0$ and $z = -1.28$.

Use the standard normal curve table to find the *z*-score closest for the given condition.

17. 4% of the total area is to the right of the *z*-score.
18. 18% of the total area is to the right of the *z*-score.
19. 21.5% of the total area is to the left of the *z*-score.
20. 57.8% of the total area is to the left of the *z*-score.
21. 3% of the total area is to the right of the *z*-score.

22. 82.9% of the total area is to the left of the z-score.

23. 33% of the total area is to the left of the z-score.

24. 33% of the total area is to the right of the z-score.

Find the z-score corresponding to the following conditions.

25. Shaded area is 0.9599 units:

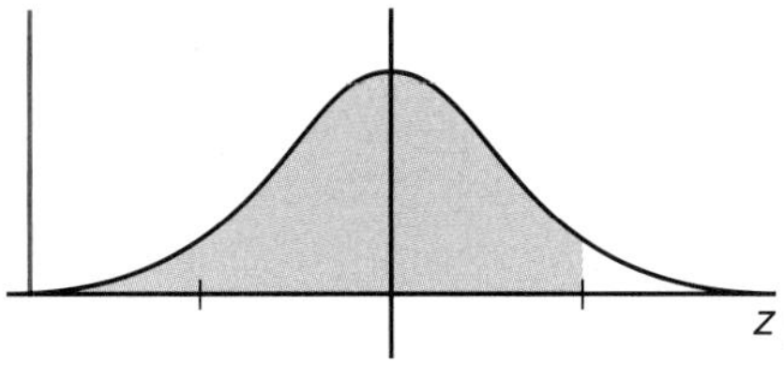

26. Shaded area is 0.4013 units:

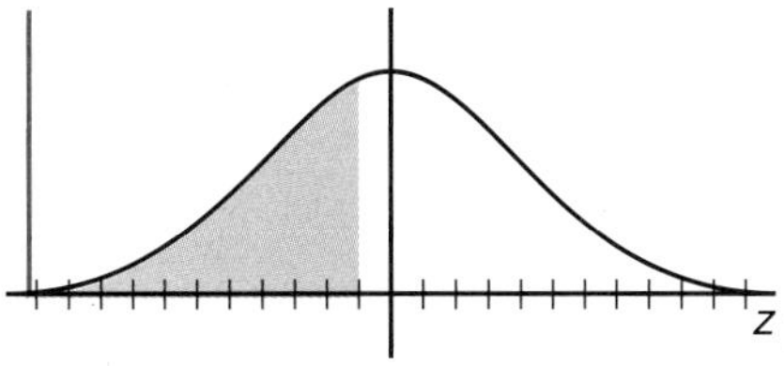

For the following, if z is a standard normal variable, find the probability.

27. The probability z lies between -2.41 and 0.

28. The probability z lies between -1.85 and 0.

29. The probability z is less than 1.13.

30. The probability z is less than 2.74.

31. The probability z is greater than 0.95.

32. The probability z is greater than -0.51.

33. $P(z < 0.97)$

34. $P(z < -1.50)$

13.6 The Binomial Distribution

Binomial Experiments and Binomial Probability Distributions

A **binomial experiment** is characterized by the following traits: (1) there are a fixed number of trials, or repetitions of the experiment, denoted by n; (2) each trial has only two possible outcomes, which we can characterize as "success" or "failure" (the probability of success on any given trial is denoted by p and the probability of failure by q); and (3) the n trials are independent of one another and are conducted under identical conditions in each case. Any experiment that satisfies only criteria 2 and 3 is referred to as a **Bernoulli trial** in honor of the great seventeenth-century mathematician Jacob Bernoulli, and thus a binomial experiment occurs when the number of successes are counted in one or more Bernoulli trials.

An example of a binomial experiment would be the following. It is known that the probability is 0.70 that students in a particular class will do their homework in time for it to be collected and graded without penalty. In a class of 20 students, what is the probability that at least 10 will do their homework on time?

The experiment is binomial because there are a fixed number of trials (20 students), there are only two possible outcomes for each trial (they do their homework or they don't), and the probability of success (doing their homework on time) does not change from student to student, and no one student's doing or not doing his or her homework influences whether another student completes his or her homework on time.

Outcomes of such an experiment are said to fit a **binomial probability distribution**, and the random variable X denotes the number of successes obtained through the n independent trials of the experiment.

In a binomial probability distribution, the mean $\mu = np$, and the standard deviation $\sigma = \sqrt{npq}$. Establishing these two results is not particularly difficult, but we'll suppress the work here, since it is not the focus of our efforts.

EXAMPLE 13.29 Imagine you participate in a game where you must either win or lose. The probability that you win the game is 0.75, and the probability you lose the game is 0.25. If you play the game 20 times, then the mean of the binomial probability distribution is $\mu = (20)(0.75) = 15$, and the standard deviation is $\sigma = \sqrt{(20)(0.75)(0.25)} = \sqrt{3.75} \approx 1.94$.

There is a standard notation in use for binomial probability distributions having n trials and probability of success p, and that is $X \sim B(n, p)$. There is no universal agreement on how this is read, but most people would read it as "X is a random variable with a binomial distribution."

Binomial probabilities are often calculated using software, but there exist tables (similar to the standard normal distribution table encountered in Section 13.5) we can use to compute the probabilities by hand (Table 13.6). Observe that this table computes the cumulative probability of obtaining at most x successes in n trials of a binomial experiment with probability of success p. If we want to determine the probability of obtaining exactly x successes, we would subtract the table entry for $(x - 1)$ successes from the entry for x successes.

EXAMPLE 13.30 Let's revisit the game described in Example 13.29, where you participate in a game where you must either win or lose. The probability that you win the game is 0.75, and the probability that you lose the game is 0.25. What is the probability that you will win the game at most 10 times?

SOLUTION

Referring to the table, we note that the number of trials is $n = 20$, so we would refer to that portion of the chart where we see 20 in the n column on the far left. Since the probability of success $p = 0.75$, we also find the column corresponding to $p = 0.75$ and the row for $x = 10$, the number of successes in which we are interested.

The corresponding table entry is seen to be 0.014, and thus we conclude that the probability of winning this game at most 10 times is 1.4%.

TABLE 13.6 The Cumulative Binomial Probability Distribution Table

n	r	.01	.05	.10	.15	.20	.25	.30	.35	.40	.45	.50	.55	.60	.65	.70	.75	.80	.85	.90	.95
2	0	.980	.902	.810	.723	.640	.563	.490	.423	.360	.303	.250	.203	.160	.123	.090	.063	.040	.023	.010	.002
	1	.020	.095	.180	.255	.320	.375	.420	.455	.480	.495	.500	.495	.480	.455	.420	.375	.320	.255	.180	.095
	2	.000	.002	.010	.023	.040	.063	.090	.123	.160	.203	.250	.303	.360	.423	.490	.563	.640	.723	.810	.902
3	0	.970	.857	.729	.614	.512	.422	.343	.275	.216	.166	.125	.091	.064	.043	.027	.016	.008	.003	.001	.000
	1	.029	.135	.243	.325	.384	.422	.441	.444	.432	.408	.375	.334	.288	.239	.189	.141	.096	.057	.027	.007
	2	.000	.007	.027	.057	.096	.141	.189	.239	.288	.334	.375	.408	.432	.444	.441	.422	.384	.325	.243	.135
	3	.000	.000	.001	.003	.008	.016	.027	.043	.064	.091	.125	.166	.216	.275	.343	.422	.512	.614	.729	.857
4	0	.961	.815	.656	.522	.410	.316	.240	.179	.130	.092	.062	.041	.026	.015	.008	.004	.002	.001	.000	.000
	1	.039	.171	.292	.368	.410	.422	.412	.384	.346	.300	.250	.200	.154	.112	.076	.047	.026	.011	.004	.000
	2	.001	.014	.049	.098	.154	.211	.265	.311	.346	.368	.375	.368	.346	.311	.265	.211	.154	.008	.049	.014
	3	.000	.000	.004	.011	.026	.047	.076	.112	.154	.200	.250	.300	.346	.384	.412	.422	.410	.368	.292	.171
	4	.000	.000	.000	.001	.002	.004	.008	.015	.026	.041	.062	.092	.130	.179	.240	.316	.410	.522	.656	.815
5	0	.951	.774	.590	.444	.328	.237	.168	.116	.078	.050	.031	.019	.010	.005	.002	.001	.000	.000	.000	.000
	1	.048	.204	.328	.392	.410	.396	.360	.312	.259	.208	.156	.113	.077	.049	.028	.015	.006	.002	.000	.000
	2	.001	.021	.073	.138	.205	.264	.309	.336	.346	.337	.312	.278	.230	.181	.132	.088	.051	.024	.008	.001
	3	.000	.001	.008	.024	.051	.083	.132	.181	.230	.276	.312	.337	.346	.336	.309	.264	.205	.138	.073	.021
	4	.000	.000	.000	.002	.008	.015	.028	.049	.077	.113	.156	.206	.259	.312	.360	.396	.410	.392	.328	.204
	5	.000	.000	.000	.000	.000	.001	.002	.005	.010	.019	.031	.050	.078	.116	.168	.237	.328	.444	.590	.774
6	0	.941	.735	.531	.377	.262	.178	.118	.075	.047	.028	.016	.008	.004	.002	.001	.000	.000	.000	.000	.000
	1	.057	.232	.354	.399	.393	.356	.303	.244	.187	.136	.094	.061	.037	.020	.010	.004	.002	.000	.000	.000
	2	.001	.031	.098	.176	.248	.297	.324	.328	.311	.278	.234	.186	.138	.095	.060	.033	.015	.006	.001	.000
	3	.000	.002	.015	.042	.082	.132	.185	.236	.276	.303	.312	.303	.276	.236	.185	.132	.082	.042	.015	.002
	4	.000	.000	.001	.006	.015	.033	.060	.095	.138	.186	.234	.278	.311	.328	.324	.297	.246	.176	.098	.031
	5	.000	.000	.000	.000	.002	.004	.010	.020	.037	.061	.094	.136	.187	.244	.303	.356	.393	.399	.354	.232
	6	.000	.000	.000	.000	.000	.000	.001	.002	.004	.008	.016	.028	.047	.075	.118	.178	.262	.377	.531	.735
7	0	.932	.698	.478	.321	.210	.133	.082	.049	.028	.015	.008	.004	.002	.001	.000	.000	.000	.000	.000	.000
	1	.066	.257	.372	.396	.367	.311	.247	.185	.131	.087	.055	.032	.017	.008	.004	.001	.000	.000	.000	.000
	2	.002	.041	.124	.210	.275	.311	.318	.299	.261	.214	.164	.117	.077	.047	.025	.012	.004	.001	.000	.000
	3	.000	.004	.023	.062	.115	.173	.227	.268	.290	.292	.273	.239	.194	.144	.097	.058	.029	.011	.003	.000

(continued)

TABLE 13.6 The Cumulative Binomial Probability Distribution Table *(Continued)*

n	r	.01	.05	.10	.15	.20	.25	.30	.35	.40	.45	.50	.55	.60	.65	.70	.75	.80	.85	.90	.95
	4	.000	.000	.003	.011	.029	.058	.097	.144	.194	.239	.273	.292	.290	.268	.227	.173	.115	.062	.023	.004
	5	.000	.000	.000	.001	.004	.012	.025	.047	.077	.117	.164	.214	.261	.299	.318	.311	.275	.210	.124	.041
	6	.000	.000	.000	.000	.000	.001	.004	.008	.017	.032	.055	.087	.131	.185	.247	.311	.367	.396	.372	.257
	7	.000	.000	.000	.000	.000	.000	.000	.001	.002	.004	.008	.015	.028	.049	.082	.133	.210	.321	.478	.698
8	0	.923	.663	.430	.272	.168	.100	.058	.032	.017	.008	.004	.002	.001	.000	.000	.000	.000	.000	.000	.000
	1	.075	.279	.383	.385	.336	.267	.198	.137	.090	.055	.031	.016	.008	.003	.001	.000	.000	.000	.000	.000
	2	.003	.051	.149	.238	.294	.311	.296	.259	.209	.157	.109	.070	.041	.022	.010	.004	.001	.000	.000	.000
	3	.000	.005	.033	.084	.147	.208	.254	.279	.279	.257	.219	.172	.124	.081	.047	.023	.009	.003	.000	.000
	4	.000	.000	.005	.018	.046	.087	.136	.188	.232	.263	.273	.263	.232	.188	.136	.087	.046	.018	.005	.000
	5	.000	.000	.000	.003	.009	.023	.047	.081	.124	.172	.219	.257	.279	.279	.254	.208	.147	.084	.033	.005
	6	.000	.000	.000	.000	.001	.004	.010	.022	.041	.070	.109	.157	.209	.259	.296	.311	.294	.238	.149	.051
	7	.000	.000	.000	.000	.000	.000	.001	.003	.008	.016	.031	.055	.090	.137	.198	.267	.336	.385	.383	.279
	8	.000	.000	.000	.000	.000	.000	.000	.000	.001	.002	.004	.008	.017	.032	.058	.100	.168	.272	.430	.663
9	0	.914	.630	.387	.232	.134	.075	.040	.021	.010	.005	.002	.001	.000	.000	.000	.000	.000	.000	.000	.000
	1	.083	.299	.387	.368	.302	.225	.156	.100	.060	.034	.018	.008	.004	.001	.000	.000	.000	.000	.000	.000
	2	.003	.063	.172	.260	.302	.300	.267	.216	.161	.111	.070	.041	.021	.010	.004	.001	.000	.000	.000	.000
	3	.000	.008	.045	.107	.176	.234	.267	.272	.251	.212	.164	.116	.074	.042	.021	.009	.003	.001	.000	.000
	4	.000	.001	.007	.028	.066	.117	.172	.219	.251	.260	.246	.213	.167	.118	.074	.039	.017	.005	.001	.000
	5	.000	.000	.001	.005	.017	.039	.074	.118	.167	.213	.246	.260	.251	.219	.172	.117	.066	.028	.007	.001
	6	.000	.000	.000	.001	.003	.009	.021	.042	.074	.116	.164	.212	.251	.272	.267	.234	.176	.107	.045	.008
	7	.000	.000	.000	.000	.000	.001	.004	.010	.021	.041	.070	.111	.161	.216	.267	.300	.302	.260	.172	.063
	8	.000	.000	.000	.000	.000	.000	.000	.001	.004	.008	.018	.034	.060	.100	.156	.225	.302	.368	.387	.299
	9	.000	.000	.000	.000	.000	.000	.000	.000	.000	.001	.002	.005	.010	.021	.040	.075	.134	.232	.387	.630
10	0	.904	.599	.349	.197	.107	.056	.028	.014	.006	.003	.001	.000	.000	.000	.000	.000	.000	.000	.000	.000
	1	.091	.315	.387	.347	.268	.188	.121	.072	.040	.021	.010	.004	.002	.000	.000	.000	.000	.000	.000	.000
	2	.004	.075	.194	.276	.302	.282	.233	.176	.121	.076	.044	.023	.011	.004	.001	.000	.000	.000	.000	.000
	3	.000	.010	.057	.130	.201	.250	.267	.252	.215	.166	.117	.075	.042	.021	.009	.003	.001	.000	.000	.000
	4	.000	.001	.011	.040	.088	.146	.200	.238	.251	.238	.205	.160	.111	.069	.037	.016	.006	.001	.000	.000
	5	.000	.000	.001	.008	.026	.058	.103	.154	.201	.234	.246	.234	.201	.154	.103	.058	.026	.008	.001	.000

	6	.000	.000	.000	.001	.006	.016	.037	.069	.111	.160	.205	.238	.251	.238	.200	.146	.088	.040	.011	.001
	7	.000	.000	.000	.000	.001	.003	.009	.021	.042	.075	.117	.166	.215	.252	.267	.250	.201	.130	.057	.010
	8	.000	.000	.000	.000	.000	.000	.001	.004	.011	.023	.044	.076	.121	.176	.233	.282	.302	.276	.194	.07
	9	.000	.000	.000	.000	.000	.000	.000	.000	.002	.004	.010	.021	.040	.072	.121	.188	.268	.347	.387	.315
	10	.000	.000	.000	.000	.000	.000	.000	.000	.000	.000	.001	. 003	. 006	. 014	. 028	. 056	. 107	. 197	. 349	.599
11	0	.895	.569	.314	.167	.086	.042	.020	.009	.004	.001	.000	.000	.000	.000	.000	.000	.000	.000	.000	.000
	1	.099	.329	.384	.325	.236	.155	.093	.052	.027	.013	.005	.002	.001	.000	.000	.000	.000	.000	.000	.000
	2	.005	.087	.213	.287	.295	.258	.200	.140	.089	.051	.027	.013	.005	.002	.001	.000	.000	.000	.000	.000
	3	.000	.014	.071	.152	.221	.258	.257	.225	.177	.126	.081	.046	.023	.010	.004	.001	.000	.000	.000	.000
	4	.000	.001	.016	.054	.111	.172	.220	.243	.236	.206	.161	.113	.070	.038	.017	.006	.002	.000	.000	.000
	5	.000	.000	.002	.013	.039	.080	.132	.183	.221	.236	.226	.193	.147	.099	.057	.027	.010	.002	.000	.000
	6	.000	.000	.000	.002	.010	.027	.057	.099	.147	.193	.226	.236	.221	.183	.132	.080	.039	.013	.002	.000
	7	.000	.000	.000	.000	.002	.006	.017	.038	.070	.113	.161	.206	.236	.243	.220	.172	.111	.054	.016	.001
	8	.000	.000	.000	.000	.000	.001	.004	.010	.023	.046	.081	.126	.177	.225	.257	.258	.221	.152	.071	.014
	9	.000	.000	.000	.000	.000	.000	.001	.002	.005	.013	.027	.051	.089	.140	.200	.258	.295	.287	.213	.087
	10	.000	.000	.000	.000	.000	.000	.000	.000	.001	.002	.005	.013	.027	.052	.093	.155	.236	.325	.384	.329
	11	.000	.000	.000	.000	.000	.000	.000	.000	.000	.000	.000	.001	.004	.009	.020	.042	.086	.167	.314	.569
12	0	.886	.540	.282	.142	.069	.032	.014	.006	.002	.001	.000	.000	.000	.000	.000	.000	.000	.000	.000	.000
	1	.107	.341	.377	.301	.206	.127	.071	.037	.017	.008	.003	.001	.000	.000	.000	.000	.000	.000	.000	.000
	2	.006	.099	.230	.292	.283	.232	.168	.109	.064	.034	.016	.007	.002	.001	.000	.000	.000	.000	.000	.000
	3	.000	.017	.085	.172	.236	.258	.240	.195	.142	.092	.054	.028	.012	.005	.001	.000	.000	.000	.000	.000
	4	.000	.002	.021	.068	.133	.194	.231	.237	.213	.170	.121	.076	.042	.020	.008	.002	.001	.000	.000	.000
	5	.000	.000	.004	.019	.053	.103	.158	.204	.227	.223	.193	.149	.101	.059	.029	.011	.003	.001	.000	.000
	6	.000	.000	.000	.004	.016	.040	.079	.128	.177	.212	.226	.212	.177	.128	.079	.040	.016	.004	.000	.000
	7	.000	.000	.000	.001	.003	.011	.029	.059	.101	.149	.193	.223	.227	.204	.158	.103	.053	.019	.004	.000
	8	.000	.000	.000	.000	.001	.002	.008	.020	.042	.076	.121	.170	.213	.237	.231	.194	.133	.068	.021	.002
	9	.000	.000	.000	.000	.000	.000	.001	.005	.012	.028	.054	.092	.142	.195	.240	.258	.236	.172	.085	.017
	10	.000	.000	.000	.000	.000	.000	.000	.001	.002	.007	.016	.034	.064	.109	.168	.232	.283	.292	.230	.099
	11	.000	.000	.000	.000	.000	.000	.000	.000	.000	.001	.003	.008	.017	.037	.071	.127	.206	.301	.377	.341

(continued)

TABLE 13.6 The Cumulative Binomial Probability Distribution Table (*Continued*)

n	*r*	.01	.05	.10	.15	.20	.25	.30	.35	.40	.45	.50	.55	.60	.65	.70	.75	.80	.85	.90	.95
	12	.000	.000	.000	.000	.000	.000	.000	.000	.000	.000	.000	.001	.002	.006	.014	.032	.069	.142	.282	.540
15	0	.860	.463	.206	.087	.035	.013	.005	.002	.000	.000	.000	.000	.000	.000	.000	.000	.000	.000	.000	.000
	1	.130	.366	.343	.231	.132	.067	.031	.013	.005	.002	.000	.000	.000	.000	.000	.000	.000	.000	.000	.000
	2	.009	.135	.267	.286	.231	.156	.092	.048	.022	.009	.003	.001	.000	.000	.000	.000	.000	.000	.000	.000
	3	.000	.031	.129	.218	.250	.225	.170	.111	.063	.032	.014	.005	.002	.000	.000	.000	.000	.000	.000	.000
	4	.000	.005	.043	.116	.188	.225	.219	.179	.127	.078	.042	.019	.007	.002	.001	.000	.000	.000	.000	.000
	5	.000	.001	.010	.045	.103	.165	.206	.212	.186	.140	.092	.051	.024	.010	.003	.001	.000	.000	.000	.000
	6	.000	.000	.002	.013	.043	.092	.147	.191	.207	.191	.153	.105	.061	.030	.012	.003	.001	.000	.000	.000
	7	.000	.000	.000	.003	.014	.039	.081	.132	.177	.201	.196	.165	.118	.071	.035	.013	.003	.001	.000	.000
	8	.000	.000	.000	.001	.003	.013	.035	.071	.118	.165	.196	.201	.177	.132	.081	.039	.014	.003	.000	.000
	9	.000	.000	.000	.000	.001	.003	.012	.030	.061	.105	.153	.191	.207	.191	.147	.092	.043	.013	.002	.000
	10	.000	.000	.000	.000	.000	.001	.003	.010	.024	.051	.092	.140	.186	.212	.206	.165	.103	.045	.010	.001
	11	.000	.000	.000	.000	.000	.000	.001	.002	.007	.019	.042	.078	.127	.179	.219	.225	.188	.116	.043	.005
	12	.000	.000	.000	.000	.000	.000	.000	.000	.002	.005	.014	.032	.063	.111	.170	.225	.250	.218	.129	.031
	13	.000	.000	.000	.000	.000	.000	.000	.000	.000	.001	.003	.009	.022	.048	.092	.156	.231	.286	.267	.135
	14	.000	.000	.000	.000	.000	.000	.000	.000	.000	.000	.000	.002	.005	.013	.031	.067	.132	.231	.343	.366
	15	.000	.000	.000	.000	.000	.000	.000	.000	.000	.000	.000	.000	.000	.002	.005	.013	.035	.087	.206	.463
16	0	.851	.440	.185	.074	.028	.010	.003	.001	.000	.000	.000	.000	.000	.000	.000	.000	.000	.000	.000	.000
	1	.138	.371	.329	.210	.113	.053	.023	.009	.003	.001	.000	.000	.000	.000	.000	.000	.000	.000	.000	.000
	2	.010	.146	.275	.277	.211	.134	.073	.035	.015	.006	.002	.001	.000	.000	.000	.000	.000	.000	.000	.000
	3	.000	.036	.142	.229	.246	.208	.146	.089	.047	.022	.009	.003	.001	.000	.000	.000	.000	.000	.000	.000
	4	.000	.006	.051	.131	.200	.225	.204	.155	.101	.057	.028	.011	.004	.001	.000	.000	.000	.000	.000	.000
	5	.000	.001	.014	.056	.120	.180	.210	.201	.162	.112	.067	.034	.014	.005	.001	.000	.000	.000	.000	.000
	6	.000	.000	.003	.018	.055	.110	.165	.198	.198	.168	.122	.075	.039	.017	.006	.001	.000	.000	.000	.000
	7	.000	.000	.000	.005	.020	.052	.101	.152	.189	.197	.175	.132	.084	.044	.019	.006	.001	.000	.000	.000
	8	.000	.000	.000	.001	.006	.020	.049	.092	.142	.181	.196	.181	.142	.092	.049	.020	.006	.001	.000	.000
	9	.000	.000	.000	.000	.001	.006	.019	.044	.084	.132	.175	.197	.189	.152	.101	.052	.020	.005	.000	.000
	10	.000	.000	.000	.000	.000	.001	.006	.017	.039	.075	.122	.168	.198	.198	.165	.110	.055	.018	.003	.000
	11	.000	.000	.000	.000	.000	.000	.001	.005	.014	.034	.067	.112	.162	.201	.210	.180	.120	.056	.014	.001

	12	.000	.000	.000	.000	.000	.000	.000	.001	.004	.011	.028	.057	.101	.155	.204	.225	.200	.131	.051	.006
	13	.000	.000	.000	.000	.000	.000	.000	.000	.001	.003	.009	.022	.047	.089	.146	.208	.246	.229	.142	.036
	14	.000	.000	.000	.000	.000	.000	.000	.000	.000	.001	.002	.006	.015	.035	.073	.134	.211	.277	.275	.146
	15	.000	.000	.000	.000	.000	.000	.000	.000	.000	.000	.000	.001	.003	.009	.023	.053	.113	.210	.329	.371
	16	.000	.000	.000	.000	.000	.000	.000	.000	.000	.000	.000	.000	.000	.001	.003	.010	.028	.074	.185	.440
20	0	.818	.358	.122	.039	.012	.003	.001	.000	.000	.000	.000	.000	.000	.000	.000	.000	.000	.000	.000	.000
	1	.165	.377	.270	.137	.058	.021	.007	.002	.000	.000	.000	.000	.000	.000	.000	.000	.000	.000	.000	.000
	2	.016	.189	.285	.229	.137	.067	.028	.010	.003	.001	.000	.000	.000	.000	.000	.000	.000	.000	.000	.000
	3	.001	.060	.190	.243	.205	.134	.072	.032	.012	.004	.001	.000	.000	.000	.000	.000	.000	.000	.000	.000
	4	.000	.013	.090	.182	.218	.190	.130	.074	.035	.014	.005	.001	.000	.000	.000	.000	.000	.000	.000	.000
	5	.000	.002	.032	.103	.175	.202	.179	.127	.075	.036	.015	.005	.001	.000	.000	.000	.000	.000	.000	.000
	6	.000	.000	.009	.045	.109	.169	.192	.171	.124	.075	.037	.015	.005	.001	.000	.000	.000	.000	.000	.000
	7	.000	.000	.002	.016	.055	.112	.164	.184	.166	.122	.074	.037	.015	.005	.001	.000	.000	.000	.000	.000
	8	.000	.000	.000	.005	.022	.061	.114	.161	.180	.162	.120	.073	.035	.014	.004	.001	.000	.000	.000	.000
	9	.000	.000	.000	.001	.007	.027	.065	.116	.160	.177	.160	.119	.071	.034	.012	.003	.000	.000	.000	.000
	10	.000	.000	.000	.000	.002	.010	.031	.069	.117	.159	.176	.159	.117	.069	.031	.010	.002	.000	.000	.000
	11	.000	.000	.000	.000	.000	.003	.012	.034	.071	.119	.160	.177	.160	.116	.065	.027	.007	.001	.000	.000
	12	.000	.000	.000	.000	.000	.001	.004	.014	.035	.073	.120	.162	.180	.161	.114	.061	.022	.005	.000	.000
	13	.000	.000	.000	.000	.000	.000	.001	.005	.015	.037	.074	.122	.166	.184	.164	.112	.055	.016	.002	.000
	14	.000	.000	.000	.000	.000	.000	.000	.001	.005	.015	.037	.075	.124	.171	.192	.169	.109	.045	.009	.000
	15	.000	.000	.000	.000	.000	.000	.000	.000	.001	.005	.015	.036	.075	.127	.179	.202	.175	.103	.032	.002
	16	.000	.000	.000	.000	.000	.000	.000	.000	.000	.001	.005	.014	.035	.074	.130	.190	.218	.182	.090	.013
	17	.000	.000	.000	.000	.000	.000	.000	.000	.000	.000	.001	.004	.012	.032	.072	.134	.205	.243	.190	.060
	18	.000	.000	.000	.000	.000	.000	.000	.000	.000	.000	.000	.001	.003	.010	.028	.067	.137	.229	.285	.189
	19	.000	.000	.000	.000	.000	.000	.000	.000	.000	.000	.000	.000	.000	.002	.007	.021	.058	.137	.270	.377
	20	.000	.000	.000	.000	.000	.000	.000	.000	.000	.000	.000	.000	.000	.000	.001	.003	.012	.039	.122	.358

At first glance, the result of Example 13.30 may seem surprising, but if we give the matter further thought, the reasonability of the result becomes evident. The game is heavily weighted toward winning, and we are wondering what the probability is that we will win at most 10 times. Considering the high level of probability of winning any individual trial, we can conclude that it would be surprising to win at most a mere 10 games of the 20 played.

EXAMPLE 13.31 A biased, or unfair, coin is tossed 15 times. The probability of obtaining heads on any particular toss of the coin is 0.3. What is the probability that the number of heads that come up will be less than 6?

SOLUTION

In this case, "success" is defined by a result of heads occurring. Here, we have a fixed number of trials, 15, and since no one flip influences the result on any other flip, the trials of the experiment are independent, and the probability of success on each individual trial is fixed, and thus we are looking at a problem whose results should follow a binomial distribution.

Since the table results for the cumulative binomial distribution tell us the probability that the number of successes is less than or equal to a particular value x, when we consider the probability that the number of heads that come up will be less than 6, we must view this as the probability that the number of heads coming up will be less than or equal to 5. Referencing our chart, with $n = 15$, we find the probability is .7216, or 72.16%, that the number of heads coming up will be less than 6.

EXAMPLE 13.32 Suppose a quality control engineer is in charge of testing whether 90% of the DVD players produced by his company conform to specifications. To do this, he randomly selects 12 DVD players from each day's production and passes the day's production if no more than one DVD player fails to meet specifications. Should this not happen, then the entire day's production is tested.

(a) What is the probability that the engineer incorrectly passes a day's production as acceptable if only 80% of the day's DVD players actually conform to specification?

(b) What is the probability that the engineer needlessly requires the entire day's production to be tested if in fact 90% of the DVD players conform to specifications?

SOLUTION

(a) In this case, we are considering a situation in which the number of trials, 12, is fixed, and success can be defined as determining whether a particular DVD player conforms to specifications. Since no DVD player being within specifications influences whether another will be within those specifications, this qualifies as a binomial experiment. In this case, the engineer will pass the day's production as acceptable if no more than one DVD player fails to meet specifications, and this is equivalent to saying that we want to find the probability that the number of successes (as we've defined them here) is at least 11. We can calculate this using our table if we compute first the probability that the number of successes will be less than or equal to 10 and then subtracting this result from 1. Referencing our chart, we find the probability of finding 10 or fewer DVD players that meet specifications to be 0.7251, and therefore the probability that 11 or more of the players will be within specifications is $1 - 0.7251 = 0.2749$. This is a good result, since it means that we have only a 27.49% chance that the engineer will incorrectly pass the day's production as acceptable under these conditions.

(b) Here, we must recall that the engineer will require the entire day's production to be tested if we should find the number of successes to be less than or equal to 10. In this example, the number of trials is still 12, but now the probability of success is 0.90. Referring to our chart, we find that the probability the engineer will needlessly require the entire day's run to be tested will be 0.3410, or 34.1%.

Relating Binomial to Normal Distributions

For values of n sufficiently large, it turns out that the binomial distribution is approximately equal to the normal distribution. For reasons beyond the scope of this text, we will point out that this generally holds if the product np is at least 10, but the higher the value of n, the greater the accuracy of the approximation. Recall that n is the number of trials and that p is the probability of success for any individual trial. The usefulness of this fact lies in the realization that there is extensive research done on the normal distribution, and the calculations in that distribution are somewhat easier than those in the binomial distribution.

Of course, in order to use the normal distribution, we will need to find a z-score for our particular data value, and for this we must know the mean and the standard deviation for binomial distributions, which we have stated are $\mu = np$ and $\sigma = \sqrt{npq}$.

Suppose we know that 10% of all men are bald (let's set aside whether this is true and the ambiguity of how one would categorize being bald). What is the probability that fewer than 100 in a sample of 818 men are bald? Our cumulative binomial probability table does not accommodate a sample size of 818, but since that is a large number, we will use a normal approximation, since $np = 81.8$ is clearly larger than 10. The value of the mean and standard deviation, according to our formulas, are 81.8 and approximately 8.5802, respectively, and the z-score for $x = 100$ would be $z = \dfrac{100 - 81.8}{8.5802} \approx 2.12$. Using the standard normal distribution chart, we find the probability that fewer than 100 men of the 818 will be bald is ≈ 0.9830, or 98.3%. Were we to have access to a sufficiently large cumulative binomial distribution table, it turns out that the table would show that the actual binomial probability is 0.9833, which is obviously very close to that which we saw in the normal distribution result.

Exercises

Use the binomial probability table to find the probability of at most x successes given the probability of a single trial is p, where n is the number of trials.

1. $n = 4, x = 3, p = 0.20$
2. $n = 6, x = 3, p = 0.30$
3. $n = 10, x = 2, p = 0.40$
4. $n = 5, x = 2, p = 0.35$
5. $n = 30, x = 10, p = 0.20$
6. $n = 12, x = 5, p = 0.25$
7. $n = 30, x = 5, p = 0.20$
8. $n = 60, x = 5, p = 0.04$

Find the mean, μ, for the binomial distribution which has the stated values of n and p.

9. $n = 36, p = 0.2$
10. $n = 67, p = 0.7$
11. $n = 22, p = 0.6$
12. $n = 2000, p = 0.5$

Find the standard deviation, σ, for the binomial distributions having the stated values of n and p.

13. $n = 50, p = 0.2$
14. $n = 47, p = 0.6$
15. $n = 700, p = 0.7$
16. $n = 500, p = 0.65$

13.7 Linear Correlation and Regression

The word "correlation" indicates the existence of a mutual relationship between two or more things, or interdependence among those things. As we'll see, the degree to which two variables correlate is expressed using a quantity called the **correlation coefficient**, which we will define shortly. **Linear regression** analyzes the manner in which two variables, X and Y, and seeks a line of "best fit" through the data when plotted on an X,Y coordinate plane. The line's equation should allow us to predict the value of Y given the value of X and vice versa.

Linear Correlation

Linear correlation refers to a relationship that exists between the two variables X and Y in the particular instance where every value of x is associated with just one value of y. An example of this concept would be the score of a particular person on the mathematics component of the SAT, paired with the corresponding score of that particular person on the verbal component of the SAT. Another would be the age of a particular individual, paired with the corresponding number of traffic tickets the individual has received.

When we say that the correlation is linear, we essentially mean that one of two situations exists: if you have more of variable X, then you have more of variable Y, or (conversely) if you have more of variable X, then you have less of variable Y. When the relationship is "more and more," that is, the first case described, we say there is a positive correlation. When the relationship is "more and less," we say there is a negative correlation between the variables.

EXAMPLE 13.33 There is a linear correlation between the number of minutes spent daydreaming in class by a particular student and the score attained by that student on a particular examination. Is this correlation positive or negative?

SOLUTION

The correlation here would likely be negative. If a student increases the amount of time she daydreams in class, her score on the examination is likely to decrease. Thus, if we increase the variable expressing the number of minutes daydreaming, we will decrease the variable expressing the score by the student on the examination.

Linear correlation between variables is something we may suspect to exist or not exist, but it is also something we can examine and quantify. Suspicion of the existence of linear correlation is often a result of examination of the graph of a particular set of data on an X, Y plane. If visual inspection of the information suggests a pattern in which the data points generally form a linear path or alignment, we might suspect that a linear correlation exists. If the points do not form a linear path (they could form some other, nonlinear path, or they may be a patternless jumble in the plane), we might suspect that **zero correlation** would exist (Figures 13.9 to 13.11).

The perfect situation (highly unlikely) of linear correlation would be the situation where the data would line up along a particular straight line in the plane. In this case, we would describe the arrangement of data as an illustration of perfect correlation, which would represent the ultimate linear correlation. Real-world situations rarely exhibit this perfection, and it is usually encountered only in situations depicting elementary physical principles, such as the relationship between voltage and current in a circuit possessing constant resistance.

Measuring Correlation

The principal method for quantifying the degree of linear correlation is the **Pearson product-moment correlation coefficient**, often referred to simply as the correlation coefficient. The correlation coefficient is a single numerical value, telling us whether a relationships exists between the variables, whether the relationship is positive or negative, and the strength of that relationship. Symbolized by the letter r, its values range from -1 to 1, inclusive. The value $r = 1$ indicates a perfect positive linear correlation, $r = -1$ a perfect negative linear correlation, and $r = 0$ an utter lack of linear correlation. Any value of

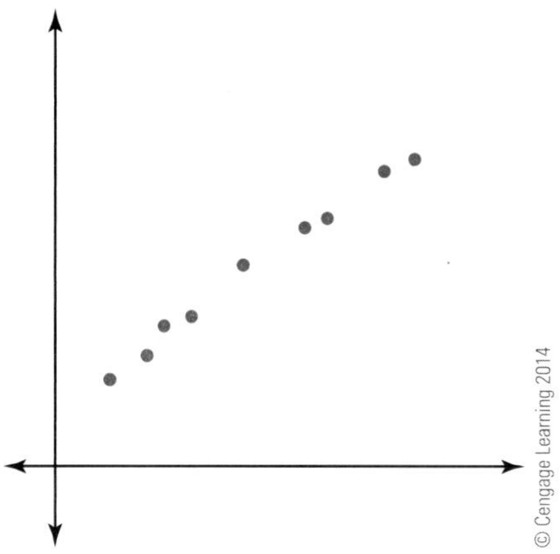

Figure 13.9 Likely positive linear correlation

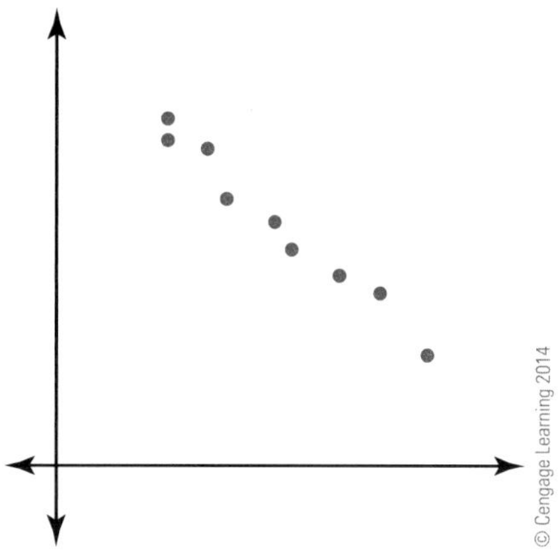

Figure 13.10 Likely negative linear correlation

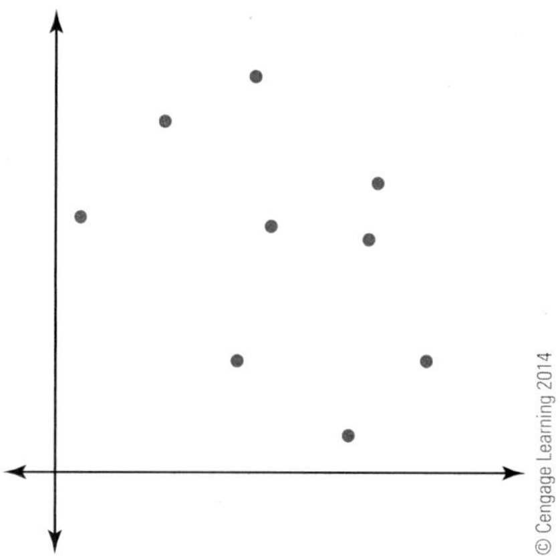

Figure 13.11 Likely zero linear correlation

r falling between 0 and 1 represents a degree of positive linear correlation, with values of r closer to 1 indicating increasing linear correlation over those values of r closer to 0. Likewise, those negative values between 0 and that are closer to -1 describe an increasing degree of negative linear correlation.

A related measure of linear correlation, the value of r^2, is called the coefficient of determination, which has nonnegative values between 0 and 1, inclusive. The benefit of using the coefficient of determination is that it offers a scale for determining the "strength" of any particular correlation, with the weakness that it masks the type (positive or negative) of the correlation. It is common to use the value $(100)(r^2)\%$ to express the strength of a particular correlation. We will exhibit this and expand on the notion once we have explained the method for calculation of r and hence of r^2. Graphing calculators and computer software can, through their built-in regression function, produce the Pearson product-moment correlation coefficient, but we should develop an understanding of where the value arises and how it is computed by hand before we lean on technology too much.

It is, quite fortunately, rare that we are asked to calculate the value of r by hand. While the formula is not complicated, it is tedious to use but lends itself easily to programming into a computer.

Mathematically, the formula for the computation of r is as follows:

$$r = \frac{n\sum xy - \left(\sum x\right)\left(\sum y\right)}{\sqrt{n\sum x^2 - \left(\sum x\right)^2}\sqrt{n\sum y^2 - \left(\sum y\right)^2}}$$

Here, n is the number of pairs of data, and x and y are the individual values of the variables X and Y. A value of $|r|$ that is greater than 0.8 is typically considered "strong," while a value of $|r|$ less than 0.5 is considered "weak." This rule is not hard and fast but is a good rule of thumb.

Consider the following situation, constructed using entirely fictitious data. It is natural to speculate that the compensation of an employee of an organization is related to the time of employment. We study the employees of the organization and find the information shown in Table 13.7. Once we have compiled our information, the next logical step would be to construct a scatter plot of the information, as shown in Figure 13.12.

TABLE 13.7 Employees' Wages and Number of Years with Organization

Employee	Years with Organization	Hourly Wage (Dollars)
A	7	52
B	12	49
C	13	50
D	18	43
E	21	39
F	24	38
G	26	30
H	30	29
I	36	19

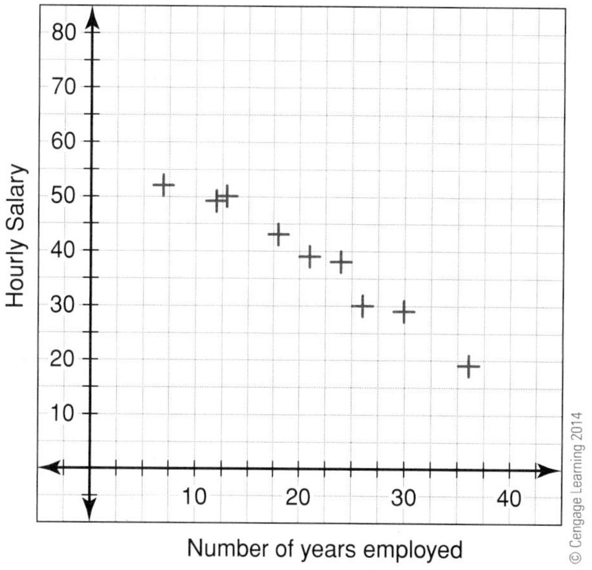

Figure 13.12 Relationship between years employed and hourly salary

Note

We must be very careful not to draw inferences from our data, such as a claim that time with the company results in lowering salaries. We cannot assume that one causes the other! All we will have established is that a relationship exists between the variables.

Observe that the points in the scatter plot appear to align linearly, falling from left to right in the graph. We are now prepared to determine if such a relationship exists and how strong that relationship would be. Observe that, if true, the situation is highly unusual and merits investigation: the workers with the longest tenure are paid the lowest salary!

Let's do the computations for this illustration. Recall that the formula for r was

$$r = \frac{n\sum xy - \left(\sum x\right)\left(\sum y\right)}{\sqrt{n\sum x^2 - \left(\sum x\right)^2}\sqrt{n\sum y^2 - \left(\sum y\right)^2}}$$

To perform the calculation, we'll arrange things as shown in Table 13.8, with columns for all the quantities we need.

TABLE 13.8 Calculating the Correlation Coefficient

x	y	x^2	y^2	xy
7	52	49	2704	364
12	49	144	2401	588
13	50	169	2500	650
18	43	324	1849	774
21	39	441	1521	819
24	38	576	1444	912
26	30	676	900	780
30	29	900	841	870
36	19	1296	361	684

We'll now find the sum of each column, which we can then insert into the formula for r. Here, we find $\sum x = 187$, $\sum y = 349$, $\sum x^2 = 4575$,

$\sum y^2 = 14521$, $\sum xy = 6441$. Since the number of data values is 9, we can now make our substitutions:

$$r = \frac{9(6441) - (187)(349)}{\sqrt{9(4575) - (187)^2}\sqrt{9(14{,}521) - (349)^2}}$$

$$r \approx \frac{-7294}{7426.91}$$

$$r \approx -0.98$$

Observe that our suspicions were correct and that there is a nearly perfect negative linear correlation between the two data sets. Moreover, we had mentioned that r^2 was the coefficient of determination, and in this case r^2 is approximately 0.96, and thus the strength of the relationship is 96%. This indicates that about 96% of the value of the salary is directly related to the time they have spent in the organization. In other words, about 96% of the total variation in the salary of the employee can be explained by the linear relationship between time of employment and salary, with the remaining 4% unexplained.

Linear Regression

Linear regression is a procedure that uses the known existence of a linear correlation between two variables to find a line of best fit that will allow you to make predictions about the two variables. Now that we have in our hands the ability to calculate the correlation coefficient and the coefficient of determination, we are able to apply those values to determine this line of best fit.

A **regression line** for variable Y in terms of variable X will be the model that allows us to perform our process of prediction. There are several methods one can use to determine a regression line, but the technique we will use here is called the least-squares regression method, which is probably the method in widest use. In any model for prediction, there will, of necessity, be some error. However, if our correlation was found to be strong (an r value greater than 0.8), we should find that the model will be surprisingly accurate. If the value of $r = 0$, remember, there is no linear correlation at all, and thus we may as well make arbitrary guesses.

Let's revisit the scatter plot we created earlier, showing the relationship between time with an organization and hourly wage (Figure 13.13). There are several facts we will need in order to obtain the equation of the regression line. Keep in mind that most of the time, you will be using a computer or graphing calculator to find the equation of the line, but we can demonstrate the method with a relatively small data set, such as our ongoing compensation example.

We will need to know r (the correlation coefficient), σ_x (the standard deviation of the X variable), σ_y (the standard deviation of the Y variable), μ_x (the mean of the X variable), and μ_y (the mean of the Y variable). With those values in hand, we have the following relationship (presented without proof!):

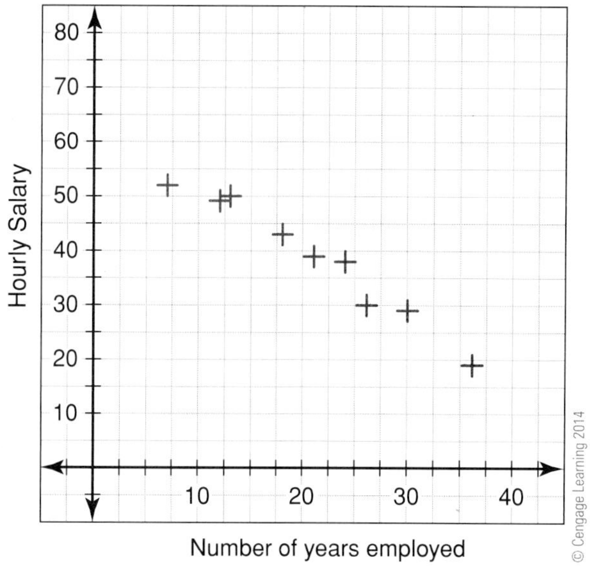

Figure 13.13 Relationship between years employed and hourly salary

$$Y - \mu_y = r\left(\frac{\sigma_y}{\sigma_x}\right)(X - \mu_x)$$

Since all the quantities involved are known to us, it is merely a matter of inserting values into the expression. Here, $r = -0.98$, $\sigma_x \approx 9.28$, $\sigma_y \approx 11.1$, $\mu_x \approx 20.78$, and $\mu_y \approx 38.78$, so we obtain the following equation for the line of regression:

$$Y - 38.78 = -1.17(X - 20.78)$$
$$Y - 38.78 = -1.17X + 24.31$$
$$Y = -1.17X + 63.09$$

This last is the regression equation for our data. The line has a number of useful properties we should mention, though we should remind you that, in practice, one uses technology to calculate the regression line. The line found in the method minimizes the sum of squared differences between the observed values of *Y* and the values that would be predicted by the model. The regression line passes through the mean of the *X* values and the mean of the *Y* values, and the slope of the regression line is the average change in the dependent variable *Y* for a one-unit change in the value of *X*.

EXAMPLE 13.34

Suppose a group of students take a composition class and then submit written papers to a national writing contest. The relationship between the student's class average in the composition class and their performance (on a scale of 100) in the writing contest is shown in Table 13.9. Find the equation of the regression line for the data.

SOLUTION

In order to find the regression line's equation, we must take the time to compile the following information: *r* (the correlation coefficient), σ_x (the standard deviation of the *X* variable), σ_y (the standard deviation of the *Y* variable),

TABLE 13.9 Students' Composition Class Averages and Writing Contest Scores

Student	Composition Class Average	Writing Contest Score
A	96	87
B	91	88
C	83	79
D	82	76
E	78	69

μ_x (the mean of the X variable), and μ_y (the mean of the Y variable). With these facts, we can substitute into the least-squares regression line formula, $Y - \mu_x = r\left(\frac{\sigma_y}{\sigma_x}\right)(X - \mu_x)$, and find our regression equation.

A straightforward but tedious set of computations reveals the following facts: $r = 0.937$, $\sigma_x \approx 7.31$, $\sigma_y \approx 7.92$, $\mu_x \approx 86$, and $\mu_y \approx 79.8$. Substitution into our formula reveals the regression equation to be $Y = 1.014X - 7.406$.

What use is this equation we have found? Since the correlation coefficient was found to be 0.937, the correlation between the two data values is strong, and therefore the regression equation should model the actual data values well. Were another student from the composition class to participate in the writing contest and that student's class grade average were 70, for example, we could predict with relative certainty that their performance on the writing contest would have been $Y = 1.014(70) - 7.406 = 63.574$. Thus, the equation allows us to make predictions, assuming that the correlation is strong.

Exercises

For the following problems, determine the linear correlation coefficient, *r*, for the ordered pairs.

1. (4,38), (3,42), (11,29), (5,31), (9,28), (6,15)
2. (3,98), (2,96), (3,88), (2,87), (4,61), (4,77)
3. (186,85), (189,85), (190,86), (193,81), (191,88)
4. (81,44), (72,59), (77,60), (86,39)
5. (2,1), (3,3), (4,2), (4,4), (5,4), (6,4), (6,6), (8,7), (10,9)
6. (32,23), (43,40), (39,43), (57,67), (71,40), (75,81), (85,82)

In each of the following, find the equation of the regression line for the data set. Note the data sets are those in problems 1–6.

7. (4,38), (3,42), (11,29), (5,31), (9,28), (6,15)
8. (3,98), (2,96), (3,88), (2,87), (4,61), (4,77)
9. (186,85), (189,85), (190,86), (193,81), (191,88)
10. (81,44), (72,59), (77,60), (86,39)
11. (2,1), (3,3), (4,2), (4,4), (5,4), (6,4), (6,6), (8,7), (10,9)
12. (32,23), (43,40), (39,43), (57,67), (71,40), (75,81), (85,82)

Use the given data to find the equation of the regression line. Round the final coefficients to three significant digits, as needed.

13.

x	y
2	7
4	11
5	13
6	20

14.

x	y
0	8
3	2
4	6
5	9
12	12

15.

x	y
6	2
8	4
20	13
28	20
36	30

16.

x	y
24	15
26	13
28	30
30	16
32	24

Find the correlation coefficient for the given data.

17. The scores of 10 students on their midterm exam (x) and their final exam (y) yielded the following data. Find the correlation coefficient for the data.

$$\sum x = 638$$
$$\sum y = 690$$
$$\sum x^2 = 43{,}572$$
$$\sum y^2 = 49{,}014$$
$$\sum xy = 44{,}636$$

18. The IQ scores (x) of 20 students whose grade point average in college (y) yielded the following data. Find the correlation coefficient for the data.

$$\sum x = 1090$$
$$\sum y = 35.2$$
$$\sum x^2 = 110{,}700$$
$$\sum y^2 = 103.9$$
$$\sum xy = 3452.7$$

Summary

In this chapter, you learned about:

- Selecting and using various types of sampling techniques.
- Constructing and interpreting statistical graphs.
- Recognizing and calculating measures of central tendency and measures of dispersion.
- Understanding and performing calculations with normal and binomial distributions.
- Establishing a linear correlation and performing linear regression.

Glossary

Bernoulli trial A binomial experiment having a fixed number of trials with each trial having only two possible outcomes, characterized as either "success" or "failure."

bimodal data set A data set having two modes.

binomial experiment An experiment having a fixed number of trials in which each trial has two possible outcomes (success or failure) and in which the trials are independent of one another.

binomial probability distribution The probability distribution of a binomial random variable.

box-and-whisker plot A graphical tool to present the concentration of the data within a sample.

cluster sample A sample in which the population is subdivided into clusters from which a random sampling is taken.

convenience sample A sample constructed using readily obtained information.

correlation coefficient A quantity that depicts how closely a best-fit regression line fits a set of data.

data Information about a population or sample.

deviation The difference between a particular data value and the mean of the data set.

empirical rule The fact that in a normal distribution, approximately 68% of all the data lies within 1 standard deviation (on either side) of the mean, approximately 95% of all the data lies within 2 standard deviations (on either side) of the mean, and approximately 99.7% of all the data lies within 3 standard deviations (on either side) of the mean.

histogram A common tool for representing data using bars depicting data frequencies.

homogeneous population A population whose membership possesses relatively uniform characteristics.

interquartile range That part of the data lying between the first and the third quartile.

linear correlation A relationship between two variables X and Y in which every X value is associated with just one value of Y.

linear regression line The graph of a linear correlation.

mean The numerical average of a set of data.

median The central value of a data set when the data have been arranged in increasing order.

mode The data value occurring most often within an overall set of data.

nonprobability sampling method A technique of sampling characterized by all members of the population not having an equal opportunity for selection.

outlier A data value lying conspicuously far from the remaining data in the set.

parameter Information gathered about an entire population.

Pearson product-moment correlation coefficient See correlation coefficient.

population The collection of objects or people studied in a statistical experiment.

random sample A sample whose members are chosen randomly from the overall population in such a manner that each member of the population has the same opportunity for being selected as a member of the sample.

random variable The characteristic of the population being studied.

range The difference between the maximum and minimum values of a data set.

regression line A line fitted to the set of paired data in a regression analysis.

relative frequency The quotient of the frequency and the cardinality of the data set.

representative sample A sample whose members' characteristics closely resemble those of the population.

sample A subset of the population.

sampling The process of creating a sample.

standard deviation A measure of dispersion in a frequency distribution, equal to the square root of the mean of the squares of the deviations from the mean of the distribution.

standard normal distribution The normalized normal distribution.

statistic Information gathered about a sample of a population.

statistical graph A visual representation of statistical data.

strata The individual subgroups within a stratified sample.

stratified sampling A sample in which the population is subdivided into strata from which a random sampling is taken.

systematic sample A sample whose members are selected using some arbitrarily designed algorithm.

variance The average of the sums of the squares of the deviations from the mean in a data set.

***z*-score** A standardized value that allows us to transform any normal distribution into what we refer to as the standard normal distribution.

zero correlation The condition where a set of data points forms no observable linear pattern.

List of Equations

Pearson product-moment correlation coefficient:

$$r = \frac{n\sum xy - \left(\sum x\right)\left(\sum y\right)}{\sqrt{n\sum x^2 - \left(\sum x\right)^2}\sqrt{n\sum y^2 - \left(\sum y\right)^2}}$$

End-of-Chapter Problems

Identify what type of sampling is used in the following problems: random, stratified, systematic, cluster, or convenience.

1. A sample consists of 39 people from Ward 1, 49 people from Ward 2, 47 people from Ward 3, and 37 people from Ward 4 of a town. The respective populations are 394, 492, 467, and 372 people.

2. A researcher takes the list of town residents and interviews every 37th resident on the list, starting with the third.

In the following exercises, construct a histogram.

3. The frequency table below shows the number of classes missed in a given semester for a class of 25 people.

Classes Missed	Frequency
0	13
1	5
2	3
3	0
4	2
5	2

4. In a survey, 55 residents of a town were asked their age. The results are summarized in the frequency table below.

Age of Resident	Number of Residents
0–10	10
11–20	8
21–30	7
31–40	8
41–50	9
51–60	8
61–70	3
71 and above	2

Construct a stem-and-leaf plot for the given data.

5. Here are the ages of 35 college students chosen at random:

   21 18 19 17 25 22 21 23 19 18 18 20 22 21

   18 22 23 21 18 19 21 18 19 19 18 17 18 21

   18 46 21 22 18 19 17

6. A "Math for IT" class has 26 students. Here are the number of students who came to each class over the course of a semester:

   33 34 35 33 27 30 31 32 31 30 17 29 31 27 31 35 33 35 28 30 32 27 30 31 34 35

Use the data to create a box-and-whisker plot.

7. A number of students were surveyed and asked how many hours of television they watched per week. The answers were as follows:

   4 5 4 8 12 8 7 3 1 0 8 2 4 5 3 7 2 4 8 4 2 3 2 2 3 4 4 3 5 2 7 8 3 2 1 0 1 1 2
   3 4 6 7 7 8 3 2 2 8 3 3 4 3 3 5 5 8 2 7 6 5 3 2 5 7 8 3 2 1 8 7 6 5 3 2 8 2 1 0
   8 2 3 4 5 5 7 2

8. A number of professors were surveyed and asked how many hours of television they watched per week. The answers were as follows:

   1 2 0 1 3 1 1 3 3 3 2 0 0 1 3 1 2 1 0 0 0 1 1 3 1 2 8 3 2 0 0 1 1 0 3 2 4 1 3 2
   3 3 2 1 0 7 0 2 3 1 1 2 4 3 2 1 0 1 1 2 3 2 2 0 3 1 1 0 0 1 2 1 1 3 3 2 2 2 1 0
   1 2 0 1 2 3 1 2

Provide an appropriate response.

9. Given the following histogram, which shows the age ranges for a group of people, how many people were between 40 and 50? 70 and above? What is the interval length?

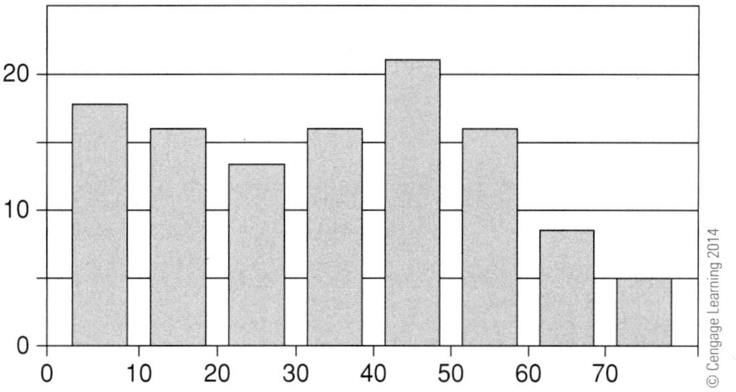

10. The following box-and-whisker plot shows the number of hours of commuting time per week of a set of people. What was the median value? What were the minimum, the maximum, the first quartile, the third quartile, and the interquartile range?

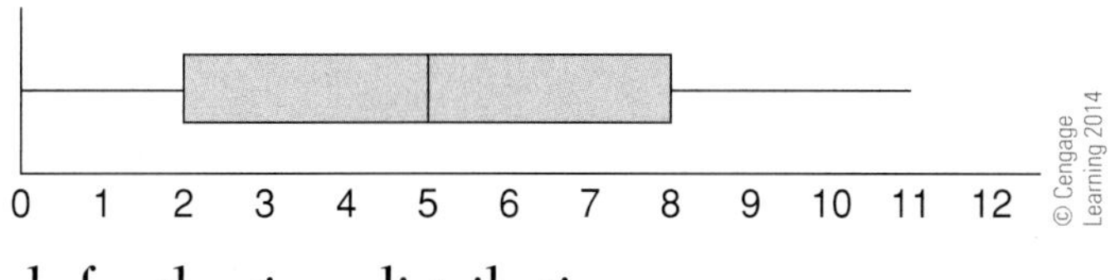

Find the mean, median, and mode for the given distributions.

11. 24, 30, 18, 30, 27, 22, 28, 18, 30, 30, 30, 22, 29, 20, 26
12. 11, 7, 10, 14, 9, 9, 12, 9, 14, 15, 12, 14, 13, 12, 8

In each of the following, find the range of the set of data.

13. 24, 30, 18, 30, 27, 22, 28, 18, 30, 30, 30, 22, 29, 20, 26
14. 11, 7, 10, 14, 9, 9, 12, 9, 14, 15, 12, 14, 13, 12, 8

In each of the following, find the variance.

15. 19, 14, 28, 22, 32, 25, 27, 16, 34, 27, 32
16. 11, 14, 13, 15, 16, 13, 14, 15, 9, 16, 16, 15, 15, 16, 13, 13, 14

In each of the following, find the sample standard deviation.

17. Use the data from Exercise 15.
18. Use the data from Exercise 16.

In each of the following, solve the problem.

19. A baseball pitcher has a selection of pitches clocked, and it is found that the pitches were at 124, 137, 119, 124, 135, 122, and 117 feet per second. What is the range, the mean, and the standard deviation?

20. A person is playing a particular computer game. She is trying to get past the first level, and on a sample of tries, it has taken her 3, 5, 4, 3, 7, 11, 3, 1, 2, and 8 attempts to make it. What are the range, mean, variance, and standard deviation?

For each of the following, find the percentage of the area under the standard normal curve.

21. $z = 0$ and $z = 1.23$

22. $z = 1.17$ and $z = 1.94$

For each of the following, use the table to find a *z*-score that fits the given condition.

23. 57% of the area under the standard normal curve is below the score.

24. 37% of the area under the standard normal curve is above the score.

For each of the following, find the area under the normal curve for the condition.

25. Find the percentage of the area under the curve between the mean and 2.37 standard deviations from the mean.

26. Find the percentage of the area under the curve between 1 standard deviation below the mean and 1 standard deviation above it.

For each of the following, use the standard normal curve table to find the closest *z*-score for the given condition.

27. 7% of the total area is to the left of z.

28. 33% of the total area is to the right of z.

For each of the following, find the *z*-score.

29. The shaded area is 0.0475.

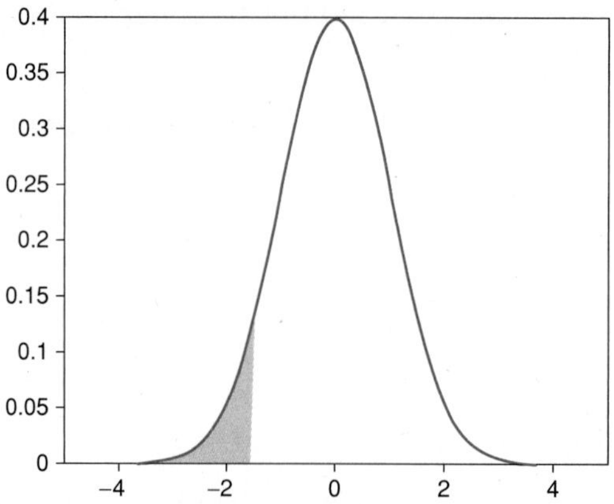

30. The shaded area is 0.7486.

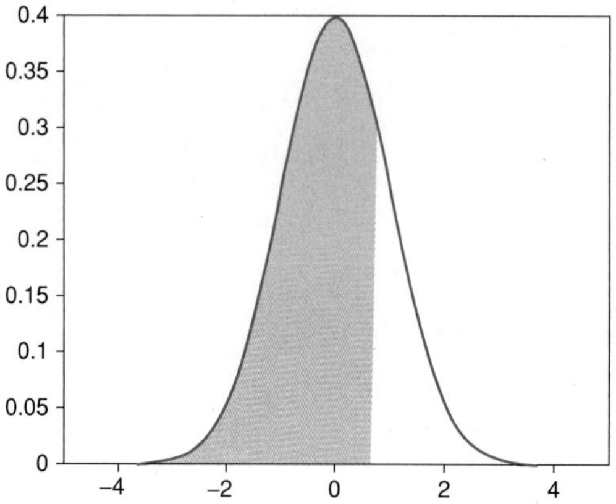

For each of the following, if z is a standard normal variable, find the probability.

31. The probability that z lies between 0 and 2.35
32. The probability that z lies between -1.34 and 0.87

For each of the following, use the cumulative binomial probability distribution table to find the probability of exactly x successes given the probability p of success on a single trial.

33. $n = 7, x = 4, p = 0.5$
34. $n = 12, x = 9, p = 0.75$

For each of the following, find the mean, μ, for the binomial distribution that has the stated values of n and p.

35. $n = 42, p = 0.43$
36. $n = 3552, p = 0.22$

For each of the following, find the standard deviation, σ, for the binomial distribution that has the stated values of n and p.

37. $n = 53, p = 0.1$
38. $n = 4276, p = 0.37$

For each of the following, find the value of the linear correlation coefficient r.

39. For the values in the following table:

x	0	0	19	14	28	22	32	25	0	0	27	16	0	34	27	32	0
y	11	14	13	15	16	13	14	15	9	16	16	15	15	16	13	13	14

40. For the values in the following table:

x	3	5	4	8	4	2	1
y	5	3	4	0	4	6	7

For each of the following, use the given data to find the equation of the regression line. Round the final values to three significant digits, if necessary.

41. Use the data from Exercise 39.
42. Use the data from Exercise 40.

For each of the following, state what kind of correlation, if any, the scatter plot indicates.

43.

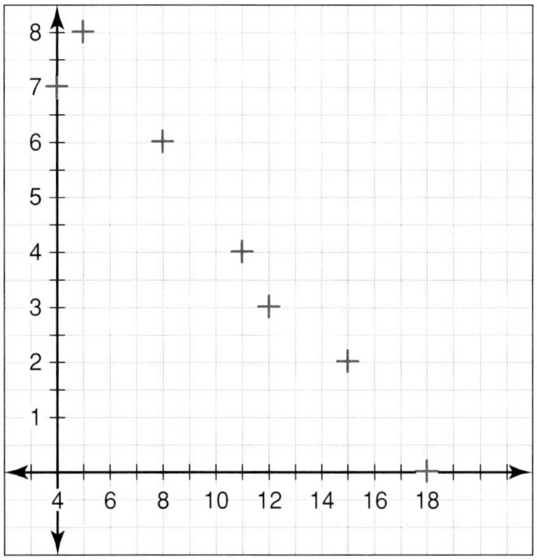

44. 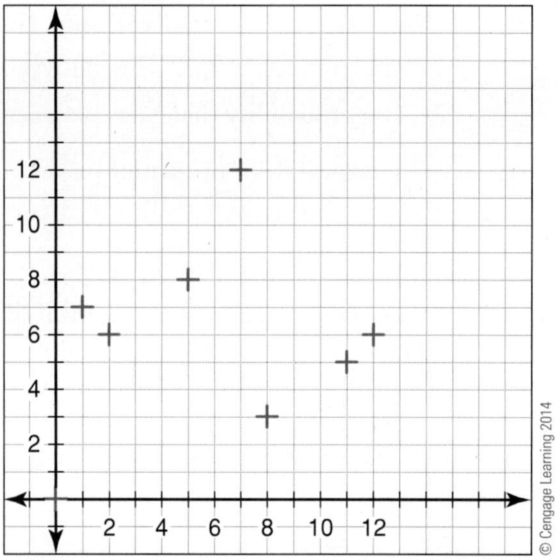

For each of the following, answer the questions.

45. In playing a particular game, the player received the following high scores. Each score consists of the money made (in millions) and the number of buildings constructed. Is there a correlation between the money made and the number of buildings constructed?

x	314	309	308	303	303	303	301	300	300	299
y	278	278	282	294	283	273	285	282	285	281

46. A summer camp takes in a certain number of campers and earns some profit. A sample of the number of campers in a given year together with the profit made (in 1000s of dollars) in that year follows. Is there a correlation between the number of campers and the profit made? If so, what is the equation of the regression line?

x	315	330	300	320	325	322	310	308	312	340
y	10	12	8	11	11	12	7	7	8	13

47. If we want to take a stratified sample of about 800 from a population of 3,000,000, where 53% identify themselves as white, 19% identify themselves as African American, 20% identify themselves as Asian American, and 8% identify themselves as Hispanic, how many of each group should be taken?

48. Given the following data, what is the correlation coefficient? If it is greater than 0.8 (or less than −0.8), what is the equation of the regression line?

x	8	7	8	3	4	5
y	5	6	4	11	9	7

49. For a female Portuguese water dog to win a point at a dog show in Division 1 (New England and New York), there must be two females entered; to win two points, there must be at least five females entered; and so

on. The numbers for males are similar; for one point there must be two males entered, for two points there must be three males entered, and so on. The table below summarizes the point schedule:

Points	1	2	3	4	5
Number of males	2	3	4	5	8
Number of females	2	5	8	12	19

Are the number of males and the number of females required to get a certain number of points correlated? If so, what is the equation of the regression line?

50. What is the equation of the regression line for the following data set?

x	11	8	12	9	13	10
y	20	17	25	17	28	19

Chapter 14 Graph Theory

In this chapter, when we refer to "graphs," we are no longer using the word as it applies to graphs of equations, say, in the Cartesian coordinate system. Rather, we are talking about a finite set of points that are connected by a line segments (or possibly curved segments, if line segments are impractical for whatever reason). Graphs, in this sense, can be used to represent real-life situations, with much of the "clutter" of extraneous information removed.

For instance, if we were to consider a salesperson traveling through a series of towns, we could obviously acquire a map of the region and examine his path on that document. We could, however, strip away the extraneous information, represent each of the towns by a point, and connect the points with line segments intended to depict his travel path (not necessarily to scale) from one town to another. As we'll see, this allows us to investigate the problem of minimizing travel distances, which would be of interest if we wanted to minimize our travel time or cost.

Objectives

When you have successfully completed the materials of this chapter, you will be able to:

- Identify graphs, and the terminology used when describing them.
- Understand the concepts of paths and circuits.
- Work with Euler paths and Euler circuits.
- Work with Hamilton paths and Hamilton circuits.
- Construct and use the specialized form of a graph: trees.

14.1 Graphs, Paths, and Circuits

Preliminaries of Graphs

A **graph**, in the sense we will consider here, is a finite set of points connected by line segments or arcs. The points are referred to as the **vertices** or **nodes** of the graph (each individual point is called a vertex), and the line segments or arcs are referred to as **edges**. Should an edge connect a vertex to itself, that edge is called a **loop**, and two vertices connected by an edge are said to be **adjacent vertices** within the graph. If a pair of vertices are connected by two distinct edges, then those edges are said to be **parallel**. The number of vertices in a particular graph is referred to as the order of the graph, and the number of edges in a graph is referred to as the graph's size.

There are many technical terms with which we can continue to specify individual features of graphs and thereby categorize them. For instance, a graph is a **complete graph** on a set of vertices if every pair of vertices are adjacent to one another. A graph that is not complete is referred to as an **incomplete graph**. A graph is called a **simple graph** if it has no loops or parallel edges. The list goes on and on.

We'll use capital letters like *A*, *B*, *C*, and so on to name the vertices of a graph and the two vertices connected by an edge to name the edges. That is, if you have the pair of vertices named *A* and *B*, then the edge connecting them would be *AB* or *BA* (unless otherwise specified, direction is immaterial in a graph, so *AB* and *BA* are considered equivalent). For loops, since the edge connects the vertex to itself, the naming convention would be *AA*, *BB*, and so on. Should two edges cross one another, the intersection is not considered a new vertex of the graph.

If we can represent a graph on a piece of paper in such a way that the edges only intersect at vertices, then the graph is called planar. When you view an integrated circuit, you will notice that the components of the chip are set up in such a manner that none of the connections cross one another, and therefore (in a sense) the chip can be viewed as a planar graph.

The utility of graphs lies in their simplicity. That is, through the use of a graph, we can remove all extraneous or potentially distracting information from our problem setup and visualize only the important aspects of the problem. Let's illustrate this with an introduction to a classic problem involving graphs, the Konigsberg bridge problem. Konigsberg was the capital city of what was once East Prussia (the territory of which was divided between Poland and the Soviet Union following World War II), and in that city (now Kaliningrad, Russia), the Pregel River was spanned by a series of bridges that connected several islands (Figure 14.1).

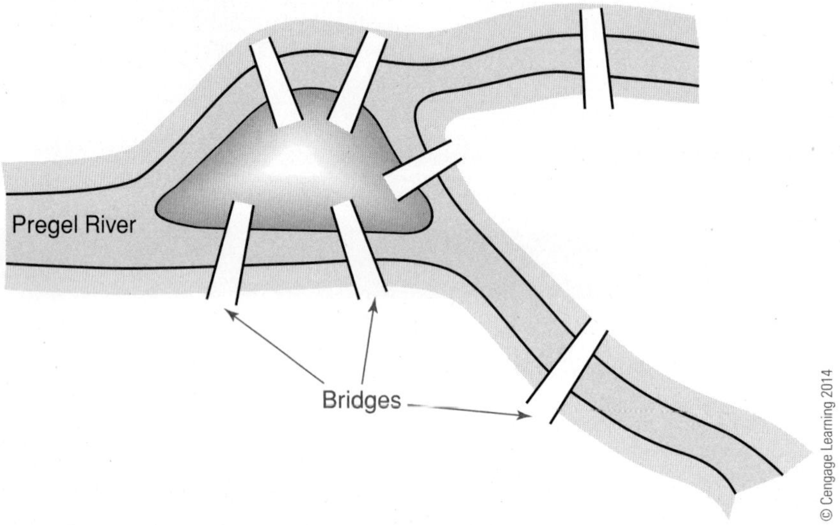

Figure 14.1 The Konigsberg bridge problem

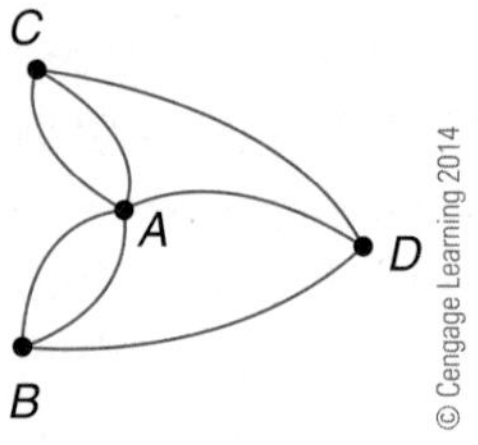

Figure 14.2 Graph depicting the Konigsberg bridge problem

The problem was to determine if anyone could traverse all seven of the bridges without crossing any of the bridges twice. The great Swiss mathematician Leonhard Euler simplified the problem to one involving a graph, as shown in Figure 14.2.

We'll revisit this problem shortly once we have a bit more information about graphs. For now, let's think of a more contemporary example and then continue with our introduction to the terminology we'll need.

EXAMPLE 14.1

Consider a communications network, such as a collection of computers used to communicate via email. The computers in the collection are considered the vertices of this graph (though they might be more appropriately thought of as "nodes" in this context), and the wiring connecting the computers is considered the edges in the graph.

The number of edges emanating from or connecting to a vertex is referred to as the degree of the vertex. This number could, if the vertex were isolated, be 0. Since a loop both emanates from and connects to a vertex, a loop counts as two edges connected to its vertex. Based on this definition of degree, we can classify vertices as being even or odd, depending on their degree. A graph is said to be a **regular graph** (or a ***k*-regular graph**) if every vertex has the same degree, k. Finally, a graph is said to be a **trivial graph** if it has only one vertex and otherwise is said to be **nontrivial graph**.

EXAMPLE 14.2

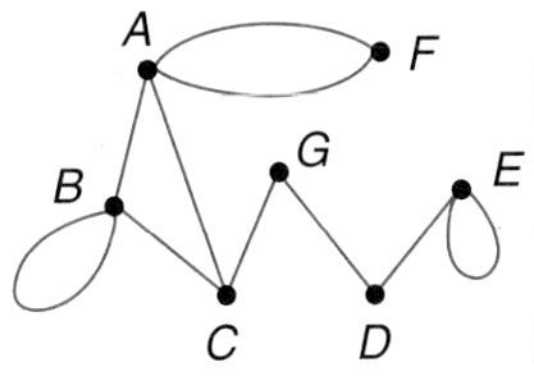

Figure 14.3 Graph for Example 14.2

Identify the degree of each vertex in the graph shown in Figure 14.3 and classify the vertices as even or odd. Also identify the order of the graph. We will consider the vertices in alphabetical order and address the questions of degree and even/odd first. In closing, we will identify the order of the graph. To keep things organized, we will express our answer in tabular form.

Vertex	Degree	Even/Odd
A	4	Even
B	4	Even
C	3	Odd
D	2	Even
E	3	Odd
F	2	Even
G	2	Even

There are a total of seven vertices, and therefore the order of the graph is 7.

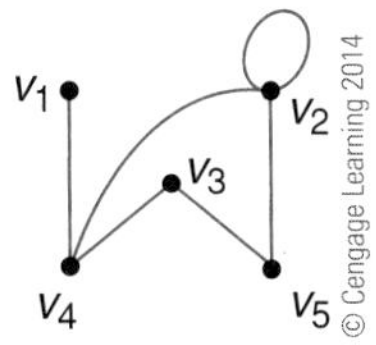

Figure 14.4 Graph for adjacency matrix

Adjacency Matrices

Because computers are particularly adept at working with matrices, it is convenient to describe graphs using such an array structure. If we were to designate the n vertices of a graph as $v_1, v_2, \ldots, v_n$, then we could construct an $n \times n$ adjacency matrix, where the i, j entry of the matrix is 1 if there exists an edge connecting v_i to v_j and 0 otherwise. Consider the graph shown in Figure 14.4. For this graph, the adjacency matrix would be

$$\begin{bmatrix} 0 & 0 & 0 & 1 & 0 \\ 0 & 1 & 0 & 1 & 1 \\ 0 & 0 & 0 & 1 & 1 \\ 1 & 1 & 1 & 0 & 0 \\ 0 & 1 & 1 & 0 & 0 \end{bmatrix}$$

Observe that an adjacency matrix for a graph that is not a directed graph must, by necessity, be a symmetric matrix. The reason this must be the case is that if vertex v_i is connected by an edge to vertex v_j, then it must also be the case that vertex v_j is connected by an edge to vertex v_i.

Here, as an illustration, what does the 1 in the (2, 2) position mean? The graph contains an edge connecting vertex v_2 to v_2. What does the 1 in the (1, 4) position mean? The graph contains an edge connecting vertex v_1 to v_4. For every entry of 1 in the matrix, identify its i, j position in the matrix and observe that an edge connects vertices v_i and v_j.

EXAMPLE 14.3

For the graph shown in Figure 14.5, produce the adjacency matrix. By inspection, we can find the adjacent vertices and insert 1s in those entries of

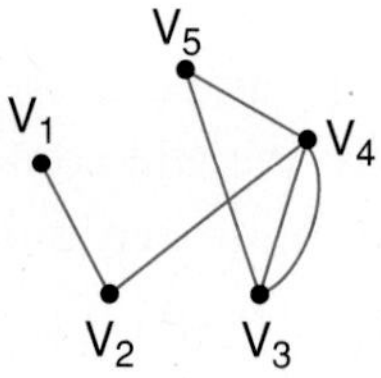

Figure 14.5
Graph for Example 14.3

the matrix whose row and column correspond to the adjacent vertices. In all other entries, we place a 0. Thus, in this situation, we obtain the matrix

$$\begin{bmatrix} 0 & 1 & 0 & 0 & 0 \\ 1 & 0 & 0 & 1 & 0 \\ 0 & 0 & 0 & 1 & 1 \\ 0 & 1 & 1 & 0 & 1 \\ 0 & 0 & 1 & 1 & 0 \end{bmatrix}$$

Paths

In a graph, a **walk** is a sequence of vertices connected by edges. There are special types of walks, called trails and paths, that are characterized by the edges they traverse and the vertices they visit. A **trail** is a walk in which no edges are repeated, and a **path** is a trail that has no repeated vertices. Thus, a path is a special type of walk wherein no edges are repeated and no vertices are revisited. A **circuit**, or cycle, is a walk that starts and ends at the same vertex but that is not a loop. The number of edges traversed by walk, trail, or path is called the **length** of the walk, trail, or path.

The term "path" is, in application, sometimes made a bit more relaxed, and revisiting of vertices is permitted. Thus, a distinction is drawn between a path and a simple path, with the term "path" permitting revisitation of vertices. In this chapter, we will not require that paths be simple, and therefore the words "path" and "walk" are synonymous, and when we intend to refer to a simple path, we will specifically use that terminology.

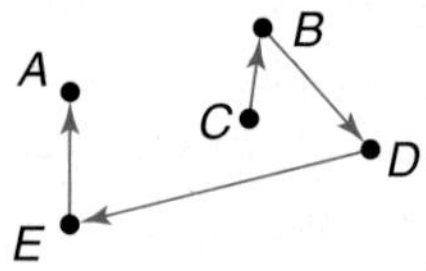

Figure 14.6
Graph depicting path C, B, D, E, A

There are several accepted conventions for the naming of paths, but we will adopt the practice of listing the vertices of the path in order of traversal as the path's name. For instance, if we have a graph with numerous vertices, among which are the vertices *A*, *B*, *C*, *D*, and *E*, we could specify the path *C*, *B*, *D*, *E*, *A*, which describes motion beginning at *C*, moving to *B*, then to *D*, then to *E*, and ending at *A* (assuming that the graph contains edges *CB*, *BD*, *DE*, and *EA*) (Figure 14.6). Any path that ends at the same vertex as it began is referred to as a circuit, and a one-edge loop path is referred to as a **pendant**. We point out that the name "circuit" may be a bit misleading and specify that a circuit (or any path, for that matter) need not include every vertex in the graph.

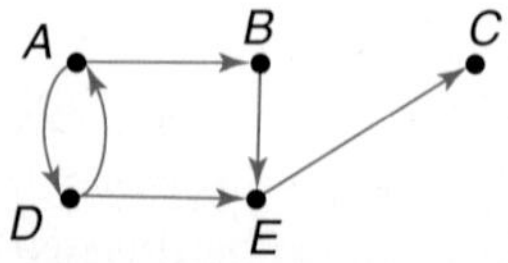

Figure 14.7
A digraph

Although we shall not explore them here, a graph is said to be a directed graph if the edges all have an indicated direction on each edge. Such graphs are sometimes called digraphs, and the direction of each edge is indicated by an arrowhead drawn on the edge (Figure 14.7). For digraphs, changing the order of vertices when naming an edge is significant, since *AB* would no longer be viewed as the same as *BA*. A graph that is not directed is sometimes called an undirected graph.

The **distance between vertices** in a graph is defined to be the number of vertices in the shortest path that connects the vertices. Should there be no paths connecting the pair of vertices in question, then we say the distance between the vertices is infinite.

A graph in which every pair of vertices can be connected by a path is called a **connected graph**, and a graph that is not connected is called a **disconnected graph**. If removal of a particular edge changes a connected graph into a disconnected graph, then that edge is referred to as a **bridge**.

EXAMPLE 14.4

Sketch a complete three-regular graph of order 4.

SOLUTION

The problem calls for a graph having four vertices (order 4), each of which has three edges entering it and every vertex connected to every other vertex. Although the graph in Figure 14.8 is not the only possibility, it has the desired properties.

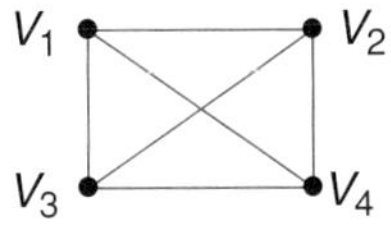

Figure 14.8 Graph for Example 14.4

Weighted Graphs and the Shortest-Path Problem

It is possible that we would want to assign numerical values to the edges in a graph. In such a case, the numerical values are said to be the **weights of an edge**, and the graph is called a **weighted graph**.

Weights appear frequently in applications of graph theory. In the so-called friendship graph, the edges would represent connections or friendships between individuals and the weights the strength of those friendship (measured on some appropriate scale). In a communications network graph, the weights could indicate the cost of maintenance or construction of the communications links between individuals or locations within the network.

We can investigate a weighted graph and a pair of connected vertices within the graph, and we can search for the minimum (or maximum) sum of weights of the edges forming a path between the vertices. If we calculate the sum of weights for any particular such path, that sum is defined to be the length of the path, and the minimum length of all such paths is defined to be the **weighted distance between vertices.** This definition coincides with the definition of the distance between vertices in an unweighted graph if we make the assumption that the weights of all edges are 1.

The shortest-path problem lies behind a very familiar Web application you may have used yourself: an online map generator that provides you with the "shortest travel distance between locations." Sound familiar? It can also be used to solve problems like network routing, where the goal is to find the shortest path for data packets to take through a switching network. It is also used in more general search algorithms for a variety of problems ranging from automated circuit layout to speech recognition.

The technique we will use here is called Dijkstra's algorithm, after the Dutch computer scientist Edsger Dijkstra, and is often used to solve routing problems. In fact, the algorithm is generally used to determine the shortest distance from a particular vertex to all other vertices in the graph, but in this case, we are interested in the shortest distance to a particular vertex, so will use a truncated version of the algorithm. We could repeat the process for any particular vertex we wished.

During the algorithm, we will designate any vertex whose minimum distance from the starting vertex is known as "solved." All other vertices shall be referred to as "unsolved." Until we know the shortest distance from the starting vertex to a particular vertex with certainty, the distance to that vertex will be referred to as a candidate distance.

Let's suppose we are interested in finding the shortest distance from vertex S to vertex T in Figure 14.9. We begin by identifying all unsolved vertices connected to S, and these are seen to be A, B, and C. The candidate distances from vertex S to these other vertices are 2, 5, and 4. The smallest of these candidates is retained, and the vertex connected to S by that distance is now considered known, and its candidate distance is its minimum distance from S.

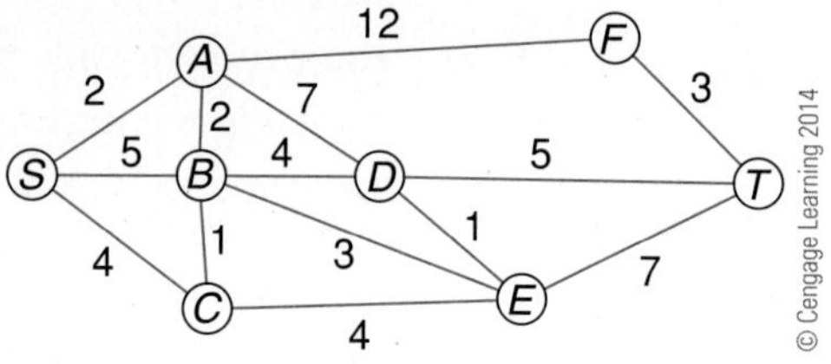

Figure 14.9 Graph for a shortest distance problem

We can keep track of the minimum distances by use of a table in which we place the minimum distance from S once it is known:

Vertex	Shortest Distance from S
S	0
A	2
B	
C	
D	
E	
F	
T	

Vertices S and A are now considered solved. Next, we consider all unsolved vertices connected by a path to a known vertex and determine their distances from S. The connected unsolved vertices here are F, D, B, and C, and we compute the candidate distances from S to these vertices. The distances are $S, A, F = 14$, $S, A, D = 9$, $S, A, B = 4$, $S, B = 5$, and $S, C = 4$.

We have a tie for the shortest distance from S, and therefore we make an arbitrary choice of which one to keep. Here, we'll choose to keep S, A, B, and thus the distance from S to B is known to have minimal distance 4:

Vertex	Shortest Distance from S
S	0
A	2
B	4
C	
D	
E	
F	
T	

Again, we consider the distance from S to all unsolved vertices connected by a vertex to a solved vertex. The unsolved vertices in question are F, D, E, and C. We compile the candidate distances from S to those vertices and again choose the shortest distance:

$$S, A, F = 14,\ S, A, D = 9,\ S, A, B, D = 8,\ S, A, B, E = 7,\ S, C = 4$$

We keep the shortest distance from the list, and thus the shortest distance from S to any unsolved vertex is 4, and the shortest distance from S to C is known:

Vertex	Shortest Distance from S
S	0
A	2
B	4
C	4
D	
E	
F	
T	

The process continues until we know the shortest distance to T. The set of unsolved vertices connected by a path to a solved vertex are D, E, and F, and we now find the candidate distances to those vertices from S. These distances are

$$S, A, F = 14, S, A, D = 9, S, A, B, D = 8, S, A, B, E = 7, S, C, E = 8$$

The shortest of these distances is 7, and thus the distance from S to vertex E is known to be 7:

Vertex	Shortest Distance from S
S	0
A	2
B	4
C	4
D	
E	7
F	
T	

The unsolved vertices now connected to a solved vertex are D, F, and T. We find the candidate distances from S to each of these vertices:

$$S, A, F = 14, S, A, D = 9, S, A, B, D = 8, S, A, B, E, D = 8, S, A, B, E, T = 14$$

The smallest of these is 8, connecting S to D. We keep one of those paths, and the distance from S to D is now known to be 8:

Vertex	Shortest Distance from S
S	0
A	2
B	4
C	4
D	8
E	7
F	
T	

Both unknown vertices are connected by an edge to a known vertex, so we list all candidate distances from S to F and to T. They are

$$S, A, F = 14, S, A, D, T = 14, S, A, B, E, D, T = 13$$

We keep the smallest distance, which is 13, and thus we know that the shortest distance from S to T is 13. Since this is the distance we were seeking, we can terminate the process. Observe that the shortest distance to F was never completely resolved, but we could determine it if we continued the process through one more iteration:

Vertex	Shortest Distance from S
S	0
A	2
B	4
C	4
D	8
E	7
F	
T	13

EXAMPLE 14.5

The given weighted graph depicts a switching network and the delay times (in nanoseconds) for data packets of a particular size to pass through the various paths of the network (Figure 14.10). Find the shortest path for a packet of data between points A and B in the network.

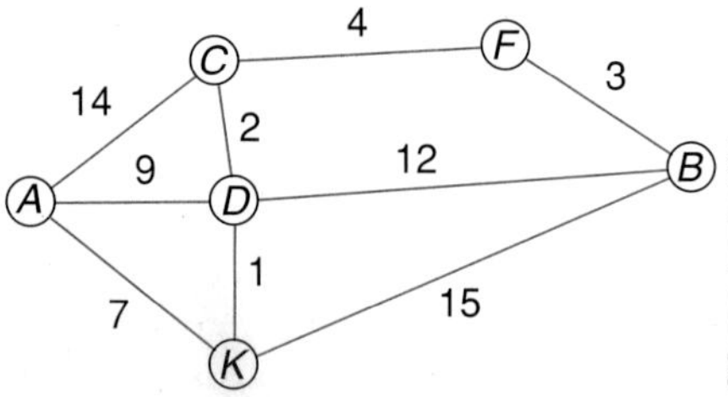

Figure 14.10 Weighted graph of a switching network

SOLUTION

As before, we construct a table showing the shortest time from vertex A to any other vertex in the network:

Vertex	Minimum Distance from A
A	0
C	
D	
E	
F	
B	

There are three unsolved vertices connected to vertex A, and of those three, the minimum distance is from vertex A to vertex E, and hence the distance from A to E is known to be 7, which we enter into the chart:

Vertex	Minimum Distance from A
A	0
C	
D	
E	7
F	
B	

The unsolved vertices connected by an edge to a solved vertex are now C, D, and B. We compute the candidate distances from A to each of these vertices:

$A, C = 14$

$A, D = 9$

$A, E, D = 8$

$A, E, B = 22$

The shortest of these distances is 8, and thus vertex D is solved, with distance 8 from vertex A:

Vertex	Minimum Distance from A
A	0
C	
D	8
E	7
F	
B	

The unsolved vertices connected to solved vertices are C and B, and we compute the candidate distances as

$A, C = 14$

$A, E, D, C = 10$

$A, E, B = 22$

$A, E, D, B = 20$

The smallest of these distances is 10, and hence the distance from A to C is now solved:

Vertex	Minimum Distance from A
A	0
C	10
D	8

Vertex	Minimum Distance from A
E	7
F	
B	

Both of the remaining unsolved vertices are connected by an edge to a solved vertex, so we can find distances from A to each of these vertices:

$A, E, D, C, F = 14$
$A, E, D, B = 20$
$A, E, B = 22$

The smallest of these distances is 14, and hence the vertex F is now solved, and its distance from A is 14:

Vertex	Minimum Distance from A
A	0
C	10
D	8
E	7
F	14
B	

We have been forced to run through all other vertex distances in this example and now can finally find the distance from A to B:

$A, E, B = 22$
$A, E, D, B = 20$
$A, E, D, C, F, B = 17$

The minimum of these distances is 17, and thus the minimum distance from A to B is 17, and we conclude that the shortest routing path for the data is A, E, D, C, F, B, which (oddly!) runs through all intermediate data transfer points:

Vertex	Minimum Distance from A
A	0
C	10
D	8
E	7
F	14
B	17

Exercises

Answer the following questions using proper grammar, complete sentences, and correct terminology.

1. Define what is meant by a mathematical graph, including the distinction between the edges and vertices of the graph.

2. What does it mean to say that a graph is connected?
3. In an unweighted graph, what is meant by the notion of distance between vertices?
4. Explain Dijkstra's algorithm.
5. What is the distinction between a path and a circuit?
6. What is a bridge in a graph? Draw an example of a graph having a bridge, identify the bridge, and then explain why that edge is a bridge.

For the following graphs, identify the degree of the vertices of the graph, classify the graph as either even or odd, and give the order of the graph.

7.
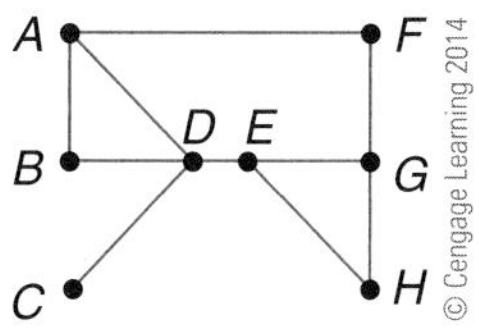

8.
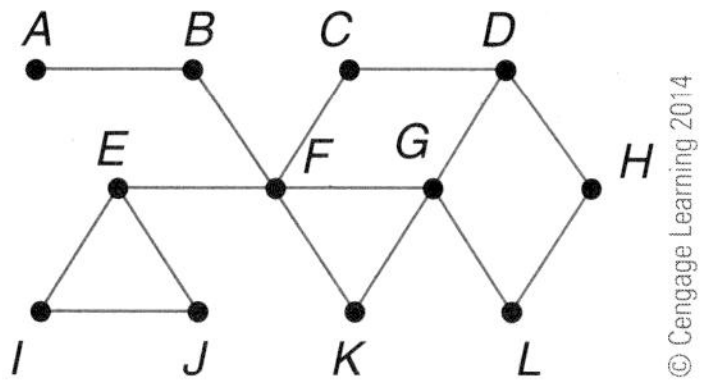

9.
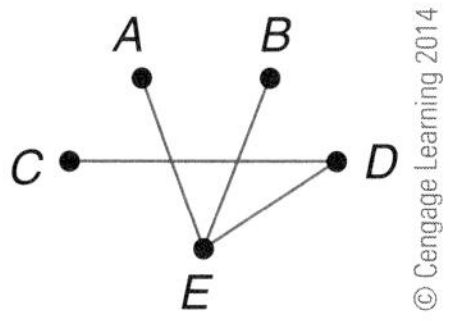

10.
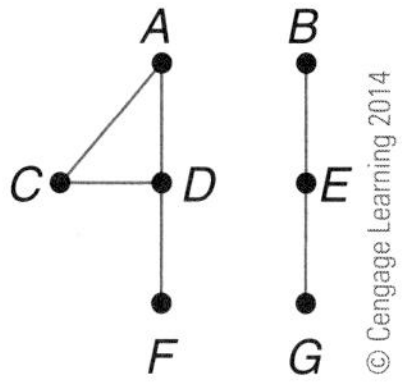

11.
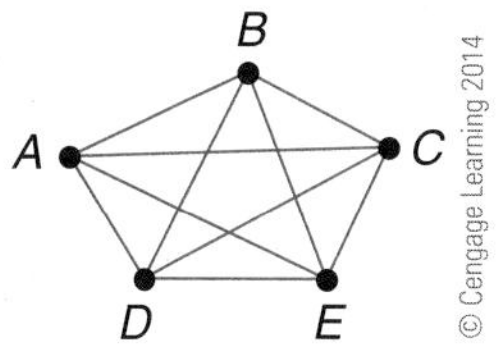

12.
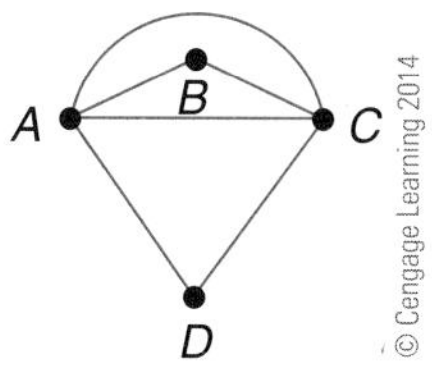

13.
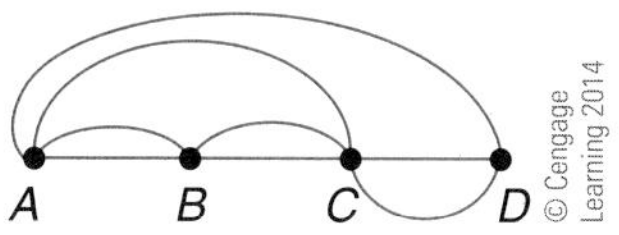

14.
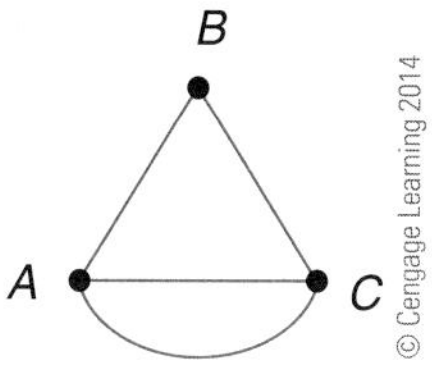

In the following problems, graphs from previous problems are referenced. Produce the adjacency matrix for each of the graphs.

15. Problem 7
16. Problem 8
17. Problem 9
18. Problem 10
19. Problem 11
20. Problem 12
21. Problem 13
22. Problem 14

In the following problems, a graph is described. Sketch an example of a graph having the indicated properties.

23. The graph has five vertices.
24. The graph has five vertices, three of which are even.
25. The graph has seven vertices with two bridges.
26. The graph has four odd vertices.

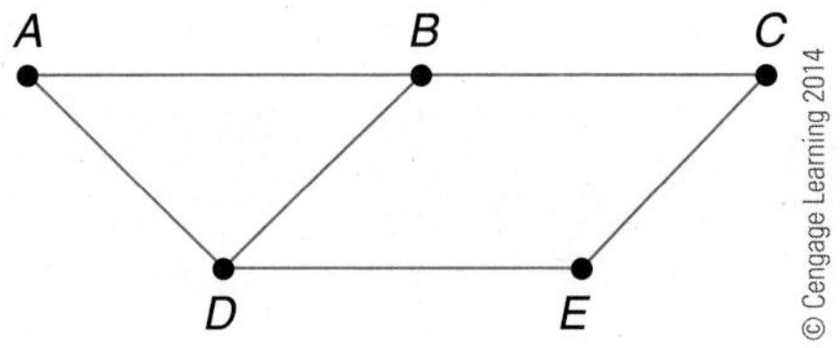

Figure 14.11

The following problems refer to the graph shown in Figure 14.11.

27. Is *A*, *B*, *C*, *D* a path within the graph?
28. Which edges of the graph are not included in the path *A*, *B*, *C*, *E*?
29. Within the graph, can we find a path that includes all the edges without using any edge twice? If so, find the path.

In the following problems, sketch a graph with the indicated properties.

30. Sketch a complete seven-regular graph of order 8.
31. Sketch a complete four-regular graph of order 5.
32. Sketch a complete five-regular graph of order 6.
33. Sketch a complete six-regular graph of order 7.

For the following graphs, determine if the graphs are connected or disconnected.

34.
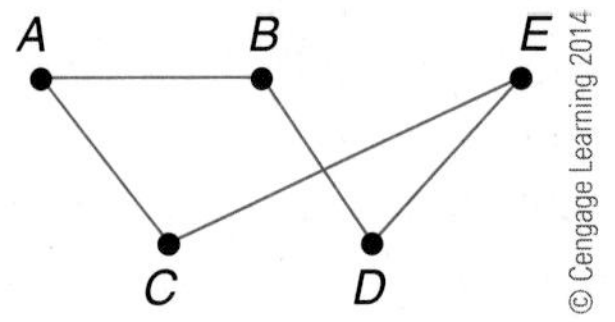

35.
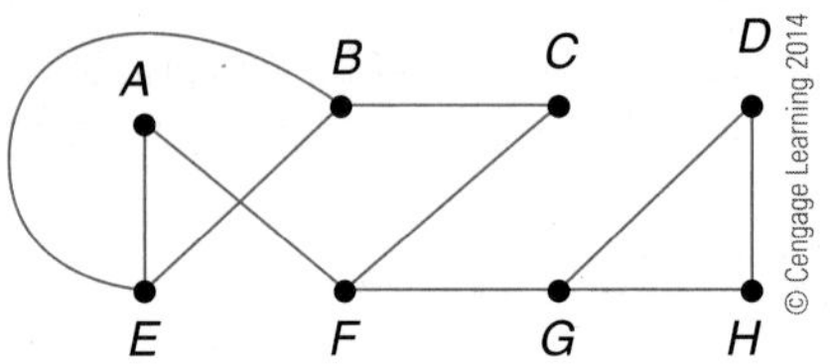

36.
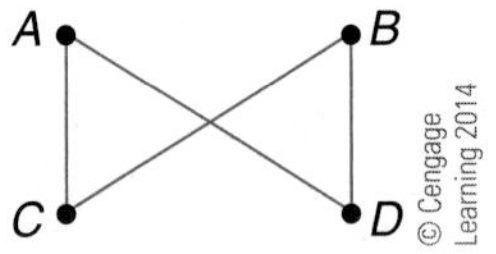

37.
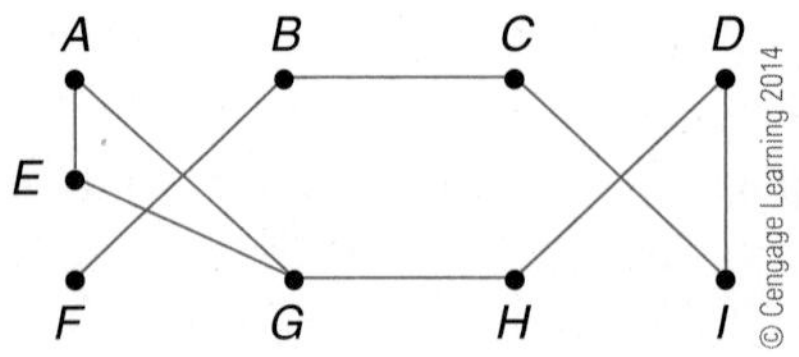

In the following weighted graphs, find the shortest path from *S* to *T*.

38.
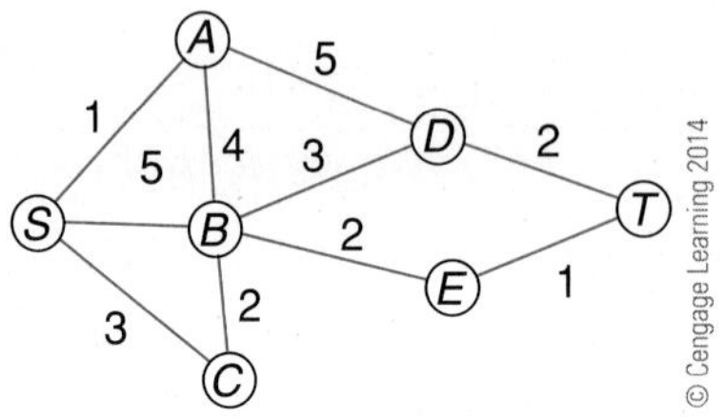

39.

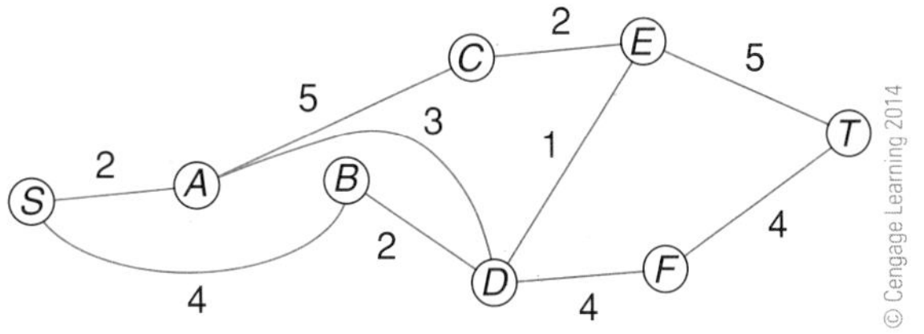

14.2 Euler Paths and Euler Circuits

In the previous section, we laid the groundwork for the fundamentals of graph theory, and now we want to delve a bit deeper into the concept. To do so, we'll summon the name of one of the true giants of mathematical history, Leonhard Euler, and introduce the concepts of the Euler path and the Euler circuit.

Euler Paths and Circuits

An **Euler path** is a path along a connected graph that connects all the graph's vertices and that traverses every edge of the graph only once. Since a circuit is a path that begins and ends at the same vertex, it follows that an **Euler circuit** is a circuit that passes through each edge of a graph exactly once. These two concepts are closely related and differ in that an Euler circuit must start and finish at the same vertex. The definition indicates, we should mention, that every Euler circuit must be an Euler path, but the converse is not true.

Our first challenge will be to determine if a given graph contains an Euler path, an Euler circuit, or neither. This can be determined by brute-force experimentation, although other, more elegant methods for resolving the question do exist, as we'll see. Let's begin with an example.

EXAMPLE 14.6

Does the graph shown in Figure 14.12 contain an Euler path? If so, find the Euler path.

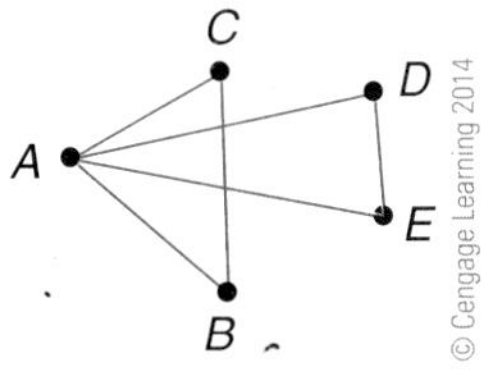

Figure 14.12 Graph for Example 14.6

SOLUTION

Without a theoretical tool (yet!) to help us, what we need to do is to rely on some trial and error to see if we can determine the answer. Beginning with vertex *A*, we proceed to vertex *B* (arbitrarily chosen). In order for us to construct an Euler path (if possible), we must reach every vertex and traverse no edge more than once. Thus, our only choice is to advance to vertex *C* and from there advance to vertex *A*. Since we must continue to blaze new trails across new edges, our choices now are somewhat limited: we can follow either *AE* or *AD*. Randomly choosing to follow *AD*, we proceed to vertex *D*, then *E*, and our Euler path is complete. (Note that in this example, we could have proceeded back along *EA* to complete an Euler circuit, but this was not asked of us.)

The method for solution in Example 14.6 was what we refer to as a proof by demonstration. That is, we established that the graph contained an Euler path by finding the path and then exhibiting it for all to see. Sometimes, this can be quite challenging, and it is better (well, easier!) to prove the existence of the item we're looking for by using theoretical results.

EXAMPLE 14.7

Does the graph shown in Figure 14.13 contain an Euler path? What about an Euler circuit? If so, find them.

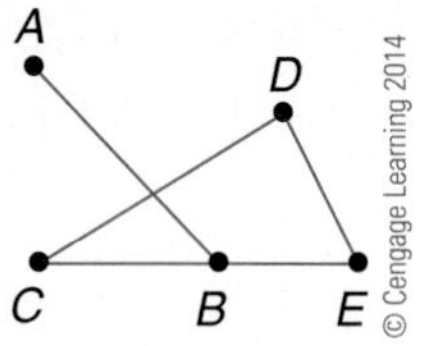

Figure 14.13 Graph for Example 14.7

SOLUTION

Again, we choose an arbitrary vertex at which to start, and (to avoid monotony) we'll select vertex *C*. Beginning at *C*, we traverse the path *C*, *D*, *E*, *B*, *A*. This has allowed us to visit all the vertices of the graph but does not qualify as an Euler path, since edge *CB* was never traversed.

Observe that this does not tell us that no Euler path exists. Hardly so! All it demonstrates is that the particular path we found, which did reach all the vertices, was not an Euler path.

Beginning at vertex *B*, we traverse the path *B*, *C*, *D*, *E*, *B*, *A*, and this is another path that reached all the vertices, and it had the property of traversing all of the edges in the graph. Therefore, this path was an Euler path! Thus, we have proved, by demonstration, the existence of an Euler path.

There can be no Euler circuit, however, because of the structure of the graph. Observe that vertex *A* sits at the end of a "dead-end road" in the graph. The only access to vertex *A* is via edge *BA*, and since this can be traversed only once, there is no way to begin and return to vertex *A* or any way to manage to reach vertex *B*, then *A*, and then another vertex without recrossing edge *BA*.

What is it about these graphs that allows the existence of an Euler path? The answer lies in a result called Euler's theorem, which tells us that the number of odd vertices within a graph controls whether an Euler path or an Euler circuit will occur.

Euler's Theorem

Suppose we have a connected graph. Then

a. A graph with no odd vertices has at least one Euler path, which will also be an Euler circuit. The circuit can be initiated at any vertex you wish and will terminate at that same vertex.

b. A graph with exactly two odd vertices has at least one Euler path but no Euler circuits. The Euler path must begin at an odd vertex, and it will end at the other odd vertex.

c. A graph with more than two odd vertices has neither an Euler path nor an Euler circuit.

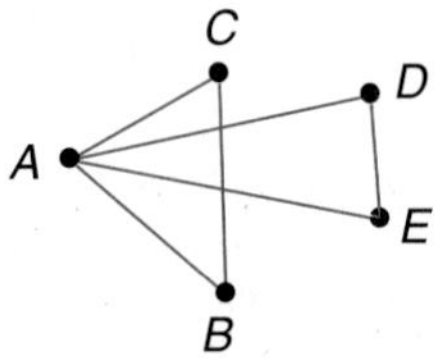

Figure 14.14 Graph of Example 14.6

Let's reexamine Example 14.6, with the weapon of Euler's theorem now cast and in our hands. Observe that every vertex is even, with degrees of either 4 (for vertex *A*) or 2 (the rest of the vertices) (Figure 14.14). By Euler's theorem, there is at least one Euler path, which will also be an Euler circuit. Recall that we constructed an Euler path and remarked that (had we proceeded one vertex further) we would have had an Euler circuit as well.

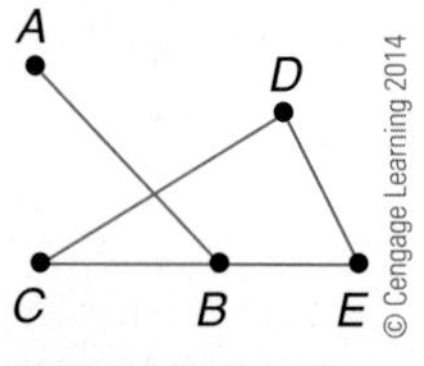

Figure 14.15 Graph of Example 14.7

What about Example 14.7? Vertex B has degree 3, so it's an odd vertex, and vertex *A* has degree 1, so it's odd (Figure 14.15). The remaining vertices are even, so we are in the situation of having exactly two odd vertices. The second statement within Euler's theorem tells us that there is at least one Euler path but no

Euler circuit, which is what we found. Moreover, the Euler paths we could find would have to begin at one odd vertex and end at the other. Recall, we initially attempted to begin our path at vertex *C*, which was an even vertex, and we failed to construct an Euler path. When we began at vertex *B*, one of the odd vertices, we were able to construct an Euler path, which ended at the other odd vertex, *A*.

EXAMPLE 14.8

Using the result of Euler's theorem, construct a connected graph that has no Euler paths or circuits.

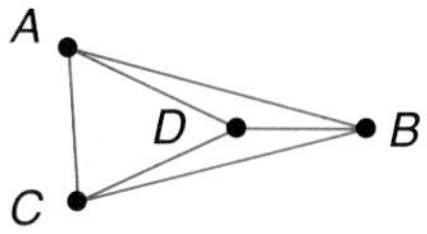

Figure 14.16 Graph for Example 14.8

SOLUTION

The third criterion from Euler's theorem tells us that we must construct a graph with more than two odd vertices. So, we will devise a graph with three odd vertices (if we can manage it), and that will serve as our example. Consider the graph shown in Figure 14.16.

Each of the vertices has degree 3, and therefore no Euler path should exist. Let's see (for purposes of investigation) what might happen if we attempt to construct an Euler path. Starting at vertex A, we can follow *A*, *C*, *B*, *A*, *D*, *C*,…, but at this point there is no way to cross edge *BD*, and we are stuck at vertex *C*. A similar experience awaits no matter which series of edges we cross in an attempt to visit all vertices while crossing all edges a single time.

EXAMPLE 14.9

Ali needs to use an electronic scanner device to read the electricity meters on all the houses in a particular neighborhood. In an effort to minimize his driving time and gas expenses, he wants to read the meters in all the houses, passing down each street only one time. A map depicting the neighborhood is shown in Figure 14.17. Is it possible for Ali to accomplish his goal?

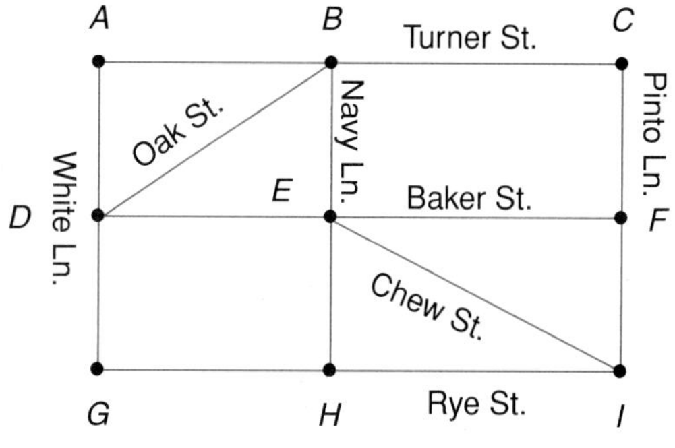

Figure 14.17 Graph for Example 14.9

SOLUTION

For Ali to achieve his objective, he must traverse all the streets in the neighborhood, passing through each a single time. Over a cup of coffee, he labels all the intersections in the neighborhoods as vertices *A* through *I* and considers the situation. Familiar with Euler's theorem, he decides to count the number of odd vertices in the graph and finds that vertices *F*, *I*, *E*, and *H* are odd vertices. Since there are more than two odd vertices, Ali realizes (with a bit of frustration!) that his task is a hopeless one, and he is certain to have to pass down at least one street a second time.

EXAMPLE 14.10 For the graph shown in Figure 14.18, does an Euler path or an Euler circuit exist? If so, give an example of such a path and/or circuit.

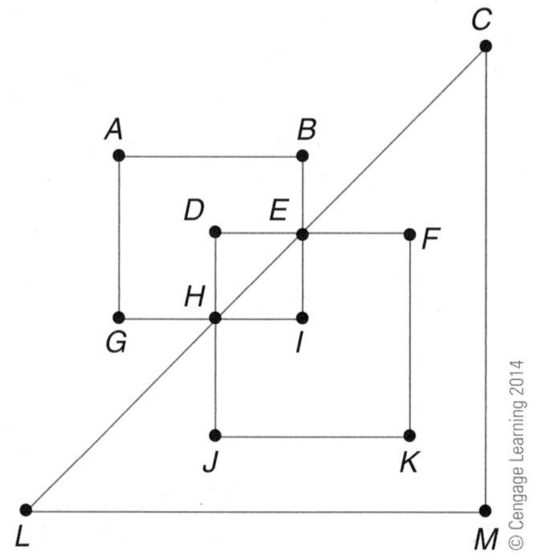

Figure 14.18 Graph for Example 14.10

SOLUTION

Examination of the graph reveals that, surprisingly, all the vertices are even. Therefore, the first statement in Euler's theorem tells us there is at least one Euler path, which will also be an Euler circuit. All we have to do now is produce one.

Beginning at vertex *A*, we will attempt to construct an Euler path, knowing that (if we are successful) we will return to vertex *A* to close the circuit. One path is

$$A, B, E, F, K, J, H, D, E, I, H, E, C, M, L, H, G, A$$

There are many others, but this one suffices.

Fleury's Algorithm

As we've seen, we can use Euler's theorem to determine the existence or nonexistence of Euler paths and circuits, and we've been able to construct such paths and circuits knowing that they exist. It would be very useful to have an algorithm to which we could appeal for the construction of Euler paths or circuits so that our construction process would be systematic rather than ad hoc.

Fleury's algorithm, a rather famous result from the 1860s, has the curious property of having an uncertain attribution. That is, though certain writings of the French mathematician Edouard Lucas refer to Fleury, nowhere is there mention of Fleury's full name. Because of the precision inherent in all mathematics, this is a truly unusual occurrence, but limitations on time prevent us from discussing the anonymity of the devisor of the algorithm we now present.

Fleury's algorithm tells us that to construct an Euler path or circuit, begin at any vertex you wish. From that vertex, pick an edge to traverse, darkening it or marking it in some manner so as to ensure that you will not recross it. *Never cross a bridge unless no other choice exists!* Traverse that edge to reach the next vertex. Repeat steps 2 and 3 until all edges have been traversed, returning to the original vertex if possible (to create an Euler circuit).

EXAMPLE 14.11

The graph in Figure 14.19 has two odd vertices: *A* and *C*. Euler's theorem tells us there is an Euler path but not an Euler circuit and that the Euler path must begin at one odd vertex and end at the other. Construct an Euler path for the graph.

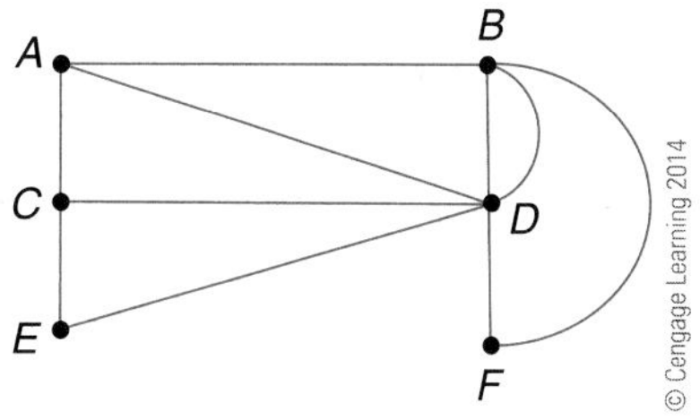

Figure 14.19 Graph for Example 14.11

SOLUTION

We must begin our path at either *A* or *C*, so we will use alphabetical preference and begin at vertex *A*. Our path can be *A*, *B*, *D*, *B*, *F*, *D*, *E*, *C*, *D*, *A*, *C*, which crosses each of the edges a single time. Observe that, as predicted by Euler's theorem, we ended at vertex *C*, the other odd vertex.

EXAMPLE 14.12

For the graph shown in Figure 14.20, if an Euler path or circuit exists, find examples of them.

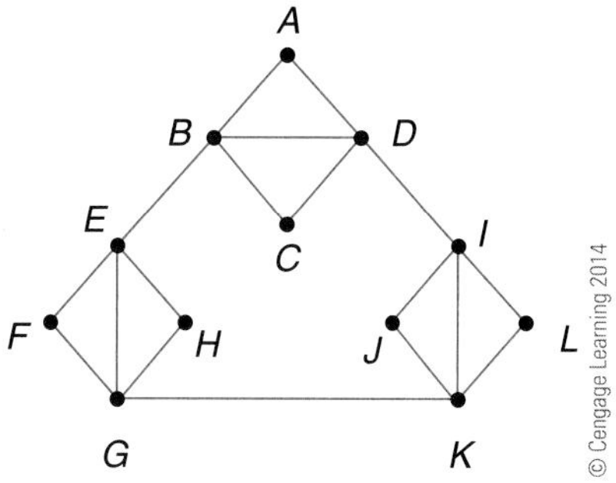

Figure 14.20 Graph for Example 14.12

SOLUTION

All the vertices in the figure are even, and therefore Euler's theorem tells us there is an Euler path that is also an Euler circuit. We arbitrarily choose to begin at vertex A and attempt to cross all edges of the graph, returning to *A*. *A*, *B*, *D*, *C*, *B*, *E*, *F*, *G*, *E*, *H*, *G*, *K*, *L*, *I*, *K*, *J*, *I*, *K*, *L*, *I*, *D*, *A* serves as the Euler path/cycle in this case.

EXAMPLE 14.13

In a small office, the relationships as confidante's among the workers is displayed in Figure 14.21. Adjacent workers (vertices) know one another and talk regularly. If Thuy tells a secret to Marc, is it possible for the secret to be relayed across all the lines of communication, with no one "repeating the gossip" to someone whom they have already told or who has told the secret to them?

SOLUTION

In order for the message to be relayed across all lines of communication, it is necessary for the secret to be transmitted along all the edges of the graph and to every vertex, and thus the question amounts to whether there exists an Euler path within

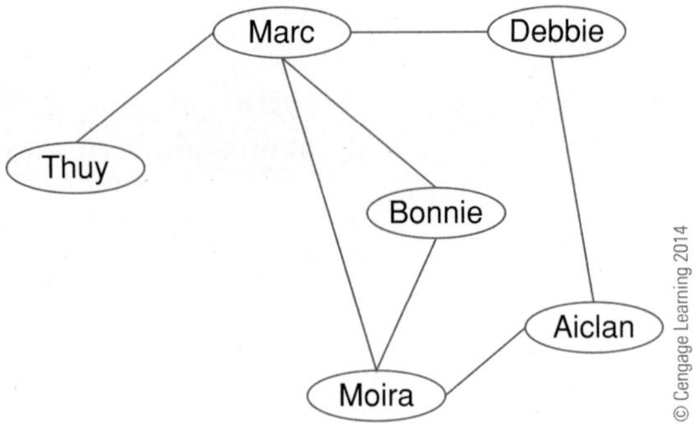

Figure 14.21 Relationship Among Workers

the communication network, starting with Thuy. We notice there are two odd vertices, Thuy and Moira, and therefore Euler's theorem guarantees the existence of an Euler path, and since it is beginning with Thuy, it must end with Moira.

The only person Thuy knows in the network is Marc, so he is told first. Marc has a choice of three individuals to whom he can tell the secret, and he arbitrarily chooses to tell Debbie, who passes the story on to Aidan, who tells it to Moira. The path is not complete because Bonnie has not yet been told Thuy's secret, and all the lines of communication have not been used.

Moira tells the secret to Bonnie, who passes it on to Marc, who then repeats it for the final time, to Moira, and the Euler path is complete.

Map Coloring

A famous problem arising from the 1800s is the so-called map-coloring problem: given *any* map, what is the minimum number of colors required so that no two countries sharing a common border have the same color? The problem's solution, which is entirely nontrivial, is four, established using a computer to perform proof by contradiction during the mid-1970s.

Given a connected set of countries on a map, we can attempt to resolve the question for a particular case by using edges to depict the geographical adjacency of neighboring countries, with the countries depicted as vertices. Such a graph is, by the way, referred to as a dual graph for the map.

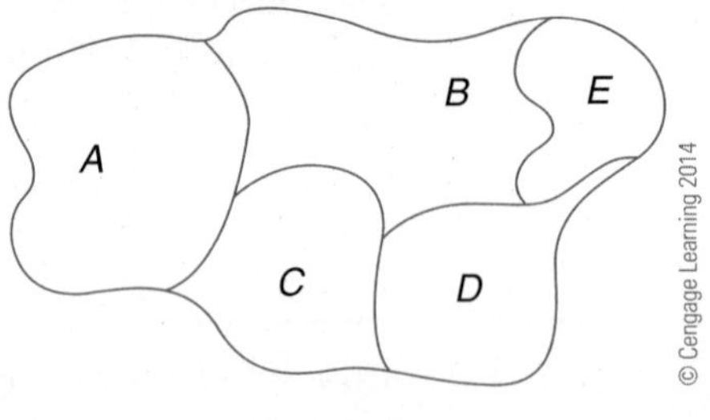

Figure 14.22 Map of five countries

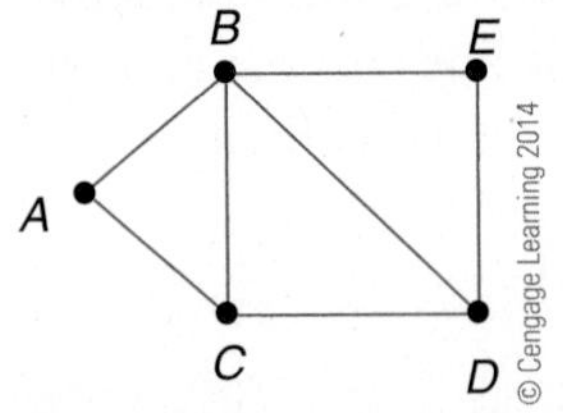

Figure 14.23 Graph for the map of five countries

The map shown in Figure 14.22 depicts five countries subdividing a large landmass. The dual graph for the map is shown in Figure 14.23. What is the minimal number of colors needed in order to ensure that no two neighboring countries are filled with the same color on the map?

Let's start our investigation with vertex A. Since *A* is adjacent to both *B* and *C* and those two vertices are, themselves, adjacent, the map requires at least three colors, possibly antique white, blue, and carmine. Are any more colors needed? From the graph, we see *E* is adjacent to both *B* and *D* but not to *A* or

C, and therefore we can color *E* the same as *A*, and therefore we still have used only three colors.

Is a fourth color needed for *D*? Vertex *D* is adjacent only to *E*, *B*, and *C*, and therefore it cannot be antique white, blue, or carmine, and thus a fourth color is needed to fill in *D*.

You may be wondering if there is a method for determining the exact number of colors required for a particular map, and the answer to that question is no. It is known that a maximum of four colors will be required, but there is no algorithm for determining how many colors a given map will need.

EXAMPLE 14.14

Draw a map for which the graph shown in Figure 14.24 would be a dual graph.

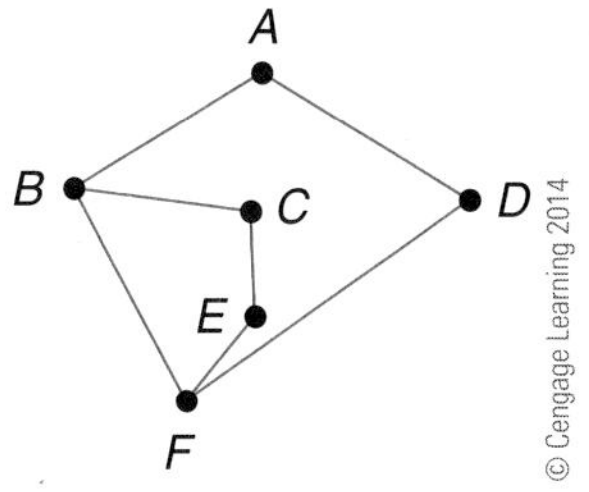

Figure 14.24 Graph for Example 14.14

SOLUTION

Many solutions are possible because of the creativity of the person drawing the map. Figure 14.25 shows one example that would suffice.

All we need to do is to construct a map in such a way that our countries touch only those with whom the graph demonstrates they share a border. In this case, we appear to have a rather wild graph, suggesting the existence of a body of water between countries *C* and *F* and also between *D* and *B*, *C*, and *E*.

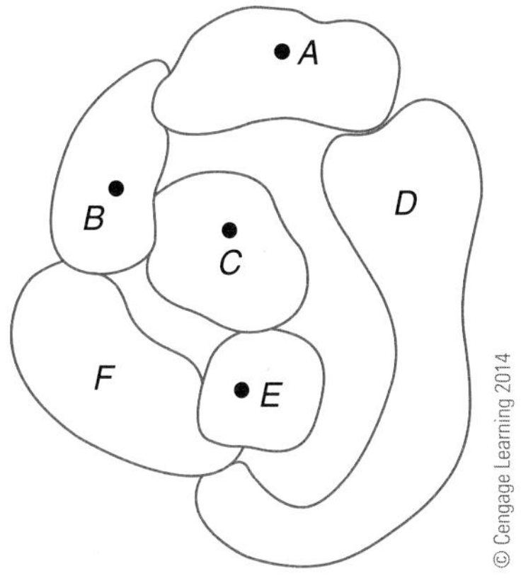

Figure 14.25 Possible Map

EXAMPLE 14.15

What is the solution to the Konigsberg bridge problem? That is, is it possible to cross all seven of the bridges spanning the Pregel River without traversing any individual bridge more than once?

SOLUTION

In the depiction of the Konigsberg bridge problem in Figures 14.26a and 14.26b, we see that the degrees of the vertices are 5 for vertex *A*, 3 for vertex *B*, 3 for vertex *C*, and 3 for vertex *D*. Thus, each of the four vertices is odd. According to Euler's theorem, a graph with more than two odd vertices can have no Euler path, which in this case is the route sought to cross all the bridges a single time. Thus, the walk described in the problem is not possible.

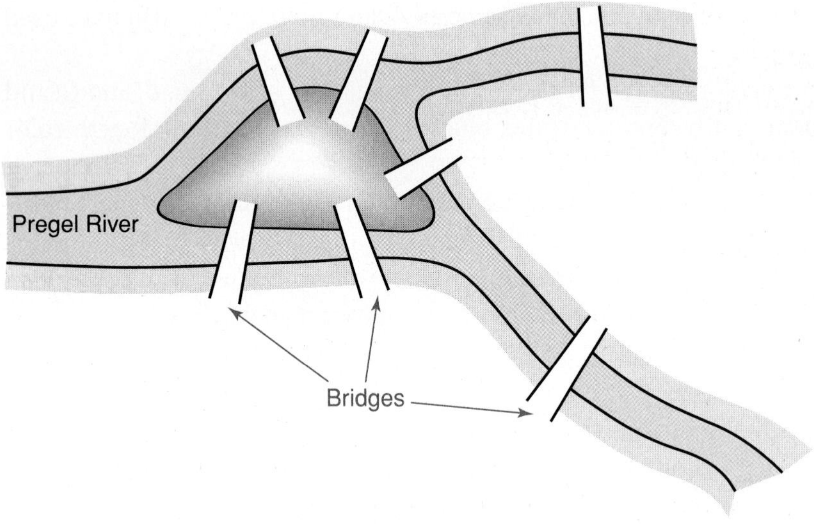

Figure 14.26a The Konigsberg Bridge Problem

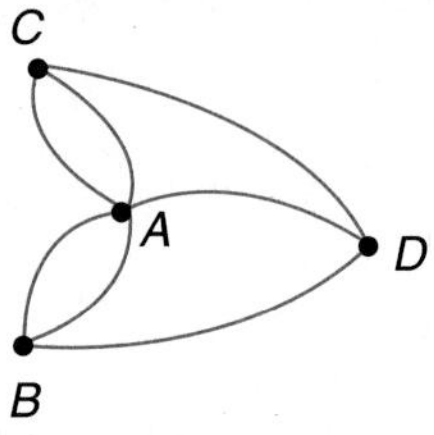

Figure 14.26b Graph for the Konigsberg Bridge Problem

Exercises

Answer the following questions using proper grammar, complete sentences, and correct terminology.

1. What is an Euler path?
2. How does an Euler circuit differ from an Euler path?
3. What is Fleury's algorithm, and for what is it used?
4. How can we use the number and type of vertices to establish the existence or nonexistence of an Euler circuit?

For the following exercises, use the graph shown in Figure 14.27.

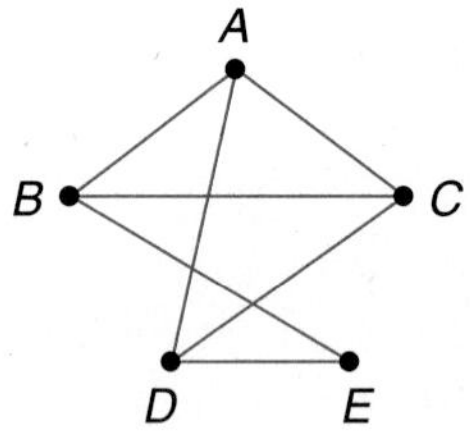

Figure 14.27

5. Find an Euler path, if one exists, that begins at vertex *A*.
6. Find an Euler path, if one exists, that begins at vertex *D*.
7. Find an Euler path, if one exists, that begins at vertex *B*.
8. Does an Euler circuit exist for this graph?

For the following exercises, use the graph shown in Figure 14.28.

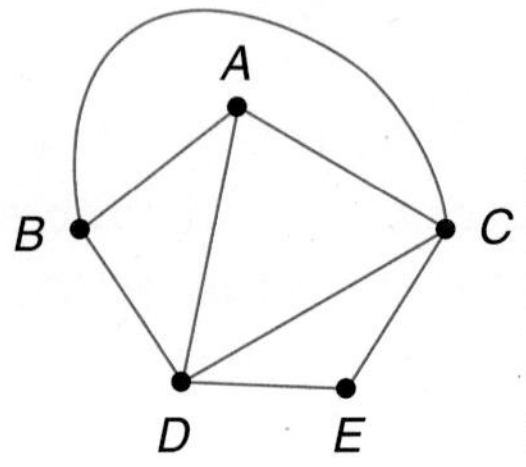

Figure 14.28

9. Find an Euler path, if one exists, that begins at vertex A.
10. Find an Euler path, if one exists, that begins at vertex C.
11. Find an Euler path, if one exists, that begins at vertex B.
12. Does an Euler circuit exist for this graph?

For the following exercises, use the graph shown in Figure 14.29.

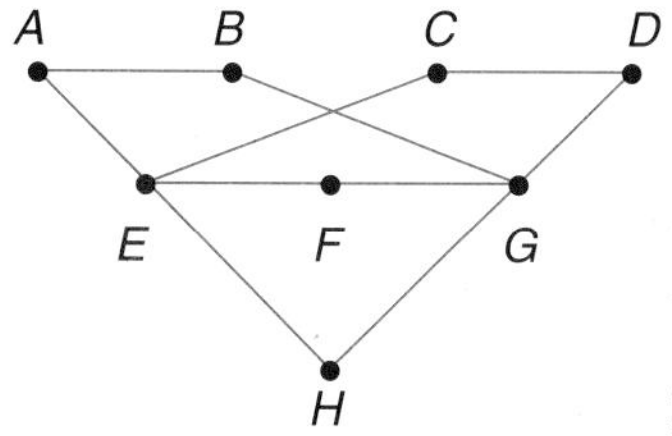

Figure 14.29

13. Find an Euler path, if one exists, that begins at vertex A.
14. Find an Euler path, if one exists, that begins at vertex C.
15. Find an Euler path, if one exists, that begins at vertex B.
16. Does an Euler circuit exist for this graph?
17. Suppose we had a graph with 1000 vertices, all of which are even (Figure 14.30). (a) Does an Euler path exist for the graph? Explain. (b) Does an Euler circuit exist for the graph? Explain.

Figure 14.30 The Western United States

18. Represent the shown map as a graph, using a vertex to represent each state, with states sharing a common border joined by an edge.
19. Does the graph described in Problem 18 possess an Euler path? Explain, and if the answer is yes, produce such a path.
20. Does the graph described in Problem 18 possess an Euler circuit? Explain, and if the answer is yes, produce such a circuit.

In the following graphs, use Fleury's algorithm to find an Euler path and an Euler circuit (if one exists).

21.
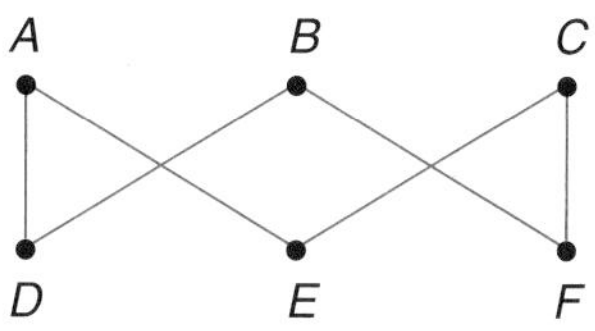

22.
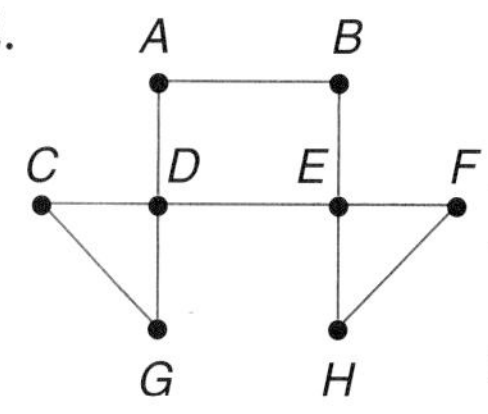

23.

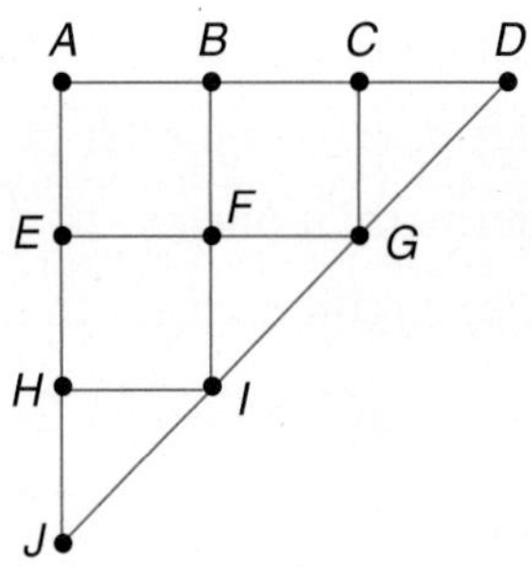

24.

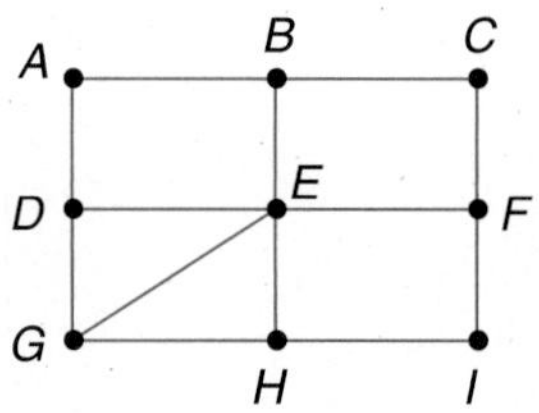

14.3 Hamiltonian Paths and Circuits

Introduction to Hamiltonian Paths

The description of a Hamiltonian path is somewhat similar to that of an Euler path, and that of a Hamiltonian circuit likewise is similar to that of an Euler circuit. A **Hamiltonian path** is a path in an undirected graph that visits each vertex exactly once. A **Hamiltonian circuit** is a Hamiltonian path that begins and ends at the same vertex, and graphs possessing such a cycle are referred to as **Hamiltonian graphs** or sometimes as traceable. At present, there is no known efficient method for determining whether or not a given graph is Hamiltonian.

Our first effort in the analysis of Hamiltonian graphs will be to discuss particular properties that such a graph must possess, and our second will be to produce some special graphs that are known to be Hamiltonian. In the latter case, we will see that the demonstration of "Hamiltonianicity" relies on the general concept that a graph having "enough" edges must be Hamiltonian (although, granted, the term "enough" is weak and imprecise).

Conditions Necessary for a Graph to Be Hamiltonian

A first condition that must hold for an undirected Hamiltonian graph is that it must be connected. That is, every pair of vertices in the graph must be capable of being connected by a path within the graph. This virtually goes without saying, since the condition of being Hamiltonian requires that a path exists by which all the vertices can be visited exactly one time.

A second condition is that every vertex must have degree two or more. This may also be apparent from the definition of being Hamiltonian, since there must be a path that visits all the vertices at least one time. However, let's do an example where a particular graph (Figure 14.31) has a vertex of degree 1 and understand what the consequences would be of a vertex of degree 1 being present within the graph.

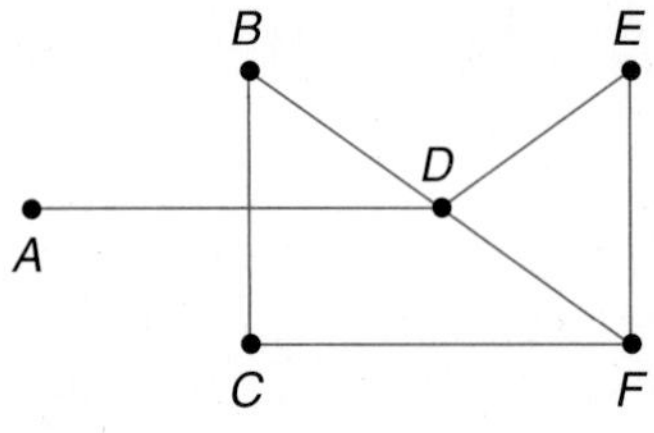

Figure 14.31 Graph with vertex of degree 1

Clearly, vertex *A* is of degree 1, with only edge *AD* connecting that vertex to the rest of the graph. If we attempt to begin our path at *A*, the only choice for our first stop in the circuit is vertex *D*. We are then free to choose any of three other vertices to which to travel from *D* to proceed with the circuit. However, to close the circuit by returning to vertex *A*, we must again pass through *D*, which is forbidden in a Hamiltonian graph. By similar reasoning, starting at any vertex other than *A*, we must eventually visit vertex *A*, meaning that the circuit must contain the sequence of vertices *D*, *A*, *D*, again violating the definition of being a Hamiltonian graph.

A third condition that must exist within a Hamiltonian graph arises if we introduce the concept of bipartite graphs. A graph is called bipartite if the vertices of the graph can be divided into two sets where every edge within the graph goes between the sets and there is no edge within either of the sets. If a graph is bipartite and Hamiltonian, then the two sets into which the graph can be split as mentioned in the definition must have the same number of vertices. The rationale for this will be left for you to discuss in the exercises, but it is not a terrifically deep result.

When particular conditions must be present in a given situation, those conditions are referred to as necessary conditions. On the other side of this concept lie sufficient conditions. When we say a set of conditions are sufficient for event E to happen, we mean that if the conditions are present, then E must occur. Recall, however, that we had said there was no efficient method for establishing that a graph is Hamiltonian. At this point, we should mention that, in all but relatively simple graphs (having a fairly small number of vertices), establishing the existence of these conditions is challenging.

Let's present some sufficient conditions to ensure Hamiltonianicity. First, if a graph having more than two vertices is undirected and complete, then it must be Hamiltonian. Why would this be the case? First of all, for completeness, every vertex must have degree at least 2, since it must be connected to all the other vertices in the graph. Therefore, we can number the vertices in any particular order we wish and then use the cycle $v_0, v_1, \ldots, v_n, v_0$. All the edges must exist because of completeness of the graph, and therefore this cycle establishes a Hamiltonian circuit.

The next sufficient condition is that if you have an undirected, loop-free graph with n vertices, and if for every pair of distinct vertices A and B it is true that $\deg(A) + \deg(B) \geq (n - 1)$, then the graph must have a Hamiltonian path. The proof of this result is a bit more challenging, and we will suppress it in this book, leaving it to a mathematical course on graph theory for its establishment. It is, however, valid.

Note the result is really saying, in another form, that if the graph has "enough" edges (there is the recurrence of our vague term "enough"), then it must have a Hamiltonian path. The graph shown in Figure 14.32 satisfies the criteria listed above. Observe that it has quite a few edges! Let's see if we can find a Hamiltonian path: A, B, F, C, E, D suffices as an illustration. Note that it is not a Hamiltonian cycle, but we were asked only for a path in this case.

There are other sufficient conditions for a graph to be Hamiltonian, which we'll list here:

- If the degree of each vertex of an undirected loop-free graph is at least $\frac{n-1}{2}$, where n is the number of vertices of the graph, then the graph has a Hamiltonian path.
- If G is undirected and loop-free with $n \geq 3$ vertices and if $\deg(x) + \deg(y) \geq n$ for all vertices x and y that are not adjacent, then G has a Hamiltonian cycle.
- If the graph is undirected and loop-free with $n \geq 3$ vertices and if $\deg(x) \geq n/2$ for all vertices x, then the graph has a Hamiltonian cycle.

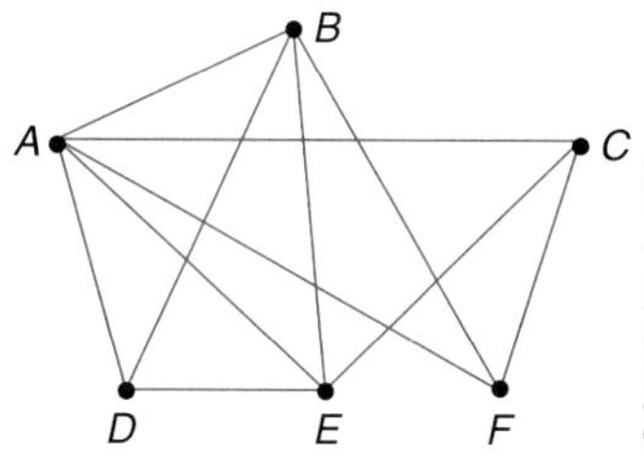

Figure 14.32 A Graph that MUST have a Hamilton Path

The Traveling Salesman Problem

Given a complete, undirected graph with weighted edges, find the Hamiltonian circuit of lowest total weight. This problem is referred to as the traveling salesman problem because, if we were to view the weights on the graph as the miles between cities, then the circuit mentioned would correspond to a route the salesman could follow where all cities are visited and the travel distance is minimized. Obviously, this would be of great interest to anyone who worked in an industry where representatives would want to visit the client list face-to-face. As was the case with the general problem on establishing that a graph is a Hamiltonian graph, there is no known (efficient) solution to the traveling salesman problem, nor is it known that no such efficient solution exists!

This problem is an example of a much larger class of mathematical problems known as combinatorial optimization problems. One could, in theory, attempt to find all possible Hamiltonian circuits and then simply choose the one having minimal total weight, but this is impractical. In one illustration of a 16-city traveling salesman problem (the problem of Homer's Ulysses attempting to visit all the cities described in *The Odyssey* exactly once, there are 653,837,184,000 distinct routes). Merely listing all the possibilities required over 90 hours of computer time on a powerful workstation, and it was from the list of possibilities that the smallest result was chosen.

Since the brute-force technique of enumeration of possibilities is clearly inadequate, there have been specific algorithms created that reduce solution time by eliminating most of the round-trips without ever having to explicitly consider and compare them. The problem finds application in computer science in the design of an idealized printed circuit board having a particular number of holes where the distances would be the amount of wire used to connect the holes.

Exercises

Answer the following questions using proper grammar, complete sentences, and correct terminology.

1. What is meant by the traveling salesman problem?
2. What is a Hamiltonian path, and what (if anything) distinguishes it from an Euler path?
3. What conditions are needed to ensure that a graph is Hamiltonian?
4. What does it mean to say that a graph is bipartite? Give an example of a graph that is bipartite.

Shown are a collection of graphs. Determine whether they possess a Hamiltonian path. If they do possess such a path, produce the path. If they do not possess such a path, explain why.

5.

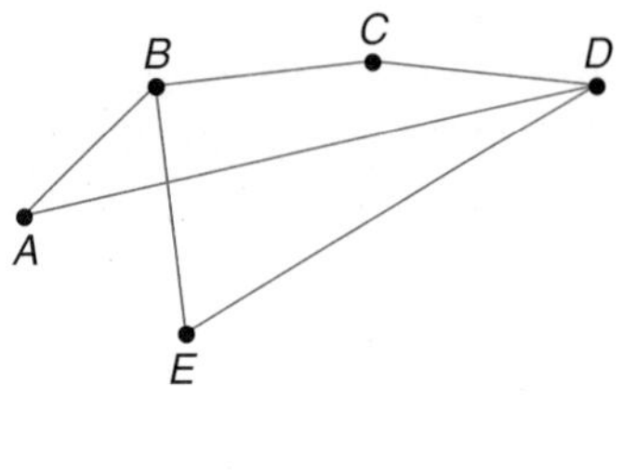

6.

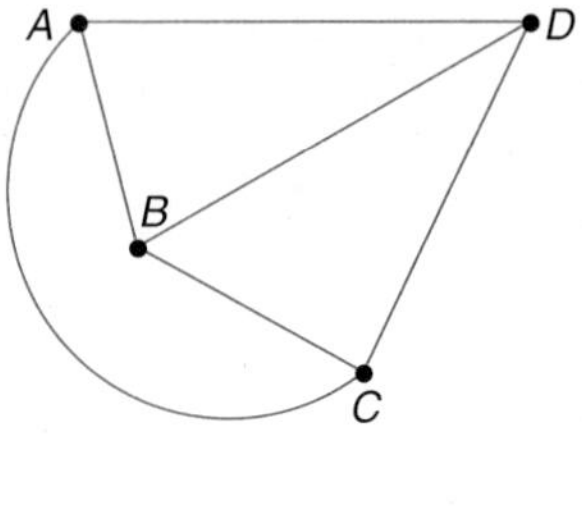

7.

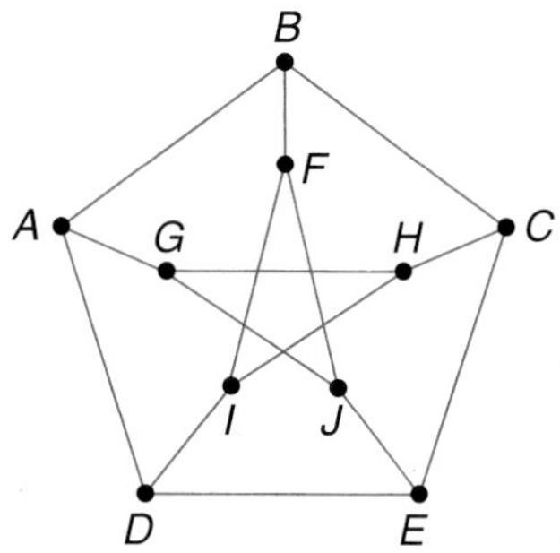

8.

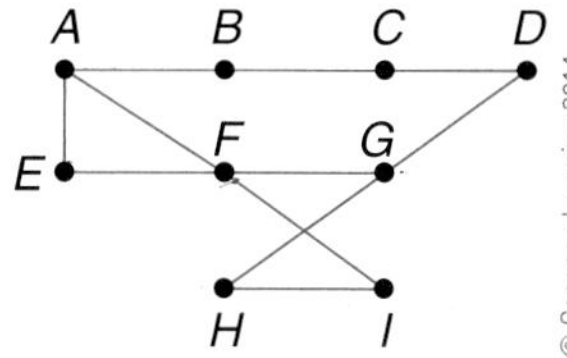

9.

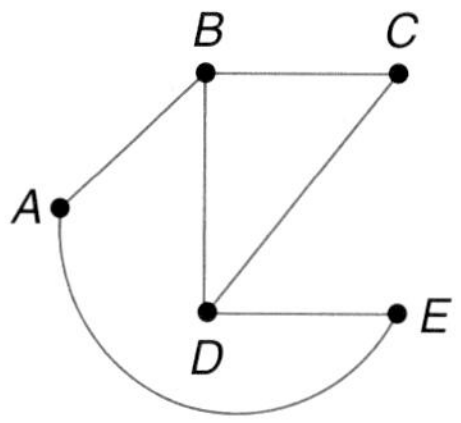

10.

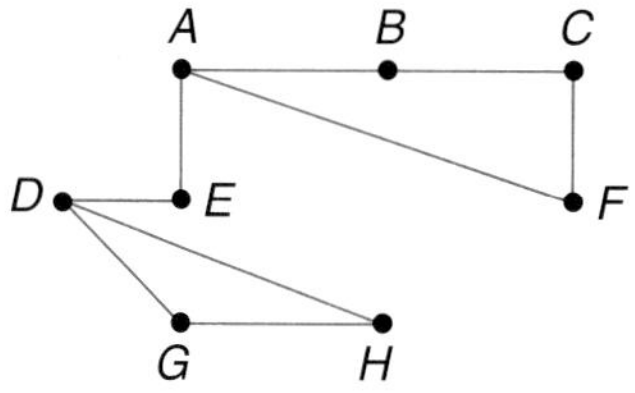

The following graphs do possess at least one Hamiltonian path. Find any particular one of these paths.

11.

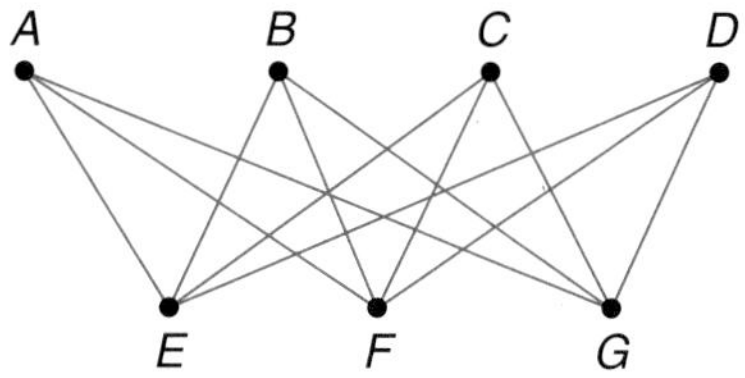

13.

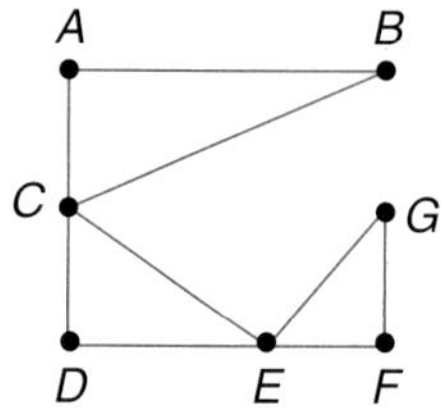

12.

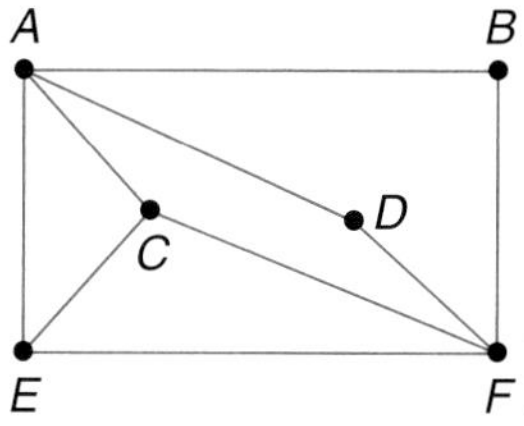

14.

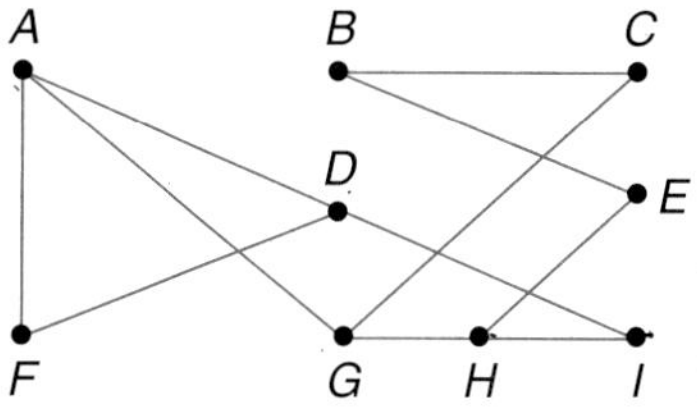

Use the associated graphs to solve the following problems.

15. Suppose a regional manager headquartered in Chicago must travel by car to visit offices in Minneapolis, Cleveland, and St. Louis. If the distances between the cities are as shown in Figure 14.33, find the Hamiltonian circuit that will allow the manager to visit all the regional offices while traveling the shortest overall distance.

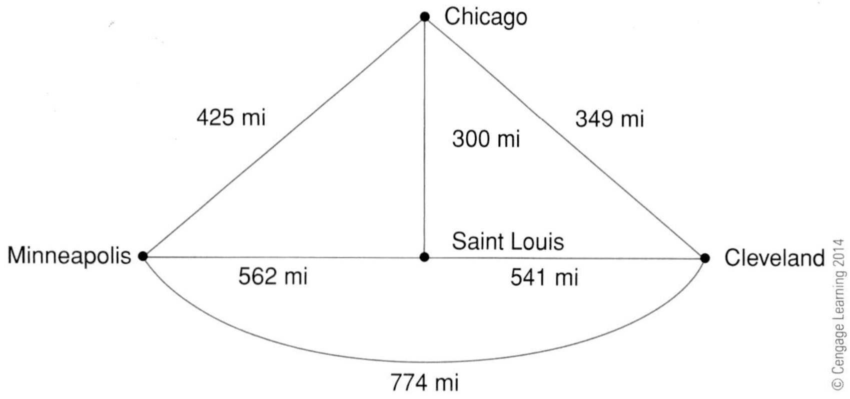

Figure 14.33

16. Repeat Problem 15, using the following "greedy" algorithm. Start at Chicago, find the nearest city to you, and travel to that first. From there, choose to visit the city closest to where you are but not Chicago. Iterate this process, making the last leg of your journey back to Chicago. Compare the result to that found in Problem 15. (This algorithm is referred to as the nearest-neighbor algorithm).

17. Suppose a videophone service will be established on an experimental basis among five cities. Figure 14.34 gives the links that are included in the network, with the value of each edge corresponding to the installation costs in millions of dollars for that link. Each city need not be directly connected because calls can be routed through a sequence of intermediate cities. Establish a Hamiltonian path for the network that minimizes the cost of construction of the videophone service.

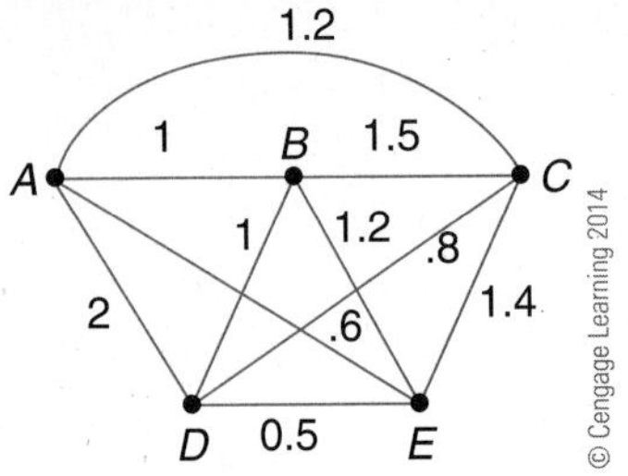

Figure 14.34 Graph of proposed videophone service

14.4 TREES

An Introduction to Trees

A graph that contains no circuits is referred to as an **acyclic graph**, "acyclic" meaning "without cycles." If a simple, connected graph is acyclic, then we call it a **tree**. The trees one can form on a graph having five vertices are shown in Figure 14.35.

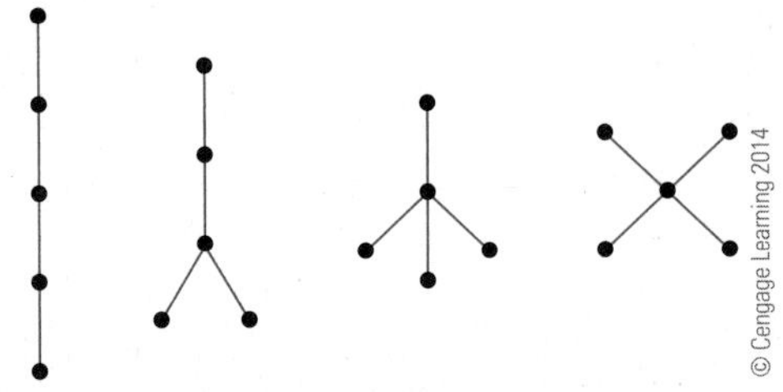

Figure 14.35 Possible trees on a graph with five vertices

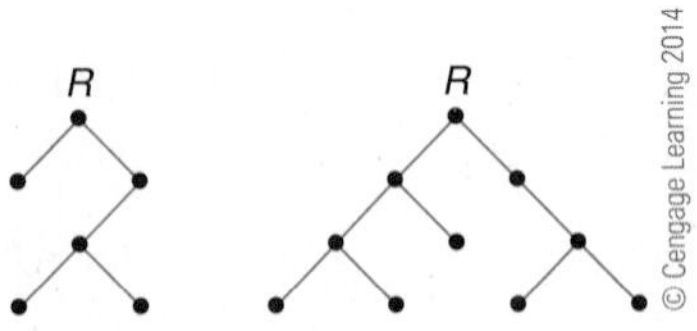

Figure 14.36 Examples of trees

Typically, one vertex of a tree is designated as the root vertex, and it is from this vertex that the tree is considered to "grow." Unlike a physical tree, however, it is common to place the root of the tree at the top of the graph, from which the edges of the tree can be followed downward (Figure 14.36). If no root is designated, we refer to the tree as nonrooted, or free.

Much like the trees we encounter in nature, graphical trees can have branches, and each of these can be considered subtrees. The vertices at which the branches split are referred to as parent vertices, and all those vertices lying beyond that branch location are considered children of the parent vertex. A vertex having no children is referred to as a leaf of the tree.

A particular type of tree, called a binary tree, is used extensively in the organization of data (Figure 14.37). A binary tree is one where each vertex has at most two children, which are designated as the left child and the right child, based on their position relative to the parent. Should a binary tree be such that all vertices except the leaves have two children and all the leaves have a

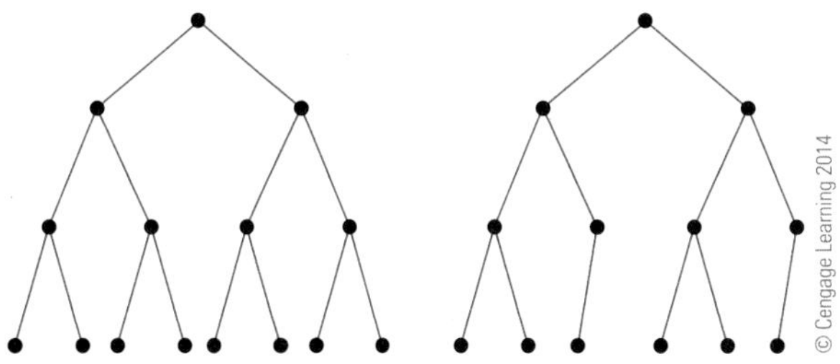

Figure 14.37 A full and a complete binary tree

common depth, then the tree is called a full binary tree. If a tree is nearly full but the bottom level is missing some leaves, we refer to the tree as a complete binary tree.

Since all trees are connected graphs (by definition), there must be a path from the root to any vertex in the tree, and since the graph is acyclic, the path is unique. The height (in some books called the depth) of a vertex in the tree is the length of the path from the root to the vertex, and the root itself is said to have height 0. The height of the tree is the maximum of all vertex depths.

EXAMPLE 14.16 For the following graph, determine the height of the tree, the right child of vertex A, and the height of vertex K (Figure 14.38).

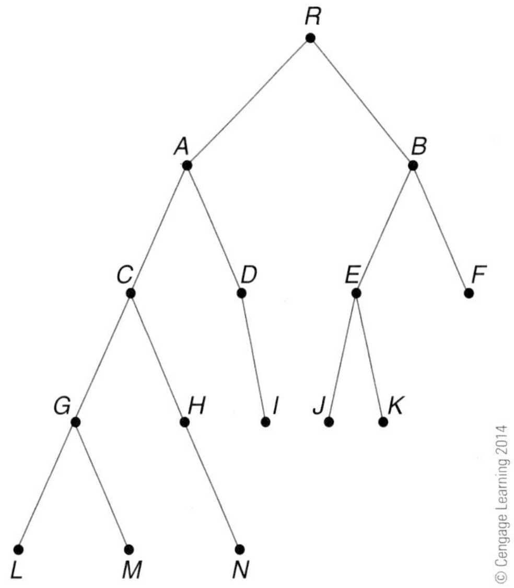

Figure 14.38 Graph for Example 14.16

SOLUTION

The greatest height of any vertex in the tree is 4, and thus the height of the tree is 4. Vertex *A* has two children, *C* and *D*, but *D* is the right child of vertex *A*. Finally, vertex *K* has height 3.

The Importance of Trees

One of the most widespread uses of trees is a corporate organization chart (Figure 14.39). This type of graph depicts the corporate hierarchical structure of an organization and allows members or outsiders to see where the various members of the organization stand in relation to one another.

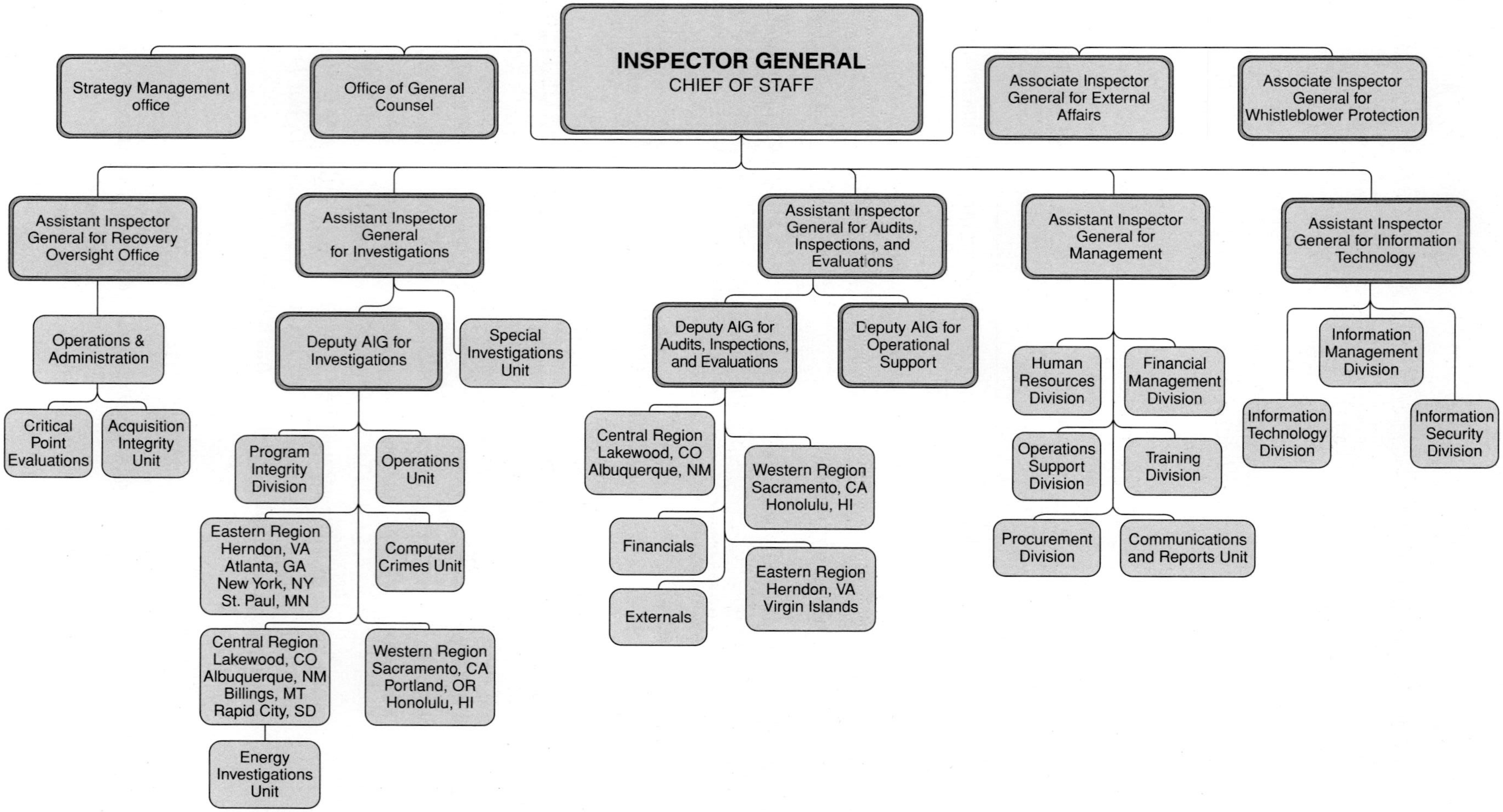

Figure 14.39 Corporate organization chart

Files on a computer are also organized in a treelike structure, which you can see listed in a vertical pattern, with the roots on the left (Figure 14.40).

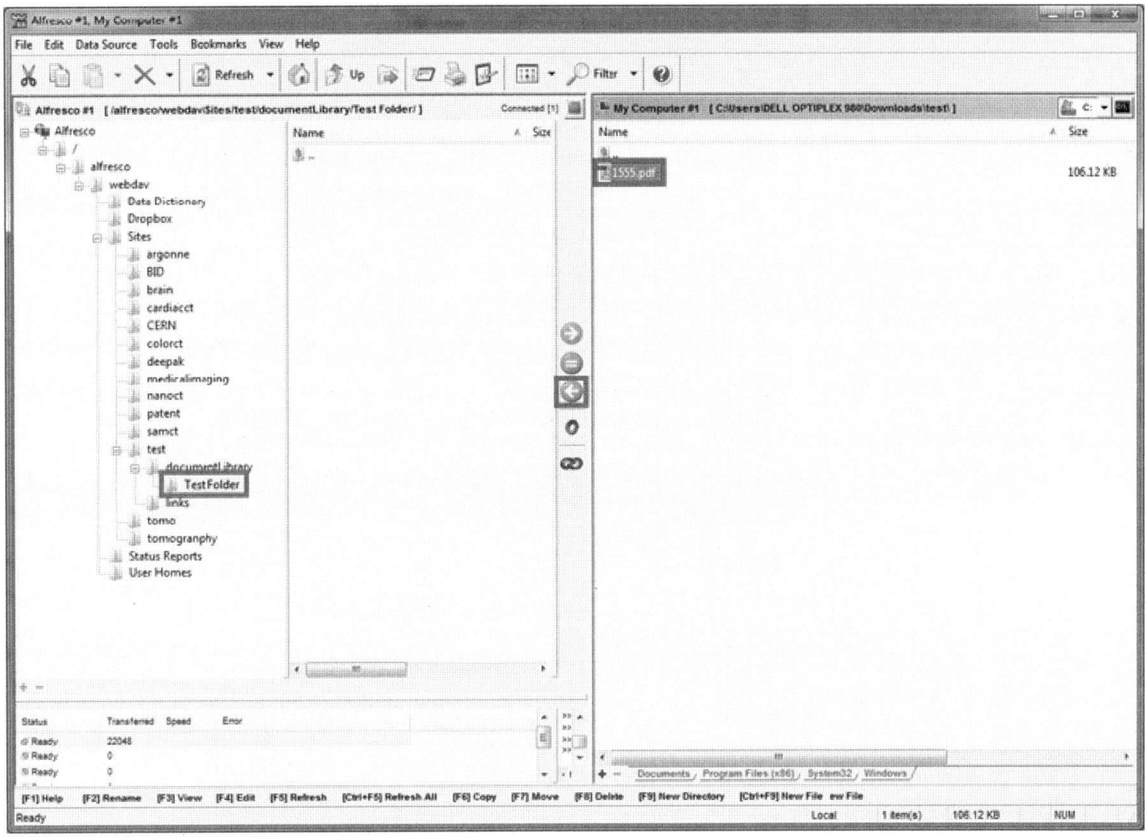

Figure 14.40 Computer file listing

The spread of a chain letter can be represented through the use of a tree, with the originator of the chain letter serving as the root (Figure 14.41). If we assume that each recipient forwards the chain letter to three friends, then each root would have three children, and hence the *n*th stage of the chain would expand the number of overall individuals involved in the chain letter process to 3^n.

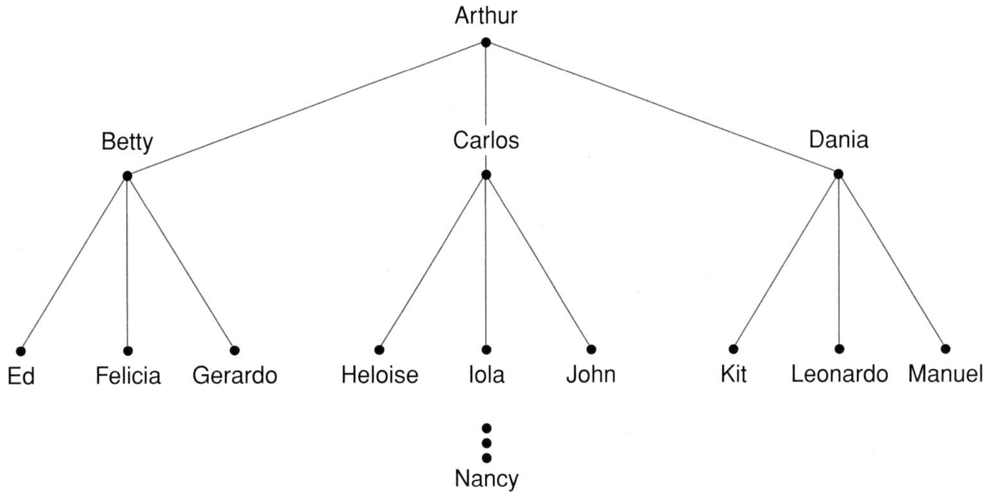

Figure 14.41 Chain letter recipient list

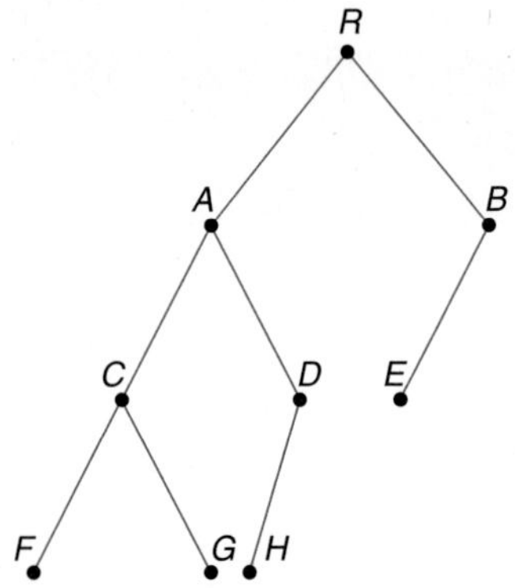

Figure 14.42 A binary tree

Array Representation of a Binary Tree

When we first introduced graphs, we presented a method by which graphs could be represented using an adjacency matrix. This technique, we pointed out, would permit us to describe the graph effectively in a manner that would capture the essential information about the relationships (if any) among the vertices in a form readily used by computers. The analog to this concept in the case of a binary tree is the array representation, which is an $n \times 3$ array, having n rows (corresponding to the n vertices of the tree) and three columns (indicating the vertex and the left or right child of the vertex).

For instance, consider the binary tree shown in Figure 14.42. The array for this tree will be constructed in the following manner. We will have a row of the array corresponding to each vertex in the graph and then an indication in the middle and right cells, the left and right children of the vertex. For the example given in Figure 14.42, we would have the following:

Vertex	Left Child	Right Child
R	A	B
A	C	D
B	E	0
C	F	G
D	H	0
E	0	0
F	0	0
G	0	0
H	0	0

Note that a 0 is used to indicate the absence of a particular type of child for any given vertex. The virtue of this construction, as was the case for the adjacency matrix, is that it is easily programmed into a computer, and thus the information contained within the tree is able to be stored in a form usable to the computer.

Tree Traversal

Another method for storing the relevant information of a tree is to describe how one would traverse the tree, visiting each of the individual vertices present. There are several methods for doing this, but the one we shall present here is called the preorder method. In this technique, we describe the branches of the tree, following them to their final leaf and then moving to the next branching.

Consider the example shown in Figure 14.43. The tree structure begins with the root r, which we will list first. We will then follow the left-most branch of the tree to its final leaf:

$$R, A, C, G, J$$

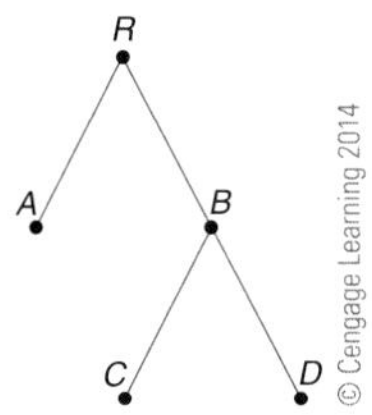

Figure 14.50

12. Give the preorder traversal for the tree given in Figure 14.50.
13. Give the array representation of the tree given in Figure 14.47.
14. Give the array representation of the tree given in Figure 14.48.
15. Give the array representation of the tree given in Figure 14.49.
16. Give the array representation of the tree given in Figure 14.50.

A pair of vertices within a tree having a common parent vertex are said to be siblings. In each of the named figures, list all sibling pairs of vertices.

17. Figure 14.47
18. Figure 14.48
19. Figure 14.49
20. Figure 14.50

Any vertex lying between a given vertex and the root vertex is said to be an ancestor of the given vertex. In each of the named figures, identify all ancestors of vertex *B* (if any).

21. Figure 14.47
22. Figure 14.48
23. Figure 14.49
24. Figure 14.50

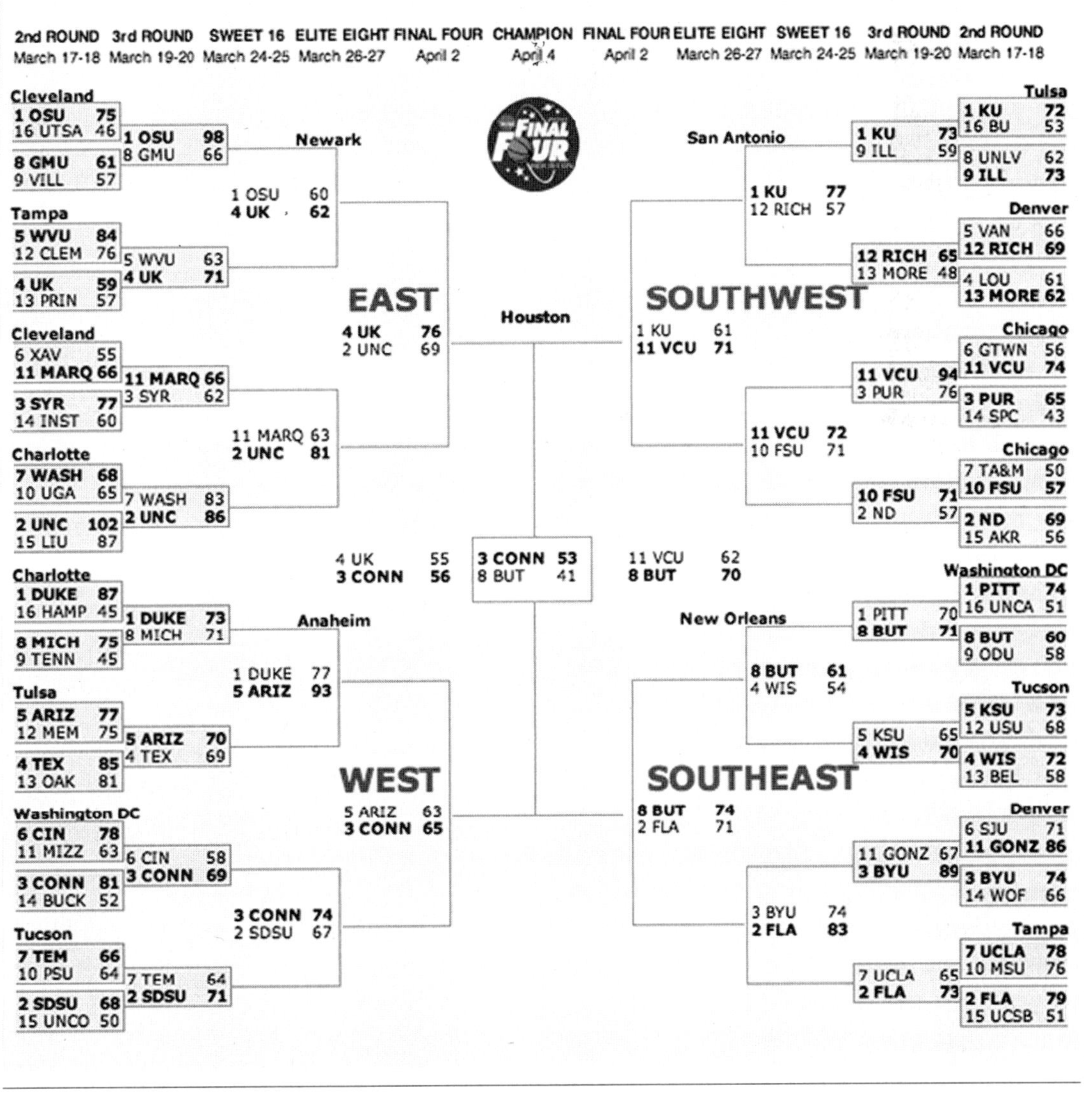

Figure 14.51 NCAA Tournament tree

Bracket			
Round 4	Quarterfinals	Semifinals	Final
C. Wozniacki (1) « S. Kuznetsova (15) 6-7(6-8),7-5,6-1	C. Wozniacki (1) « A. Petkovic (10) 6-1,7-6(7-5)		
A. Petkovic (10) « C. Suarez Navarro 6-1,6-4		C. Wozniacki (1) S. Williams (28) « 2-6,4-6	
S. Williams (28) « A. Ivanovic (16) 6-3,6-4	S. Williams (28) « A. Pavlyuchenkova (17) 7-5,6-1		
A. Pavlyuchenkova (17) « F. Schiavone (7) 5-7,6-3,6-4			S. Williams (28) S. Stosur (9) « 2-6,3-6
M. Niculescu A. Kerber« 4-6,3-6	A. Kerber « F. Pennetta (26) 6-4,4-6,6-3		
P. Shuai (13) F. Pennetta (26) « 4-6,6-7(6-8)		A. Kerber S. Stosur (9) « 3-6,6-2,2-6	
M. Kirilenko (25) S. Stosur (9) « 2-6,7(17)-6(15),3-6	S. Stosur (9) « V. Zvonareva (2) 6-3,6-3		
S. Lisicki (22) V. Zvonareva (2) « 2-6,3-6			

Figure 14.52 U.S. Open tree

A common application of binary trees in the sporting world is the **tournament tree**, which presents a binary tree in a horizontal manner, with the root tree at the center, representing the tournament champion. Such a tree is common in the NCAA basketball tournament or a professional tennis tournament. (See figures 14.51 and 14.52)

25. What is the height of the tournament tree shown for the NCAA tournament?
26. What is the height of the tournament tree in the U.S. Open tennis tournament?
27. How many ancestors does the NCAA championship team possess?
28. How many ancestors does the U.S. Open tennis tournament champion possess?

A vertex within a tree that is not a leaf is called an **internal vertex** of the tree. In the following, we consider how many internal vertices a full binary tree with height k will possess.

29. How many internal vertices does a full binary tree of height 25 possess?
30. How many leaves does a full binary tree of height k possess?
31. How many leaves does a full binary tree of height 25 possess?

A **ternary** tree is analogous to a binary tree, differing in that every vertex possesses 0 or 3 branches.

32. How many leaves will a full ternary tree of height k possess?
33. How many leaves will a full ternary tree of height 25 possess?

Most data are stored in a binary system using a fixed length binary string. ASCII is a fixed length format, using seven (or eight) bits to identify specific characters. **Huffman codes** are special binary trees intended to encode data. By using a string of 1s and 0s, a user or program can locate a specific data element. The primary purpose is to allow paths of varying lengths. The leaves of the tree represent letters of the alphabet, and each branch of the tree is assigned a value of 0 or 1 systematically (here, every left branch will be assigned value 1 and every right branch value 0), and letters are assigned a binary string corresponding to the path from the root to the letter's leaf.

In Figure 14.53, we see an example of a Huffman tree. Note that more commonly used letters, such as vowels, are positioned in such a manner that they can be represented by shorter binary strings. Here, the string 111 corresponds to the letter T, 01 to the letter I, and 001 to the letter P. Thus, the binary string 11101001 corresponds to the representation of the word TIP.

Using the Huffman tree in Figure 14.53, produce the binary strings corresponding to the following English words.

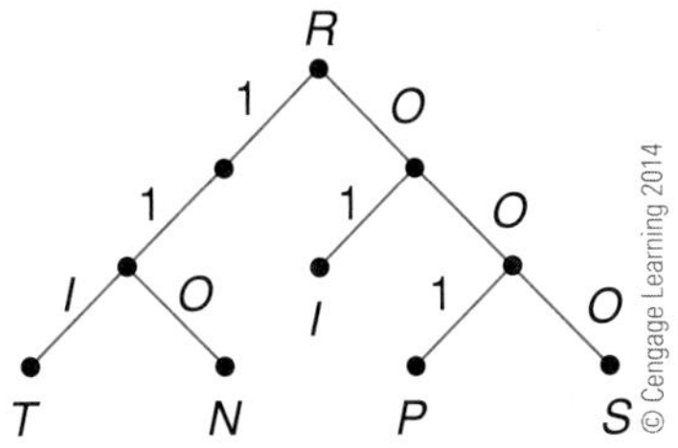

Figure 14.53 A Huffman tree

34. NIT
35. TIN
36. TINS
37. PIN
38. NIP
39. SPIN
40. PINS
41. SNIP
42. TIPS

Summary

In this chapter, you learned about:

- graphs, paths, and circuits.
- the terminology of graphs.
- Euler paths and Euler circuits.
- Hamiltonian paths and Hamiltonian circuits.
- the construction and use of trees.

Glossary

acyclic graph A graph containing no circuits.

adjacent vertices Two vertices connected by an edge.

ancestor Any vertex lying between a given vertex and its root vertex.

bridge An edge that, if removed, changes a connected graph into a disconnected graph.

circuit A walk that starts and ends at the same vertex.

complete graph A graph in which every pair of vertices forms an adjacent pair.

connected graph A graph in which every pair of vertices can be connected by a path.

disconnected graph A graph that is not connected.

distance between vertices The number of vertices in the shortest path that connects the vertices.

edges The line segments connecting vertices in a graph.

Euler circuit An Euler path that forms a circuit.

Euler path A path along a connected graph that connects all the vertices and that traverses every edge only once.

Fleury's algorithm A step-by-step process for the construction of an Euler circuit: Begin at any vertex you wish.

1. From that vertex, pick an edge to traverse, darkening it or marking it in some manner so as to ensure you will not recross it. *Never cross a bridge unless no other choice exists!*
2. Traverse that edge to reach the next vertex.
3. Repeat steps 2 and 3 until all edges have been traversed, returning to the original vertex if possible (to create an Euler circuit).

graph A finite set of points connected by line segments (or curved segments, if line segments are impractical to use).

Hamiltonian circuit A Hamiltonian path that forms a circuit.

Hamiltonian graph A graph possessing a Hamiltonian circuit.

Hamiltonian path A path in an undirected graph that visits each vertex exactly once.

Huffman code Special binary trees intended to code data.

incomplete graph A graph that is not complete.

internal vertex A vertex within a tree that is not a leaf.

***k*-regular graph** A graph in which every vertex has the same degree.

length The number of edges traversed by a walk, trail, or path.

loop An edge that connects a vertex to itself.

nodes Another term for the vertices of a graph.

nontrivial graph A graph with more than one vertex.

parallel vertices Two vertices connected by two distinct edges.

path A trail having no repeated vertices.

pendant A path with a one-edge loop.

regular graph A graph in which each vertex has the same degree.

siblings A pair of vertices within a tree having a common parent vertex.

simple graph A graph having no loops or parallel edges.

ternary tree A tree in which ever vertex has 0 or 3 branches.

tournament tree A binary tree commonly used in the sporting world to represent tournament team seeding.

trail A walk in which no edges are repeated.

tree A simple, connected acyclic graph.

trivial graph A graph with one vertex.

vertices The points in a graph.

walk A sequence of vertices in a graph that are connected by edges.

weight of an edge A numerical value assigned to an edge of a graph.

weighted distance between vertices The minimal sum of weights of the edges in all possible paths connecting two vertices.

weighted graph A graph in which all the edges have been assigned weights.

End-of-Chapter Problems

For the following graphs, identify the degree of the vertices of the graph, classify the vertices as either even or odd and give the order of the graph.

1.

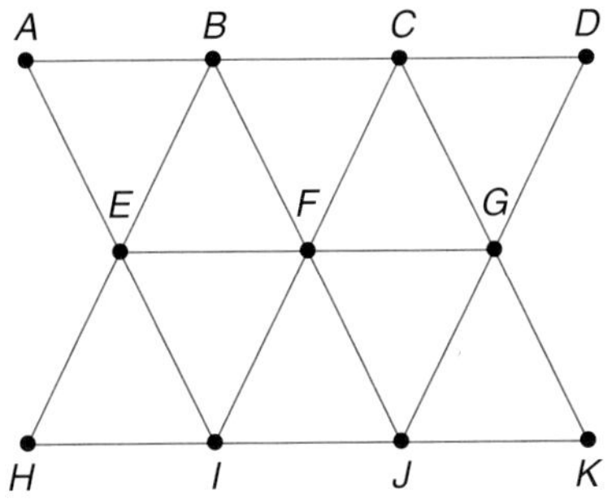

2.

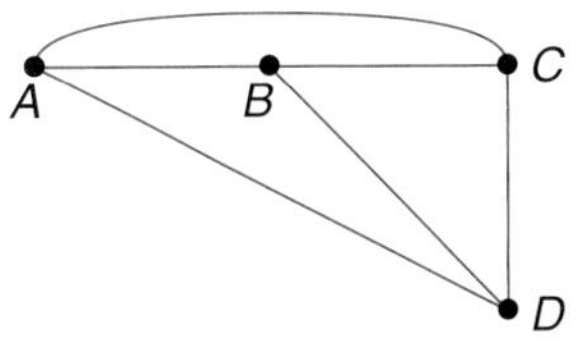

3.

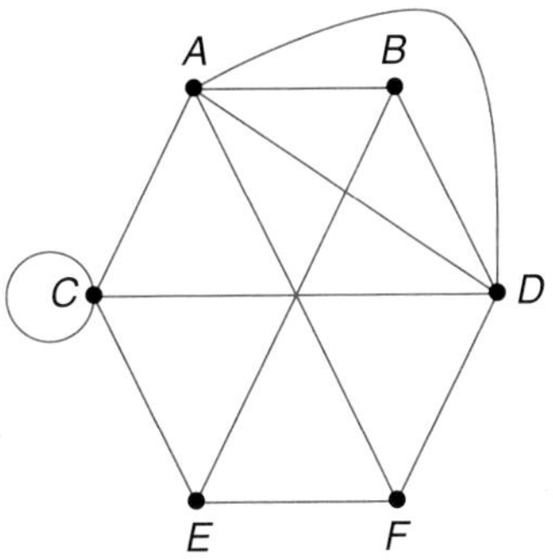

4.

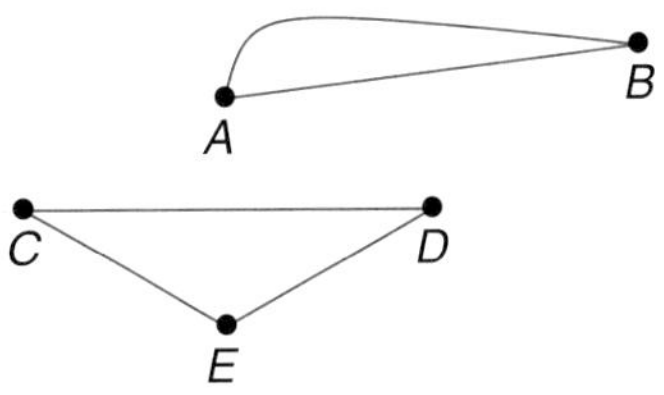

In the following problems, graphs from previous problems are referenced. Produce the adjacency matrix for each of the graphs.

5. Problem 1
6. Problem 2
7. Problem 3
8. Problem 4

In the following problems, a graph is described. Sketch an example of a graph having the indicated properties.

9. The graph has six vertices, four of which are even.
10. The graph has six odd vertices and one bridge.

The following problems refer to the graph shown next:

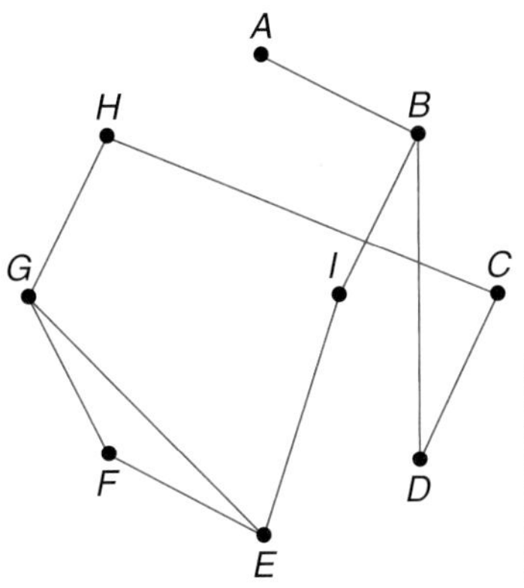

11. Is A, B, C, D a path within the graph?
12. Which edges of the graph are not included in the path A, B, D, E?
13. Is the graph connected or disconnected?

In the following problem, sketch a graph with the indicated properties.

14. Sketch a three-regular graph of order 6.
15. In the following weighted graph, find the shortest path from S to T.

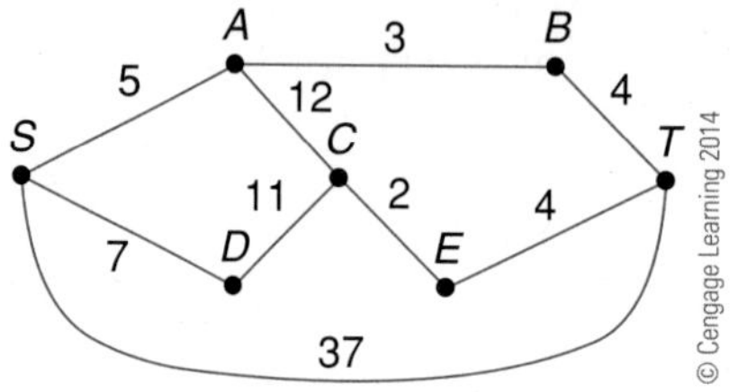

For the following exercises, use the graph shown here.

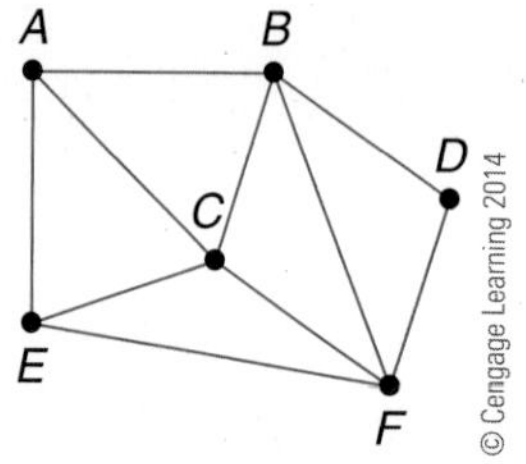

16. Find an Euler path, if one exists, that begins at vertex A.
17. Find an Euler path, if one exists, that begins at vertex B.
18. Find an Euler path, if one exists, that begins at vertex E.
19. Does an Euler circuit exist for this graph?
20. Suppose we had a graph with 872 vertices, of which exactly two are odd. (a) Does an Euler path exist for the graph? Explain. (b) Does an Euler circuit exist for the graph? Explain.

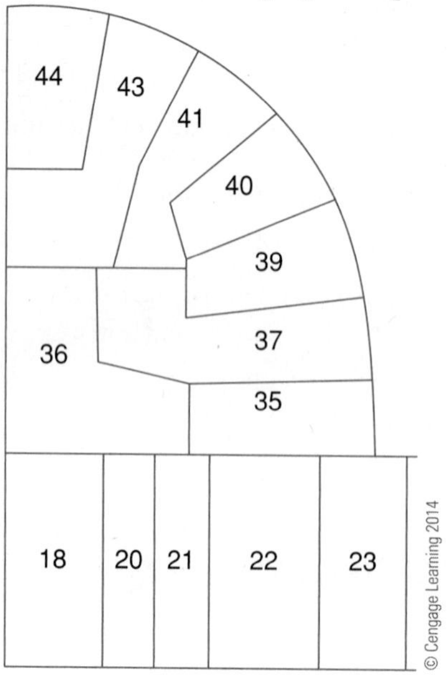

21. The picture above is a portion of a (simplified) assessor's map, showing the layouts of some lots. Represent the map as a graph, using a vertex to represent each lot, with lots sharing a common border joined by an edge.
22. Does the graph described in Problem 21 possess an Euler path? Explain, and if the answer is yes, produce such a path.
23. Does the graph described in Problem 21 possess an Euler circuit? Explain, and if the answer is yes, produce such a circuit.

In the following graphs, use Fleury's algorithm to find an Euler path and an Euler circuit (if one exists).

24.

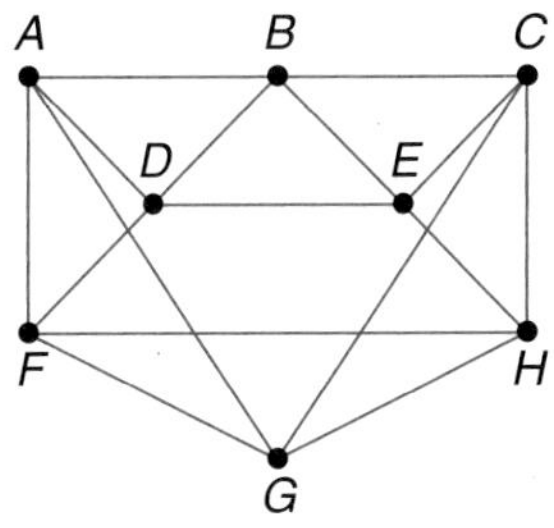

25.

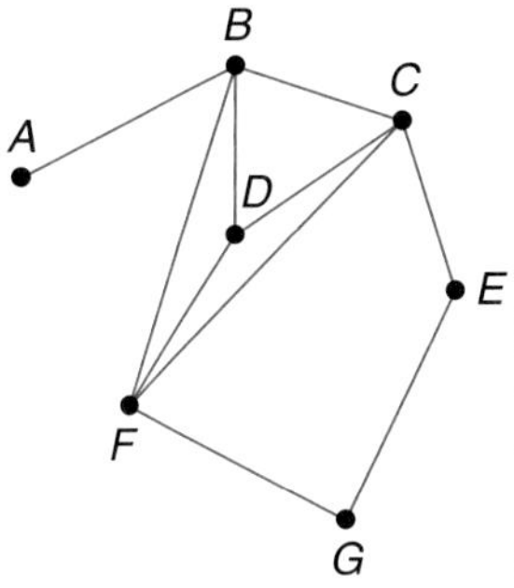

Shown are a collection of graphs. Determine whether they possess a Hamiltonian path. If they do possess such a path, produce the path. If they do not possess such a path, explain why.

26.

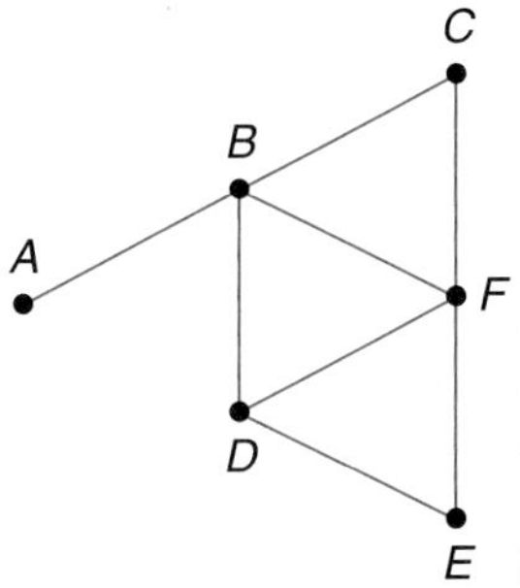

27.

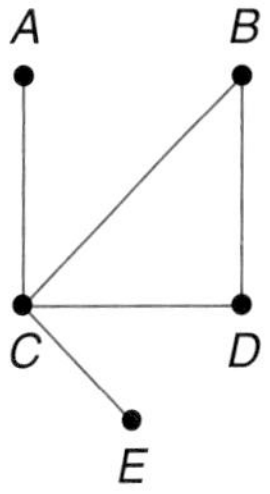

28.

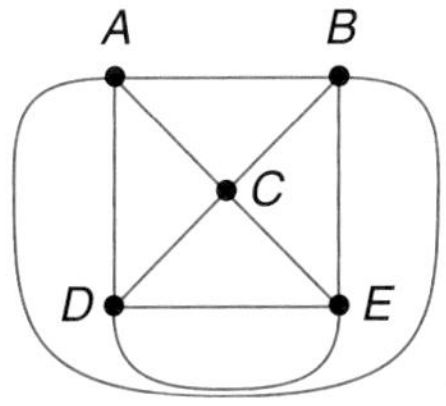

The following graphs do possess at least one Hamiltonian path. Find any particular one of these paths.

29.

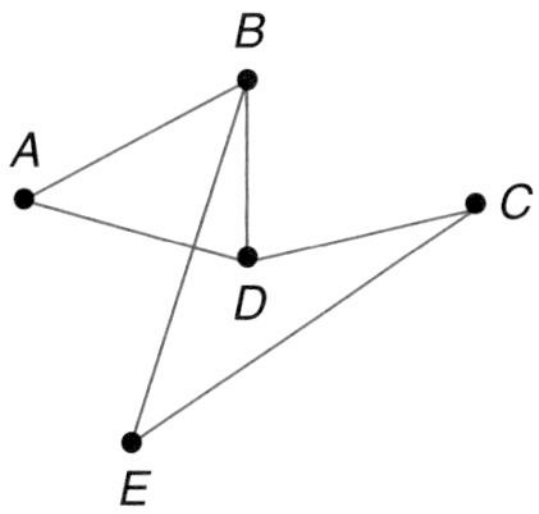

30.

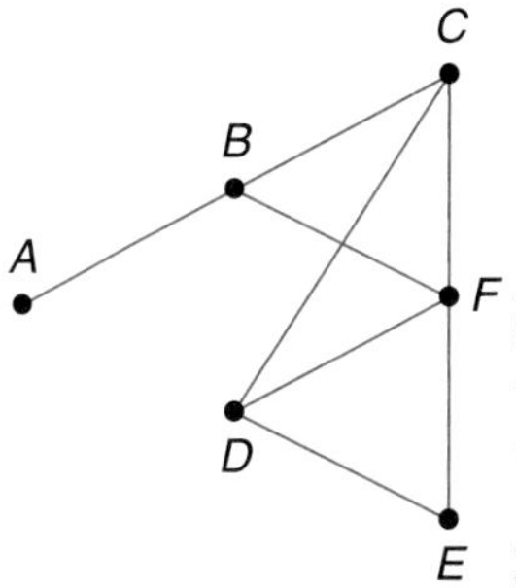

31. Suppose you live in Boston, Massachusetts, and want to visit the capitals of each of the six New England states (Boston, Massachusetts; Montpelier, Vermont; Hartford, Connecticut; Providence, Rhode Island; Concord, New Hampshire; and Augusta, Maine) exactly once, ending up back at Boston. The following graph shows the distance for a trip between each of the capitals. Find the Hamiltonian circuit in the graph that minimizes the total distance traveled.

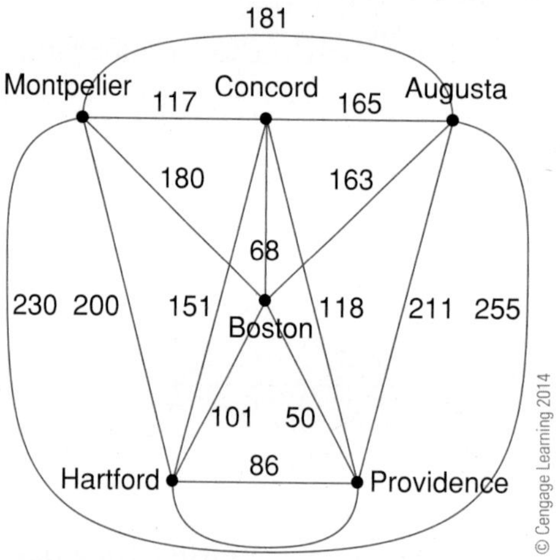

Problems 32 to 42 reference the following trees:

a.

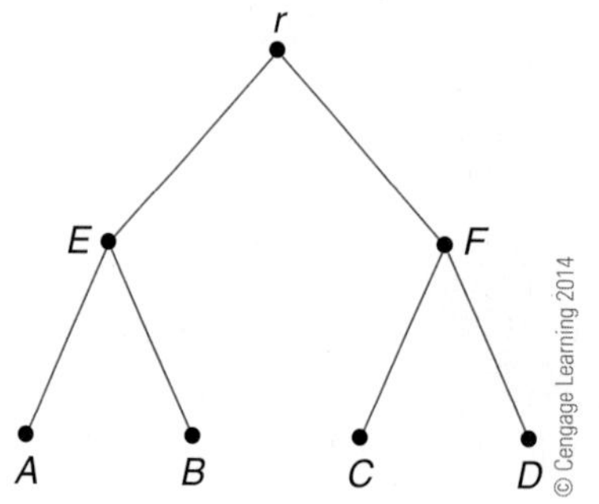

b.

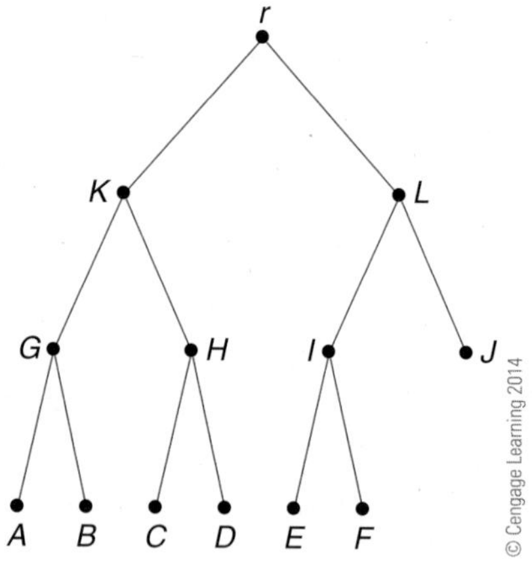

c.

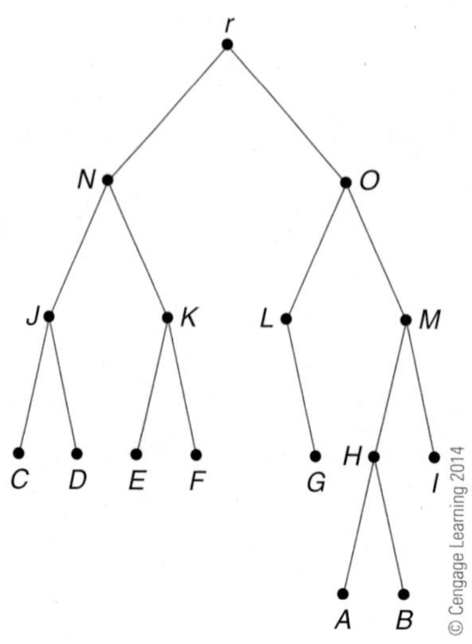

d.

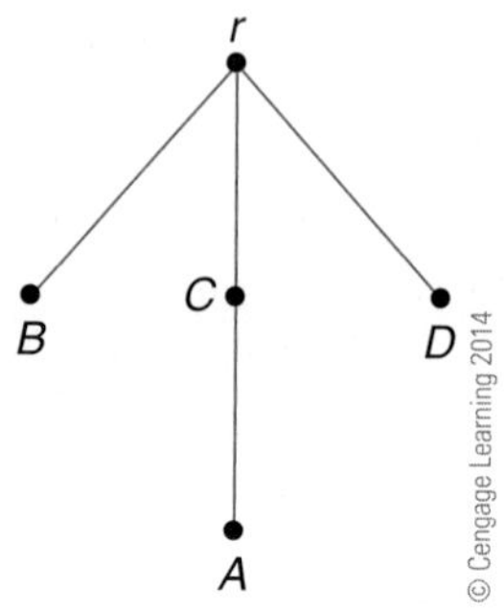

32. For each tree shown in the above figures, determine the height of the tree.
33. For each tree shown in the above figures, determine whether the tree is full binary.
34. For the tree shown in item b, find the height of vertex E and vertex L.
35. Give the preorder traversal for the tree given in item a above.
36. Give the preorder traversal for the tree given in item b above.
37. Give the array representation of the tree given in item a above.
38. Give the array representation of the tree given in item b above.

Remember that a pair of vertices within a tree having a common parent vertex are said to be siblings. In each of the named figures, list all sibling pairs of vertices.

39. Item b above
40. Item c above

Remember that any vertex lying between a given vertex and the root vertex is said to be an ancestor of the given vertex. In each of the named figures, identify all ancestors of vertex B (if any).

41. Item c above
42. Item d above

For Problems 43 and 44, you will consider the following tournament tree.

An Invented Chess Tournament

<table>
<tr><td>Joe Jones
John Smith</td><td rowspan="2">John Smith
Wilma
Flimisione</td><td rowspan="4">Wilma
Flimisione
Jim Jacobs</td><td rowspan="8">Wilma
Flimisione
Sally Forth</td><td rowspan="8">Sally
Forth</td></tr>
<tr><td>George Jelson
Wilma Flimisione</td></tr>
<tr><td>John Doe
Billy Smith</td><td rowspan="2">John Doe
Jim Jacobs</td></tr>
<tr><td>Jim Jacobs Fred
Flimisione</td></tr>
<tr><td>Sally Forth
Jill Platt</td><td rowspan="2">Sally Forth
Jesse Jones</td><td rowspan="4">Sally Forth
Bill Jenkins</td></tr>
<tr><td>George Fiedler
Jesse Jones</td></tr>
<tr><td>Cynthia Pusey
Jiminy Cricket</td><td rowspan="2">Cynthia Pusey
Bill Jenkins</td></tr>
<tr><td>Bill Jenkins
Sarah McNab</td></tr>
</table>

43. What is the height of the tournament tree?
44. How many ancestors does the tournament winner possess?
45. Remember that a vertex within a tree that is not a leaf is called an internal vertex of the tree. How many internal vertices does a full binary tree with height k possess?

46. How many leaves does a full binary tree of height k possess?

Remember that a ternary tree is analogous to a binary tree, differing in that every vertex possesses between zero and three branches.

47. How many internal vertices does a full ternary tree of height k possess?
48. How many leaves will a full ternary tree of height k possess?

Consider the Huffman tree given in Figure 14.53. If we are given a binary string and told that it is encoded with this tree, then we can decode it simply by following the branches from the root of the tree labeled with the appropriate bit value. When we reach a leaf, we have decoded a character, and we start over. For instance, if we are given the bit string 01110, we decode it as follows. From the root follow the right branch, consuming the 0 bit, then follow the left branch consuming the 1 bit. We are at the leaf labeled "I," so we have an "I," and we start over. From the root, we follow the left branch, labeled 1, and consume the 1 bit, then again follow the left branch labeled 1, consume the 1 bit, and finally follow the right branch, labeled 0, and consume the 0 bit. This leaves us at a leaf labeled "N," and we have reached the end of the bit string, so the decoded string is "IN." Using the Huffman tree in Figure 14.53 (see page 537), produce the English words corresponding to the following binary strings.

49. 00000101110
50. 00101110111000

Glossary

abscissa That element of an ordered pair corresponding to position relative to the horizontal axis of the plane.
acyclic graph A graph containing no circuits.
additive identity matrix A matrix all of whose entries are 0s. Also called a zero matrix.
adjacent vertices Two vertices connected by an edge.
ancestor Any vertex lying between a given vertex and its root vertex.
angle The span between two rays sharing a common initial point.
angle of depression The angle the line of sight makes with the top of an object at a lower level.
angle of elevation The angle the line of sight makes with the top of an object at a higher level.
antilogarithm of *N*, base *A* The number *B* in the equation $\log_A B = N$.
Argand diagram A Cartesian-type plane structure in which the *x*-axis is designated as the real axis, the *y*-axis is designated as the imaginary axis, and the complex number $a + bi$ is associated with the point (a, b).
argument A collection of statements alleged to give support to a logical conclusion.
argument of a complex number The value of theta in $r\angle\theta$.
arithmetic progression The pattern followed by the terms of an arithmetic sequence.
arithmetic sequence A sequence in which the difference between successive terms is constant.
ASCII (American Standard Code for Information Interchange) An encryption device for the English alphabet using binary numbers.
associative A reorganization of mathematical operations using symbols of grouping, which does not change its outcome.

base The number of distinct characters employed for numeration within a particular number system. Also called the radix.
base-10 number system The decimal number system.
base-16 number system A system of numeration using 16 distinct symbols: 0, 1, 2, 3, 4, 5, 6, 7, 8, 9, A, B, C, D, E, and F. Also called the hexadecimal number system.
base of an exponential function In a function $f(x) = ca^x$, the value of *a*.
BCD (binary-coded decimal) code Any of the codes used to express the decimal digits (0 through 9) using a nibble.
8421 BCD code The particular binary decimal code using the binary equivalents of the numbers 0 through 9 as the encryption code.
Bernoulli trial A binomial experiment having a fixed number of trials with each trial having only two possible outcomes, characterized as either "success" or "failure."
biconditional statement A compound statement involving the "if and only if" logical connective form.
bimodal data set A data set having two modes.
binary number system Also called base-2, a numeration system in which all numbers are represented using the digits 0 and 1 only.
binary point The analog to the decimal point in base-10; the point of demarcation between the whole number and fractional parts of a binary number. Also called the binary separator.
binary separator The analog to the decimal separator in the binary system. Also called the binary point.
binomial coefficients The coefficients of a binomial expansion.
binomial expansion The simplified expression resulting from calculating $(a + b)^n$, for any values *a* and *b*, and any whole number *n*.
binomial experiment An experiment having a fixed number of trials in which each trial has two possible outcomes (success or failure) and in which the trials are independent of one another.
binomial probability distribution The probability distribution of a binomial random variable.
binomial theorem A formula with which one can calculate the binomial expansion of an expression.
bit A two-digit pairing in the binary number system.
box-and-whisker plot A graphical tool to present the concentration of the data within a sample.
bridge An edge that, if removed, changes a connected graph into a disconnected graph.
byte An eight-bit grouping in the binary number system.

cardinality (of a set) The quantity of members within a set; represented as $n(S)$.
Cartesian plane A two-dimensional construct allowing us to systematize location within a mathematical plane.

certainty An event whose probability of occurrence is 1.

characteristic property (characteristic trait) A quality or feature of an object that makes it both distinctive and identifiable.

circuit A walk that starts and ends at the same vertex.

cluster sample A sample in which the population is subdivided into clusters from which a random sampling is taken.

coefficient matrix A matrix created using the coefficients of the variables of a linear system of equations.

cofactor expansion A technique for calculating the determinant of a matrix.

color codes Six-digit hex number HTML designations that identify the text color in a document.

column of a matrix Any of the vertical alignments of entries of a matrix.

common difference The difference between successive members of an arithmetic sequence.

common ratio The ratio of one term of a geometric sequence to its predecessor.

commutative An operation whose order can be reversed without changing its outcome.

complement (of a set *S*) Those elements of the universal set that are not elements of *S*.

complement The set of all outcomes in the sample space not in the event.

complete graph A graph in which every pair of vertices forms an adjacent pair.

complex fraction A fraction whose numerator, denominator, or both contains one or more fractions.

complex plane An Argand diagram.

compound event A combination of a pair of events *A* and *B* using the words "and" or "or."

compound statement A statement conveying two or more ideas.

conclusion A final statement in a logical argument alleged to follow from given premises.

conditional statement A compound statement involving the "if-then" logical connective form.

conjugate pairs Two binomial expressions differing only in the sign between the terms of the expressions.

connected graph A graph in which every pair of vertices can be connected by a path.

consistent A system of equations having a finite number of solutions.

consistent and dependent A system of equations having infinitely many solutions.

contradiction A compound statement that, under all conditions, is exclusively false.

contrapositive The logical equivalent form ~B → ~A to A → B.

convenience sample A sample constructed using readily obtained information.

correlation coefficient A quantity that depicts how closely a best-fit regression line fits a set of data.

cosine of an angle The ratio $\frac{x}{r}$, where x is the x-coordinate of an arbitrary point on the terminal side of the angle in standard position, and r is the distance from the origin to that point.

countable set (countability) A set that is finite or that can be placed in one-to-one correspondence with the set of natural numbers.

data Information about a population or sample.

decimal number system A system of numeration using 10 distinct symbols: 0, 1, 2, 3, 4, 5, 6, 7, 8, and 9.

decimal separator A dot used to indicate the separation from whole number to fraction number places within a number's base-10 representation.

decision problem A question posed by Hilbert that asked if a standard procedure could be developed to determine if a particular statement could be proven.

DeMorgan's laws $(A \cup B)^c = A^c \cap B^c$ and $(A \cap B)^c = A^c \cup B^c$.

dependent events Two events, the occurrence or nonoccurrence of one having no effect of the occurrence or nonoccurrence of the other.

determinant of a matrix A numerical value corresponding to a particular square matrix, used to determine invertibility.

deviation The difference between a particular data value and the mean of the data set.

diagonal matrix A square matrix whose entries not on the main diagonal are all 0.

directed distance A distance measurement in which sign indicates direction.

directed line segment A line segment assigned a direction, indicated by an arrowhead at one end of the segment.

disconnected graph A graph that is not connected.

disjoint sets Sets whose intersection is empty.

distance between vertices The number of vertices in the shortest path that connects the vertices.

distance formula (distance between two points) Formula to calculate the length of the line segment connecting the point (x_1, y_1) to the point (x_2, y_2).

$$D = \sqrt{(y_2 - y_1)^2 + (x_2 - x_1)^2} = \sqrt{|y_2 - y_1|^2 + |x_2 - x_1|^2}$$

DMS notation An expression of the measure of an angle using degrees, minutes, and seconds.

domain The set of input values of a function for which the function is defined.

double negation A succession of two "not" symbols in a logical expression.

edges The line segments connecting vertices in a graph.
element An object that is a member of a set.
elementary row operations Adding a multiple of any row of a matrix to another, multiplication of any row of a matrix by a nonzero scalar, and interchanging of rows of a matrix.
elimination One of the techniques of solving systems of equations.
ellipsis A series of three dots, indicating a progression of numbers continues according to a demonstrated pattern.
empirical probability of event *A* Probability of occurrence based on observation of a succession of trials. If n trials are performed and event A occurs k times, the empirical probability of occurrence is k/n.
empirical rule The fact that in a normal distribution, approximately 68% of all the data lies within 1 standard deviation (on either side) of the mean, approximately 95% of all the data lies within 2 standard deviations (on either side) of the mean, and approximately 99.7% of all the data lies within 3 standard deviations (on either side) of the mean.
empty set A set having no members.
end point The starting point of a ray.
entries The numbers within a matrix.
equally likely events Two events whose probabilities of occurrence are equal.
equal matrices Two matrices having the same order and identical entries in corresponding locations.
equal sets Two sets having precisely the same set of members.
equivalent sets Two sets having precisely the same number of members.
Euler circuit An Euler path that forms a circuit.
Euler diagram A Venn diagram used to determine validity of reasoning.
Euler path A path along a connected graph that connects all the vertices and that traverses every edge only once.
even-odd identities A set of trigonometric identities expressing relationships between functions reliant on the properties of symmetry displayed by their graphs. Also called symmetry identities.
event A subset of the sample space of an experiment.
excess-3 (XS-3) codes A BCD code and numeral system using a prespecified value N as a bias value.
expanded form The decomposition of a number as a sum where each term displays a product of a digit of the number's representation with its place value.
expected value The expected return in a game of chance or in an experiment.
experiment A planned operation conducted under controlled conditions.
exploded form Another term for the expanded form of a number.
exponential form of a complex number $re^{i\theta}$, where r is the modulus and θ is the argument of the complex number.
exponential form of an equation $A^N = B$, where A and B are positive constants, with A not 1.
exponential function A function of the form $f(x) = ca^x$.

factorial An abbreviation for a particular mathematical product; $n!$ represents the product of the natural number n with all lesser natural numbers.
factorial notation For a natural number n, the notation $n!$, defined to be the product of n with all smaller natural numbers.
fallacy An argument form known to be invalid.
fallacy of the converse $A \to B$
B
$\therefore A$
fallacy of the inverse $A \to B$
$\sim A$
$\therefore \sim B$
finite set A set having a limited number of members.
Fleury's algorithm A step-by-step process for the construction of an Euler circuit: Begin at any vertex you wish.
1. From that vertex, pick an edge to traverse, darkening it or marking it in some manner so as to ensure you will not recross it. *Never cross a bridge unless no other choice exists!*
2. Traverse that edge to reach the next vertex.
3. Repeat steps 2 and 3 until all edges have been traversed, returning to the original vertex if possible (to create an Euler circuit).

function A rule of assignment that associates with each element of one set (the domain) exactly one element of a second set (the range).

general form of the equation of a line $Ax + By + C = 0$, where A, B, and C are integers having no common factors, with A nonnegative and at least one of A or B nonzero.
general term of a sequence A formula with which the individual terms of a sequence can be calculated as a function of the order of occurrence.
geometric progression The pattern followed by the terms of a geometric sequence.
geometric sequence A sequence in which each term is a constant multiple of its predecessor.
graph A finite set of points connected by line segments (or curved segments, if line segments are impractical to use).
gray codes A binary numeral system where two successive values differ only in one bit.

Hamiltonian circuit A Hamiltonian path that forms a circuit.
Hamiltonian graph A graph possessing a Hamiltonian circuit.
Hamiltonian path A path in an undirected graph that visits each vertex exactly once.
hexadecimal number system Also called base-16, a numeration system in which all numbers are represented using the symbols 0, 1, 2, 3, 4, 5, 6, 7, 8, 9, A, B, C, D, E, and F only.
hexadecimal point The analog to the decimal point in base-16; the point of demarcation between the whole number and fractional parts of a base-16 number. Also called the hexadecimal separator.
hexadecimal separator The analog to the decimal point in the hexadecimal number system.
histogram A common tool for representing data using bars depicting data frequencies.
homogeneous population A population whose membership possesses relatively uniform characteristics.
Huffman code Special binary trees intended to code data.

identity A true statement of equality.
identity matrix A square matrix having 1s in all main diagonal entries and 0s elsewhere.
imaginary axis The vertical axis in an Argand diagram.
imaginary number Any number whose square is a negative real number. Equivalent to pure imaginary number.
imaginary part In an imaginary number $a + bi$, the number b.
imaginary unit The mathematical concept number i, defined by the rule $i = \sqrt{-1}$.
impossible event An event whose probability of occurrence is 0.
inclusive or That usage of the word "or" taken to mean "either or both."
incomplete graph A graph that is not complete.
inconsistent A system of equations having no solution.
inconsistent system A system of equations having no solution.
independent events Two events, one of whose occurrence does not influence the occurrence or nonoccurrence of the other.
index of summation The variable used within the formula of sigma notation.
infinite set A set having an unlimited number of members.
internal vertex A vertex within a tree that is not a leaf.
interquartile range That part of the data lying between the first and the third quartile.
intersection (of sets) The elements common to two given sets.
invalid argument A logical argument that is not valid.
invertible matrix A square matrix for which a multiplicative inverse exists.

***k*-regular graph** A graph in which every vertex has the same degree.

law of contraposition $A \rightarrow B$
$\underline{\sim B}$
$\therefore \sim A$
law of disjunctive syllogism $A \rightarrow B$
$A \vee B$
$\sim A$
$\therefore B$
law of large numbers That law stating that if an experiment is repeated a great number of times, the empirical probability will approach the theoretical probability for the event.
law of syllogism $A \rightarrow B$
$B \rightarrow C$
$\therefore A \rightarrow C$
least significant binary digit (LSB) The base-2 instance of the least significant digit. Also called the least significant bit.
least significant digit (LSD) The nonzero digit farthest right within a number's representation.
length The number of edges traversed by a walk, trail, or path.
length of a sequence The number of terms in a sequence.
Liar's paradox A classic statement ("I am lying") that cannot be either true or false.
linear correlation A relationship between two variables X and Y in which every X value is associated with just one value of Y.
linear equation An equation whose variables appear to the power of 1 only, capable of being rewritten in the general form $Ax + By + C = 0$.
linear regression line The graph of a linear correlation.
line segment A connected portion of a straight line, having a specific starting and ending point.
logarithmic expressions Expressions involving logarithms with any particular base.
logarithmic form of an equation $N = \log_A B$, where A and B are positive constants, with A not 1.
logarithm If N is a number such that $A^N = B$, where A is positive but not 1, then we say that N is the logarithm, base A, of the number B.

logical connectives Words that connect one simple statement to another.
logically equivalent statements Two statements having the same truth-functional value.
loop An edge that connects a vertex to itself.

main diagonal In a square matrix, those entries occurring in row *k*, column *k*.
mathematical induction One technique of mathematical proof.
matrix An array of real numbers.
mean The numerical average of a set of data.
median The central value of a data set when the data have been arranged in increasing order.
member An object belonging to a set.
members of a sequence The numbers occurring within a sequence.
midpoint formula The formula used to calculate the midpoint of the points (x_1, y_1) and (x_2, y_2) $\left(\frac{x_1+x_2}{2}, \frac{y_1+y_2}{2}\right)$.
midpoint That point lying in the exact center of a line segment connecting two specified points in the plane.
mode The data value occurring most often within an overall set of data.
modulus of a complex number The value of *r* in the polar form $r\angle\theta$.
most significant binary digit (MSB) The base-2 instance of the most significant digit. Also called the most significant bit.
most significant digit (MSD) The nonzero digit farthest left within a number's representation.
multiplication rule for dependent events If *A* and *B* are dependent events, then $P(A \text{ and } B) = P(A) \times P(B|A)$.
multiplication rule for independent events If *A* and *B* are independent events, then $P(A \text{ and } B) = P(A) \times P(B)$.
multiplicative inverse of a square matrix *A* A second square matrix, named A^{-1}, having the same size as matrix *A*, such that $AA^{-1} = A^{-1}A = I$.
mutually exclusive events Two events *A* and *B* such that $P(A \text{ and } B) = 0$.

natural exponential base The mathematical constant *e*, whose approximate value is 2.71828182845904523560287.
natural exponential function An exponential function whose base is the mathematical constant *e*.
natural logarithm A logarithm whose base value is the mathematical constant *e*.
nibble A four-bit grouping in the binary number system.
nodes Another term for the vertices of a graph.
noninvertible matrix A matrix for which no multiplicative inverse exists.
nonprobability sampling method A technique of sampling characterized by all members of the population not having an equal opportunity for selection.
nonsingular matrix An invertible matrix.
nontrivial graph A graph with more than one vertex.
nonweighted codes Codes that employ a weighted positioning system.
null set A set having no members.

octal number system Also called base-8, a numeration system in which all numbers are represented using the symbols 0, 1, 2, 3, 4, 5, 6, and 7 only.
octal separator The analog to the decimal point in the octal number system.
odds against an event The ratio *b*:*a*, where *a* is the number of favorable outcomes for the event and *b* is the number of unfavorable outcomes of the event.
odds in favor of an event The ratio *a*:*b*, where *a* is the number of favorable outcomes for the event and *b* is the number of unfavorable outcomes of the event.
one degree $\frac{1}{360}$ of one revolution around a point.
one radian The measurement of an angle determined by the sweep of an arc of length equal to the radius of an arbitrary circle whose center lies at the common initial point of the two involved rays.
one-to-one correspondence A relationship between two sets that associates to each member in one set a unique element in the other set and vice versa.
ordered triple A three-coordinate expression identifying a location in three-dimensional real space.
order The size of a matrix, given as $m \times n$, where m is the number of rows and n the number of columns within the matrix.
ordered pairs A coupled pair of numbers having the form (a, b), where *a* and *b* are real numbers. The ordered pair specifies the point in the plane at the intersection of the vertical line $x = a$ and the horizontal line $y = b$.
ordinate That element of an ordered pair corresponding to position relative to the vertical axis of the plane.
origin The intersection of the horizontal and vertical axes of the plane.
outcome The result of an experiment.
outlier A data value lying conspicuously far from the remaining data in the set.

parallel vertices Two vertices connected by two distinct edges.
parameter Information gathered about an entire population.

Pascal triangle A shortcut method for determining the binomial coefficients of a binomial expansion.
path A trail having no repeated vertices.
pearson product-moment correlation coefficient See correlation coefficient.
pendant A path with a one-edge loop.
permutation A rearrangement of the members of a set.
place value The value associated with a particular digit location in a numerical representation.
point-slope form of the equation of a line The form of a line's equation given by $y - y_1 = m(x - x_1)$, where m is the line's slope and (x_1, y_1) is a point lying on the line.
polar coordinates A system of representing location within the plane using a directed distance and an angle measured off a polar axis.
polar form of a complex number The notation $r\angle\theta$, where r is the directed distance from the origin to the point representing the complex number in an Argand diagram, and θ is the angle between the polar axis and the line segment connecting the origin to the point representing the complex number in that Argand diagram.
population The collection of objects or people studied in a statistical experiment.
power set (of a set) The set of all subsets of a given set.
premises Statements within a logical argument that allege to imply a particular conclusion statement.
principal square root The nonnegative square root of a number.
proper subset A subset containing some but not all of the members of a particular set.
properties of logarithms A set of equivalent forms of specific logarithmic expressions used to rewrite logarithmic expressions in alternative forms.
pure imaginary A complex number whose real part is zero, or (equivalently) any number whose square is a negative real number. Another term for imaginary number.
Pythagorean trigonometric identity $\cos^2\theta + \sin^2\theta = 1$

quadrantal points Points lying on either of the coordinate axes in the plane.
quadrants The four regions into which the plane is subdivided by the horizontal and vertical axes.
quantifying A word (such as "all," "none," or "some") that expresses the portion of a set possessing a particular property.

radian system of measurement A means of measuring angles whose basis is a circle of arbitrary radius centered at the origin.
radicand A number occurring within a radical symbol.
radix The base of a number system.
random sample A sample whose members are chosen randomly from the overall population in such a manner that each member of the population has the same opportunity for being selected as a member of the sample.
random variable The characteristic of the population being studied.
range The difference between the maximum and minimum values of a data set.
ray A half line having an initial point and infinite extension in one direction only.
real axis The horizontal axis of an Argand diagram.
real number An imaginary number whose imaginary part is zero.
real part In an imaginary number $a + bi$, the number a.
reciprocal identities A set of trigonometric identities expressing relationships between functions reliant on the reciprocal relationships.
rectangular components of a vector The x- and y-components of a vector.
rectangular form of an imaginary number The mathematical form $a + bi$. Also called the standard form.
rectangular form of a vector $\langle a, b\rangle$, where (a, b) is the terminal point of the vector when in standard position.
recursive relationship A relationship in which the value of a term in a sequence depends on earlier terms in the sequence.
regression line A line fitted to the set of paired data in a regression analysis.
regular graph A graph in which each vertex has the same degree.
relation A rule of assignment that associates to each element of one set one or more members of a second set.
relative frequency The quotient of the frequency and the cardinality of the data set.
representative sample A sample whose members' characteristics closely resemble those of the population.
resultant The sum of two vectors.
RGB The red/green/blue ratios used to construct colors in HTML; also called color codes.
right triangle A triangle containing a 90° angle at one vertex.
rise The vertical change in position within the plane when moving along a straight-line path connecting one point to another.
rise-over-run method A counting technique used to calculate the value of the slope of a line.
roster notation A listing of all the members of a set.
row of a matrix Any of the horizontal alignments of entries within a matrix.

run The horizontal change in position within the plane when moving along a straight-line path connecting one point to another.

sample A subset of the population.
sample space The set of all possible outcomes of an experiment.
sampling The process of creating a sample.
scalar A number not appearing within any particular mathematical context.
scatter plot A graph depicting with distinct dots the locations of distinct points within the plane.
sequence A progression of real numbers whose ordering cannot be changed.
series The sum of the terms of a sequence.
set A well-defined, unordered collection of objects having no duplicate members.
set-builder notation Representation of set membership, including an arbitrary variable used as a generic set member, together with a characteristic property/trait describing the set members.
sexagesimal system Subdivision of degree measure into minutes and seconds.
siblings A pair of vertices within a tree having a common parent vertex.
sigma notation A compact notation representing the sum of the terms of a sequence. Also called summation notation.
simple event A particular outcome in a sample space.
simple graph A graph having no loops or parallel edges.
simple statement A statement conveying a single idea.
sine of an angle The ratio $\frac{y}{r}$, where y is the y-coordinate of an arbitrary point on the terminal side of the angle in standard position, and r is the distance from the origin to that point.
singleton A set whose cardinality is 1.
singular matrix A square matrix having no inverse.
slope A numerical value quantifying how far a line is tilted from horizontal.
slope intercept form of the equation of a line The form of a line's equation $y = mx + b$, where m is the slope of the line, and the point $(0, b)$ is the y-intercept of the line. Also called the "$y = mx + b$" form.
square matrix A matrix having as many rows as columns.
standard deviation A measure of dispersion in a frequency distribution, equal to the square root of the mean of the squares of the deviations from the mean of the distribution.
standard form of an imaginary number The mathematical form $a + bi$. Also called the rectangular form.
standard normal distribution The normalized normal distribution.
standard position of a vector A sketch of the vector where its initial point is at the origin.
statement A declarative sentence that is either true or false.
statistical graph A visual representation of statistical data.
statistic Information gathered about a sample of a population.
straight line A path in the plane having no curvature, defined as a geometric primitive.
strata The individual subgroups within a stratified sample.
stratified sampling A sample in which the population is subdivided into strata from which a random sampling is taken.
subset A collection of objects taken from a particular, specified set.
substitution One of the techniques of solving systems of equations.
superset A larger set containing a given set as a subset.
syllogism A set of premises followed by a conclusion.
symmetric property of equality The property stating that A = B is equivalent to B = A.
symmetry identities A set of trigonometric identities expressing relationships between functions reliant on the properties of symmetry displayed by their graphs. Also called even-odd identities.
system Two or more linear equations, considered as a group.
systematic sample A sample whose members are selected using some arbitrarily designed algorithm.

tangent of an angle The ratio $\frac{y}{r}$, where x and y are the coordinates of an arbitrary point on the terminal side of the angle in standard position.
tautology A compound statement that, under all conditions, is exclusively true.
terms of a sequence The numbers occurring within a sequence.
ternary tree A tree in which ever vertex has 0 or 3 branches.
theoretical probability of event *A* The number of outcomes for event A divided by the total number of outcomes within the sample space.
tip-to-tail method of vector addition A visual method of calculating the resultant of two vectors.
tournament tree A binary tree commonly used in the sporting world to represent tournament team seeding.
trail A walk in which no edges are repeated.
tree A simple, connected acyclic graph.
trigonometric form of a complex number $(r\cos\theta) + (r\sin\theta)i$, where r and theta are the modulus and argument of the complex number, respectively.
trigonometric identity A statement expressing the equality of two trigonometric expressions.
trivial graph A graph with one vertex.

truth table A device we can use to investigate a statement containing a finite number of simple statements in combination.

truth-value The truth or falsity of a statement.

turing machines A theoretical device that would manipulate symbols on a spool of tape in accordance with a set of predesignated rules.

two-intercept form of the equation of a line The form of a line's equation $\frac{x}{a}+\frac{y}{b}=1$, where $(a, 0)$ and $(0, b)$ are the x- and y-intercepts of the line, respectively.

two-point form of the equation of a line The form of a line's equation $y-y_1=\left(\frac{y_2-y_1}{x_2-x_1}\right)(x-x_1)$, where (x_1, y_1) and (x_2, y_2) are points on the line.

union (of sets) A combination of two sets.

universal set An overall set specified in a particular problem within which other sets are specified.

upper limit of summation The maximum value of the index of summation in sigma notation.

valid argument An argument such that, if the given premises are true, the conclusion must be true.

variable exponent The expression in the exponent of an exponential function.

variable matrix A column matrix whose entries are the variables of a linear system of equations.

variance The average of the sums of the squares of the deviations from the mean in a data set.

vector A quantity possessing both direction and magnitude.

Venn diagram A visual tool used to demonstrate the relationships between sets within a universal set.

verbal description (of a set) A nonambiguous written description of the members of a set.

verification of identities A process of algebraic manipulation and substitution through which one can establish the equality of two trigonometric expressions.

vertices The points in a graph.

walk A sequence of vertices in a graph that are connected by edges.

weighted code a code in which representation of the base-10 digits is performed using a four-bit combination for which each position within the nibble carries a particular weight.

weighted distance between vertices The minimal sum of weights of the edges in all possible paths connecting two vertices.

weighted graph A graph in which all the edges have been assigned weights.

weight of an edge A numerical value assigned to an edge of a graph.

well-defined set A collection of objects membership within which is unambiguous.

x-axis The horizontal axis in the Cartesian plane.

x-component of a vector $R_x = (|\vec{v}| \cos\theta)\,\vec{i}$, where $\vec{i}$ is the standard unit vector $<1, 0>$ and θ is the angle between the positive x-axis and the vector when drawn in standard position.

x-coordinate The abscissa in an ordered pair in the Cartesian plane.

x-intercept The point on the x-axis intersected by a nonhorizontal line.

$y = mx + b$ form of the equation of a line The form of a line's equation $y = mx + b$, where m is the slope of the line, and the point $(0, b)$ is the y-intercept of the line. Also called "slope-intercept" form.

y-axis The vertical axis in the Cartesian plane.

y-component of a vector $R_y = (|\vec{v}| \sin\theta)\,\vec{j}$, where $\vec{j}$ is the standard unit vector $<0, 1>$ and θ is the angle between the positive x-axis and the vector when drawn in standard position.

y-coordinate The ordinate in an ordered pair in the Cartesian plane.

y-intercept The point on the y-axis intersected by a nonvertical line.

zero correlation The condition where a set of data points forms no observable linear pattern.

zero matrix A matrix all of whose entries are 0s. Also called an additive identity matrix.

z-score A standardized value that allows us to transform any normal distribution into what we refer to as the standard normal distribution.

Index

Page numbers followed by "f" indicate figures; page numbers followed by "t" indicate tables.

T

U

V

W

X

Y

Z